How Physicians Can Fix Health Care:
One Innovation at a Time

Adjunct Professor of Business Administration

Tuck School of Business at Dartmouth
100 Tuck Hall
Hanover, NH 03755
(603) 646 0463
chris.trimble@dartmouth.edu

American Association for
PHYSICIAN
LEADERSHIP®
Inspiring Change. Together.

Toll-free: 800-562-8088
Fax: 813-287-8993
Web: physicianleaders.org
Email: info@physicianleaders.org

400 N. Ashley Drive, Suite 400 · Tampa, Florida 33602-4322

ii

ISBN: 978-0-9825482-9-5

Library of Congress Card Number: 2015952884

Printed in the United States of America by Lightning Source.

"Trimble understands physicians ... the way we are feeling and the obstacles we are facing. This is a clear, practical, and engaging guide to making innovation happen. Trimble's work has been enormously helpful to me."

—ERIC ISSELBACHER, **MD**
Staff Cardiologist and Director of the Healthcare Transformation Lab,
Massachusetts General Hospital

"I wish I had read this book fourteen years ago when I started the Camden Coalition of Healthcare Providers. I would be much further along in my work and the journey wouldn't have been so difficult."

—JEFFREY BRENNER, **MD**
Executive Director, Camden Coalition of Healthcare Providers

"I'm excited to see Chris Trimble bring his formidable expertise on innovation to health care delivery. With engaging and inspiring stories, Trimble reveals a path forward, toward better outcomes and experiences for both providers and patients."

—ROY ROSIN
Chief Innovation Officer, University of Pennsylvania Health System

"Trimble offers an thoughtful and provocative challenge to physicians and senior leaders ... to tackle the big issues requiring innovation head on and to discard incrementalism."

—PAUL KECKLEY
Managing Director, Navigant Center for Healthcare Research and Policy Analysis

"Health care is in desperate need of innovation and Trimble's book provides a well-timed prescription. His collaborative approach with defined roles for clinicians and leaders will drive results and improve morale."

—CRAIG WRIGHT, **MD**
Senior Vice President, Physician Services, Providence Health and Services

"At a time when many doctors are considering leaving medicine, Trimble offers a voice of hope and an innovative approach for how doctors can lead the way to a more collaborative, effective, and sustainable system."

—HEATHER FORK, **MD**
Founder, Doctor's Crossing

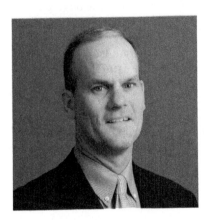

Chris Trimble has dedicated more than fifteen years to studying the best practices for making innovation happen within established organizations.

His past works, with co-author Vijay Govindarajan, include the New York Times bestseller *Reverse Innovation: Create Far From Home, Win Everywhere,* which focuses on innovation in emerging markets, and *How Stella Saved the Farm: A Tale About Making Innovation Happen,* which instigates productive conversations about the organizational dynamics of innovation.

Chris is on the faculty at Dartmouth. This is his sixth book.

www.chris-trimble.com

Table of Contents

....

Foreword

....

THE CLICK OF THE DOOR CLOSING. FROM THE MOMENT I BECAME A physician and for the 25 years that I practiced, that sound had a special meaning for me. A gantlet of training, acculturation and personal growth culminated in a lovely instant of transition, as my patient and I entered the examination room, a sanctuary of trust and duty: the patient's trust, my duty. In the formative moments of my professional development, my teachers' often repeated the message: "This is your patient, Dr. Berwick. Sit down. Pay attention." Click.

And so, the door closed, and we were together. At my best, I focused every nerve and sense on my single task: help. The world receded.

But the door, of course, leaked. One way it leaked was from the inside out. I often found myself not up to the task alone. I needed to open the door to reach out to specialists for advice, books for reminders, nurses and social workers for follow-up and support. I was dependent. The heroic sense of self-sufficiency groomed in residency (in part by the "guess-what-I'm-thinking" faux-Socratic "teaching methods" on too many medical rounds) did not survive in the vast world of real practice. At first, it felt like a loss. Was I not, after all, supposed to be Superman? But, I was lucky enough to be working in a superb group of pediatricians in a fine Health Maintenance Organization where helping each other was not a power trip but a source of joy. That turned out fine.

But there was another leakage, from the outside in, and it felt different and far worse. Its main feature was "scrutiny." As the years passed, more and more hidden cameras and invisible periscopes seemed to be watching. I was given rules for coding my encounters for billing purposes. My patterns of X-ray use were studied and compared to others. Lawyers coached me on how not to get sued. Our HMO's quality of care was rated by new scoring systems. Within a short time, I was running some of that scoring as the first "vice president for quality-of-care measurement" in the HMO, a newly created corporate officer position for doing what its title said: scrutinizing and reporting on the care.

It was interesting but nasty work. I was walking through the hallway of my pediatric practice group one evening and found my colleague and friend, I'll call him Steve, sitting at his desk with his head in his hands.

"What's wrong, Steve?" I asked.

He slowly looked up, raised a piece of paper in his hand, and said, in a monotone, "I feel so misunderstood."

The piece of paper was a check from the HMO, Steve's quarterly performance bonus of a few hundred dollars, his reward for the department's hitting some benchmarks of quality and productivity (over which, by the way, he had little personal control).

"Do these people really think that a few bucks are going to make me try harder to be a good doctor?" asked this very good pediatrician. "If so, they just don't know me."

Steve's sadness and his question have haunted me for the nearly 30 years that I have since devoted to working on and in the very "system" that Steve experienced as so crass and insouciant. He was, of course, exactly right. Perhaps some very junior and doctrinaire economist could believe that the blunt mechanisms of accountability, incentives, metrics and goal setting that were, even then, leaking into Steve's consulting room sanctuary would help him help patients, but no one who spent an hour getting to know him could continue in that belief.

There lies a chasm between the romantic, mutually treasured image of the "doctor-patient relationship" as the sanctuary door clicks shut and the ever-growing, ever-more-technically-enabled enterprise of trying to get a troubled, too-costly, too-risky, too-variable health care industry under control. Today's doctors in legions echo Steve – *I feel so misunderstood* – and the evidence of their demoralization makes Steve's case seem mild.

Enter Professor Chris Trimble. He gets it. He has spent a lot of time with Steve's descendants in their despair and he minces no words in describing what he has found. He writes, "I don't mean that, as a group, physicians are feeling a bit down. It's much worse than that. ... One departmental leader told me that most doctors are 'precariously balanced on the edge of burnout.' "

But Trimble brings hope. With decades of experience in fostering innovations in other industries, he now carries to health care new lenses and new tools. His central premise is one that I hold with equal certainty. Physicians have a choice. They can either allow the outside-in forces continue to try alone to concoct solutions to out-of-control health care cost and quality problems. Or they can take control themselves, not to defend the present, but to create the new care. "Let me be clear," he writes. "Only physicians can fix the system."

May I quibble here? Instinctively, I would have preferred "clinicians" to "physicians." Indeed, I have seen throughout my career that all health care professionals can participate in and lead innovation efforts. Trimble has made a calculated choice to narrow his focus to physicians, believing that by doing so his work will have greater impact. Maybe he has made a sensible choice, given his focus on larger and more ambitious innovation efforts — those that involve building new teams from the ground up and thus require a certain level of organizational influence. Nonetheless, I hope that non-physicians also read and act upon this important book.

It is easy to say that physicians should do the fixing instead of being fixed, but it is devilishly hard to do without a method to take control. Had I told Steve to take control, he would simply have laughed at me. He had no idea where the control levers were.

That's where this book could not be more helpful or better timed. It seeks to give willing doctors a set of guidelines for innovating within their span of their control to accomplish two goals: to create better care and to feel better themselves. Trimble's approach to innovation is simple and elegant, and the implications are profound: new roles, strong teams, new financial plans, new levels of patient and family involvement in decisions, and more. Most of all, he counsels, it is crucial to identify and extinguish a collection of time-honored but time-worn beliefs that are inconsistent with, in fact toxic to, the innovations needed. "Start with a clean slate," he advises, "Forget everything you know about the system we have."

For each component of action — the most crucial, in his view, being a new and bold approach to creating and nurturing teams — Trimble offers a roadmap grounded in his experience in other industries and informed by his extensive fieldwork in health care. What he advises is not easy, but his book is a wonk-free zone using clear and relaxed prose and a medley of vivid case studies to make both the challenges and the possibilities fully comprehensible.

Importantly, Trimble addresses his advice directly not just to doctors but also to senior leaders. Each chapter closes with a summary of points for each. To me, the wisdom in this is clear. Doctors may feel misunderstood, but, I assure you, health care executives feel pretty much the same. In my decades of work with hospital and health system leaders, far and away the most frequent question they ask me, with furrowed brows and strong frustration, is: How do we get the doctors involved? Their tactics, like Steve's performance bonus, may miss the point, but trust me, they feel misunderstood, too.

So Trimble, with experience in both camps now — the world of physicians and the world of executives — tries to build a bridge of shared endeavor — doctors to remake care, and executives to set in place the conditions to support and allow them to do so.

Would this book have helped Steve? I frankly don't know. So much of what Trimble counsels depends on context, and I don't know if Steve would have found the personal reserves, the local encouragement, the slack time, and the senior leadership trust to make it seem worthwhile to try.

But then I recall another part of Steve's story. Maybe from pure frustration, or maybe from a more positive source of motive, Steve became the best expert in the HMO on one topic that impassioned him: the fight against lead poisoning. He developed and ran a new lead poisoning care model, began teaching instructional programs to increase the skills of the entire staff, and became the local "go-to" consultant for us all. In other words, Steve didn't just work in the system, he worked on it. And that, exactly, is Trimble's clear-eyed invitation in this beautifully grounded book.

If you are like Steve, don't give up. Trimble offers you an alternative. Read on.

Donald M. Berwick, MD
President Emeritus and Senior Fellow
Institute for Healthcare Improvement

A Note to Readers

....

IT WILL TAKE THE DEDICATED EFFORTS OF MANY TO FIX HEALTH care — physicians, yes, but also policy makers, payers, administrators, researchers, technologists, and clinicians of all kinds. All can be innovators. All can create positive change — improved quality, safer hospitals, greater efficiency and more. I hope all will find value in these pages.

Nonetheless, I have made the deliberate choice to write *directly* for physicians and *indirectly* for all other readers. From the first draft of this book that I shared, I learned that my emphasis on physician leadership can evoke charged reactions from some. I have listened to many voices of concern, but I remain firm in my choice.

My rationale is straightforward. The focus in this book is *not* the totality of all innovation work. Rather, it is a subset of innovation initiatives that I believe are particularly crucial to fixing health care. These initiatives have the following characteristics:

1. They require a certain resource commitment — at least a few people, full time, working on a single initiative.

2. They involve the redesign of care, from scratch, for a particular patient population.

3. They only thrive with steadfast support and ongoing engagement from senior leaders.

4. They generally have a substantial impact on the work lives of the physicians involved.

In my judgment, for this subset of innovation initiatives, the odds of success are much higher when physicians lead. Physician acquiescence is not enough. Physician acknowledgment is not enough. Physician buy-in is not enough. What is needed is physician leadership.

Note that I did not say that physicians must lead the charge alone. I have seen several examples, a few of which are profiled in this book, in which physicians have co-led an innovation effort with a non-physician. Typically, the physician is a medical director and the non-physician is an operations director. Furthermore, the time commitment for the physician tends to decline over time, such that only the non-physician holds a permanent full-time leadership role.

Also, every rule has exceptions. I have seen a small number of initiatives of the type that I describe that have been successful without physician leadership. These initiatives tend to be led by non-physicians with unusual access to resources. They also tend to be the initiatives that do not much affect physicians, perhaps only making their work lives easier. I suspect that such initiatives represent no more than 10 percent of the work that needs to be done.

§

I have received the strongest negative reactions to the notion of physicians doing the fixing from policy experts who hold a particular point of view on why the system is broken in the first place. That view, presented in the starkest possible terms, is as follows: Physicians have too much power. Furthermore, physician education inculcates a level of arrogance that leads to subtle abuse of that power. It encourages physicians to operate too much, prescribe too often, and spend too carelessly. It invites physicians to design care not for patients but for their own convenience. It tempts physicians to view themselves as individual superheroes and their non-physicians colleagues as diminutive minions. For those that hold this point of view, any effort to empower physicians, such as this book, is readily dismissed as pouring fuel on the fire.

I acknowledge that there is an element of truth in this point of view. Simultaneously, however, I hold physicians in extremely high regard. I believe that the medical profession is unrivaled in its combination of talent, intelligence, energy, and commitment to a larger purpose. I question the wisdom of any approach to fixing health care that posits physicians as objects to be manipulated or barriers to be overcome rather professionals to be engaged.

I am sure that there exist problem physicians. I have no interest, however, in figuring out how to constrain or control them. Instead, I'm interested in inspiring and guiding the best that the profession has to offer.

So, who can be the best? Who can help fix the system? What exactly does a physician innovator look like?

The very first requirement, in my view, is humility. Indeed, the physician innovators who will fix health care will recognize that innovation is a team sport. The creation of multidisciplinary teams — those designed in such a way that every health professional is elevated to the top of their license — lies at the very core of the recommendations developed in this book.

Are there any physicians out there who are eager to take on the challenge of leading innovation, redesigning care, and building outstanding teams from scratch? If you want to know what I think, simply observe that I have bet my career on a hunch that there are not just one or ten such physicians out there, but tens of thousands.

If you are a non-physician, I hope you will read on. I acknowledge that you may find the passages so clearly aimed at physicians somewhat distracting. Still, I suspect you will agree with the overall prescription for fixing health care presented in the first few chapters. And, as you proceed more deeply into the book, the wide variety of innovation roles that you might play will become ever more evident.

If you are a senior leader of a health care institution, you may be less interested in leading innovation yourself than in creating an environment that encourages and supports others. I have appended chapters 4-10 with an "Action List for Senior Leaders."

I invite all readers to share reactions by writing me at chris.trimble@ dartmouth.edu. I am certain to benefit from your thoughts, wisdom and experience. Thank you in advance.

Introduction — A Call to Action

NEARLY A CENTURY AGO, AS THE UNITED STATES PREPARED TO enter World War I, James Montgomery Flagg designed what is arguably U.S. history's most famous poster. An intent Uncle Sam, clad in red, white and blue, pointed directly at the viewer. The caption: *I want YOU for U.S. Army.* The government published more than 4 million copies. The campaign was so successful that the same poster called the next generation to action during World War II.

We are fortunate to live in a more peaceful era. Nonetheless, we are confronted by a massive social and economic challenge. Health care — one-sixth of the economy, crucial to our well-being and host to some of life's most vivid moments — is broken.

The good news is that health care can be fixed. The central thesis in this book is that doing so will require the mobilization of tens of thousands of *physician innovators* who change the way care is delivered. And *I want YOU* to be one of those physician innovators.

To be clear, this is *not* a job for a single physician-hero, or even a small number physician innovators. Instead, it is a job for tens of thousands — each working in their own geography, their own hospital or clinic, and their own medical domain.

It is also *not* a job for physicians who wish to work alone. Indeed, the job calls on physicians to design, build, and lead or support *multidisciplinary teams* that put health care providers of all stripes to work in new ways. Further, the mission calls on physician innovators to work collectively, sharing ideas and spreading best practices.

If you are like most physicians, you have experienced health reform to date primarily as things done *to* you. But health care reform is moving into a new phase. Now, it will be about things done *by* you.

I hope this book will help light the path. I hope that it represents at least a modest contribution to a story that all of you can write together — a story about the United States of America and how it fixed its health care system.

§

At a recent conference, I met a physician who was more galvanized to fix health care than any other I've met. Her inspiration was a cancer patient, a father of grown children in his early 50s. The patient was self-employed, hard-working and neither rich nor poor. Although the disease was advanced when diagnosed, there was a high likelihood that it could be held in check for years to come with proper care.

Sadly, in a calculated choice, the patient, who had decent insurance, refused treatment. He felt it was unfair to burden his family with debt. He wanted to leave at least a limited financial cushion.

On our current trajectory, fraught tradeoffs between physical health and financial health can only become commonplace. Indeed, health care has become so expensive that one cannot care for a patient's overall well-being without keeping an eye on the wallet.

The tradeoffs are becoming ever more visible in aggregate statistics. For years, low-wage workers have seen escalating health care costs gobble up nearly every penny of increased earnings. Health care expenses are also crowding out public and private spending on goods and services that may have a stronger connection to overall well-being, such as education, infrastructure and the environment. And, many now agree that health care is the top threat to the long-term solvency of the federal government.

The problem with health care? To put it bluntly: It costs too much.

But I am not interested in imposing cost cuts. I'm *only* interested in pursuing innovations that promise a double win — better outcomes *and* lower costs.

And I want physicians to lead the charge.

§

In recent years, adversaries in the health care debate have fought primarily over the merits of the Affordable Care Act. Who should get coverage? What services should be covered? Who should pay for it all? What government interventions in health care are necessary and appropriate?

These are contentious issues, but these are waters in which this book will not swim. Instead, we will only be interested in what can be done to simultaneously improve outcomes and reduce costs. As sharp as the rhetoric about the ACA has been, certainly this is an objective that all can share.

Because public attention is inevitably drawn to conflict, the ACA's most crucial achievement, the one that directly serves this objective, may also be its quietest. The new law has catalyzed the proliferation of new contracts between payers and providers. These contracts move away from paying for each service delivered — that is, paying for *volume*. Instead, they reward increased *value* — better outcomes and lower costs.

This form of contracting is complicated. There are many ways to go about defining the patients, medical conditions, services, outcome measures and cost benchmarks that are included in any agreement. Adding to the complexity, these contracts go by several names, such as accountable care, bundled payments, risk sharing and capitation. Despite the challenges, these contracts are gaining momentum. This is unambiguous good news. There can be no going back.

Value-based contracts (I will use only this general term) throw the doors wide open for innovations that have long been stymied by perverse incentives. Would-be physician innovators have long found themselves in an impossible position. Doing the *right* thing for the system and the *right* thing for patients has far too often meant doing the *wrong* thing for the bottom line.

New contracts change the game. They make it possible for physicians to fix health care.

§

This book is presented in three parts: Foundations, Toolkit and Action. In Part I, I will more fully describe the nature of the challenge to fix health care and lay out just a few first principles for innovation. Part II is the book's workhorse. You'll find step-by-step advice that can be applied to any innovation in health care delivery. In Part III, I will share numerous additional examples and develop specific guidance for each of four innovation categories.

In Part I, I will draw heavily on observations that I made while shadowing physicians. As I promised confidentiality in these settings, my descriptions will be general and I will not use names. By the end of the book, however, I will have introduced you to more than 20 physician innovators who have agreed to openly appear here.

I wish I could profile every innovator in the industry. There are hundreds. In fact, it wasn't long into the work on this book that I began to feel that it seemed impossible to walk more than 10 yards in health care without stumbling upon an interesting innovation. Still, there aren't nearly enough initiatives in progress to fix the system. We have hundreds, we need tens of thousands.

The initiatives that I studied ranged widely. Some were near the beginning, some near the end. Some experienced smooth sailing, others faced considerable headwinds throughout. Some were clear successes, others more ambiguous.

I have not endeavored to somehow evaluate all that is happening and put only the best, the first or the most original innovations on display. I am not in the business of awarding innovation trophies. Instead, my desire is to illustrate bedrock principles for managing innovation across a wide range of medical contexts and disciplines. Particularly in Part III, I hope that you will find one or more hooks that make you think: "I could do something very similar."

All of the physician innovators I met while working on this book were generous with their time, transparent, tolerant of my dumb questions and energetically willing to help. I am grateful to each and every one.

In particular, I am grateful for the example they have set. They have illustrated the giant strides that are possible when physicians lead innovation in health care delivery.

PART I
FOUNDATIONS

Chapter 1
The Power Is in Your Hands

....

B EFORE WE GET DOWN TO THE BLOCKING AND TACKLING, SHALL we take a few minutes to get to know each other? Let's make that our focus for this first chapter.

I have been on the faculty here at the Tuck School of Business at Dartmouth for nearly 15 years. Before this book, I published five books with Vijay Govindarajan focused on the best practices for managing innovation inside of established organizations. I've taught classes, led workshops and given speeches around the world in industries ranging from paper to publishing, energy to education and computers to communications.

By 2011, however, it had become clear to me that the work I was doing on innovation in a general context was approaching a natural end point. I knew that I would soon feel that I had done all I could do, to the best of my ability. It would be time to turn the page and start a new chapter. I vaguely knew that I wanted to tackle something less abstract, something more applied. I wanted to focus on a social challenge confronting the world, one in which innovation was a critical piece of the solution.

One day, Al Mulley knocked on my door. A Dartmouth graduate, Mulley had spent nearly four decades at Harvard, becoming chief of general medicine at Massachusetts General Hospital. He returned to Dartmouth to lead a newly created research initiative. Mulley invited me to join and apply my past work to the specific challenge of innovation in health care delivery — not new treatments but new models of care. I'd also have the opportunity to join forces with a long list of distinguished Dartmouth researchers who had profoundly influenced health policy by challenging conventional wisdom.

Talk about a no-brainer. I immediately accepted.

§

In assessing the work I had agreed to take on, I immediately saw two distinct challenges — one technical, the other translational. The technical aspect of the job that I accepted was figuring out how health care is different from other industries, and, therefore, how the managerial guidance I'd developed for any context could be tailored to health care.

My initial instinct was that there would be at least a modest level of effort required here. In truth, I underestimated. There was a *great deal* of work to be done. Health care, as an industry, has more than a few peculiarities. The more I learned, the more I saw that the generic guidance I offer in any context could be substantially improved for health care. That discovery was energizing, as it elevated my sense that the work I was doing was worthwhile.

The translational challenge, by contrast, I knew would be steep. For the first time in my career, I was going to need to communicate across professions. I had spent more than a decade communicating with MBA-types, like myself. Now, MDs?

I was acutely aware that at no point in my life had I had any significant exposure to medicine. My impressions were limited to what I had seen as a patient, friend or family member. One of these exposures was both vivid and recent, and it was effectively my starting point in learning about your profession.

§

My father's diagnosis was hypertrophic cardiomyopathy. To my family and me, this was nothing more than 11 intimidating syllables. We all felt better when we understood what it meant: Too much muscle in the heart. It seemed an entirely fitting diagnosis for a career fighter pilot in the Marine Corps.

In the hour before the surgery, to lighten the air, Dad and I traded wisecracks. At one point, I wondered aloud if the surgeon would give Dad the excised heart muscle, the way kids sometimes get to come home with their tonsils. Dad replied, "That's a very interesting thought, Chris. And what do you suppose we could do with it?" As we are both fans of the movie *The Silence of the Lambs*, I had no trouble picking up right on cue, mimicking Hannibal Lecter to the best of my ability. "Perhaps we could serve it with some fava beans and a nice Chianti," I suggested.

During the surgery, Mom and I tried to think of anything but surgery. Meanwhile, we received routine updates — on the heart-lung bypass machine, off the heart-lung bypass machine — that sort of thing.

Soon after the procedure, we had a brief interaction with the surgeon. He said: "We went in, we removed the excess muscle and we closed him up."

I suspect that the surgeon and I experienced those words quite differently. For him, it was "I stopped at the store, I picked up the milk and I came home." For me, he might as well have said: "I turned lead into gold, changed water into wine and pulled the sword from the stone."

Mom and I were by Dad's side as he regained consciousness. He looked at me, looked at Mom, he looked back at me and said, "Chris! Did you get the meat?"

Dad was well on his way to a full recovery.

§

I came into this project holding the U.S. health care system in high esteem, especially on the front lines, where doctors and patients interact. In this way, I was not that different from most Americans. I have seen that many doctors today feel that the status of the profession is in decline, and that may very well be, but from where I sit, it still looks quite high. Americans may express any number of frustrations with health care, but, by and large, they live in awe of doctors and the wonders of medical technology. Some blithely imagine that medicine can solve any problem, mend any injury and cure any illness. If there is a problem with health care, many think, it must have something to do with politics, policies and payments — not with providers or the care they deliver.

This mind-set has long been reinforced by television shows like *E.R.,* *House* and *Grey's Anatomy*. Medicine provides authors and screenwriters an abundance of terrific storytelling material. The stakes are life and death, and doctors, more often than not, are portrayed as heroes.

Having spent most of my professional life in MBA circles, I must admit to a bit of envy. As far as I am aware, there has never been a popular TV show about management consultants or investment bankers. We can, however, claim at least one terrific movie, though it is now nearly 30 years old and hardly casts business types as heroes. One protagonist falls into a metaphorical abyss and ends up in jail; the other, appropriately named after a reptile, makes a fortune trading on inside information while taking joy in laying people off. The movie, of course, is *Wall Street*.

§

To develop a richer and more nuanced understanding of the lives of MDs, one of my first steps was to read as many medical memoirs as I possibly could. Thankfully, there is something of a cottage industry here. Who knew that so many doctors were also writers?

By and large, these memoirs are full of terrific stories, well told. As I read, I felt daunted by the issues you confront every day — illness and health, life and death — which are quite a bit more weighty than the MBA fare of margins and profits, interest rates and growth rates. Frankly, reading a stack of medical memoirs made me feel small.

I also spent dozens of hours shadowing physicians at work. These up close and personal exposures to routine days in the lives of doctors provided an important counterbalance to the memoirs, which of course only highlighted medicine at its most vivid. One of the lasting impressions from shadowing was the incredible amount of time that you all spend typing in electronic medical records. The first time I shadowed in the ER, I expected drama. Mostly, I sat behind a doctor as he worked at a laptop. I learned just how hard it is to *read* what is captured in an EMR and quickly assimilate the small fraction of information that is clinically relevant.

Shadowing revealed to me just how diverse medicine is — the patients, the physicians and the nature of the work that physicians do. I watched, for example, a primary care physician pivot, chameleon-like, at 15-minute intervals, to engage patients from all walks of life — a psychiatrist, a construction worker, a nervous 12-year-old, an elderly woman. The physician's agility in doing so impressed me every bit as much as a two-hour GI procedure.

Along the way, I've tried to learn at least a bit of medicine. This was exciting at first. I've now made just enough progress for it to be entirely clear to me just how much I *don't* know. It does appear that I'm a bit better than a layperson now. In one recent visit (as a patient) to a doctor that I had not met before, she paused midway through our conversation and asked, "Wait, are you medically trained?" That made me chuckle. I suppose I now speak medicine roughly as well as I speak Spanish. I can make a good first impression, but I struggle to sustain a conversation.

This is really my way of asking your forgiveness in advance for any errors of a medical nature that I've made in these pages. I've worked hard to avoid it, and I've asked for help from people who know more than I. Naturally, I take full responsibility for any errors that remain.

§

I've now spent a few years immersed in the U.S. health care system — as a scholar, not a patient. I think it is safe to say that my understanding of your profession has evolved. One surprise stands out: I had no idea on the day I began my work that physician morale was so low. As you can imagine, given my sunny starting point, this was a jarring discovery.

I don't mean that as a group, physicians are feeling a bit down. It's much worse than that. Many of you are just trying to get through the day. One department leader told me that most doctors are "precariously balanced on the edge of burnout." Another mentioned that he estimated that 80 percent of his staff was unhappy.[1]

Will you advise your children to enter the profession? Many of you are actively discouraging it these days. I was startled to read this headline for a physician's blog post: *Why I Still Love My Job.* You have to justify loving your work to your peers now? Apparently you do.

The reasons for low morale are numerous and complex. I don't imagine that I can fully capture them, but the proximal cause appears straightforward: Too much to do, too little time. There is precious little slack in your schedule, if there is any at all. One physician told me that he is well aware the system is broken, but he is so busy that he rarely has more than a fleeting moment to reflect on how he might try to improve it. A financial planner, one who has many physicians as clients, told me that comments like "I love doctoring, but the schedule is just killing me!" are increasingly common, even from young physicians.

Much of the increase in your workload is administrative. You are ever more burdened with documentation and data entry, and this pulls you further from what satisfies you most —caring for patients. But your clinical workload is also increasing. You are up against chronic disease more often, for example, where there are no quick and satisfying victories. And you are under constant pressure to increase productivity to generate more revenue for your practice or institution.

There are other sources of low morale. You may feel bitter disgust toward the waste that surrounds you every day. Some of this waste is tied directly to the push for productivity. Shorter appointments lead to mistakes, hasty diagnoses and overreliance on expensive diagnostic tests. And sometimes greater "productivity" is waste in disguise — when it generates more revenue for the institution without delivering anything of value to the patient.

Indeed, some of you are feeling like mere pawns in a giant money-making game. Your institutions push you to do more, the pharma, device and biotech industries induce you to do it with their products, and payers try to resist by tightening rules and regulations. Many of you, over the course of your careers, have seen your ability to make independent treatment decisions narrow considerably. Some of you have seen your personal incomes slashed by a sudden change in payment policy that you had no control of whatsoever.

As if this weren't enough, in health care — indeed, in any service industry — low morale is self-reinforcing, through daily interactions between service providers and their customers. Put simply, when you are unhappy, it rubs off on patients; when patients are unhappy, they take it out on you. The misery spirals ever downward.

These are the daily pressures of medicine, but there also may be more fundamental sources of distress connected to lifelong aspirations and core values. Happiness is at least partly a function of expectations, and it seems that many of you entered medical school with incredibly high hopes. Perhaps you chose medicine imagining yourself saving lives. You were later influenced and trained by a generation of doctors who lived through a golden age of medicine, when doctors truly had it all — wealth, status, independence and a stunning expansion in medicine's capabilities that resulted in boundless gratitude from patients. You worked hard and sacrificed greatly through medical school, internship, residency and fellowship, and you did it in anticipation of a certain reward. Reality fell short, and that's painful for anyone.

On top of all else, how often do you find your daily work to be in conflict with the ideals that brought you into medicine in the first place? At a young age, you felt called to serve. Have you now become a doctor that you never imagined you'd be? Under relentless time pressure, are you giving patients less attention or empathy than you wish you could? Are you finding that your day-to-day behavior is more often driven by financial considerations than you care to admit?

For me, low morale was a surprising discovery. On closer examination, I get it. Perhaps you'd welcome some good news: The more I study health care, the more firmly I believe that the pathway to fixing the U.S. health care system and the pathway to fixing your profession are one and the same. Furthermore, you have far more power to do both than you likely imagine.

My guess is that many of you don't feel particularly powerful. Please stick with me just a bit longer.

§

While endeavoring to understand your profession, I have drawn on my own professional experiences when possible. My work in business and academia has not proven nearly as useful in this regard as my first career step. I was a submarine officer in the Navy.

As a high school student, I was powerfully attracted to mathematics and the physical sciences. Engineering school was an obvious choice, and I went on to study mechanical and nuclear engineering at the University of Virginia.

As much as I enjoyed *learning* engineering, I concluded after two summer jobs that I really had no interest in *practicing* engineering. The pace was too slow. I was young; I wanted excitement and adventure. A conversation with a Navy recruiter at a job fair helped me connect the dots. I could put my engineering knowledge to work on a submarine. As I had grown up in a military family, it was a comfortable choice for me. It didn't hurt that Tom Clancy published his first book, *The Hunt for Red October,* right around that time.

Though I did not dedicate a full career to it, the Navy was, overall, a good experience. I liked walking onto the sub each day — giant cranes rolling down the pier, the zaps and buzzes of welding, diesel fumes mixing with salt air. One pier had an enormous sign above the entrance that read "You Defend Freedom." I liked what it felt like to work for a mission-driven organization.

One doesn't just walk onto a submarine and start operating it. First comes a process that is similar to, though shorter than, medical school and residency. In Navy classrooms, all submarine officers go through the rough equivalent of a master's degree program in nuclear engineering. Then there is an intensive period of training on a land-based prototype of a submarine nuclear reactor. You learn to operate every system while closely supervised. Starting up the nuclear reactor for the first time (it's a bit more complicated than turning the key in your car) is a coming-of-age moment that I won't forget.

Training is completed at sea and includes all submarine systems. Beyond the power plant, I learned sonar, navigation, weapon systems, periscopes and life-sustaining systems such as the machinery for producing pure air and water. On the submarine on which I served, it was possible for more than 100 people to remain submerged for weeks on end, until the food ran out.

There is a final qualification that could be viewed as roughly analogous to passing your boards. Complete the exam and you are awarded gold dolphins to pin on your uniform, indicating to all that you are fully qualified to take charge of the ship, in port or at sea.

I was very proud that day. I had mastered an encyclopedia of submarine science, technology, rules and procedures. I had learned the location, design and operation of every valve, every pump and every control. I could find most anything on the sub blindfolded. I could react instinctively in most any imaginable crisis.

At one point, while thinking about this book, I imagined what it would have felt like if someone, soon after I pinned on the gold dolphins, said to me: "Chris, great job. You've learned how to operate the sub. That's terrific, but I now have a different job for you. I want you to fix the Navy." I'm certain that I would have looked at that person as though he had three heads. So I suspect that I have at least some sense of how it might sound to you to hear me say, "I want you to innovate to fix health care."

Nonetheless, that is *exactly* what I'm asking you to do.

Nobody ever asked me to fix the Navy, but there were more than a few things that I wanted to change. In fact, I began to lose interest in the Navy when I realized that I'd have to be promoted again and again, for at least 20 years, to have even a small chance at pursuing some of the improvements that I thought would be sensible (sometimes naïvely so). That's what prompted the career change that sent me off to business school and ultimately pushed me toward many years of studying innovation.

Because of those years of study, I can say the following with certainty: You are in a far better position to innovate in health care than I ever was in the Navy. The Navy is, appropriately, a highly centralized, command-and-control operation. Health care is, appropriately, exactly the opposite. Most of the power is on the front lines, in the hands of physicians.

It is estimated that physicians make the decisions that account for 80 percent of all health care spending.[2] As I recall, the peer of mine on the sub who had the greatest direct financial responsibility managed the ship's recreation fund, which paid for the annual picnic and softball game.

Indeed, there has always been a great deal of power in your hands, but set against that power has been a single mammoth counterforce: the perverse incentives of fee-for-service reimbursements. At last, that barrier is crumbling. New forms of contracting are spreading. Now, the job of fixing health care is primarily in your hands.

§

"My job is to serve the patient in front of me. Fixing the system? That's someone else's problem." My guess is you've heard a comment like that more than once. You may have even heard it during your residency, from a mentor whom you respected deeply.

Let me be clear. *Only* physicians can fix the system.

Policymakers, payers and administrators can help create the right incentives and right conditions, but they cannot finish the job. Their core tools are too blunt. Consider payers. Their most potent levers are choosing whom they will work with, what they will pay for and how much they will pay. These choices may help contain costs, but they almost inevitably do so by trading costs for something else of value — for patients, it's quality, access or out-of-pocket expenses; for physicians, it's income or work conditions.

Having given up on payers, some people are now wondering if *employers* can fix the system. They are paying the bill, the thinking goes, therefore they have the greatest incentive to get the job done. To me, the rising energy behind this line of attack shows just how far we have strayed. Do you really think that a corporation that is best in the world at, say, building jet engines is likely to also be adept at managing a network of primary care clinics?

Enough. The action is on the front lines. We need leadership on the front lines.

I want you to step up to the plate and lead innovation in health care delivery. If you do so, you accept a challenge. You make a bet that *you can build a better system* — not for the entire country, of course, but within your own geography, your own clinic or hospital, and your own medical domain.

You won't get it done alone. You will need to build a team. Furthermore, you will need engagement and support from the senior leaders in your organization. Or, if you work in a small physician practice, you may need to partner with the larger health care delivery organizations or health plans in your local area to gain similar support.

I will specify the specific help you need throughout this book. Sometimes it will be near at hand, sometimes you'll need to be patient — though hopefully not for long. As value-based contracts continue to spread, your ability to effect change will only strengthen.

.... § ....

Here, specifically, is the challenge for you (which the rest of this book will be dedicated to addressing):

1. Choose a patient population.
2. Be certain that you understand these patients' wants and needs.
3. Design from scratch a team that can meet those needs.
4. Define the roles and responsibilities of each team member.
5. Hire and train each team member.
6. Invent the schedules and operating routines for your team.
7. Accept responsibility for both outcomes and costs for the population.

Note that *designing, building and leading a team* is the crux of innovation in health care delivery. There has been a lot of chatter of late about the importance of team-based medicine. Less talked about is the reality that these teams by and large do not yet exist. I'm asking you to build them. My bet is that many of you — tens of thousands of you out of the nearly 1 million physicians in the country — are eager to call this challenge your own.

I acknowledge that for many of you the list above looks daunting. Note, however, that there is nothing technically or scientifically complicated on that list. I give my MBA students a deep dive into material similar to what is in this book (though without a health care focus) in a course with just nine 90-minute sessions. The course has no prerequisites. Though I ask my students to work hard, it is well understood that any student who does the work will do well.

You've demonstrated the academic horsepower and discipline to emerge through many years of medical school and residency. My instinct: If you have at least a year or two of clinical experience, then you likely have what it takes to lead innovation in health care delivery.

.... § ....

Not long ago, I attended a conference for primary care physicians. I went to sit quietly, listen and learn. Many of the presentations that day were excellent. There was, however, one true stinker — the one that was more about management than medicine. At first, I wondered if I was less interested only because I already knew plenty about management. As I looked around the room, however, it became painfully clear that I was not the only one who was bored.

I was reminded, in fact, of my days in engineering school, where I was exposed to a small dose of coursework that was managerial in nature. My passion at the time was applied mathematics. On any given day, I was as likely to pick up a calculus textbook as I was to pick up the new issue of *Sports Illustrated* — and I was then, and remain today, an avid sports fan.

Compared to the glitter of mathematics, the stuff on management seemed dull in the extreme. My attitude was that if it didn't involve partial derivatives and double integrals, it couldn't possibly be all that interesting or even relevant. Furthermore, the management concepts felt an awful lot like common sense, only dressed up with contrived terminology. Today I know that there was a lot of truth in that youthful assessment. There is way too much incentive in management scholarship to be needlessly inventive with language and to make straightforward ideas overly complex.

I promise not to do that to you in this book. I can't provide any cutting-edge bioscience or epidemiology. I can guarantee, however, that the straightforward ideas in this book will be presented in like manner. Though it took years of research to nail down the central concepts in the pages that follow, there is nothing here that is difficult to understand.

The major challenge that you'll face in leading innovation in health care delivery is *not* getting your mind around the ideas presented in this book. The challenge will be one of leadership. In particular, you will need to be able to communicate what you are doing and why, so that everyone involved in your initiative moves in the same direction as a cohesive unit. This won't be as easy as it sounds because, as we will see, innovation and conflict go hand in hand. Many of the people on your team will experience opposing pressures, subtle and otherwise. You'll need to help everyone anticipate these conflicts and pressures, mitigate them and guide the path forward.

§

If you say *yes* to the challenge of leading innovation in health care delivery, two aspects of your work life change — *accountability* and *autonomy*.

You get more of each. You accept that you are responsible for the patient in front of you *and* for system performance (as it pertains to a specific and local patient population). In return, you get more freedom. *You* hire the team. *You* design the care. *You* set the schedule. You refuse to be infantilized by rules that specify what you can and cannot do and what you will and will not be paid for. You refuse to run endlessly on a hamster wheel designed by others. Instead, *you* create the apparatus — hamster wheel or otherwise.

When you are successful, you will feel *much* better. You will feel more empowered, you'll have a broader span of control, and, given the incentives built into new contracts, your income will improve or you'll be able to justify a higher salary. In short, you'll diminish most of the pressures that are driving your profession's morale into an abyss.

Of course, it's not all about you. There is a larger mission here. By answering the call to lead innovation in health care delivery, you are taking on a new kind of patient — a very ill one — the U.S. health care system. And, you are taking on the challenge of building new teams that can deliver better value for the individual patients you see every day.

Not every physician will answer this call to action. A few docs will continue to brazenly work the system. You, however, are not one of them. If you were, you wouldn't be reading this book.

You want to innovate. Value-based contracts are opening doors. The opportunities are nearer at hand than ever.

Chapter 2
Doctors and Dollars

....

I F YOU DON'T MIND, I'D LIKE TO START THIS CHAPTER IN MY ABDOMEN. I'm lucky to be able to report that I've lived a healthy life. My one significant health episode was way back in high school, when my appendix ruptured. The ensuing infections and the need for industrial-strength antibiotics kept me hospitalized for more than two weeks.

That was a long time ago, but I have a few vivid recollections. I remember that I lost so much weight that a friend remarked that he could count my ribs from 20 yards. I remember my mom fattening me up with an endless supply of milkshakes. And I recall the surgeon. I can still visualize what he looked like, and I can remember a particular comment that he made to me: "You know," he said, "not that long ago, people died from this."

As you might imagine, that was a formative experience. I was surprised and intrigued. What I could not do at so young an age, however, was share the surgeon's perspective on the velocity of medical advance. He'd lived through an era in which medical breakthroughs must have seemed routine.

Consider the remarkable progress that the science of medicine has delivered. The 1940s brought us penicillin; the 1950s, a vaccine for polio. In the 1960s, doctors attempted the first liver, lung and heart transplants. The 1970s saw the introduction of CT and MR images; the 1980s, angioplasty and stents. In the 1990s, medicine turned the tide against AIDS in the developed world, and in the 2000s, scientists identified discrete molecular subtypes of breast cancer.

All of this has happened in the lifetimes of many who are currently on Medicare. That's jaw-dropping progress. It has been an age of medical miracles.

There's a nagging question, however, that must be addressed: *At what cost?*

Medicine and costs are, at best, uncomfortable cousins.

Indeed, it can be challenging to discuss costs at all in medicine. Cost reduction can quickly be equated with rationing.

But this is a chasm that must be bridged. Innovations in health care delivery have a *dual* purpose — better outcomes *and* reduced costs. If we avoid talking about costs, we ignore half the job. We must embrace *both* objectives.

My goal in this chapter is simply to explore the fraught relationship between doctors and dollars.

§

A dose of historical perspective is a helpful starting point. I won't be doing much with graphs in this book, but at this moment one is clearly indicated — one that shows both outcomes and costs — and the movement of these variables over time. I'll be using variations of this graph throughout the book. Here, my intent is to show the *frontier of what's medically possible* and how that frontier has advanced.

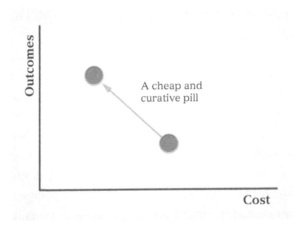

Figure 1

On the graph's vertical axis is a hypothetical aggregate measure of outcomes — longevity, function, absence of disease and discomfort — and on the horizontal axis, per capita spending on health care. At a given moment in time, the graph would show just one point, representing the current medical frontier — outcomes and costs.

The role of innovation, of course, is to improve what's possible. In the best-case scenario, an innovation pushes the point on the graph upward and to the left, toward better outcomes at lower costs. A cheap and curative new pill, for example, would do exactly that, by making patients healthier more quickly and simultaneously avoiding expensive alternative treatments. See Figure 1.

Now imagine for a moment where the frontier was in, say, 1940. We didn't spend much on health care back then; there simply wasn't much that medicine could do. On the graph, we were at a point that was quite low and to the left.

What did innovation do over time? While it is nice to imagine cheap and curative pills, most of the innovations in the biosciences haven't been quite so good to us. (A richer discussion of the historical trajectory of outcomes and costs would include advances in public health. For simplicity here, I will focus only on medical advances.) In aggregate, innovation in the biosciences has moved the frontier not upward and to the left, but upward and to the right — better outcomes but also higher costs. Back in the days of breakthroughs like penicillin and polio vaccines, the innovation trajectory was quite steep. Sure, innovation raised costs somewhat, but we were more than happy to pay because the gains in medical outcomes were enormous. As time has passed, however, the curve has flattened, as illustrated in Figure 2. Cost increases have become more dramatic; medical gains less so.

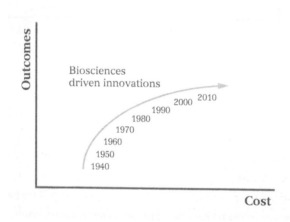

Figure 2

This makes health care like most any other industry. Innovation trajectories tend to have a decreasing slope. One sign of the flattening of the curve in health care is an increasing frequency of medical reversals, in which early research supports a new therapy but subsequent research reverses the recommendation.[3] Another sign: I'll bet you can think of more than one recently released pharmaceutical that offers, at best, a marginal improvement over the prior best option but costs much more.

To complete the story, we need to add one more element to the graph: a cost constraint — that's the vertical line to the right in Figure 3. Today we are clearly running up against a societal limit on how much we are willing to spend on health care. Thus, the most pressing question in health care today is: *How can we deliver health care to everyone without going broke?*

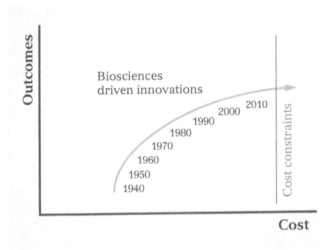

Figure 3

Quite clearly, we cannot simply continue along the same trajectory. And yet, I do not want, nor do I suspect that you want, to stop pursuing the development of new diagnostics and new therapies. What now, then?

The solution is innovation. Step one, however, is to recognize that we need a completely different *type* of innovation. What got us here is not going to get us where we need to go. We can no longer focus solely on innovation in the biosciences; we must now simultaneously pursue innovation in health care delivery.

And on that note, here is some really good news: *There are plenty of innovations in health care delivery that are like the cheap and curative pill.* Instead of pushing the frontier upward and to the *right,* they push it upward and to the *left* — equal or better outcomes at *lower* cost, as shown in Figure 4.

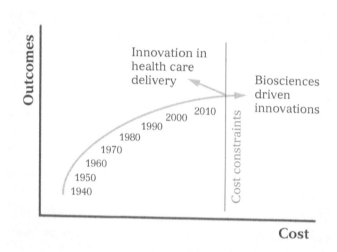

Figure 4

This may sound too good to be true. It's not. Thanks primarily to the inducements of fee-for-service medicine, innovations in health care delivery have been overlooked for decades. As a result, opportunities are everywhere. The fruit is both low-hanging and abundant. There are ways to simultaneously improve outcomes and reduce costs in every specialty, every institution and every geography.

Because of ever-tightening cost constraints in health care, we can only *earn the right* to continue our vigorous pursuit of advances in the biosciences through the *equally vigorous pursuit* of innovation in health care delivery. We need both.

§

As you know, I want you to lead innovation in health care delivery. I want you to fix health care. There is, however, a prerequisite. First, you must decide that you care — truly care — about costs. Do you?

I'm afraid I've had several exposures to physicians that indicate an almost complete lack of interest in cost. The stories that follow are not, of course, representative of *all* physicians. But I think you'll agree that they are not outliers, either.

§

While shadowing, I witnessed a conversation between physician and patient that I imagine made the physician's week, month and year. The patient could express nothing but gratitude. It had been a long battle with a complex disease and the patient was all but cured. Life had gone from fear and discomfort back to near normal.

I must admit I was jealous. It will not surprise you to hear that I've never taught a class or given a speech that induced a similar level of appreciation. Several physicians I have met have called their work a privilege. I agree. One even used the words *sacred* and *intimate* to describe the work. These are, I can assure you, descriptors that are not often used by MBAs.

I admire your passion. I also, however, see your passion as a tremendous potential barrier to the work of fixing health care. To put it bluntly, why would a profession oriented to a mission as intoxicating as *saving lives* ever give a rat's tail about cost?

§

A few years ago, during a casual conversation, a physician I had just met asked me to share a bit about my research. At one point during my description, she furrowed her brow and said to me, "So you're just studying how health care is organized?"

There was quite a bit packed into those few words. First of all, she was right. Studying how health care is organized (and its relationship to cost) was exactly what I was doing. But don't overlook that word *just*. She had called my work trivial, but she had meant no harm. It was an honest and impromptu reaction. To her, innovation in health care meant *cutting-edge science that saves lives.* And I get it. How could *how care is organized* or *how much care costs* compete with that?

So far, it's not competing very well, at least if you use the federal budget as an indicator. The National Institutes of Health has a $30 billion annual budget, while the recently formed Center for Medicare and Medicaid Innovation spends less than one-tenth that amount each year. It's also not competing well for public attention. In our community, we have a massive annual fundraiser to raise money for cancer research. I'm not yet aware of any similar efforts to raise money to support innovation in health care delivery. Are you?

Cost has never been sexy. It has never been exciting. It has never been a priority ... but it must be now.

§

At one sizable primary care practice that I visited, a physician told me that it was a cherished element of the clinic's culture that medicine was practiced without an eye toward what anything costs. Note that she did not simply say that she didn't concern herself much with costs; she described her indifference as a *cherished element* of the clinic's culture.

I'm afraid that just a short time later I came across a similar sentiment at a much higher level of authority. I interviewed a surgeon who had recently become chief. Of the many topics we discussed, one was the nature of his new job responsibilities. I wanted to explore, in particular, the financial aspects of his work. As it turned out, there wasn't much to discuss.

He told me that before he accepted the position, he wanted to be sure he wouldn't be in over his head on finances. He was confident with surgery, not with budgeting. So he asked questions. Bottom line: His boss, the chief executive, also a surgeon, assured him that he need not worry about it. "We have a talented staff for managing the finances. Leave it to them."

Reading between the lines, but only just a bit, I'd say that the surgeon felt that worrying about money was simply beneath him. There were other hospital employees who could cook the food, take out the trash and add up the numbers.

This is, I'm afraid, a lousy place from which to launch a quest to fix health care.

§

Among the many medical memoirs I read were two by Emily Transue, a primary care physician in Seattle. In one chapter, she described her very first few days on the job in a private clinic, just after she had completed her residency. She received something of a crash course in coding and billing. After seven years of medical school and residency with barely a word about money, her introduction to the topic was a hurried blast of arcane rules. Straightforward visit with a new patient? Code that a 99201. If it's an established patient, that's a 99211. And if it's a longer or more comprehensive visit, well, that might be a 99212, 99213, 99214 or 99215. And don't forget that the financial viability of the clinic depends on accurate coding and the documentation to back it up.

So this is how doctors are introduced to dollars, I thought as I read. Dollars are red tape and aggravation. No wonder docs care so little.

§

I saw that emotions can run much deeper than indifference when I met a psychiatrist who shared his worry about one of his patients who was at risk of suicide. The psychiatrist was trembling as he spoke with me about it. Underlying his fear for the patient was a deep anger that was directed at the insurer and its procedures for prior authorization. The insurer was refusing to pay, and the patient was too poor to pay out of pocket. Tangling with insurance companies, he went on to say, was a part of the job. The task was impersonal and frustrating on good days, intolerable on others. As I listened to his story, I imagined him on hold with the insurer, internal monologue something like: "I'm trying to save a life, and this *asshole* on the other end of the phone is trying to save a buck!"

The bean counters, thus, are the evil enemy. And who would want to join *that* team?

§

One final enlightening interaction was with a physician innovator whom I interviewed and then followed up with by email. I asked for some additional data to more fully understand the financial impact of the work she was doing. I got more than I asked for. In addition to sharing additional data, she wrote that she had attended a funeral that weekend and stood with the family whose loved one she had treated in the last week of life. It was a good death, she wrote, or at least as good as a death can be, in that it was expected, non-traumatic and peaceful.

And then she wrote: *How can you put a price on that?*

§

I've studied lots of industries. In general, cost and quality compete on equal terms. Some companies emphasize cost, others quality, but both are always in the picture.

And then there is health care. Uniquely among industries, health care has evolved with minimal attention to cost. Forces of cost constraint have historically been so weak as to be irrelevant. With fee-for-service reimbursement and with most patients paying only a small fraction of the incremental cost of each service, only payers have had the incentive to try to control costs. Most have struggled to intervene productively, and many have simply passed as much risk as possible on to employers.

I don't enjoy being the bearer of bad news, but that era of obliviousness to cost has definitively come to a close. We can mourn its passing, but we must move on.

Cost matters.

Want to help fix health care?

First you have to care about costs.

§

I want you to be part of the solution. I want you to innovate to fix health care. On the other hand, I'm pretty sure that you didn't choose medicine as your profession because you were keenly interested in solving abstract macroeconomic problems such as the burden of escalating health costs on the federal budget. So here are some motivators that perhaps will hit closer to home.

For starters, you might choose to care about costs because you care about your patients. Among the many screening questions you ask, how often do you ask your patients whether they are concerned about their ability to pay their medical bills, or even to refill their prescriptions? Even when the Affordable Care Act is fully implemented, many patients will have insurance that leaves a substantial financial burden on their shoulders. Are you doing patients any favors if you lessen their back pain but simultaneously render them unable to pay their rent?

That's a direct connection between cost and the overall well-being of the patient in front of you. In most cases, however, the connections are less immediate and more diffuse. Indeed, it may feel like the only clear and immediate benefit of cutting costs might be to fatten some large organization's bottom line or to make a tiny dent in the Medicare budget. Seen that way, money is a filthy commodity. It is a number, devoid of any human virtue.

What if, however, we viewed money through a longer-term and broader lens — and not as an *end* in and of itself, but as an *intermediator* between real human choices? It may be helpful to pull out of the world of medicine for a just a moment to the simpler world of personal finance. A quick example: Much of my income comes from self-employment — from traveling and giving speeches and workshops about innovation. In good years, I have been faced with the happy problem of having more opportunities than I could fulfill. I was confronted with the question Just how hard do I want to work?

I never found it very helpful to ask: "Will my family be happier if I earn X more dollars this year?" That's too difficult a question to answer. It's just a number on a tax return. Instead, I found it helpful to ask questions on the other side of money. "Will my family be happier if I'm away from home a bit more, but we can go on a longer vacation?" Or, "Will we be happier if I'm away from home a bit more, but we save more and reduce the odds that my wife and I will ever become a financial burden on our children?"

So what's on the other side of money in health care? One innovation I studied — a simple one, but one that required the formation of a new kind of team — saved roughly $400,000 annually. That's just a number, and not a huge one in a $3 trillion industry. On the other hand, it's a number that's easy to convert to a human scale. It's enough to pay for health insurance for nearly 50 average Americans. That's 50 people who no longer have to worry about the possibility of an illness forcing a choice between treatment and personal bankruptcy.

And, honestly, how can you put a price on that?

A wasted dollar is not a triviality for someone else to worry about. It has real consequences. It makes health care more expensive for everyone. It increases the number of uninsured and underinsured. It squeezes budgets everywhere. It reduces what individual families can direct to good living, what governments can spend on education or infrastructure, and how much businesses, both small and large, can invest in growth and job creation.

A wasted dollar does harm. It is not as obvious or as direct as the harm you might inadvertently cause with, say, a misdiagnosis, but it is every bit as real.

One physician I have spoken with is deeply engaged in the cause of innovation in health care delivery and likes to talk about developing an *ethic of conservation* in medicine. To me, such a principle dovetails naturally with a more familiar one: *First, do no harm.*

§

There is one more reason that I think you might choose to care about costs. Put simply: You won't like what happens to your work life if you don't. Physician morale may be low today, but I'd wager a fair sum that it can go lower still.

Costs are going to be contained one way or another. The only question is how. If physicians do not engage and lead the effort, a solution will be imposed from above. As I write this, I'm reminded of a lie I was told when I was hospitalized back in high school. Just before inserting a tube in my nose, down the back of my throat, through my esophagus and into my stomach, a provider said to me, "This might be a little bit uncomfortable."

I respect you too much to say something like that. If a solution is imposed from above, you will be miserable. You will hate it. You will kick and scream, even more than you do now.

Payers and policymakers, when they work alone, work with sledgehammers, not scalpels. Think about the regulations, requirements and cost controls that you probably already detest. You want more? There will be more rules about what is and what is not covered, more administrative hoops for approvals, more reductions in reimbursement rates, more unhappy patients and more frustrated doctors.

It appears to me that one of the most satisfying aspects of being a doctor is the degree of independence you enjoy. Yes, it is less than what it once was, but it is still substantial. Glyn Elwyn, a colleague of mine at Dartmouth, likes to say that only in health care do millions of board meetings take place every day. He's talking about the conversations you have with patients. That's where almost all of the spending decisions take place. That's where the power is — and in my view, where it *ought* to be.

If physicians don't fix health care, however, that power will be further eroded. You'll go to work every day feeling a bit smaller than the day before.

I want you instead to feel large. I want you to take control.

The thought of pure business (or policy) types trying to fix health care on their own is, for me at least, a bit scary. I know businesspeople a little bit too well. They tend to underappreciate the complexity of health care, and, as a result, their solutions tend to be too simplistic and too sweeping. Furthermore, bottom-line incentives can be strong — too strong for an industry so rife with ethical conflict — and I don't much like where that might lead.

To the extent that I get a vote, I want physicians in charge.

.... § ....

Someone once said to me that when there is an elephant in the room, you ought to introduce it.

By now, many of you have considered the reality that what looks like *cost* to one person looks like *salary* to someone else. Won't controlling health care costs lead to fewer jobs for health care providers — people like you?

Let's address this head on. To do so, I'm going to engage in some armchair speculation, looking out several years into the future and assuming we make dramatic progress in containing costs. Will there be losers? Certainly. Here are my top four:

Hospital construction and capital equipment. Study the physics of business cycles and you'll quickly see that a supplier of capital goods lives in a far more volatile world than the industry it serves. If we have any success at all in containing health care costs, we're likely to find that we don't need another hospital and we don't want that new MRI machine. The impact on the companies that provide these goods and services could be severe.

The pharmaceutical, biotech and medical device industries. If we start thinking about value in health care much more critically, we're going to see that the track record of these industries in recent years doesn't look nearly as good as it once did. We're going to demand a lot fewer of the "breakthrough" treatment options that these industries have been selling.

Hospitals. Hospital care is expensive. Many of the positive steps that can be taken to reduce costs amount to doing a better job of discerning when hospital care is truly needed or can be avoided through prevention and proactive care. Nonetheless, the hospital business will be lifted by other forces, especially growth in the population, the aging of the population and the expansion in the number of insured patients under the ACA. I suspect that at worst, hospitals face a gradual decline, not a sharp and sudden one. Some hospitals will find themselves vulnerable, and a small fraction may fail.

Specialist care. I hope that there are more than a few specialists reading these pages, so this is worth a bit more discussion. Should you be concerned about your livelihood? Yes, you should, though not acutely so. The forces affecting specialists are similar to those affecting hospitals, in that many innovations in health care delivery will amount to making better choices about when specialist care is truly needed. Any impact from such innovations will be offset by all of the same forces lifting demand for hospitals. A gradual decline in demand looks like a worst-case scenario.

Note, however, that in responding to a decline in demand, individual specialists have greater flexibility than institutions. Your options, roughly in the order of impact on the system, from best to worst:

Lead the charge. Become like one of the physician innovators you'll meet later in this book. Change the way care is delivered to improve value. By doing so, you'll prove your mettle in the new world of health care. You'll make yourself indispensable.

Relocate. Health care spending is far from uniformly distributed across the country.[4] If you are in a geographic region in which your specialty is overrepresented and you start to see a slackening in demand, one option is simply to move to a part of the country where your services are not as readily available. And, if exploring the world is your cup of tea, consider the emerging economies, where health industries are under construction and will be growing for decades to come.

Just keep doing what you are doing. Specialist care is hardly going to disappear. If you are good at what you do, your services will remain in demand.

Fight like hell to protect the status quo. When confronted with change, resistance is a natural first instinct. But do you really want to follow that instinct full steam? High costs are harming patients, communities, businesses and governments. If you do not want to step up to the plate to lead innovation in health care delivery, wouldn't you feel better about yourself if you at least simply stood aside, out of the way, while others worked to fix the system?

.... § ....

As you know by now, I want you to engage in fixing health care by think-ing about costs, by driving for improved value and by being something of an entrepreneur — a physician innovator. There's a bit of a paradox here. Earlier, I said that I'm far more comfortable with physicians, not businesspeople, taking the lead in fixing health care. Yet simultaneously, I'm asking you to think and act a bit more like businesspeople.

In part, this simply means embracing closer partnerships with talented people with business training. The nature of innovation in health care delivery and the omnipresence of cost reduction as an objective demands it. You are going to need good analytical help.

To date, however, I haven't seen physicians place much value on business acumen. This is particularly evident in hiring practices. As I have traveled around the country and visited various hospitals, I have often noticed pairings of physician leaders and managers that look like Superman coupled with a third-grader. The physician is a nationally known specialist with degrees from the most prestigious institutions; the business partner is a recent graduate of the local community college.

Closer engagement between physicians and businesspeople, I recognize, is difficult. Distrust of business runs deep in the medical profession, to the point that it is common to hear physicians express a low opinion of other physicians who are just a little bit *too* successful on the business side.

Historically, doctors have zealously defended their independence from business interests, intolerant of the possibility that the drive for profits might distort decision-making about patient care. The longstanding tradition of physician self-employment is a direct result. The business of running a hos-pital, in turn, is quite peculiar by the standards of any other industry. Most of the spending decisions have traditionally been made by people who are not even employees.

While I see the rationale, I can't imagine fixing health care without closer engagement between MDs and MBAs, or, more broadly, between providers and administrators. Physicians are increasingly choosing hospital employ-ment over self-employment, and this can only help, but I do not view it as strictly necessary. An alternative is the creation of partnerships in the form of a co-management agreements that, in effect, turn over one aspect of a hospital's operations to a physician or practice. Lighter-weight partnerships that simply allow self-employed physicians to work closely with hospital business analysts to assess outcomes and costs are also certainly possible.

§

I want you to do more, however, than just partner with someone with strong business skills. I want you to *embrace* the role of physician innovator — and, perhaps, your inner businessperson. To help you do so, I'd like to share two simple exercises I have used in the classroom.

The first is one that I used with fourth-year medical students. I wanted to get a sense of how the students felt about business in general, so I asked them to share the first word that popped into their heads based on the following prompts I put on the screen:

MBA.

Hospital Executive.

Insurance Company CEO.

There was a bit of awkward silence at first. Then someone said "greedy." That got the ball rolling, and there wasn't much ambiguity or dissention from there.

It was only an exercise and it was somewhat contrived. Nonetheless, I suspect that those called to the profession of medicine generally believe that their purpose lies on a higher plane than those who choose business. I have no problem with that. Not only that, I don't think too many of my MBA students would disagree.

Not all business activity is equal, however. Sometimes, the work of business lies on a high plane indeed. To get my MBA students thinking about this, I ask them to complete the following SAT-like analogy:

Medicine / Health

Science / Truth

Law / Justice

Business / ?

I wish I could say that my students acquit themselves well during this exercise, but by and large they fall prey to the obvious temptation to just say "profit" or "shareholder value." Most have a hard time articulating how business success improves the world.

It doesn't always. There are many ways to drive profits up. I find it useful to put the possibilities into two categories — those that *shift value* from one entity to another and those that *create value*. In the former category, consider a company's effort to renegotiate with employees, customers, partners or suppliers. The direct impact of any new contract terms is simply to shift money from one entity's pocket to another. Doing so is neutral for society at large.[5]

Businesses *create value,* on the other hand, when they innovate. They do so by developing new offerings that consumers want — faster computers, smarter phones, softer fabrics, you name it — or by figuring out how to produce and deliver existing offerings at lower cost. These are the activities that drive the world forward. This is what *progress* looks like. And progress, I believe, is the best one-word answer to the question above that I pose in my MBA classroom.

Most efforts to innovate — and certainly the efforts you will read about in this book — are efforts to create value. They are efforts to move the world forward. This connection has helped motivate my work on innovation for many years. There are a great many business issues that I would have a hard time getting the least bit excited about, but innovation is not one of them. Innovation is business at its best. It is through innovation that businesses solve the unsolved, create jobs and improve lives.

To the extent that you engage in *value creation,* you can feel confident that your work is honorable and will lead to societal progress. You can lead innovation in health care delivery, pay close attention to costs and *feel great about it!*

Take note that what you've seen in your industry, for many years, is the dominance of strategies that emphasize value shifting, not value creation. Payers and providers, for example, have been deeply absorbed in strategies focused on size.

Indeed, chances are quite good that at some point in your career, and maybe even quite recently, you have seen a merger or acquisition in your local region. Chances are also quite good that the strongest business rationale for that merger or acquisition was to achieve greater negotiating power. For example, when a health system makes an acquisition, the primary goal is often to amass enough heft in the local region that payers simply cannot live without the health system and still offer desirable or credible insurance. Payers then have little choice but to acquiesce to whatever prices the health system wishes to charge. It works exactly the same way in reverse. Mergers among payers are driven by the same logic.

This strategy is not complicated. It's the rough equivalent to two heavy-weight boxers punching the hell out of each other. In some regions, the payers have the upper hand; in others, the providers do. Either way, blood is shed at the negotiating table. One bottom line goes up a bit, the other goes down a bit and the direct impact on society is neutral. It is business as usual on the front lines. Outcomes and costs remain unchanged.[6]

By and large, businesses opt for and stick with value-shifting strategies for as long as the bottom-line results are acceptable. Value shifting takes less skill and is less risky. If shareholders are happy, why attempt more?

But frankly, I'm not at all interested in having you engage in the game of value shifting. It's not worthy of your considerable talents or your considerable commitment to doing good in the world. I am only interested in your engagement in the work of value creation.

In other words, I'm not asking you to *join* the businesspeople that have shaped the trajectory of your industry. I'm asking you to reach higher.

Chapter 3
The Gold Lies Just Beyond the Wall

F ROM TIME TO TIME, I AM INTRODUCED TO PEOPLE WHO WONDER how one really goes about writing an *entire book*. I usually tell people that writing is not that different from bricklaying. You show up, spend several hours laying bricks, come back the next day, and repeat. In other words, while the entire task is large, each step is nothing more than nouns and verbs, sentences and paragraphs. More than anything else, completing a book is matter of showing up.

Fixing health care is, obviously, a bigger task than writing a book. Similarly, however, it can be broken down into manageable pieces. What I'm asking you to do is to take charge of *laying one brick*. If you and tens of thousands of others take this on, we all win. Collectively, the book about how physicians fixed health care will be written.

Laying a brick will, in all likelihood, take you into unfamiliar territory. It may take you well outside your comfort zone. But it is achievable. These are bricks that individual physicians can lift.

§

In the introduction, I briefly described the nature of the innovation initiatives that I believe are necessary to fix health care. I called for tens of thousands of initiatives, each involving a handful of people — a few to a few dozen — brought together in a new team structure.

You may wonder how I arrived at this prescription. I must say I was surprised by it myself. I did not anticipate calling for a large number of small initiatives, but this is the conclusion my research led me to. (Are initiatives involving a few people to a few dozen really "small"? Well, they are tiny against the backdrop of an industry that employs nearly 20 million, though possibly quite substantial within a single department in a single hospital or clinic.)

I was surprised because in my work in other industries, I have had to work hard to explain the *limitations* of such approaches. I spend a lot of time doing so, because the "let a thousand flowers bloom" innovation mind-set is far more popular than it is practical. It is popular because it dovetails nicely with the happy notion that everyone can be an innovator. The problem is that companies in other industries generally pursue strategies that call for a more concentrated approach — a small number of ambitious initiatives.

Health care, again, proves to be different. It is a massively heterogeneous and fragmented industry, composed of tens of thousands of small and local distinct business operations, dozens of which are often under the roof of a single institution. Each small and local business needs innovation on a small and local scale.

It is important to get the size of each innovation initiative right. Small does not mean "as small as possible." Indeed, it's possible both to undershoot and overshoot. My purpose in this chapter is to further describe and illustrate the sweet spot in the middle. This is where, I believe, far too little work is being done. We need a massive scale-up of the work that is in progress.

§

By calling your attention to the sweet spot, I'm not trying to stop or detract from either larger or smaller initiatives. There is plenty of valuable work in progress on either side of the sweet spot.

Efforts on the larger side include those to launch new companies, build new institutions, offer new health plans, merge, acquire, restructure or consolidate. You might think of these as "CEO-sized" initiatives. A colleague here at Dartmouth, Elizabeth Teisberg, and her co-author, Michael Porter, advocate for a specific restructuring of health care delivery, one in which the traditional departmental structure is abandoned in favor of units focused on medical conditions. Such a change, done well, would help catalyze several of the innovations in health care delivery that will be described in this book.

The potential risk with large-scale changes is that they may force an overly sweeping solution. The initiative that wins in cardiology will not necessarily look much like the one that wins in primary care. Each specialty and each patient population faces distinct challenges. Therefore, CEO-level initiatives should focus on enabling innovation without over specifying it.

In the smaller category are the projects that can be executed *on the job* and within people's *slack time*. Such initiatives are the primary thrust, though not the exclusive thrust, of quality improvement programs. The strength of such efforts is that they have the potential to draw in the creativity, talent, and energy of all employees. Everyone can be an innovator. Typical changes that result from this work include the use of safety checklists in the operating room, improved scheduling to increase access or reduce waiting time, and even added vigilance in washing hands. The mind-set is this: The car has been designed, let's make it run faster and more efficiently, and with less waste and fewer errors. Much has been gained through such endeavors in the past, and I suspect that this will remain the case for years to come.

But consider two limitations of this approach. The first is the practical reality that people are already very busy. There is only so far that an effort to squeeze innovation work into people's slack time can go. Any project that is larger than what one person, or at most a small handful of people, can execute within their slivers of slack time are bound to get stuck. Second, quality improvement efforts work *within* the existing organizational model. They do *not* reimagine job roles and team structures. They take what exists as a given.

These two limiting factors are not soft constraints. Together, they constitute a concrete wall. What all of the examples in this book collectively illustrate is that there is a pot of gold just beyond the wall.

The gold is *just* beyond, not *well* beyond the wall. It is not necessary for health systems to "bet the farm" by making enormous bets on speculative innovation concepts. It is not necessary to pursue the risky initiatives described above that can only be led by chief executives. Instead, the innovation agenda is composed of modest bets on small teams — a few people to a few dozen — that have been redesigned from scratch. I like to think of these initiatives, shown in Figure 5, as "physician sized."

Note that ideas for physician-sized initiatives might result from efforts that are labeled either innovation or improvement. Regardless of the source of the ideas, however, the work cannot stop with small projects that can be jammed into the slack time in the existing organizational structure. Indeed, beyond further pursuing and expanding value based contracts, the most important change that senior executive teams in hospital systems can make to fix health care is to make more bets on physician-sized initiatives.

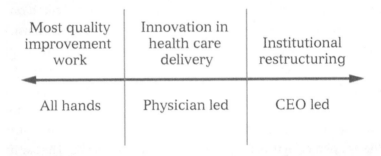

Most quality improvement work	Innovation in health care delivery	Institutional restructuring
All hands	Physician led	CEO led

Figure 5

Then, your role is to design and lead the team. The task, of course, is not quite so simple as "your role is to design and lead the team" makes it sound. But is it achievable?

Absolutely.

§

I know it is achievable because I have seen so many terrific examples already. There are many to come in this book.

Because I have touted the critical nature of new types of contracts between payers and providers, it is important that I highlight, even before presenting the first illustration, that most of the examples in this book took place *in absence of* a new value-based contract. Some were grant funded, for example, and some were supported by senior executives who were willing to lose money temporarily while getting ready for a future of value-based care. Others, as we will see, found a variety of other creative options for getting started even without a value-based contract.

Such methods for financing innovation are limiting and difficult to sustain. Nonetheless, these initiatives help light the path forward. They give future leaders of innovation in health care delivery the opportunity to learn from those who preceded them.

They also make us all hungry for what becomes possible as new contracts proliferate — the scale-up of innovation in health care delivery across the country. One of the happiest pieces of news I received while writing this book, from an insurance industry insider, is that in 2014 roughly one in three health care institutions in the United States was at least experimenting with new forms of contracting. And there were no signs of reversal.

§

Of many innovations I studied, I was particularly intrigued by what I observed in Salt Lake City, Utah, at Primary Children's Hospital, which is owned and managed by Intermountain Healthcare and staffed by pediatricians and surgeons of the University of Utah School of Medicine. Primary Children's chief medical officer, Ed Clark, has been anticipating a future of value-based contracting for years. He has been pursuing a variety of improvements, so that when that future arrives, Primary Children's is already several steps ahead.

Clark's work includes both quality improvement projects that are small enough to be squeezed into slack time and innovations in health care delivery that involve building new teams from scratch. To further reinforce the distinction between the two, one example of each follows, starting with a quality improvement effort.

Several years ago, Clark and his team began working to improve the treatment of respiratory syncytial virus. The infection can lead to pneumonia or bronchiolitis, and, in many cases, hospitalization. Each winter, Primary Children's experienced unpredictable spikes in admissions of children with pulmonary complications. This created hospital crowding that was difficult to manage and often led to delays of elective or semi-elective procedures.

Infants with RSV typically needed to be supported with fluids and supplemental oxygen until the disease ran its course. In general, RSV patients were discharged only when they were no longer in need of supplemental oxygen. Clark believed that patients and their families would be better served if they could be discharged and cared for at home sooner. He and his team surveyed hospital practices within Primary Children's for treating RSV and discovered wide variations. Some physicians kept patients hospitalized substantially longer than others.

The team gathered and hammered out a standard protocol, seeking to establish a clear baseline against which subsequent improvements could be measured. With the new protocol in place, average length of stay actually went *up,* but only temporarily. The team began systematically testing departures from protocol, assessing results and making quick progress. Most critically, the team began discharging the lowest-risk patients sooner, sending them home on supplemental oxygen.

Before discharge, families and caregivers needed a bit of training on supplemental oxygen plus a period of observation to make sure they could manage feeding and secretions on their own. With just that much, however, most families did quite well. So Clark and his team kept pushing forward, discharging more patients sooner. They even set up a rapid-treatment center for RSV that enabled lower-acuity patients to avoid admission altogether.

With time and experience, Clark's team became more accurate in its predictions of which patients would succeed at home on supplemental oxygen and which would struggle. The team became more cautious about discharging patients who lived at high altitude in the nearby mountains, who were not fluent in English, or who did not have transportation to get to the hospital quickly if needed. Also, the team began consulting social workers to help judge whether the patients' family structure was sufficiently stable and capable of managing supplemental oxygen.

§

Clark's efforts with RSV illustrate classic methods for quality improvement and continuous learning. In particular, note that the improvements were *within* the existing organizational structure. There is plenty of progress that can and should be made in this manner.

Again, however, my interest is in the larger leaps that are achievable with modestly larger initiatives — those that lie just beyond the wall — and those that involve the construction of new kinds of teams. Another of Clark's efforts proceeded in exactly that way.

Like many around the country, Clark is keenly interested in what has become known as *hot spotting* — efforts to anticipate which individual patients or segments of the patient population are most likely to require intensive medical care in the not-too-distant future, and intervening early to minimize the need. It is the specificity of the prediction and the focus on the avoidance of near-term acute care episodes that makes hot spotting different from traditional prevention.

Clark and his team have extensively mined hospital data and even genealogical records to learn as much as possible about where pediatric hot spots lie. Clark is able to rattle off statistics based on his team's analysis, such as these: The most expensive 0.1 percent of pediatric patients account for 13 percent of total pediatric costs (based on a sample of commercially insured families). And, 40 percent of health care costs in the 1- to 9-year-old segment are attributable to children with a prior stay in the neonatal intensive care unit.

Clark is quick to point out that pediatric hot spotting does not look much like adult hot spotting. The latter is mostly about the major chronic diseases — diabetes, cardiovascular disease and respiratory diseases — and the behavior change that is necessary for adults to stay as healthy as possible. Pediatric hot spotting is more of a mixed bag. It takes on a variety of forms to match a range of pediatric conditions.

Despite this complexity, Clark reasoned that a simple bet was likely a winner. His specific hypothesis was that for a small subset of the patients with the greatest health care needs, a dramatically increased investment in primary care was highly likely to both improve outcomes and reduce costs. Clark knew someone at Primary Children's who he thought was perfect for the job of building and leading a primary care clinic focused on this population: Nancy Murphy. Clark and Murphy launched the clinic in 2007.

Murphy had a deep passion for serving the population often referred to in the literature as children with complex medical conditions — those, for example, with congenital abnormalities, genetic syndromes, brain injuries, spinal cord injuries and neuromuscular disorders. (Murphy's clinic did not serve all children with complex needs. Some were already served by other special programs at Primary Children's, including those with spina bifida and cystic fibrosis.)

In general, Murphy's patients had tragic life stories. Many did not survive childhood. They also had intensive and expensive health care needs, some reaching more than $100,000 per year in care. Typically, across the country, care for such patients was poorly coordinated. Families were often overwhelmed and confused by multiple and sometimes conflicting directives from various specialists.

Murphy's team worked to improve care through more thoughtful and longer-term care planning, more careful coordination and closer engagement with families in the form of lengthier appointments and as-frequent-as-needed check-ins by phone and email. In setting up her clinic, Murphy had the flexibility to follow her instincts on team structure and scheduling. She soon had a full-time team of four — herself, a nurse practitioner, a nurse care manager and a medical assistant — that by 2014 was serving approximately 600 patients.

Murphy devised a schedule in which appointments were slotted into 60- or 90-minute blocks, with up to four patients seen by her team during any one block. A full one-third of the schedule was left free of appointments altogether. Murphy and her team used the "free" time for care planning, care coordination, and phone and email contacts with patients' families.

Let's pause for a moment so you can reflect on your own practice. Imagine what you might be able to do if you had Murphy's degree of autonomy. Imagine you could hire your own team. Imagine you could redefine each team member's roles and responsibilities from scratch. Imagine you could reinvent your clinic's schedule — how much time with patients, how much time meeting with the team, how much time for email and phone calls to patients and the other providers who treat them — all without respect to what payers are and are not willing to pay for. Could you improve outcomes and reduce costs? As value-based contracts proliferate, it becomes ever more likely that you will have the opportunity to show that you *can* do so.

Before going on with the story, it is also worthwhile to make a few observations about Murphy's work. First, her initiative is far from enormous. It is a small team. Second, at a high level, the concepts at play are not radical, breakthrough, or even original. They are common sense. Indeed, there are initiatives based on similar principles moving forward elsewhere in the country. Finally, note that Murphy's work is focused on a small and unusual population of patients and thus could not readily be generalized for more mainstream settings.

Given these observations, you may reasonably wonder why I've chosen to present Murphy's work first. It's not because I anticipate an over-the-top *wow* reaction from readers (though Murphy's work with her patients did hit hard with me personally). Instead, I chose this example because I see it as emblematic of the nature of the innovation work we need much more of. For every Nancy Murphy we have today, we need 100 more tomorrow, each leading a physician-sized initiative based on common-sense principles, each building a small team that delivers better care at lower costs for a local patient population with a specific set of needs.

We don't need a handful of Silicon Valley style innovation home runs. Instead, we need tens of thousands of base hits.

Now, back to the story. In defining how her team would operate, Murphy needed to address who did what. She did not start with all of the answers —

only a philosophy that she wanted to delegate as much as possible to be as efficient as possible, such that every team member was working at the top of their capability and licensure. Of significance, this meant that Murphy herself would spend less time doctoring and more time managing and supervising — a not-always-easy transition that those who lead innovation in health care delivery must frequently confront. Over time, Murphy's team members worked out their routines. Because they all worked together in the same small room every day, they had plenty of face-to-face contact to make it happen.

During an appointment, patients and their families typically had a lengthy conversation with one provider and shorter conversations with one or more additional team members. In addition to the four full-time team members, several others joined the team part time as circumstances demanded. On most days, one or two pediatricians joined the team, and on days when patients who required ventilators to breathe were seen, an ENT, pulmonary specialist and rehabilitation doctor joined. Murphy's team also frequently called upon the hospital's social workers to help develop a fuller understanding of what it would take to keep patients at home and as healthy as possible.

At the end of each scheduled 60- or 90-minute appointment block, the team gathered to quickly discuss care plans for each of the patients who had just been seen and to assign follow-up actions. Every team member had the opportunity to offer input. When the team could not quickly agree on the best course of action, the discussions continued at the end of the day.

During half-day intervals without appointments, the team responded to a steady stream of emails from families. Some could be answered quickly by email; many required phone conversations. After particularly traumatic appointments, families often needed to hear the care plan a second time to fully absorb it.

The families that the team served were highly engaged and rarely missed a step in caring for their children. On occasion, Murphy's team followed up on missed appointments or unfilled prescriptions. The families did not find the reminders intrusive; they appreciated the help. Indeed, Murphy believed that such calls helped to cement the partnership between the clinic and the family.

In 2013, Murphy submitted a report describing the clinic's performance to the Center for Medicare and Medicaid Innovation. The most concrete sign of improvement: emergency room visits were down 14 percent and hospitalizations down 25 percent. Furthermore, surveys showed that families were more satisfied. Murphy and her team had worked diligently, through conversations and through a survey, to discover what families valued most: keeping their child's pain at bay and keeping them comfortably in school. Families were so satisfied, in fact, that the clinic's panel grew quickly through word of mouth. Murphy never needed to make any effort to spread the news of the clinic's opening. In fact, by 2014, Murphy was hiring to double the size of her team and patient panel.

Outcomes up? Check. Total system costs down? Check — and by an estimated 12 percent. Success!

§

There is one sour aspect of Clark and Murphy's story. In 2014, Primary Children's was still overwhelmingly paid on a fee-for-service schedule, and a substantial proportion of the clinic's work was not reimbursable. As such, Murphy's clinic operated at a financial loss. Her team's ability to cut system costs accrued to the payer, not to the hospital.

Fortunately, Primary Children's is in a privileged position, one that affords Clark an unusual degree of flexibility. Primary Children's is a pediatric referral center for an enormous geographic area in the Intermountain West, within which the hospital has minimal competition. As a result, Primary Children's income statement is strong and the institution can make financial sacrifices in pursuit of its mission. It can do the right thing for patients and for the system, even when doing so diminishes the hospital's income somewhat.

The management team also wants to be ready for the new world of value-based contracting before it arrives. A happy sign of the shifting times: By 2014, Clark and the hospital team had negotiated a new value-based contract with one payer and was in talks with others.

Aware of the special circumstances that supported her work, Murphy felt grateful for the privilege of running the clinic. She also worked diligently to run as tight a ship as possible, taking every possible step to maximize allowable billings and to minimize costs.

§

I've spent most of my professional life rubbing shoulders with MBAs, not MDs. As such, it is second nature for me to presume that successful innovators will pursue geographic growth. If all physician innovators thought this way, then perhaps we'd just need a few dozen of them to fix the system, rather than tens of thousands.

Not so fast. I have asked most every physician innovator I have spoken with some version of the following question: *Are you considering national expansion?* Honestly, most seemed to regard my question as strange or at least surprising. The question may reveal the starkest divide between MBAs and MDs. To an MBA, the answer is "Duh." To an MD, it's "Huh?" If you show an MBA something that works locally, the very next instinct is "How can we grow this business?" The typical MBA dream is to be the next Jeff Bezos or Steve Jobs — to build the next mammoth global corporation that becomes an indispensable part of everyone's daily lives. For MDs, the canonical youthful dream is quite different. It is to save a life.

A few MDs may have the ambition for expansion, but it is worthwhile to think through what the pathway to national expansion would look like. First, the physician innovator's work life would change dramatically, and through several predictable phases of increasing managerial abstraction — from doing the work to training others to do the work, to measuring performance and holding leaders accountable, to being a guardian of the culture of a national organization. Along the way, there would be far fewer face-to-face visits with patients and far more time in airports. This is a sequence that seems quite normal to MBAs, quite less so to MDs.

There is more to consider. Most innovators in health care delivery recognize that the U.S. health care system is already overbuilt. As a result, adding capacity through national expansion is a dubious step in the drive to fix health care. It doesn't make sense unless you are simultaneously putting someone else out of work — a messy, bloody process.

Now in MBA land, you can say, "Put the inefficient bastards out of business," and it's regarded as sporting or at least normal. I'm not sure most MDs are ready to turn on each other to that degree. Even if some are, the fight strikes me as particularly delicate in health care because doing well in any market requires some degree of trust and collaboration with other local providers — at the most basic level, to attract referrals. That may be hard if it is clear that your intent is to put local rivals out of business.

So, even as we are getting more and more contracts between payers and providers that create the right incentives, I don't see it as highly likely that a substantial fraction of the work that needs to be done will be tackled by single business entities spreading innovations in health care delivery nationwide.[7] We might see great ideas systematically spread within the nation's largest health systems — but, as you know, even the largest account for only a small fraction of the industry.

And that leaves us with a need for tens of thousands of initiatives.

One of the reasons I am optimistic that this approach will work is that most of the physician innovators I have spoken with are willing, if not eager, to share what they are doing with anyone who wants to learn. Clark and Murphy, for example, enjoy hosting visitors and routinely present their work at conferences. Other physician innovators I have spoken with aspire to create nationally recognized "centers of excellence" for their new model of care, hosting visitors, sharing job descriptions, creating training materials, publishing results — doing everything possible to spread the model short of building a nationwide organization to deliver it themselves.

This openness highlights another stark difference between MDs and MBAs. Such open sharing of innovations is anathema to the MBA set, but it will be a powerful expedient in the effort to fix health care. The collective and collaborative drive of tens of thousands of physician innovators is exactly what it will take to get the job done.

I have seen that giant leaps in performance are possible through innovation in health care delivery. One initiative I studied, in its first year, generated savings equal to five times the money invested, while improving outcomes. That's remarkable. In most industries, returns on investment are measured in percentages, not multiples!

Given the potential gains, it seemed odd to me that payers have not sooner engaged in spreading new models of care through value-based contracts. In most industries, if there is a buck to be made, somebody is going to make it. Why hasn't payment reform proceeded sooner?

It could be that the level of distrust between payers and providers has proven, for many years running, to be too great to overcome. As one investor said to me, "If an idea requires payers and providers to work together, forget about it, no matter how worthy it might be." Furthermore, through the 2000s, payers have by and large been delivering solid financial results and feeling secure enough in their futures without taking risks on innovation. Motivation was simply too low because times were "good enough."

That all changed with the Affordable Care Act. The uncertainty in the legislative process made payers very *in*secure about their futures. The forward-thinking payers are now working aggressively to change payment models, to be part of the solution and to demonstrate that they can add value. Most critically, perhaps, payers have learned from the failure of 1990s health reform. They are recognizing that the value they bring, beyond signing contracts that align incentives, is not in the form of controlling or limiting what physicians and patients do; rather, it is in the form of supporting physicians in doing their jobs better, especially by sharing data and analytics that physicians and patients can use to guide care.

It will take time for payers to reach all delivery organizations with these new approaches. In all likelihood, the smallest practices will wait the longest. Nonetheless, the trend is promising. The pieces are falling together, such that a new generation of physician innovators can take control and do what needs to be done.

Chapter 4
Walk and Chew Gum

....

TO GET STARTED, ALL YOU NEED IS A GOOD IDEA. PLEASE DON'T spend too much time thinking about it. Are you surprised to hear me say so? A brief analogy will put this in perspective.

One of my favorite recreational activities is hiking. This hardly makes me unusual in my home community here in Vermont. In fact, compared to some of my neighbors, I'm quite the casual hiker. I am rarely gone for more than one day. I have a small backpack, I know how to put lunch in it, and that's about it.

Nonetheless, I have often contemplated the possibility of taking on a genuine mountaineering challenge — high altitudes, snow and ice, climbing equipment and so forth. I haven't done it yet, but I've talked to a few people who have. From those conversations, I've gleaned an important lesson. *Only a novice celebrates at the summit.*

The expert, on the other hand, knows that the real dangers are on the descent. As the day goes on, the sun gets higher, the temperatures rise and the snow gets softer and more vulnerable to collapse. Meanwhile, the climbers get more fatigued and more likely to make a big mistake. The expert understands that you never celebrate until you are safely back down to the bottom of the mountain.

Innovation is similar. The "summit" is that moment when you say, "That's a fantastic idea! Let's make it happen." If you examine the books and articles that have been written about innovation, or look at what companies do when they decide to increase their emphasis on innovation, what you will see is that almost all energies are focused on finding that game-changing idea. The implicit assumption is that if somehow we can generate a flash of inspiration, the rest will be a snap.

Far from it. A great idea is only a great beginning. As it is with mountaineering, the hidden dangers are on the descent.

I have been using the analogy between innovation and mountaineering for more than five years now. It seems to be an effective way to shift at least some attention from the front end to the back end of the innovation challenge — from ideas to execution.

This is not an easy shift to achieve. These two sides of the challenge are night and day, and it's not too hard to see why the front end garners all of the attention. The front end is about ideation and creativity, science and technology, dreams and possibilities. The back end, by stark contrast, is about the blood, sweat and tears of getting the work done.

Furthermore, there is a strong tendency to underestimate the degree of difficulty of innovation execution. Many people think, "Hey, we can execute. We execute every day! If we can just come up with a great idea ..." But comparing day-to-day execution to innovation execution is like comparing a simple somersault to a triple flip with a quadruple twist.

I don't want to overstate the point. There are a few contexts in which the quality of the idea is everything and execution is second nature. Book publishing strikes me as a reasonable example. Find a winning manuscript and the rest is routine. A more relevant example for our purposes is biosciences-driven innovation in health care. There, the ideas are indescribably complex. In drug development, for example, you may need a PhD in molecular biology just to join in the conversation. Once you create that new drug, however, a highly evolved industrial system for taking it to market awaits.

Innovation in health care delivery, however, is *not* an exception to the rule. In fact, I have never seen a context in which the mountaineering metaphor is more apt. The ideas are not at all complicated. Indeed, the following four-point innovation agenda (which we will examine much more closely in subsequent chapters) likely captures more than 90 percent of the work that needs to be done.

1. Standardize and delegate

2. Coordinate

3. Prevent

4. Improve treatment decisions

Yes, that's the entire list. Almost all innovations in health care delivery that I've examined employ one or some combination of these strategies for delivering better outcomes at lower costs. And, no, you do not need a PhD in anything to grasp any of them.

That's not to say that there isn't some nuance in further specifying these ideas for a particular health care context. For example, considering the first agenda item, I do not think, nor do I imagine that you believe, that all health care can be converted into standardized processes. Health care operations are not like airline operations or automobile manufacturing; they are messier and more human. (Heard in a hospital hallway: "I'm so sick of consultants coming in here and telling us to run this place like Toyota." Exactly right.) But there are certain aspects of health care delivery that can and should be standardized, and the productivity gains can be enormous.

Still, the core point remains. The ideas in innovation in health care delivery are the easy part. This is not a game where one should study, debate and refine an idea before even getting started. Don't polish the cannonball. Just get to work. Launch and learn. Adjust as you go.

Your idea needs to be neither perfect nor original. Remember, what we need to fix health care is tens of thousands of initiatives. Most of them will be similar to initiatives that have already been tried elsewhere, with some adjustments for the local context and the specific medical domain.

If you don't have any good ideas, go to a conference this year that is more focused on health care delivery than new treatment options. If you don't sleepwalk through the conference, you'll come back with a dozen ideas. Or network with your colleagues at other hospitals and learn from what they are trying. Or visit the Agency for Healthcare Research and Quality website, where you'll find nearly 1,000 reports about innovation initiatives around the country, along with contact information for the people who lead them.

Finally, as innovation in health care delivery is heavily focused on cost, you might consider looking at what providers in low-resource environments are doing, both domestically and in the emerging economies. Precisely because they have always faced tight resource constraints, these organizations have had no choice but to innovate — and in ways that may suggest directions that you should go in as well. Later in the book we'll look, for example, at a clinic that serves the homeless.

Bottom line: The idea is the easy part.

-------------------------------- § --------------------------------

My goal in the remainder of this final chapter of Part I is to provide an overview of the challenge of executing an innovation initiative. In Part II, we'll get into the details.

Execution is the hard part, and the difficulty starts here: Innovation execution is not one job, it is two. It requires simultaneously building something new *and* sustaining excellence in what already exists. Not *just* innovation, but innovation *and* ongoing operations.

As it turns out, this pairing is quite a bit harder than walking and chewing gum. Innovation and ongoing operations are not only very different from one another, they are inevitably in conflict with each other. It is crucial to elaborate on this point, but to do so I need to divert our attention *away* from health care for a few paragraphs.

Most companies, in most industries, like to view themselves as innovative. The reality, however, is that most organizations evolve not toward excellence in innovation but excellence in ongoing operations. They become what I have called in past books Performance Engines.

Well-run Performance Engines are masters of many tasks. They excel at serving existing customers and fighting existing rivals. They are experts in efficiency. They run on time, on budget and on spec every single day. And, of course, they consistently deliver bottom line profits.

Already, it is not hard to sense the conflicts with innovation. Nonetheless, a well-run Performance Engine is highly desirable. The Performance Engine is *not* the enemy. Great organizations build great Performance Engines. I'd even go so far as to say that by mastering the construction and operation of Performance Engines, the management profession has raised living standards throughout the world.

But excellence in ongoing operations creates complications for innovation. The conflicts lie in the method of the Performance Engine, which is to make every task and every process as *repeatable* and *predictable* as possible. Each is powerful in the drive for steady and growing profitability. The rub, of course, is that innovation is exactly the opposite. Every innovation initiative is, by definition, *non-routine* and *uncertain*. These are fundamental incompatibilities, and they drive right at the heart of how managers are trained and how organizations are designed.

Therefore, to succeed at innovation execution, you somehow have to simultaneously master two deeply opposing activities. Adding a bit more spice to the challenge is the reality that in most organizations, nearly all resources are allocated to ongoing operations.

This is a difficult challenge, but it is one that *can* be mastered. This is the challenge that has been the focus of all of my prior books.

§

Now let's get back to health care delivery.

The way I just framed the innovation challenge has worked well across dozens of industries and around the world. Once again, however, health care proves unique. Health care delivery organizations are simply not accurately described as Performance Engines. They are less driven by efficiency, repeatability or predictability — and by far — than any other organizations I've ever observed.

Health care delivery organizations have been driven by a different ideal — to deliver the best possible care for any medical condition, regardless of cost. Furthermore, the industry as a whole has aspired to advance the science of medicine, so that even better care — new diagnostics, drugs, devices and therapies — can be delivered tomorrow than is possible today.

Measured against these objectives, U.S. health care delivery organizations would have to be regarded as supremely successful, bringing cutting-edge science and technology to bear against the full spectrum of human diseases. As a result, we are all surviving what would have been deadly maladies just a few decades ago. This is an awesome achievement, though a different kind of achievement than creating and managing a Performance Engine.

There are fundamental conflicts for us to deal with in health care, but the problem is not well characterized as innovation versus efficiency. Instead, innovation in health care delivery versus innovation in the biosciences is closer to the mark. Indeed, the health care system we have seems perfectly aligned for advancing the biosciences yet perfectly *mis*aligned for innovation in health care delivery.

To see the conflicts fully and in sharp relief, consider what defines "best possible care." Over the years, a number of assumptions have become deeply entrenched in the mind-sets of providers and patients alike. All of these assumptions looked better a few decades ago than they do today.

More care is better. With fee-for-service payments, weak forces of cost containment and a history of spectacular medical break-throughs, there has been little reason to question this assumption. With the era of nearly unlimited spending behind us, however, this mind-set must be challenged.

High tech care is better. The golden age of medicine has shaped everyone's thinking. Almost inevitably, the latest and greatest treatment gets the benefit of the doubt over the tried-and-true alternative. It shouldn't.

Every patient deserves custom care. It's hard to argue with the ideal of treating every patient as a unique individual, but some pockets of health care are quite amenable to standardization, and in ways that yield better outcomes.

To fix is better than to prevent. Everyone has heard the old axiom that an ounce of prevention is better than a pound of cure, but preventive steps never generate the excitement that even the possibility of a breakthrough cure can. As a result, powerful opportunities in targeted prevention have been overlooked for years.

Physicians should be autonomous. If the goal is to deliver the right individualized care for each patient, then physicians *must* be able to act autonomously — to design the right care, one patient at a time. Further, physicians are all conditioned, during residencies, to feel complete responsibility for the patient in front of you, to the extent that you comfortably speak of "my patient" and "his patient." This mind-set must also be challenged. The examples throughout this book illustrate the power of *team* care.

Specialists deliver cutting-edge care. The desire to drive the frontiers of medicine forward has naturally led to an organizational model that prizes narrow specialization. In this way, the organizational model for health care delivery is quite similar to that of academia. Health care professionals specialize in ever tighter domains and tend to be most tightly networked with others in the same specialty. This makes a great deal of sense when the overarching objective is to advance knowledge. It is also, however, at obvious odds with the goal to treat patients as whole persons, not parts.

Innovations in health care delivery go against the grain by questioning one or several of these assumptions. They shift the emphasis in care delivery along the following dimensions:

FROM		TO
Volume	to	Value
High-tech	to	Low-tech
Custom	to	Standard
Cure	to	Prevention
Autonomous physicians	to	Teams
Specialization	to	Whole patients

I do not share these stark contrasts as an indictment of an "old way." Far from it. Earlier, I said that great companies build great Performance Engines. Similarly, we ought to honor and take pride in what the health care industry has accomplished to date. Further, we ought to be mindful that in many segments of health care, these assumptions are likely still serving us well. In my view, much of what is written and said about health care reform takes on too strong a tone and too hard a line about what's broken.

The goal is not to tear down what we've built. The goal is to sustain it while simultaneously advancing innovations in health care delivery. We must walk and chew gum. We must continue to deliver the best possible care to each patient while working aggressively to improve value for the dollar. We must take on two sets of activities that are not just distinct from each other but in conflict with each other.

And it can be done.

§

One of the biggest enemies of innovation within established organizations is a widespread myth. Put simply, it is the belief that innovation is primarily a function of brilliant individuals with brilliant ideas — superheroes who are so committed to those ideas that they do whatever it takes to move them forward. They break rules, they work underground, they create little conspiracies. And, because they are so supremely talented, they succeed.

This may not be the way you personally think about innovation, but I can assure you that the myth is widespread. Two common questions I am asked when I speak about innovation are: How can I identify the real innovators in my organization? And, how can I knock down at least some of the barriers that I know they are going to face? It seems completely natural to frame the innovation challenge as a battle between brilliant individuals and bureaucratic behemoths.

We must do better than that.

I believe that the myth's persistence has everything to do with the overemphasis on innovation's front end — the hunt for the big idea. Yes, one person really can come up with a great idea. But can one person execute? Can one person simultaneously build something new and sustain excellence in what already exists?

Clearly not. Indeed, innovation is a team sport.

Furthermore, good old-fashioned teamwork is insufficient. Executing both innovation and ongoing operations simultaneously requires a particular team structure.

To show why, a quick analogy. I had a friend way back in high school, I'll call him Tom. People both loved Tom and loathed him, because he seemed to be good at everything. He was a rock-star student and a rock-star athlete. He was also exceedingly popular, both with the guys and the girls.

There really do exist a handful of individuals who are (annoyingly) multitalented. However, groups of individuals — teams — are an entirely different story. Inevitably, getting teams to function well together requires specifying roles — that is, who does what and when. That makes teams inherently less flexible, and definitively less multitalented, than individuals. Design a team for Activity A and it is not likely to be adept at Activity B — especially if there are deep incompatibilities between A and B. You are much better off with a separate and independent team for B, one with a distinct design.

So how then can your hospital, clinic or team succeed at both ongoing operations and innovation simultaneously? It is quite possible, but it does require some organizational fancy footwork.

First, it requires some separation. You cannot ask the same groups of people to be fully immersed in both activities at the same time. Doing so is too conflicting and too confusing. Instead, there must be teams for ongoing operations and teams for innovation in health care delivery.

The complication is that both will need to exist within the same larger organization and there will inevitably be some overlaps. As you might imagine, these overlaps tend to be trouble spots. They need to be defined and managed carefully.

The details await in Part II.

§

To reinforce the most important insights and recommendations from Part I, I provide the summary below. Similar summaries are included at the end of each chapter in Part II.

Note in particular the list of action steps for senior executives within health care delivery institutions. For staff physicians, part of the purpose of this list is to give you a clear picture of the support you will likely need to succeed. Independent practices and physicians will sometimes need to partner with larger institutions to move forward with innovation in health care delivery. In such cases, the list will help give you a sense of what you need from the partnership.

Chapters 1-4: Summary of Observations and Recommendations

1. Physicians can fix health care by leading innovation in health care delivery. We need an army of tens of thousands of physician innovators to get the job done.

2. The expansion of value-based contracts is opening the door for innovations that have long been stymied by perverse incentives.

3. If you want to help fix the system, you must care about costs.

4. There are countless opportunities for innovation in health care delivery that are just a bit too large to be executed within typical quality improvement programs. They require the creation of small, full-time teams ranging in size from a few people to a few dozen.

5. The core challenge for physician innovators is designing, building and leading new teams that can deliver better outcomes at lower costs.

6. The idea is the easy part. Focus your energies on execution, and specifically on the complexities of building something new inside of an established organization.

Action Steps for Senior Executives

1. Continue pursuing and expanding value-based contracts.

2. Make more bets on physician-sized innovation initiatives — that is, on small, full-time teams that can redesign care from scratch to deliver better outcomes at lower costs.

PART II
TOOLKIT

Chapter 5
Solidify Your Idea

In 2011, Shreya Kangovi hired her first two community health workers, or CHWs, at the University of Pennsylvania Health System. Their role was to develop partnerships with low-income hospitalized patients who were at high risk of readmission. The CHWs ensured that the patients understood their discharge instructions, bridged communication gaps between the patients and providers, and informed providers about socioeconomic challenges that could affect care. After discharge, the CHWs assisted patients in taking care of their health until they had gained access and transitioned to primary care.

Three years later, the results had proven powerful. Readmission rates were down, patient satisfaction was up and access to primary care was up. UPHS had created the Center for Community Health Workers, under Kangovi's direction. The center was hiring to expand its staff to 30 people and broadening the range of situations in which CHWs worked. Kangovi's experience will serve as the primary illustration in this chapter and the next.

I have said that the idea is the easy part, but it is certainly worthy of a brief discussion. In this chapter, I will focus on developing a kernel of an idea into a more complete concept so that you can share it with others, build momentum and move forward to execute. In the next chapter, we will take our first look at execution, and, specifically, team design.

§

You don't need a perfect idea to get started, and you certainly do not need an original one. Indeed, just four fundamental ideas account for nearly all of the work that needs to be done in innovation in health care delivery. They are:

Standardize and delegate. Some aspects of care are highly amenable to standardization. If you can specify a care process, step by step, and make it routine, productivity will rise and errors will fall. System costs will fall.

To further reduce costs, it may be possible to shift tasks from providers with high salaries and many years of education to providers with lower salaries and less education. Standardization and delegation are generally paired, because it is easier to delegate tasks that are scripted and consistent. These steps are the basics of industrialization, practiced by Henry Ford in automobiles a more than a century ago, as well as many others before him.

Coordinate. From the patient's perspective, fighting illness is a continuous process that stretches out over weeks, months or even years — and the patient is intimately familiar with every minute. From the provider's perspective, health care is a collection of discrete interactions with the patient, spread out across time, space and multiple providers. In a perfect world, these interactions collectively would constitute a coherent, planned, well-coordinated and well-executed health care intervention, one that adapts as circumstances change.

Much easier said than done, as you are already all too aware. Fragmented, poorly coordinated care is commonplace, and it leads to confusion, contradictory care plans, missed care steps, wasteful duplication and errors. It can frustrate patients to the point that they disengage or give up, which of course leads to deteriorating health, more care and more waste. Improve the coordination of care, and outcomes go up while costs go down.

Prevent. Is your patient population at significant risk of needing expensive medical care in the not-too-distant future? Can you intervene to prevent it? If your prediction is accurate and the cost of the intervention is low, outcomes go up and costs go down.

Improve treatment decisions. Few treatment decisions are black and white. There are typically multiple treatment options and uncertainty about which is best. An ideal decision is a deliberate one. Both patient and doctor are clear about the latest evidence regarding the benefits, risks and side effects of each treatment option. Both are clear about what matters most to the patient.

Practically speaking, such idealized decisions are difficult to achieve. Is it possible, however, to at least move closer to the ideal? Outcomes improve when the outcomes that matter most to patients are identified and considered during the decision-making process. Costs fall in cases in which well-informed patients choose less care.

One of these four pathways to improved care is typically dominant for an innovation in health care delivery. Kangovi's central thrust was preventing patients from falling ill again and potentially needing to be rehospitalized.

Her initiative, as is commonly the case, touched on other pathways as well. She and her colleagues have *standardized* work routines for CHWs and *delegated* as much work as possible to them. CHWs endeavor to *coordinate* hospital care and primary care. And, they help patients and physicians *improve treatment decisions* in the form of more realistic care plans and goals.

§

Your job, as a physician innovator, is rarely to come up with a fantastic new-to-the-world breakthrough concept that nobody has ever contemplated. It is instead to apply and refine one or more of these four fundamental ideas for a specific context.

A strong first step in doing so is to develop clear and crisp one-phrase or one-sentence answers to five questions. Here are the first three:

1.   Who are the patients you will serve?

2.   Which outcomes will you improve for these patients?

3.   How will you do it?

These questions are straight from the standard MBA playbook. They are tried and true. In MBA jargon, the questions are: *Who is your customer? What is the value proposition? How will you deliver?* Starting with these questions will help ensure that your initiative is patient-centered, as of course it should be.

Every entrepreneur is trained to develop an "elevator pitch" that answers these questions quickly — in as little time as it takes to ride an elevator. A good pitch demonstrates that the entrepreneur has clarity and simplicity of purpose. It also opens the door for conversations with potential customers, partners, suppliers and funders. The feedback from these conversations

...... the entrepreneur sharpen the idea, or, in some cases, abandon it before investing too much time and energy.

You'll benefit from a similar process. By developing a quick pitch and sharing it — with patients you want to serve, other providers you may collaborate with and institutional leaders who may support you — you'll learn a great deal, and, potentially, sidestep a great deal of future heartache.

Inevitably these conversations will turn to money, and that brings us to the final two questions:

4. How will you reduce system costs?

5. What financial impact will you have on your practice or institution?

Innovators outside the health care industry have it a bit easier, in that there is no system to fix. There is only question five — generically, *Can you make money?* An innovation in health care delivery, however, must do both — save money for the system and prove financially sustainable for the practice or institution.

Let's now look a bit more closely at each of the five questions, keeping in mind that what is called for at this early stage is simply a crisp and clear one-sentence or one-phrase answer to each.

.... § ....

First, who are the patients you will serve? How old are they? Are they male or female? Where do they live or work? What is their income level? How are they insured? What is their socioeconomic status? Do they have a particular medical condition? Chronic or acute? Where are they in their overall care process?

In general, more specificity in your answer will serve you well, as long as you do not go so narrow that there is an inadequate number of patients in your service area to keep your small team busy. Note that the patients you choose do not all need to have the same medical condition. Instead, they ought to have similar needs. Kangovi's initiative served patients who had a low income, were soon to be discharged from the hospital and were at high risk of readmission.

You want to choose a patient population that you are particularly passionate about serving. It may very well be just a subset of the patients that you are serving today. Quite likely, it is a group of patients that the system is failing in specific ways that leave both you and the patients deeply frustrated.

.... § ....

Once you've defined a patient population, the next question is: Which outcomes will you improve for these patients? Here are a few related questions that may stimulate your thinking: What does success look like for these patients? What do these patients value? What barriers to better health do they face? How is the system failing these patients today?

These may sound like straightforward questions, but it is easy to go astray even at this early stage. The key to success is to be certain that you are truly taking the patient's perspective, not your own or that of your institution.

No MBA student graduates without examining at least one case study that makes this same point. The subject company might be an information technology company that understands circuits and code much better than it understands customers and their needs. The downfall is predictable. The company launches a new gadget that is technologically whiz-bang and appeals to a few sophisticated customers, but most have little idea what to do with it.

Your daily language is organ systems, lab test results, imaging studies, tissue cultures and genetic sequences. These concepts, however, are foreign to mainstream patients. What's interesting to you may be irrelevant to your patients. Professional market researchers know that they have to be detached when deciphering what people want. You need the same discipline.

Mnemonic devices are widely used in medical school; here is one for helping to ensure that you are getting outside of your own head and into the heads of your patients. The five D's of patient-centered outcomes are death, disease, discomfort, dysfunction and dissatisfaction. (And if you want a bonus for an age of out-of-control health costs, how about a sixth D, for destitution?)[8] At least at a high level, these are the outcomes that patients care about.

In figuring out what your customers want, you have a significant advantage over, say, software engineers, in that you interact with your patients often. As such, your intuition about their needs ought to be at least in the ballpark. Remember, however, that you are exposed to patients during just a sliver of their lives. You are hardly getting a panoramic view.

Furthermore, recognize that your intuition about patients' needs is most likely to be accurate when you consider patients whose lives are similar to your own — of similar age and socioeconomic status, in particular. One geriatrician I interviewed shared that his favorite moment when training his residents is taking them on a house call. It satisfies him to watch as residents suddenly gain tremendous new insight into the full context of patients' lives

— in particular, why patients might struggle to stay healthy or follow medical advice even when they have the best of intentions.

Before Kangovi hired her first CHW, she made an extensive effort to study the lives of the patients that she intended to serve. She had entered the Robert Wood Johnson Clinical Scholars program; the foundation funded portions of her research. She partnered with a member of the low-income Philadelphia community who conducted more than 100 interviews with patients to study the key barriers to staying healthy after a hospital stay. One interesting insight from the research, right on point: The patients found the hospital providers to be friendly and professional, but lacking any depth of understanding about the real-life challenges the patients faced, such as food insecurity and homelessness.

The research also revealed several common pathways to readmission to the hospital. Post-discharge, patients often had a hard time remembering what had happened in the hospital or what they were supposed to do next. Furthermore, they had such a difficult time accessing primary care that they frequently received no follow-up care unless they felt sick enough to go to the emergency room. Finally, patients reported in the interviews that providers were setting goals for them that were simply unrealistic. When they failed, they felt incapable of taking care of themselves. This led, in turn, to a negative cascade of declining health and motivation.

Kangovi's endeavor to understand patients' needs (in MBA circles, we'd call it market research) was exceptional. The work paid off. She designed her CHW interventions to address precisely the problems that the research revealed.

Not every innovation effort must begin with research that is as extensive as Kangovi's. Most innovation efforts in health care delivery are modest in size, and, in general, research need not be out of proportion to the scale of the innovation effort. But *some* research is a good idea. Even if your intuition about patient needs is quite strong, it is best to be cautious and humble by checking your assumptions. You can interview a handful of patients, seek insights from other providers in your area who treat the same patients, and, possibly, compare your insights with colleagues in other parts of the country. When you get to the point that you no longer think that you are learning anything that would change your approach, it is time to move on.

§

So now you have identified which patients you want to serve and how you'd like to improve their lives. Next: *How will you do it?*

Here, it is imperative that you start with a clean slate. Forget everything you know about the system we have. Instead, think: *How would I design care if I could do it from scratch?* What would care look like if we had a system that emphasized value, not volume? Low tech, not high tech? Teams, not autonomous physicians? Whole patients, not body parts?

Consider also the four fundamental ideas presented at the beginning of this chapter. How much improvement is possible through standardization and delegation? Better coordination? Prevention? Improved treatment decisions?

This seems an opportune moment to re-emphasize that originality counts for nothing. Copy joyfully! If you can locate someone elsewhere in the country who has already started pursuing an initiative similar to your own, by all means get on the phone and learn everything you can from that person. If you are unable to find an initiative that is highly similar to your own, consider anything close. Hundreds of physician innovators around the country are hard at work. Even if you can't find one who is addressing exactly what you hope to improve, you can, no doubt, learn from related efforts — and there are many more still to come in this book.

More distant analogies — outside health care, even outside the United States — might also help you. Kangovi, who grew up in India, was inspired, in part, by Mumbai's network of *tiffin wallahs,* lunchbox delivery people. More than 100,000 lunches are delivered daily, and each might pass through several hands before being delivered. What impressed Kangovi was that the tiffin wallahs, many of whom are illiterate, almost never make a mistake, despite the fact that they use no technology. It made her wonder what a low-tech team that included laypeople — community health workers — could accomplish in health care.

A second point of inspiration for Kangovi was her knowledge of the role of CHWs in the treatment of tuberculosis. To recover, patients took medication with very unpleasant side effects for about six months. Some patients lapsed, putting not just themselves at risk but those around them. Public health authorities sometimes pursued a policy of insisting that patients come to clinics to take their medication under direct observation each day. An alternative approach—sending CHWs to patients' homes — proved more effective and satisfying to patients.

Keep in mind that at this point, you are not looking for a detailed care model, but a high-level sketch —a sentence or two — that describes how you will get the job done. We will tackle the next level of detail in the next chapter.

§

Now that you have a clear story about patients, outcomes and how to make it happen, let's shift our focus to costs, starting with system costs.

Analyzing system costs, at least in principle, is straightforward. You get credit for reducing the cost of care that continues to be delivered and you get credit for any care that is avoided. If you add any *new* services, such as those of Kangovi's CHWs, you deduct the cost of these services from the savings. That's as far as you need to go at this point.

§

The last of the five questions is: What financial impact will you have on our practice or institution? While cost analysis is at least conceptually straightforward at the system level, it can be far more complicated at the institutional level. The U.S. health care system is devilishly complex. There are many payers, some public and some private, and each has its own reimbursement formulas, rules and procedures.

Unfortunately, under traditional fee-for-service rules, the deal for innovators is usually this: You do the work, the system wins, the patient wins, the payer wins and you/your practice/your institution lose. It's a lousy deal, one that has thwarted countless sensible initiatives.

Fortunately, payment models are very much in flux around the country. New value-based contracts are being signed daily. You will want to be sure that you understand your current local reality.

In the best-case scenario, your organization has been aggressive about signing value-based contracts. Beyond a certain tipping point in the transition, likely somewhere near the point at which nearly half of patients are under new contracts, institutions will start behaving as though *all* patients are under value-based contracts.

As of early 2015, few institutions had reached this point. That should not deter you, however. There are several possibilities for getting your innovations off the ground even before this point is reached.

First, you could focus on a subset of patients that *is* under a value-based contract. In particular, your institution may be self-insuring its employees, or, as is increasingly common, it may own a health plan. (You could thin. either of these arrangements as a special form of value-based contracting, between your institution and itself.) If you work with either population — employees or members of the health plan — then any cost savings you generate remains in house. This approach can in some cases be a practical way of getting your initiative started and gaining experience while waiting for the further proliferation of value-based contracts.

Next, you might look for certain programs initiated by payers that fit what you are trying to do. These programs, by and large, are in pursuit of an alternative model for funding innovation in health care delivery. Rather than value-based contracting, it is expanded fee-for-service. Policymakers or payers try to do the thinking and the innovating centrally. They make decisions to pay for services have not traditionally been reimbursable — a bit more for coordinating care for challenging patients, a bit more for extra work to prevent avoidable acute episodes, a bit more for meeting patient centered medical home standards, and so forth.

A brief aside: Honestly, I do not favor this approach. Health care is probably the most heterogeneous industry I've ever examined. Every city tells its own story, with its own state and local policies, its own health care organizations and its own patient populations. It feels to me like these efforts paint with too broad a brush.

And is fee-for-service on steroids really what you want? It does appeal to some physicians because it reduces financial risk, but knowing how medical school trained you as strong individualists, it is hard for me to understand why you would favor this route. Do you really like the idea of someone else deciding when you get paid a small biscuit and when you get paid a bigger one? Someone else designing the hamster wheel so that you can just do the running?

All of this said, it is quite possible that such programs can help you get started. They may carry you along while you await the further expansion of value-based contracts.

Failing either of the above two options, there are still a few situations in which, even under traditional payment rules, the savings accrue to providers, not payers. Three common ones are:

> **By reducing the length of stay of an inpatient episode, or otherwise reducing its cost.** Medicare (along with many other payers)

reimburses not for every discrete service but by the admission, based on the diagnosis.

By avoiding loss-making hospitalizations. A hospital that is operating at capacity can sometimes benefit financially by helping a poorly insured or uninsured patient stay healthy. By doing so, it frees capacity in the hospital for better insured patients.

By avoiding readmission penalties. With the Affordable Care Act came new penalties for readmissions within 30 days.

Note that Kangovi's initiative benefited from the second and third of these. The University of Pennsylvania Health System also received value-based payments related to the program. By 2014, UPHS's internal analysis showed that for every dollar spent on the program, it saved a handsome $1.80.

If, considering all of the above, you find that the simple reality is that your initiative, while good for the system and good for patients, will financially harm you or your organization, then I have the following advice. Go outside. Scream at the top of your lungs. Let it all out. What I'm saying is — I'm with you. I'm frustrated, too.

But don't give up.

Consider seeking grant funding from the Centers for Medicare and Medicaid Innovation or the many other private grant-making institutions that are interested in innovation in health care delivery. Or consider networking your way into a phone call with a payer. Let the payer know of your interest. Get a sense of what special contracts might be possible, and in what timeframe. Thousands of such phone calls might just help payers move even faster than they are today.

Finally, consider pushing your proposal internally, making any or all of the following arguments about why your initiative makes sense even in the face of near-term losses. You just might be surprised, especially if your institution is financially healthy. Say:

- It will prepare us for the future of value-based contracting.

- We will get ahead of the game, relative to the competition.

- We will become the go-to provider in our region for payers that want to hammer out value-based contracts.

- Because this is good for patients, we will attract donations from our wealthiest patrons.

- It will help attract well-insured patients and turn them into loyal customers of the health system.

- It will help us attract the next generation of top-notch physicians.

- We can attract new referrals, especially from other provider institutions that have already signed value-based contracts.

- It will enhance our brand image as a caregiver who cares about value, and as an innovator. It will make us more attractive to the press.

- Given current trends, this is a smaller, smarter and less risky investment than adding a new wing or buying the latest imaging machine.

- It is in line with our stated mission.

- It's just the right thing to do.

With answers to the five key questions, you have a sufficient skeleton proposal to socialize your idea. Kangovi's elevator pitch might have included the five quick answers presented in the table below.

Who are the patients?	Hospitalized low-income patients with a high risk of readmission.
In what way will you improve their lives?	Hasten recovery; reduce the likelihood of readmission. Improve access to primary care.
How will you do it?	Each patient will partner with a CHW who is trained to help guide the discharge, recovery and transition to primary care.
How will it reduce system costs?	By preventing avoidable hospital readmissions.
What financial impact will it have on the institution?	We will reduce readmission penalties and, potentially, avoid loss-making admissions. This may be sufficient to offset the cost of additional staff.

Your aim with your elevator pitch is to receive some feedback and to get some encouragement and support to develop a more detailed proposal — one that will win you the funding you need to move forward. The remaining chapters in Part II will help you complete that proposal and guide your work once you get the green light.

Remember that elevator pitches and even full proposals are simply hypotheses. Few innovations find their way to success without a few missteps along the way. You'll launch, learn, redirect, and then learn and redirect some more, and you'll do so over a period of many months or even a few years. Several physician innovators have emphasized to me the critical nature of continuing to listen to patients describe their needs and evaluate how well those needs are being met. Some have even gone so far as to add patients formally to the team.

Solidify Your Idea: Summary of Observations and Recommendations

1. Develop concise answers to these five questions:

   a. *Which patients will you serve?* It is generally sensible to define the population of patients you plan to serve narrowly, provided you have adequate volume to keep a small team busy.

   b. *Which outcomes will you improve for these patients?* Be certain that you understand what your patients value and the barriers they face to better health.

   c. *How will you do it?* How would you design care for these patients if you could build a team from scratch? Be sure to learn as much as you can about similar efforts in other parts of the country or world.

   d. *How will you reduce system costs?* Will you reduce the cost of delivering care? Avoid the need for some care? Will these savings be offset by new care steps that you are adding?

   e. *What will be the financial impact on your practice or institution?* Analysis is complicated by the complexities of many payers with distinct reimbursement policies. The picture becomes much clearer when value-based contracts are in place.

2. Originality is not required. Copy joyfully.

3. If your initiative is a money loser under fee-for-service rules, look for opportunities to get started at least on a small scale, perhaps by serving employees of your health system or members of a health plan owned by your health system.

Action Steps for Senior Executives

1. Encourage your staff physicians and physicians in affiliated community practices to develop elevator pitches for innovation in health care delivery.

2. Identify the most enthusiastic physicians with the most promising ideas. Ask them to pursue their plans further. In particular, ask them to do some "scouting," that is, to identify similar initiatives elsewhere in the country and learn as much as possible from them.

3. Communicate at least the basics of the current contracting environment for your institution and how it is evolving. Make sure physicians understand where the financially sustainable opportunities for innovation in health care delivery lie. Signal your willingness and ability to invest in innovation in anticipation of value-based contracting.

Chapter 6
Design Your Team

B Y 2014, KANGOVI HAD TAKEN AN IMPORTANT STEP BEYOND assembling a team of community health workers. She had refined her methods for recruiting, training and managing her team, and she'd carefully defined each team member's role. She'd developed an organizational formula and she was using it to guide the program's growth. She was even eagerly sharing the formula with other physicians around the country who wished to build similar programs.

Team design can at first feel like an exercise in trial and error. Kangovi's program illustrates what it looks like when the team design task is complete.

§

Team design is the central aspect of innovation in health care delivery. We will dedicate the next three chapters to it. In this first of three, we will make a simplifying assumption — that the team will operate in isolation, as opposed to within an established organization. We will expand the analysis to consider the full organizational context in the subsequent two chapters.

The first intuitive questions when thinking about team design tend to focus on individual team members. What skills are needed? Who has them? We will get to these questions shortly. There is a crucial preliminary step, one that is analogous to erasing a chalkboard before drawing on it. In team design, you must start from a clean slate. Before you *create* you must *forget*.

Across all industries and throughout all of my years of study, failure to forget when designing teams has been a stealthy nemesis that has knocked back countless innovation efforts. The mistake, more specifically, is to create a team for an innovation effort that is too similar in design to the organization that already exists. This is an easy trap to fall into. Most people, especially those who have worked in the same organization or the same industry for many years or for their full careers, are hardly even aware of the norms of their organization.

I know of no better illustration of this problem than one of the first companies I studied. This takes us back to a heady era, the dot-com boom in the late 1990s, and the story of the first few years of The New York Times Company's endeavor to build its Internet unit, New York Times Digital.

Just two facts are needed to get straight to the point. First, New York Times Digital created very little of its own content. Almost all of its content came from either the newspaper or a third party. New York Times Digital then used the unique capabilities of the Internet to enhance the content; for example, by complementing an article with an interactive chart. New York Times Digital also operated a continuous news desk, which updated the key stories throughout the day.

Second, as is true of all companies, there was a pecking order within the organization, one that had been in place for decades before the launch of the website. At the top were the most senior journalists and editors. Quite a bit farther down the line you'd find the information technology staff. The IT team was clearly in a support role, building computer systems that automated data flow and improved the efficiency of the operation as a whole.

Now consider what pecking order would make the most sense for New York Times Digital — a group that does very little journalism but must be on the cutting edge technologically. The desirable hierarchy is upside-down from that of a traditional newspaper organization. In initially staffing the team, however, the company chose mostly insiders — a handful of journalists and a handful of technologists. As could be expected, the journalists were dominant, as they always had been.

Not long into the venture's life, the leaders of the effort saw the result: New York Times Digital's product was too similar to the newspaper itself. Meanwhile, the team was bypassing a variety of promising nontraditional opportunities that were emerging online. The team simply wasn't moving quickly enough to capture the venture's full potential.

Over the next two years, the company addressed the problem by hiring a small army of Internet media technologists, many of whom had already worked for another dot-com startup. By 2000, nearly three out of four New York Times Digital employees had come from outside the company. New York Times Digital's pecking order was literally flipped upside-down. It achieved a trajectory of rapid growth, and, by 2002, profitability.

§

I have yet to see a team redesign in health care delivery that is quite as dramatic as the New York Times Digital example, wherein the hierarchy was turned upside-down. My purpose in sharing the example is to show just how sharply small innovation teams can depart from historical practice.

So, what is "historical practice" in health care? What are the organizational norms that are practically taken for granted? Are there well-established notions about who does what work? About how physicians interact with each other? About the relationships between physicians and non-physician providers? Between physicians and administrators?

Of course there are. Here are a few, stated provocatively:

* Physicians are autonomous.

* Physicians are fully and individually responsible for "their" patients, until they choose to refer the patient to a different provider.

* To the extent that physicians work in teams, those teams are temporary, convened only for a particular case.

* The pecking order among physicians is based on years of training and depth of specialized knowledge. The more you have of each, the more powerful you are.

* Non-physician providers are task-oriented and take direction from physicians.

* The primary role of administrators is to provide the resources that physicians need to deliver top-notch care.

A discussion about the appropriateness of these norms in modern medicine would be a healthy one. But I generally take a humble and conservative point of view on the need for wholesale change across entire organizations. I do not advocate taking a wrecking ball to what exists. What I *do* favor is deliberately and purposefully experimenting with new types of team structures and subsequently expanding on the successes.

To do this well, you cannot default to doctors acting like doctors, nurses acting like nurses and so forth. Indeed, you must start with no preconceived notions about who does what or what the working relationships between team members might be. Before you create, you must forget. This start-from-scratch mind-set when designing teams is what most separates the innovation I'm describing in this book from typical process improvement efforts.

Everyone involved in your initiative might experience dramatic change in what they do at work each day. To illustrate, here are some general trends in the way physician roles are shifting as a result of innovation in health care delivery.

First, I have seen that doctors involved in innovation efforts are spending less time doctoring, less time maintaining close contact with high-needs patients and less time bogged down in administrative work such as managing electronic medical records. These tasks are being delegated to other team members.

Meanwhile, physicians are spending more time on clinical tasks that cannot be delegated. They are spending more time on long-term-care planning for patients with complicated care needs, more time discussing cases with other physicians treating the same patients, more time discussing treatment options with patients confronted with consequential medical decisions, and more time carefully evaluating the performance of competing specialists before making referrals.

Doctors are also spending more time in managerial roles, especially designing, building and leading new types of teams, redefining roles for health professionals, and, in some cases, even inventing jobs for new types of health workers. Then, once the teams are in place, doctors are analyzing patient health data, improving team performance, recruiting and training new team members, evaluating the individual performance of team members and perfecting work processes.

Some of you may not be excited by this increase in the managerial workload. However, as time passes and as teams begin to prove themselves, it is likely that you will be able to delegate more and more of the managerial work to another team member, such as a nurse manager, business partner or administrative partner. Indeed, some physicians choose to sustain only a part-time oversight role. You won't be able to disappear entirely, however. Some managerial tasks are difficult to delegate, particularly navigating organizational politics to acquire and protect resources for your teams and nudging fellow physicians to move with you toward models for higher-value care.

Taken as a whole, this is dramatic change in the daily work lives of physicians. Though change can be difficult, I suspect that most of you would welcome these shifts. This is, in broad strokes, what the pathway to higher-value health care looks like.

There are similarly dramatic shifts for everyone involved. For many, the changes will be received positively.

§

Now that you understand the critical importance of starting from a blank slate and at least a taste of the possible directions in which you might go with team design, let's move on to consider how to design a team for a particular innovation initiative. You want to begin with two intuitive questions:

1. What skills are needed to get the work done?

2. Who already has the skills or can readily be trained?

These sound basic, but the doors are wide open here. As you get started, you are in essence doing something analogous to both inventing a new sport and defining what each player on the team will be doing. If you were inventing American football, for example, you'd be thinking about the skills and attributes you'd need in various positions — for example, strong linemen, running backs who could change direction quickly, quarterbacks who could throw accurately and speedy receivers.

Do you have a potential initiative that you are already contemplating? If so, why not take a first cut at team design right now? What are the skills and attributes of the team members that you want on your roster? Sometimes a helpful starting point is to create a short list of action verbs that describe the core work of the team, such as engaging patients, planning care, analyzing population needs, predicting which patients are most likely to experience an acute episode, and so forth. Who already has the skills to do the work? Or who can quickly be trained?

As you know, there is a long list of health workers with roles and capabilities that are reasonably well understood and well defined, especially compared to most industries. A list would include generalist physicians, dozens of different types of specialist physicians, physician assistants, nurse practitioners, psychologists, physical therapists, occupational therapists, physiatrists, pharmacists, advanced practice nurses, registered nurses, licensed practical nurses, nurse managers, chaplains, public health professionals, laboratory technologists, radiological technicians, medical assistants, counselors and home health aides.

Depending on your past experience and the nature of the innovation you are pursuing, it may be worthwhile for you to become more familiar with the skills and capabilities of various health care professions. Many look rather underused as we move to higher-value care. Which of these health professionals comes closest to having the necessary skills to do the job? What skills do they already have and what would they have to learn?

In some cases, one of the existing health professionals might be perfect, but, again, you do *not* want to limit yourself. You may want to invent a new role or two from scratch and recruit from outside the industry altogether.

For example, Kangovi initiated her endeavor with a specific interest in community health workers. Although there is plenty of talk these days about the potential roles that CHWs can play, there is little consistency in how they are trained or in how their roles are defined. Kangovi created a CHW role from scratch for her own specific purpose and hired from outside the industry.

As we have entered an unprecedented era of cost constraints, you want to be thinking even from this early stage about delegation. Wherever possible, you want to shift tasks from highly paid and highly trained providers to less highly paid and less highly trained ones. Start with your own role, or that of physicians on your team more broadly. Which are the tasks that *only* physicians handle?

§

So far, we have tackled team design tasks that sound simple. *Be sure to start with a clean slate. List the needed skills. Identify the people who can do the job.* We are not quite ready to move on to the next level of detail just yet, however.

There is a common stumbling block at this stage that poisons many innovation efforts. Specifically, unless you are careful, you may unintentionally refill your clean slate with traditional assumptions and habits. This can happen in a few ways.

Choosing who is readily available rather than who has the right skills. It can be tempting to jump to the question of *Who?* before giving careful thought to the question of *What skills?* When hastily-chosen team members struggle with the new tasks you hope they will learn, they will naturally gravitate to what they already know well. Traditional assumptions and habits reassert themselves.

For example, the central ambition of one initiative I studied was to engage patients with multiple chronic conditions in taking better care of their health. In an early iteration of the effort, nurses were asked to coach the patients. The nurses received little training; it was simply assumed that they could figure it out.

While nurses were a convenient choice, their prior education offered little preparation for engaging and coaching patients, and the core nursing skills that they *were* quite good at were not really relevant to the new challenge. Indeed, the leaders of the effort later concluded that the most important success factor for a health coach was simply having come from a similar life context as the patient — same race, culture, nationality, socioeconomic status. Also needed were some foundational coaching skills such as motivational interviewing that are not necessarily in the nursing curriculum. The team subsequently developed and refined its practices for recruiting and training its coaches.

Using traditional titles and job descriptions. Many innovation efforts will use health care insiders — people who already have particular titles, credentials and job descriptions. There is danger here. Remember, the central task in innovation in health care delivery is creating new kinds of teams with newly defined roles. Everyone on the team is likely to take on at least some new work, and some will have their jobs almost entirely reimagined. The pharmacist's job on the innovation team may be quite different from the pharmacist's job as traditionally understood.

You may believe that writing job descriptions is mindless administrative make-work. Perhaps in some contexts it is, but never is the task more meaningful than during the formation of innovation teams. Prior understandings about who does what must be released, new ones created. Lazily using existing job descriptions can get in the way.

Changing professional titles can be as powerful as writing new job descriptions. After all, titles tend to be received as shorthand job descriptions. Further, traditional titles tend to be anchored to traditional understandings about who does what. Therefore, for innovation teams, the best titles are ones that have never been used before. Unusual titles spark healthy discussions within the team about roles and responsibilities. One possibility is to use both a credential and a job title, as in Mark Jones, RN, CHF counselor. The former is traditional, the latter is custom created for the innovation effort.

Choosing too many insiders, especially people who already work together closely. Redefining an individual's role from scratch is hard; redefining how a pair of people work together is harder still. Work relationships are sticky, especially hierarchy. Hiring a few team members (say, at least one in four) from outside your institution helps avoid this trap. Outsiders are natural expedients in the healthy process of breaking down pre-existing work relationships and building new ones from scratch.

Recruiting people who love their current (traditional) job. You want people joining your team to be excited about doing something innovative, comfortable with uncertainty and eager to learn new skills. You want volunteers. Otherwise, your team will naturally gravitate back to traditional habits and assumptions.

Bear in mind that hiring for newly defined roles is inevitably risky. People have no way of truly understanding what they are getting themselves into. Most of the innovation leaders I have spoken with have experienced at least some turnover within their teams. Usually, those team members simply realize that they prefer their prior role.

Departure of a team member or two is a normal aspect of the innovation process. You want to be gracious in your team member's departure. Perhaps they will know of someone else that is better suited for the job.

Being trapped by traditional institutional policies. Some institutions will resist making exceptions for innovation efforts. You may be pressured to recruit using established job descriptions and compensation policies, for example, or to report performance using standard metrics and formats. Either will reinforce traditional assumptions and habits. Try to get ahead of the problem by building awareness among institutional leaders that simple "one size fits all" policymaking, while seemingly efficient and fair, can be a steep barrier to innovation in health care delivery.

Failing to guard against the re-emergence of traditional habits, norms, behaviors and assumptions. Even if you do everything right in initially forming your team, you must remain on your toes. Traditional health care can easily reassert itself. Role expectations and behavioral norms are deeply ingrained from the very first day of medical training. As such, it is not enough to write down job descriptions. The new model of care must constantly be reinforced. (Because Kangovi's team was composed almost entirely of newcomers to health care, the risk in her initiative was greatly diminished. For most innovations in health care delivery, however, the risk is substantial.)

For example, one innovation leader, trying to corral a group of physicians into new team behavior, took the sensible step of finding a physical space where everyone could sit together. She remarked to me that even when right next to each other, they at first continued to engage in what she described as "parallel play." I knew what she meant. I had heard the term used to describe the behavior of 1-year-olds in nursery school. There is not much depth of social interaction at that age; everyone is doing their own thing, but in one big room. Absent constant reinforcement of a new team model, physicians will tend to gravitate toward autonomy. A physician described the phenomenon to me this way: *Physicians working alone, together.*

§

With these cautions attended to, we can move on to the next level of detail. We've discussed *who* should be on the team but not *how many*. To size your team, you will want to make the best effort you can, as early as you can, to specify task by task the work that each team member will do. Doing so may require additional research with patients or other health workers to learn as much as you can about what it will really take to get the job done.

You want to be as fine-grained as you can, right down to the length of each interaction between provider and patient and the length of each team meeting. If you had perfect information or foresight, you'd be able to anticipate the exact number of people you'd need on your team and even schedule a typical week. You'd know, in advance, your team's staff ratios (say, number of nurses for each doctor) and panel sizes.

This may feel like an impossible amount of guesswork to you. And, in fact, you may only be able to learn the appropriate staffing ratios and panel sizes through trial and error.

Remember, however, that experience is a slow and painful tutor. I simply can't say it too many times: *Originality is not required.* There is probably someone somewhere in the country who has already launched an initiative that is similar to your own. If they have been at it for some time, then they have some hard-won knowledge that could be valuable to you. You want to understand, in particular, team structure, the roles of each member of the team, whether each team member is full or part time, staffing ratios, panel sizes, recruiting strategies and training programs.

If you are truly the pioneer, then recognize that you will have more to learn than most. Start small, evaluate results, and redirect as needed. Over time, you will develop ever more specific job descriptions, hiring practices and training programs for your team members.

Kangovi's extensive research guided her initial effort to describe exactly what the community health worker's job would be. For example, because patients so often struggled to remember what had happened in the hospital or what they would need to do next, she concluded that CHWs and patients would have to begin working together while the patient was still in the hospital. The CHWs would need to stand alongside the patient during critical conversations with doctors. They would need to take notes and to interrupt, when appropriate, to ask doctors to slow down and repeat their instructions. She also reviewed the literature to learn about the best practices for discharge communications and worked elements of those practices into the CHW job description.

Kangovi tested many of her ideas for potential CHW roles by asking patients for feedback. For example, post-discharge, she thought it might be a good idea for CHWs to pick up their patients' prescriptions. Patients warned against it, pointing out that if word got out that CHWs were routinely carrying prescription drugs, they might become a target for theft or violence.

Despite her research, Kangovi knew there was still much to be learned through experience. She started with two CHWs and with only high-level job descriptions. From there, part of Kangovi's learning process was shadowing CHWs and observing what they did as they followed their intuition. Every time a CHW did something that she hadn't put in the job description, she made a note in a red pen, and every time a CHW did *not* do something that *was* in the manual, she made a note in a blue pen. Her objective was not to get the CHWs to conform to her initial ideas of what they ought to be doing, but to systematically discover what worked and to perfect the job descriptions as quickly as possible.

By 2014, Kangovi had published manuals that outlined not just the roles of the community health workers themselves but the supervisory structure — one manager for every eight CHWs, plus a medical director for the program as a whole. She described her objective as ensuring that "if an alien from Mars landed on Earth" and picked up her manuals,[9] it would be able to replicate the program. And, indeed, the information in these manuals is detailed. To give just a flavor, Kangovi describes CHWs as trained laypeople who share life circumstances (race, language, health status, socioeconomic status) with the patients they serve. They are natural helpers, and their most important attribute is their ability to connect deeply with patients and gain their trust. Just as important, CHWs engage providers in productive exchanges. They are polite but pushy, respectful but demanding.

Kangovi's attention to detail in specifying how her team operates is noteworthy. Remember, creating new teams and defining the roles of each team member is the central aspect of innovation in health care delivery. It is a task worthy of every ounce of energy that Kangovi put into it.

Kangovi's work as a physician innovator rewarded her with handsome career dividends. At the age of 34, she was in front of what her institution viewed as a strategically critical initiative. Her work had attracted positive coverage from the media, including National Public Radio and the *New York Times*.

§

Some time ago, I was searching for a light read before a long flight. As I enjoy watching basketball, I picked up Phil Jackson's *Eleven Rings,* which chronicles this head coach's journey to 11 NBA championships. The book delivered on my expectations. I found it entertaining.

About two-thirds of the way through, however, I reflected on the broader relevance of the team-building principles that Jackson described. For a basketball team to be effective, Jackson believed that there had to be:

* Commitment to a shared goal.

* A clear structure.

* Empathy and mutual respect among team members.

I'd suggest that if you want a list of just three absolute must-haves if you want to be successful in building teams for innovation in health care delivery, you could certainly do a lot worse than these three.

In Phil Jackson's case, the shared goal is of course winning the NBA championship. In your case, it will be a specific vision for improving outcomes while lowering costs. For Jackson, the clear structure is the triangle offense (which I can't explain even after reading *Eleven Rings*). For you, the team structure is what you invent following the principles in this chapter. For Jackson, empathy and mutual respect meant that even stars like Michael Jordan and Kobe Bryant could walk in the shoes of the newest rookie on the team. For you, it means that all members of your team work hard to understand not just their own roles, but also what the challenges of daily work look like from the perspective of every other team member.

I subsequently discovered Jody Hoffer Gittell's extensive research on team performance in health care, which has highlighted nearly identical principles: shared goals, shared knowledge and mutual respect. For a deeper look, I recommend her book *High Performance Healthcare.*

Attention to these three principles is a constant necessity. Many innovation leaders have opted for a daily huddle or weekly meeting, using the time to reinforce goals, to build clarity around the (evolving) team structure, and to develop empathy and mutual respect.

§

In this chapter, we stepped through the process of developing a specific team design for a specific initiative. In the next chapter, we abandon the pretense that your team exists in isolation and start our dive into the best practices for building your team inside an established organization. This will require adding a generic team overlay to your team design — a partnership between two groups, a Dedicated Team, filled almost exclusively of full-time team members, and a Shared Staff, composed of part-time contributors.

The three principles just described can be applied to the overlay. The shared goal is dual excellence — building something new while sustaining excellence in what already exists. The clear structure is the partnership between Dedicated Team and Shared Staff. And empathy and mutual respect is all the more essential, given the inevitable conflicts between innovation and ongoing operations.

Design Your Team: Summary of Observations and Recommendations

1. Before you create you must forget. Team design is a clean slate endeavor. Start with no preconceived notions about who does what or what the working relationships among team members might be.

2. Originality is not required. As you get started, learn as much as you can from similar innovation efforts elsewhere in the country.

3. First ask: What skills are needed to get the work done? Then ask: Who possesses those skills or can readily be trained? Start with an eye toward delegating as much as possible.

4. Invent new titles and job descriptions for your team members.

5. Be equally open to hiring insiders and outsiders. Avoid teams composed entirely of insiders, especially those who have a history of working closely together or otherwise already know each other well.

6. Guard against the re-emergence of traditional norms, behaviors, habits and assumptions.

7. Over time, refine your team design, including who does what

and your estimates of the appropriate panel sizes and staffing ratios.

8.   Once you know what works, consider growth. Also, consider sharing your knowledge with other leaders of innovation in health care delivery.

9.   High-performing teams are characterized by shared goals, shared understanding of the roles of all team members and mutual respect.

Action Steps for Senior Executives

1.   Talk through team design with physician innovators. In particular, question traditional assumptions about who takes on which tasks on the team.

2.   If possible, help physician innovators identify and learn from similar efforts elsewhere in the country.

3.   Ensure that innovation leaders have the flexibility they need to define new roles and hire into those roles. If necessary, accelerate the process for approving new job descriptions.

4.   Communicate to your organization why you have approved the innovation initiative and why you are hiring new people for it, especially if you are not approving new hires elsewhere in the organization.

Chapter 7
Deploy Part-Time Team Members Carefully

AT ESSENTIA HEALTH CARE IN DULUTH, MINNESOTA, PHYSICIANS referred patients who had been hospitalized with congestive heart failure for the first time to an outpatient care management program designed to help patients meet their health goals, manage their symptoms and minimize hospitalizations and emergency room visits. A full-time team of nurse practitioners, physician assistants and nurses saw patients regularly, stayed in touch by phone and by email, and monitored body weight of the least healthy patients via a home tele-scale. The program improved use of medications, improved patients' ability to function, reduced readmission rates, and for those who were readmitted to the hospital, reduced length of stay.

Not far away, the Gundersen Lutheran Health System in La Crosse, Wisconsin, developed and refined an advance care planning program in which trained facilitators engaged patients in a series of structured conversations about treatment decisions near end of life. The facilitators — nurses, social workers and chaplains — generally allocated 10 to 20 percent of their time to this work. The program increased the number of patients who had advance directives and reduced the frequency of tense dilemmas in which consequential end-of-life decisions needed to be made while the patient's wishes remained unclear.

§

In this chapter, we will explore these two stories in more detail. We will do so with particular attention to a detail that may seem small: At Essentia, the core team all worked on the initiative full time; at Gundersen, part time.

In both cases, the team members were full-time *employees*. At Gundersen, however, they only spent a fraction of their time working on the innovation initiative. They otherwise remained engaged in their normal work responsibilities.

To spend an entire chapter discussing full time versus part time may appear to make a mountain out of a molehill, but the wrong choices can sink an otherwise promising initiative. Of greatest importance are some easily overlooked constraints on what part-timers can realistically accomplish.

For convenience, I will refer to the group working on an innovation initiative full time as the *Dedicated Team*; those working part time as the *Shared Staff*. As an innovation leader, it is critical that you are equally attentive to both groups. Your *team* is the *partnership* between the Dedicated Team and the Shared Staff, not just the Dedicated Team.

I want to be precise about my use of the adjective *dedicated* in Dedicated Team. It only signifies that a team member is spending all (or very nearly all) of his or her time on a single innovation initiative. Ideally, the Shared Staff is just as dedicated in the sense that they are hard workers who want to see the initiative succeed.

§

Some foundations must be laid for this discussion, starting with a simple observation: Innovation requires resources. In service industries in particular, the most crucial resource tends to be *people* and their *time*.

But how much of this precious resource is available inside an established organization? I have found that a chart that is rectangular but otherwise similar to a pie chart is quite useful for considering this question. The people in the organization are on the horizontal axis, from 0 to 100 percent, and their time is on the vertical axis, from 0 to 100 percent.

Inside established organizations, the immediate challenge that confronts innovators is that the pie is nearly fully claimed by ongoing operations, as shown in Figure 6. Everyone is already busy, though not quite completely consumed. Like the crumbs left over after a pack of voracious teens devours a chocolate cake, there are small slivers of *slack time* left over. The slack time shows up on the pie chart as a thin horizontal band at the top.

Organizations that are relentless about efficiency try to squeeze as much slack out of the system as possible. The fewer crumbs, the better. Health care is less relentless about efficiency than many industries; still, it is hardly the case that there are huge pockets of idleness looking for a purpose. What fraction of your time each week can you plausibly direct to innovation? Five percent? Two percent? One half of 1 percent? What would your colleagues say?

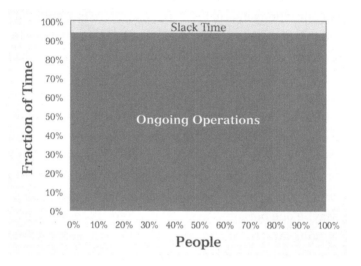

Figure 6

So, how can innovation be squeezed in?

Many organizations seem not to give the question much thought. A typical phenomenon is for executives to tout a "culture of innovation" and to encourage everyone to think creatively, share ideas, write proposals and take initiative, but to defer for as long as possible any serious consideration of where the resources are going to come from for executing all of those terrific ideas. By default, what this means is trying to make progress by squeezing innovation into slack time.

Despite the reality that slivers of slack time are quite small, they can indeed be put to work on behalf of innovation, up to a point. Innovation programs that fly under monikers like quality improvement, Six Sigma, Lean, continuous improvement and so forth try to make full use of this resource. The idea is to engage every employee in an effort to search for opportunities for improvement and to implement solutions.

There are some significant limitations to this approach, however. For starters, it can be challenging to motivate employees to direct their slack time to innovation efforts. After all, they already have full-time jobs, and now they are being asked to go the extra mile with the last little sliver of free time that they have. You may have heard the quality improvement mantra, "Everyone has two jobs, to do their work and to figure out how to do their work better."

Also, slack time is not reliably available. In most organizations, the tempo of ongoing operations fluctuates. One week there is ample slack time, the next week people are working overtime and still just barely able to keep their heads above water. The unpredictability of the availability of slack time can make it difficult to sustain any innovation momentum.

The most significant limitation, however, is a physical one: project size. The sum of all of an organization's slack time does add up, at least mathematically, to a sizable resource. However, any initiative that requires more than one person's slack time, or, at most, a few people's slack, is simply too large to be practical. The problem is that it is nearly impossible to aggregate, coordinate and manage little slivers of sometimes-available-sometimes-not slack time. Maybe one person is available on Tuesday mornings, another on Friday afternoons. Maybe one person is lightly loaded this week but fully consumed next week, while another faces exactly the opposite demands. If you have ever been excited about the potential of a new initiative, even had the buy-in of several people who wanted to be involved, but then found it was nearly impossible to schedule the first meeting, then you have run head first into this concrete wall limitation.

When your objective is to move forward with a large number of small and narrow initiatives — and this describes most activity in quality improvement programs — then slack time is powerful. As project size goes up, however, the utility of those little slivers of slack goes down quickly. And, as you know, we are focused in this book on initiatives that lie beyond — though not far beyond — the limitation of slack time.

I have made many presentations, in health care and beyond, in which the insight about the limitations of slack time has hit hard. Comments like "We've been trying to squeeze way too much into slack time" are common. Right. When executives encourage a "culture of innovation," it is soon clear that it is far easier to generate intriguing ideas than it is to pursue them. Most organizations have too many ideas — too many good ideas, even — chasing too few resources.

As you contemplate your own initiative, it is best to be clear-eyed about what is possible with slack. You may be able to work with a colleague or two to do a bit of research or write a proposal for a larger innovation initiative, but that is probably about as far as slack time will take you.

.... § ....

Beyond slack time, you have two distinct options — full and part time — Dedicated Team and Shared Staff. As illustrated in Figure 7, full-time team members show up on the chart as a vertical stripe, while part-time team members appear as a horizontal stripe. Most innovations in health care delivery, as it turns out, require a combination of both. You cannot, however, make the choice lightly. It makes an enormous difference which tasks you try to squeeze into the horizontal stripe and which you try to squeeze into the vertical stripe.

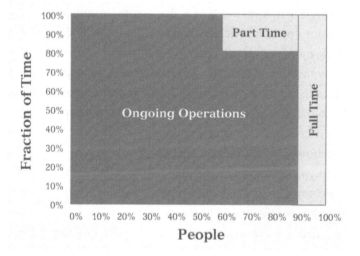

Figure 7

Let's first tackle the advantages and limitations of full-time team members. Some are immediately apparent. First, a full-time team member is a sizable resource for a nascent initiative. Second, a full-time team member is not vulnerable to unexpected short-term demands from ongoing operations. He or she is fully dedicated and thus protected from other needs. These are the advantages that innovation leaders generally have their eyes on when they make a request to hire full-time team members.

There is another advantage that, while subtle, is extraordinarily powerful. *You can reorganize the Dedicated Team.* You can redefine each team member's role, write new job descriptions, establish new work relationships and create a new team structure from the ground up. As I have been highlighting throughout the book, work of this kind lies at the very core of innovation in health care delivery. It's all about the creation of new kinds of teams. The ability to design a team from scratch is simply a "must-have" for leaders of innovation in health care delivery.

The challenge, of course, is that resources are tight and full-time team members can be hard to get. Budgets in your organization may be so tight that persuading decision-makers to create new positions for your initiative may seem well beyond "hard" to "impossible." In some organizations, because of the differences between the capital budgeting process and the process for approving new positions, it seems easier to approve a $50 million building expansion than it is to approve a new five-person team.

If this is your concern, stick with me a bit longer. There are some limited circumstances, more common in health care than in other industries, in which the crucial advantages of full-time teams are present even with part-timers. This "special case," in which a part-time team member can be considered part of the Dedicated Team, can be important in getting an initiative launched; I will describe it near the end of the chapter.

You may, however, be more successful in arguing for full-time team members than you anticipate. Let's take a look at the most common points of resistance and some responses that might at least lead you to a more specific diagnosis of what really stands in the way.

The most frequent objection boils down to cost. Full time? Too expensive!

But is it really? Or, does it just seem that way? Objectively, there is no difference in cost between staffing an initiative that requires five full-time equivalents with five full-time people or with 25 people who each contribute one-fifth of their time. Either way, the total number of person-hours expended is the same.

Sometimes the issue is not the actual resource expenditure but how it appears to others in the organization. The addition of a full-time team member is very visible. There are announcements when people are hired or transferred into new positions. Furthermore, full-time teams show up easily on internal financial statements. By contrast, there is little fanfare when someone joins a team part time. And, depending on the level of detail in the internal accounting system, part-time contributions might be invisible. If the problem is appearances, you may be able to further determine whose reactions will be problematic and come up with a strategy for dealing with those reactions.

In some cases, the leader who can approve your Dedicated Team may worry that approving your request will bring many more requests for full-time team members out of the woodwork. It is unlikely that you are the only one angling for new full-time staff, after all. Can you help the leader articulate a strategy that makes a decision to support your initiative but not others sensible and defensible?

Sometimes the anxiety is grounded in anticipation of the human resources headaches that can accompany the announcement of a Dedicated Team. Some staff will immediately start angling to be put on the new team, thinking that it sounds exciting, fun, innovative and potentially career enhancing. Others will immediately run for the hills, worried that being assigned to the team will pose a major career risk if the initiative fails. If you can help articulate, from the beginning, the specific skills and experiences you need on the team, you may be able to calm the waters a bit. You want your initiative to look exciting enough to attract the right people, but not to look so glamorous that there is a backlash of jealously and resentment from those who are not involved but wish they were.

A final potential source of concern is rooted in the notion that successful organizations have a single consistent culture, and the creation of a Dedicated Team that operates differently is a threat to that culture. The problem with organizations that have a uniform culture is that they are the least able to adapt. Given the enormous transformational challenge that the health care industry is facing, it is the last that can afford an insistence on uniformity.

Part-timers, like full-timers, bring their own mix of advantages and disadvantages. Most obviously, they cost less and thus are easier to get. Just as important, it is through the Shared Staff that you gain access to the resources and capabilities of your institution. Indeed, if you tried to launch an innovation with only a Dedicated Team, you would in effect be launching a startup. That means *everything* would have to be built from the ground up. It is much more sensible, when possible, to take advantage of what already exists. If some of the work you will be doing already lies well within the routine capabilities of your organization, there is no point in reinventing the wheel.

Against these advantages weigh some major constraints. The first is simply time. On the day that an innovation initiative is launched, the Shared Staff suddenly has *more work to do.* These part-time team members do double duty — both innovation and ongoing operations. Innovation may be the exciting new kid on the block, but the demands of ongoing operations never go away. There is no letup. Sometimes, unrealistically, the Shared Staff is expected to just squeeze the new work into the slack time. This can prove deadly to progress. If the available slack time is inadequate, then additional staff must be added to free up the needed time for innovation.

Even when the necessary resources for double duty are available, there is another, even more intractable constraint. *You cannot reorganize the Shared Staff.* You cannot redefine jobs, work relationships, hierarchy or reporting structure, because doing so would disrupt ongoing operations.

Remember, our aspiration is not just to succeed at innovation. It is dual excellence — simultaneously building something new and sustaining performance in what already exists. We cannot, in the name of innovation, damage ongoing operations. *First, do no harm.*

A misguided reorganization can be extremely harmful. Recall that while the capabilities of individuals may be broad, the capabilities of established organizations are narrow. Organizations are fine-tuned for a specific purpose, and you simply cannot organize the same group of people for two distinct purposes (ongoing operations and innovation) simultaneously.

So then, what *can* you ask part-timers to do? Simply put, you can ask the Shared Staff to do *more* work, but not *different* work. To be more specific, the tasks you ask the Shared Staff to take on must:

- Already be familiar, or require just a bit of training — say, a few days.
- Fit naturally into existing work processes.
- Fit naturally into existing roles and work relationships with colleagues.

It may be helpful to imagine that part-timers, because of their continuing obligations to ongoing operations, are members of a marching band. They are playing a tune at a certain tempo while marching in formation. If you ask a few members of the band to take on an additional performance element — a new countermelody, perhaps — that keeps them marching in the same direction, at the same pace and using the same instrument, then probably everything will go fine. On the other hand, if you ask them to take on added performance elements that call for marching in a different direction, playing at a different tempo or using a different instrument, not only will they likely fall far short of doing both jobs well, they'll probably trip others and create chaos.

A brief aside for a somewhat technical point: Just because a task meets the criteria for assignment to the Shared Staff does not mean it *must* be assigned to the Shared Staff. When there is a sufficient volume of work to keep at least one person busy full time *and* that person operates essentially as an individual performer, then it is better to assign that person to the Dedicated Team. Nothing is lost by doing so, and there are big potential benefits. Any competition for the worker's time and attention is eliminated. Furthermore, the team member may feel much more included and therefore become more committed to the success of the initiative.

The essential point, however, is this: *The Dedicated Team has enormous flexibility; the Shared Staff has nearly none.* Given adequate resources, the Shared Staff can do more work, but not different work.

With such tight constraints, it may seem like the Shared Staff would have relatively little to do. In fact, however, the division of labor between the Dedicated Team and Shared Staff can vary widely. I've seen 90 percent Dedicated/10 percent Shared, 10 percent Dedicated/90 percent Shared, and everything in between. It all depends on the nature of the innovation and the capabilities of the established organization.

Properly dividing the labor between Dedicated Team and Shared Staff is crucial. The biggest threat to getting it right is anxiety about the expense associated with full-time team members. Because of this anxiety, the default tends to be "when in doubt, give it to the Shared Staff."

I encourage exactly the opposite bias. It is all too easy to overestimate the flexibility and availability of part-time team members. Their lack of flexibility and availability is not their fault; it's because of the never-stop-even-for-a-moment demands of ongoing operations.

In larger organizations contemplating multiple innovation initiatives, sometimes the key to success is to cut back on the number of innovation initiatives that are in process at any one moment. It is better to adequately resource a small number of innovation initiatives than it is to proclaim an enterprise-wide commitment to innovation, say "yes" to many initiatives, and then leave them all starved of the resources they need. A small Dedicated Team is the minimum resource allocation for success in innovation in health care delivery. (Again, special case still to come.)

§

The team design overlay that I am describing — a partnership between the Dedicated Team and the Shared Staff — is relevant for most any innovation in health care delivery. The partnership is depicted graphically in Figure 8.

Note that the Shared Staff is represented by a square while the Dedicated Team is depicted by a diamond — a square turned on its side. This is meant to reinforce the design-from-scratch mind-set that is so imperative for the Dedicated Team.

Indeed, as the table below shows, the characteristics of the Shared Staff and Dedicated Team contrast sharply.

SHARED STAFF (PART TIME)	DEDICATED TEAM (FULL TIME*)
Familiar Tasks	Unfamiliar Tasks
Existing Workflows	New Workflows
Same Roles	Custom Roles
Same Structure	New Structure
Same Hierarchy	New Hierarchy
Nearly No Flexibility	Enormous Flexibility
No Team Redesign at All	Team Design from Scratch

* In limited special cases, members of the Dedicated Team can work only part time on the innovation initiative, as described later in this chapter.

§

The congestive heart failure program at Essentia, designed to keep patients as healthy as possible and out of the hospital, nicely illustrates a partnership between a Dedicated Team and a Shared Staff. Rather than focus strictly on the division of labor between full- and part-time team members, I'll offer a somewhat fuller telling of the story, one that will also allow some reinforcement of the principles of the prior two chapters.

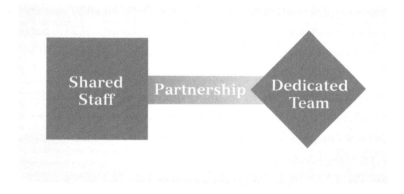

Figure 8

Note that Essentia's ambitions were similar to those of Shreya Kangovi. Both aimed to prevent readmissions. Because Essentia's focus was on a very different patient population, however, the program looked quite different operationally. The architects of Essentia's program viewed the following activities as the keys to success:

1. *Educating patients.* Following discharge from a first CHF hospitalization, patients typically had a poor understanding of heart failure. Most had forgotten or never truly understood what they had been told while hospitalized. Many only heard the words "heart failure" and perceived their condition as grave, not understanding that CHF can be managed, and often for years. With just a little bit of information, patients were far more likely to engage in taking good care of themselves.

2. *Keeping in touch to ensure that patients stay on proper medications.* Appropriate medicines following well-researched national guidelines are crucial to CHF management, but many patients had a poor understanding of the purpose of each med-icine. They stopped taking them, either out of confusion or out of displeasure with the side effects. Close contact with patients during the first few weeks could pay tremendous dividends. For some patients, it helped to ramp up doses slowly; some needed adjustments based on their diet or activity levels.

3. ***Encouraging patients to watch their diets.*** Healthy dietary habits are essential to preventing exacerbation of CHF, particularly reduced salt and fluid intake. Patients needed coaching to develop good habits.

4. ***Responding quickly to changes in symptoms.*** To fend off avoidable hospitalizations or emergency room visits, CHF patients needed to be trained to ask quickly for help whenever they experienced a change in symptoms.

5. ***Building relationships with patients.*** None of the keys to success above were likely to be achieved without a trusting and continuous relationship between patients and providers.

Before reading on, take a minute to think about how you would design a team for this program, starting from a clean slate. What is the work that needs to be done? Who would be on the team? What should their roles be? Which tasks could be handled by the Dedicated Team and which by the Shared Staff? How would you schedule the day-to-day work for the Dedicated Team? Which tasks could be handled only by physicians? Which could be delegated to other providers?

It is simultaneously energizing and daunting to be presented with a blank slate, isn't it? In my presentations, some physicians have responded with this objection: *But we haven't been trained to design teams.* That may be true, but this isn't molecular biology. It's not even rocket science. Get to know the skills and capabilities of all the health professions, then trust your intuition. It will serve you well.

Sometimes team design feels daunting because there is no obvious right answer and it seems important to get to the optimal team structure from day one. Such a high expectation is generally unrealistic. There are probably several reasonable hypotheses about the optimal structure for a CHF program based on the principles above. Most initiatives start with a solid guess — often following the example of a similar program elsewhere — and then evolve based on experience.

The CHF program at Essentia was launched in 1998 by the chief of cardiology, Carl Heltne. Linda Wick, a nurse practitioner, led the program full time starting in 2001. Right from the beginning, there was a Dedicated Team. Within it, Heltne and Wick had complete flexibility to reorganize and to create new roles from scratch. They took advantage of it. They created a team structure that bore little resemblance to Essentia's other clinics.

The Dedicated Team was composed of nurses, nurse practitioners and physician assistants. The NPs and PAs had identical roles, and each NP/PA took responsibility for about 300 patients. By 2013, eight NPs/PAs served approximately 2,400 patients. The team included four nurses, each of whom supported two NPs/PAs, bringing the total size of the Dedicated Team to 12, plus administrative support. (These panel sizes and staffing ratios are valuable bits of information for anyone wishing to start a similar program.)

A panel size of 300 will seem tiny to most primary care physicians. These patients, however, typically came into the program's care within days after their first CHF hospitalization and they needed fairly intensive interaction to stay on track. They were shaken by their hospital experience. It was an ideal moment for building trust and engaging patients in their own care.

Initial appointments lasted for one hour. A full physical was included, but it wasn't really the point of the visit. The heavy emphasis was on education and relationship building. Wick wanted patients to feel that they were being cared for by people they knew and people who knew them. It was crucial that patients became comfortable both with the NP/PA and the nurse. The nurse became the most frequent point of contact and focused on diet and lifestyle coaching. The NPs/PAs handled overall care planning and conversations about medications.

Wick's team endeavored to learn as much as it could about the day-to-day life of each patient, and, in particular, to understand what was important to him or her. Going for walks? Spending time with grandchildren? Patients were most likely to respond to coaching when it was directed at outcomes that mattered to them. Furthermore, a close relationship was crucial in encouraging patients to call to report any change in symptoms. Patients were far more likely to call someone they knew personally than "some dial-a-nurse," as Wick put it. She wanted patients to see the CHF program as a first line of response, far preferable to a visit to the ER.

Not long after the initial appointment, Wick's team followed up by phone and by email about medications, diet and symptoms. The patients were, of course, accustomed to being the ones who initiated contact with the health system, not the other way around. Receiving a call was a new experience, one that cemented trust. New patients also typically had four face-to-face

twice/m++

appointments with their NP/PA in the first two months in the program. In
the early years of the program, these appointments were scheduled for 45
minutes, but over time it became clear that 30 minutes was sufficient. During
patients' first few weeks in the program the staff focused on improving their
ability to monitor their own symptoms and properly act on minor changes.

For NP/PAs, patient appointments consumed roughly 75 percent of the
schedule. The remaining 25 percent of their day, plus some of the nurse's time,
was adequate for remaining tasks, including reviewing the panel to prioritize
which patients were in greatest need of attention, making email and phone
followups and responding quickly when patients called to report a change in
symptoms. Typical responses to such reports included changing the medication dosage, coaching about diet or scheduling a face-to-face appointment.
The program loaned home tele-scales to the subset of patients who were most
vulnerable. All the patient had to do each day was step on the scale. Wick's
team received the data and could respond to unexpected weight gain.

The Shared Staff consisted of Essentia's cardiologists. Recall that as a rule
the Shared Staff has exceedingly little flexibility; it can do more work but not
different work. For the cardiologists involved in the CHF program, that worked
just fine. The cardiologists sustained their normal work routines, scheduling
and seeing patients in the CHF program just as they would any other patient.
They saw each as needed, or, at minimum, once per year. Cardiologists were
most likely to be involved when symptoms escalated dramatically or when
there were challenging co-morbidities. One cardiologist served as medical
director of the program.

As often is the case, there were some tensions between Dedicated Team
and Shared Staff. Some cardiologists were reluctant to refer their patients to
the program, believing they could do a better job managing CHF patients on
their own. Wick found, however, that the cardiologists were responsive to data
that showed that the program increased the likelihood that patients would stay
on their medications. Wick's view was that the cardiologists, while meaning
well, simply didn't have the necessary time required to build close relationships

with patients and coach them. Ultimately, the cardiologists not only saw that the
program was good for patients, they saw that it freed their schedules, allowing
them to tackle medical challenges that called more fully upon their expertise.

Over the years, Wick worked to expand the program's impact. In particular, she endeavored to more closely coordinate care with other providers

who typically treated the same patients. Depression was a common malady faced by CHF patients, for example, and Wick also improved coordination with hospice care.

A nervous interval for Wick was a period in which costs came under close scrutiny throughout the hospital. Heltne had initially used his substantial influence to launch the program over objections about its potential negative impact on profitability. Some of the program's services could not be reimbursed.

As it turned out, the financial analysis actually helped the program. It showed that most CHF hospitalizations were money losers under Medicare's reimbursement schedule. By minimizing these hospitalizations, the program's financial impact was positive despite the significant unreimbursed work it involved. As Essentia moved toward new payment models in 2013, Wick anticipated that the program would only look better.

By 2014, Wick had been lured away from Essentia by the University of Minnesota health system. She was hard at work both building a similar CHF program and overseeing several similar initiatives that spanned a wide range of health conditions.

§

And now, as promised, a look at a special case. Under limited circumstances, part-time team members bring the best of both worlds. They don't cost as much as full-time team members, yet they *do* have the flexibility of full-timers. Their roles, work processes and work relationships with other team members can be redefined from scratch — without hurting ongoing operations. Because of this flexibility, we will consider such team members to be part of the Dedicated Team even though they are only working part time. They are part of the diamond, not the square.

This bonanza is possible when the part-time team members' two jobs are completely independent of and isolated from each other. Perhaps at some point in your life you held down two part-time jobs simultaneously. Student and bartender? Such combinations — hairdresser and tailor, retail salesperson and landscaper — are primarily a matter of making the scheduling work.

It is not quite so simple when both jobs are within the same organization. Conflicts between the demands of the two jobs are more likely, and on multiple dimensions. For the two jobs to be completely independent and isolated from one another, they must avoid four types of conflicts: scheduling conflicts, coordination conflicts, role conflicts and relationship conflicts.

Scheduling conflicts: This is the most straightforward of the four. There can be no overlaps in time between the two jobs. This is more readily achieved in health care than in most industries. It is common practice to slice and dice providers' days into small and discrete time blocks that can be scheduled in any number of ways. A single provider could schedule all appointments for one job in the morning and all of the appointments for the second job in the afternoon.

This assumes that neither of the jobs requires the provider to be available on demand in a way that is disruptive to the other job. Part of Essentia's CHF model was for a patient's most trusted provider to respond quickly any time a patient reported a change in symptoms. This alone severely limits the possibility of part-time NP/PAs or nurses in that program.

Coordination conflicts: It is not enough for team members to have adequate time to do their individual work; they must also be available to interact with other team members. The future promises much more team medicine; logically, we'll see a rise in the amount of time providers spend talking to one another, both in formal meetings and ad hoc collaborations.

More than once, I've run across team meetings that take place at, say, 6:30 a.m. because there is simply no other option. More teamwork will require more time and more convenient time slots for the coordination of efforts. Clearly, the presence of part time team members makes team coordination more difficult to achieve.

Role conflicts: Is it realistic to expect the same person to do both jobs? Even if one person has the basic skills for each, there may be subtle conflicts at the level of behaviors and biases. Taking a momentary step outside health care, could the same customer service representative work part time for a luxury hotel that wanted to lavish each customer with special attention and part time for a software company that wanted to avoid phone calls in favor of self-guided troubleshooting on the Internet?

There are similarly stark differences between today's mainstream health care and innovations in health care delivery. For example, would it be reasonable to expect a physician to seamlessly go back and forth between two jobs when one incentivizes volume and the other value?

Relationship conflicts: When any pair of part-timers has regular interactions in both of their jobs, the two work relationships must be consistent with each other. You can't expect two people to flip a switch every time they go back and forth between one job and the other.

Work relationships are sticky. Even when a pair moves on to an innovation effort full time, it can be challenging to dissolve the prior work relationship and define a new one. When the pair only takes on the innovation effort part time, changing the relationship is next to impossible. The original work relationship continues to be reinforced every day.

Hierarchy is the stickiest dimension of work relationships. For example, if A works for B in one part-time job, do you think it is realistic to imagine that B will simultaneously be happy to work for A in a second part-time job? At most, such a back-and-forth might last through a temporary special project.

§

These four conflicts, collectively, are sufficiently suffocating that the "special case" is hardly worth mentioning in most industries. They are not quite as limiting in health care delivery, and this can help get some innovation initiatives off the ground.

For the advance care planning program at Gundersen, for example, the Dedicated Team was almost exclusively composed of part-timers. A review of the evolution of the program highlights many of the central points in this chapter.

Bud Hammes, with a PhD in philosophy, joined the Gundersen Lutheran Health System in the 1980s as an ethics expert. He took a particular interest in end-of-life care after being drawn into several tense and emotional dilemmas. In each, patients were unable to make decisions on their own, and their families could not agree on what to do and weren't sure of the patients' wishes. To Hammes, the situation was appalling because it was so manifestly avoidable. All it took was a little bit of advance planning.

Hammes started by urging Gundersen's physicians and residents to take the time to meet with hospitalized patients and their families, so that if the patient had to be hospitalized in the future there would be a care plan in place. Physicians responded with enthusiasm; it sounded like a great idea. And then nothing happened.

Hoping to further motivate physicians, Hammes engineered a change in the formal discharge procedure at Gundersen, adding advance care planning to the checklist. This time, physicians responded more forcefully — not with action necessarily, but with tons of questions. How exactly do we get this done? What goes in the discharge summary? When do I meet with families? For how long?

At this point in the program, Hammes was trying to squeeze advance care planning into the slack time in the system. Making advance care planning a formal part of the discharge procedure was a clever maneuver. The work of the Shared Staff, especially work squeezed into slack time, is always more likely to get done when smoothly integrated into the normal day-to-day workflow.

Despite Hammes' efforts to train physicians for the role, the program made little headway. It became clear that advance care planning conversations needed to be lengthy and deliberate and sometimes needed to be continued across multiple interactions separated by days or weeks. Hammes concluded that it simply wasn't practical to ask doctors to squeeze this work into their busy schedules. Their slack time was inadequate, and it wasn't easy to liberate more time because advance care planning could not be reimbursed.

In parallel with his work with physicians, Hammes proceeded with a pilot project with nurses in a dialysis unit. He hypothesized that because the nurses saw these patients regularly and in many cases developed close relationships with them, they were in an ideal position to engage patients in difficult advance care planning conversations. To prepare nurses for the role, Hammes developed a training program that emphasized particular interview skills such as posing open-ended questions that prompted patients to reflect on what they valued most.

This time, the program made rapid progress. The nurses in the unit, collectively, had just enough slack time to schedule the necessary meetings with patients and their families. Before long, more than half of the patients in the dialysis unit had advance care plans in place.

Hammes was eager to expand the program beyond the dialysis unit. To do so, he formally defined a new role from scratch, that of advance care planning facilitator, and he offered it to nurses, social workers and chaplains. As we have seen, when designing new roles and new teams from scratch, creating a Dedicated Team composed of full-time team members is powerful. It confers much-needed flexibility. Nonetheless, it would have been difficult for Hammes to get a full-time staff, given that the program had no budget and no promise of reimbursement from payers.

Fortunately, the "special case" applied. Before long, facilitators were dedicating 10 to 20 percent of their time to advance care planning conversations, engaging two to four patients and their families each week — a level of commitment that is both far beyond slack time and far short of full time. The special case applied because the advance care planning facilitator role could be isolated from and independent of the facilitators' other work. None of the four conflicts was present:

1. **Scheduling conflicts.** The advance care planning facilitators that Hammes selected had sufficient scheduling flexibility so that they could fully commit to advance care planning appointments with patients.

2. **Coordination conflicts.** Little coordination was needed, as no more than one facilitator worked with any patient. There were only occasional meetings to guide the program as a whole and to share experiences between facilitators.

3. **Role conflicts.** Hammes selected facilitators whose "other" jobs were similar, not conflicting, as is evident in his selection of social workers and chaplains to join nurses on the roster of facilitators.

4. **Relationship conflicts.** There weren't any. Hammes' team structure did not call for close collaboration among facilitators.

Hammes and his team learned a great deal from experience. Because advance care planning conversations could be difficult and emotionally charged, Hammes worried about how long it would take for new facilitators to become effective, and he worried about burnout.

As it turned out, facilitators generally felt comfortable in their role after the first 10 to 15 conversations with patients, and many reported that the work was the most meaningful that they had ever done. Hammes also learned that a close prior relationship with the patient was helpful but far less important than he initially imagined. More important was basic knowledge of the natural progression of the patient's disease. Such knowledge became part of the facilitator training.

Facilitators did not have the medical expertise to address all of the questions that came up in their conversations with patients. They would not, for example, make predictions about the likely course of a patient's illness. When additional expertise was needed, facilitators scheduled consultations with physicians — the Shared Staff for the initiative. Such consultations were less frequent than Hammes initially anticipated. Patients did occasionally need medical expertise, but their appetite for assistance in reflecting upon what they valued at end of life was far greater.

Over the years, Hammes managed to expand the program and carve out as much as 20 percent of each facilitator's schedule without ever having to compose a formal budget. In his view, advance care planning was a preventive measure similar in nature to infection control. Whether it was reimbursable or not, whether it generated more patient volume or not, it was clearly the right thing to do. A narrow-minded budget-cutting effort, he speculated, might cut the program and then reduce the involved staff by 10 to 20 percent, but the program had become sufficiently well regarded by the health system that he did not believe there was a significant risk of such an occurrence.

Indeed, by 2014, Hammes was busy working to expand the program to other parts of the country. With a colleague, Linda Briggs, he had developed a branded program, Respecting Choices, that included management and training materials to make it much easier for others to replicate the program and gain the full benefit of the many lessons that Hammes had gained only through years of experience. Hammes was cautiously optimistic that the spread of value-based contracting would accelerate the adoption of Respecting Choices.

Deploy Part-Time Team Members Carefully: Summary of Observations and Recommendations

1. Only small initiatives can be squeezed into the slack time in an organization — those small enough for one person, or at most a few people, to execute in their spare time.

2. Designing new types of teams lies at the core of innovation in health care delivery.

3. The high-level organizational model for innovation in health care delivery is a partnership between a Dedicated Team and a Shared Staff.

4. The Dedicated Team has enormous flexibility. It can accommodate entirely new roles and team designs.

5. The Shared Staff has nearly no flexibility. It can only take on tasks that: a) are already familiar or can be quickly learned, and b) fit naturally into existing roles and work processes.

6. The Dedicated Team is composed of people working full time or very nearly full time on the innovation initiative. All other team members are on the Shared Staff. There is an exception: A part-time team member can be considered part of the Dedicated Team if that team member's two jobs are completely independent of and isolated from each other.

Action Steps for Senior Executives

1. Streamline the process for approving small Dedicated Teams for innovation initiatives.

2. Talk through the distinction between Shared Staff and Dedicated Team with physician innovators. Ensure that you are comfortable that the tasks being assigned to the Shared Staff fit naturally into ongoing operations.

Chapter 8
Manage the Partnership

....

DARTMOUTH-HITCHCOCK MEDICAL CENTER, NEAR MY HOME, IS an academic medical center that serves sparsely populated regions of Vermont, New Hampshire and Maine. In 2012, to extend access to critical care, DHMC implemented a telestroke program. When stroke patients arrive in remote community hospitals, emergency room staffers are connected via videoconference to Dartmouth-Hitchcock, where an on-call vascular neurologist makes a diagnosis and treatment recommendation.

DHMC's Center for Telehealth, led by Sarah Pletcher, was advancing telehealth capabilities across other specialties as well. In a teleconsult program, for example, DHMC physicians met with a remote clinician and patient by video conference. Dermatology, rheumatology, orthopedics and nephrology were the first four specialties to join the program.

Pletcher's team structure is immediately recognizable as a partnership between a Dedicated Team and a Shared Staff. By 2014, her Dedicated Team numbered eight, including herself, a business leader, a program leader for telestroke and two technology experts. The Shared Staff included, most notably, the specialists who offered virtual consults.

.... § ....

This is the third of three chapters focused on team design. Chapter 6 discussed guidelines for building teams under the pretense that the team would operate in isolation. Chapter 7 painted a more realistic picture by examining the complications of building a team within an established organization; in particular, the crucial choice of which team members work on the initiative full time and which contribute part time — that is, who is on the Dedicated Team and who is on the Shared Staff. In this chapter, using the Center for Telehealth at DHMC as the primary example, I complete the discussion of team design by considering what it takes to sustain a healthy partnership between the Dedicated Team and the Shared Staff.

Inevitably, the partnership between the Dedicated Team and the Shared Staff is challenging. Tensions, rooted in the incompatibilities between innovation and ongoing operations, can be sharp. The Shared Staff often lives with one foot in each of two conflicted worlds.

Despite the strains, the partnership can be managed well, and indeed it must be. To give up on the partnership is to give up on innovation itself. In this chapter, we will look at three simple steps to success. They are:

- Ensure that the Shared Staff has adequate resources.

- Take a positive and collaborative approach. Never antagonize the Shared Staff.

- Get help from senior leaders when you need it.

Ensure that the Shared Staff is has adequate resources. Recall that the Shared Staff used to have just one job, ongoing operations; now it has two, ongoing operations plus innovation. But does the Shared Staff have enough time for both jobs? Does it have any additional resources for the new work?

If not, you are in essence asking for volunteer contributions. People like to volunteer, but let's be realistic. Health care professionals are busy; they have limited capacity for pro-bono contributions. When you ask for volunteer work from the Shared Staff, the resource that is available to you is slack time and nothing more. Furthermore, you will be competing with any number of other claims on those slivers of slack. If you are persuasive, you may garner enough of a response to get your initiative under way, but in most cases only that.

To get beyond volunteer contributions, you need to figure out some way to increase the resources available to the Shared Staff. You are responsible for the budget of the Dedicated Team *and* the fraction of the Shared Staff's time that is directed toward your effort. Your team is the partnership, not just the Dedicated Team.

The Shared Staff's budget is often overlooked. It may be formulated and approved through a process that is completely separate from that of the Dedicated Team. One fix is to create some kind of internal accounting transfer that compensates the Shared Staff for their contribution from the budget for the innovation initiative. Once that is done, both parties become well aware of the needs and activities of the other. In many contexts it may be difficult to calculate the proper amount to transfer, but the amount may be less important than the symbolism. Instead of viewing the innovation effort as a distraction, the Shared Staff views the initiative like a regular paying customer.

Ensuring that the Shared Staff has adequate resources is easiest when the Shared Staff provides services that can be reimbursed by a payer. (Isn't it nice when the fee-for-service system actually works in your favor as a health care delivery innovator?) For Pletcher at DHMC, it was a mixed bag. For telestroke and teleconsult, the reimbursement system was on her side, but in other telehealth initiatives, it was not.

Innovation in stroke care using telehealth technology is the most dramatic example I have seen of simultaneous improvements in both outcomes and costs. You may already be familiar with the details. Stroke is among the leading causes of death and disability in the United States. Treating a stroke patient requires quick and consequential decision-making, starting with an accurate diagnosis. In particular, it is crucial to differentiate strokes caused by a clotted blood vessel and those that result from a ruptured vessel. Within about a three-hour window, a clot-busting medication can dramatically improve outcomes if the vessel is clotted, but it can be fatal if the vessel is ruptured. Because of the risk of fatality, only a small fraction of stroke patients who could benefit from clot-busting medication actually receive it. Vascular neurologists, the specialists who can accurately diagnose the stroke type, are few in number; most community hospitals do not have one on staff. Telestroke programs help avoid death and disability by connecting these specialists to remote regions. They also cut avoidable care expenditures, such as pricey but relatively ineffective helicopter transfers and extensive post-stroke care. They dramatically improve outcomes and reduce costs.

Fortunately, physician services delivered remotely are reimbursable by public payers, provided that the patient is in a clinical facility and in the presence of a clinician. It was no problem for Pletcher to expand the capacity of the Shared Staff for the telestroke program. Indeed, to ensure the availability of vascular neurologists for telestroke consults, DHMC has partnered with the Mayo Clinic to expand the on-call roster. Pletcher's teleconsult program is similarly blessed. The patients go to a clinical facility, meet face to face with a provider, typically a nurse, and then engage with a DHMC specialist via videoconference.

In other ways, however, the frontiers of telehealth are constrained by the rules for reimbursements. Why shouldn't a doctor be able to provide services via videoconference to patients in their homes, for example? Certainly, there are contexts in which doing so would make a great deal of sense.

Pletcher is moving forward with such a program, but it is an uphill struggle. She relies on a few physicians who are willing to meet with patients in this way despite that the services won't be paid for.[10] Some physicians are willing to do so because they know it is the right thing for the patient (or, in some cases, they have some grant-funded research that involves seeing patients at home via videoconference), but this can only go so far.

Expecting too much of the Shared Staff *without* ensuring adequate resources is the most common fissure in the partnership between Dedicated Team and Shared Staff. There is a closely related concern. In addition to thinking about how your initiative will affect the Shared Staff's budget, you'll want to be attentive to how it impacts performance. For example, because the innovation work is new and experimental, there may be a few "bumps in the road" that cause a decline in patient satisfaction ratings. To sustain the full support of the Shared Staff, every effort should be made to break out these scores separately for innovation and ongoing operations, so that the Shared Staff can demonstrate that it is meeting its routine performance expectations.

§

Take a positive and collaborative approach. Never antagonize the Shared Staff. As the leader of an innovation effort — and the leader of a delicate partnership between the Dedicated Team and the Shared Staff — the last thing you want to do is make the partnership any more challenging than it already is. Unfortunately, innovation leaders are all too often infected by a "lone warrior" myth, in which the innovation hero "fights the system" and somehow slays the giant, insufferable bureaucratic octopus.

Let's please dispense with this toxic myth. It makes for a nice story, and may even work once in a blue moon, but it is far from a reliable formula for success. The best innovation leaders do not fight the system, they go out of their way to build a bridge to the Shared Staff. They engage in a positive and collaborative manner. They work hard to understand the daily pressures that the Shared Staff is already facing. They share credit with the Shared Staff at every opportunity.

There were no traces of the myth in Pletcher's approach at DHMC. She endeavored to make it as easy as possible for each specialist to participate in the virtual consult program. In particular, she endeavored to customize the technology setup so that virtual consults fit as naturally as possible into normal work routines. All virtual consults were scheduled exactly as normal consults were scheduled. If the physician preferred to take the video call from a laptop while sitting in the office, no problem — the necessary software and hardware was installed on the laptop. If, on the other hand, the physician typically was joined by other providers — a nurse, a resident — for a consult, then Pletcher's team set up a large screen and camera in a separate room.

Recall that the key when dividing the labor between the Dedicated Team and the Shared Staff is to assign to the latter only those tasks that fit naturally within existing roles and work processes. This is essential to success. Breaks from the normal routine can be disruptive to ongoing operations, which is almost always the much larger of the Shared Staff's two jobs (ongoing operations plus innovation). Pletcher went the extra mile, ensuring that even minor disruptions from routine were avoided whenever possible.

The best innovation leaders are also sensitive to antagonisms that can build up slowly. These conflicts are typically rooted in how the Dedicated Team and Shared Staff view each other. In some cases, the innovation initiative may be viewed as the "future" and those not directly involved in it as "old" or as "dinosaurs." Such feelings might be heightened if the innovation initiative seems to be getting special treatment. Maybe it has priority access to certain resources, gets better compensation or faces performance expectations that seem light or loose and far from the strict accountability imposed on ongoing operations. Alternatively, the innovation initiative may be viewed as a quirky experiment while those not directly involved in it view themselves as at the center of the real action.

Such emotions are the norm, not the exception. They are not to be taken lightly. An "us versus them" atmosphere can sink an innovation effort. You want to do everything you can to stave off any emergence of such a schism.

Be particularly wary of how you might unintentionally feed this monster by talking too frequently or too loudly about how you are trying to do something "innovative" and "different" and how stridently you want to "break away from past practice." Framed poorly, such statements can sound dismissive or condescending to the terrific work that the Shared Staff is committed to doing every day.

One case that I studied, an effort to set up a high-volume surgical center inside a hospital, fell into this trap early on. Vocal and exuberant in their desire to boldly reinvent and dramatically improve the way surgery was done, the leaders of the effort alienated those who were not directly involved, to the point that employees who were invited to join the effort reported that they felt that they had to "choose a side." To their credit, the leaders involved quickly diagnosed their mistake, worked to improve the situation, and went on to a major success.

Instead of highlighting just how different you are, you must preach dual excellence — simultaneous success in both innovation and ongoing operations. The message, roughly, is this: The future of health care is unclear. To be fully prepared, we need to experiment with new models of care while continuing to sustain high levels of performance in what we already do well. One is not more important than the other. As an organization, we must do both.

Your goal is to create an environment of empathy and mutual respect that extends throughout the partnership between the Dedicated Team and the Shared Staff. Doing so is so important that in 2013, with my co-author, Vijay Govindarajan, I published a parable, *How Stella Saved the Farm*, that is designed for exactly this job. It is a simple story about a farm in trouble and how it innovates to get out of trouble. Reading and discussing the story as a group helps everyone "walk in the shoes" of everyone else involved — both Dedicated Team and Shared Staff.

§

Get help from senior leaders when you need it. Even when you do everything right, there will be tension between the Dedicated Team and the Shared Staff. Sometimes conflicts will escalate. You will then discover what all innovation leaders find at one point or another: When you engage in a fight with ongoing operations, the odds against you are steep.

Here's why. Leaders of ongoing operations are usually more senior and more powerful than you. They have more connections at the top of the organization. They have history and habit on their side. And, on top of all that, their needs are generally both more urgent and more predictable than yours.

As a result, even the best innovation leaders sometimes need help from above. They need a senior executive as an ally — one who has sufficient authority to override the immediate needs of ongoing operations when doing so is in the best long-term interests of the organization as a whole. This senior executive must do more than simply provide verbal support of innovation. He or she must be prepared to directly engage in conflict resolution, even when conflict arises at levels in the organization that lie below the senior executive's typical zone of direct involvement.

Most conflicts between Dedicated Team and Shared Staff are over resource allocations. As such, the help you are looking for from senior executives most often takes the form of overriding subordinates to redirect resources in your favor. You won't win every such battle, of course, but without this type of support from above, you may not win any.

Senior executives can employ several additional techniques to support you. They can, for example, help you in reinforcing the goal of dual excellence. They also might consider creating special incentives or special compensation arrangements for the Shared Staff that incentivize more energetic cooperation with your initiative. Or they can alter expectations of the Shared Staff, to make it realistic to hit their targets for both ongoing operations and innovation. Finally, senior executives might consider adding a specific evaluation to individual performance reviews for those on the Shared Staff, one that focuses on each individual's energy and skill in supporting innovation.

At DHMC, Pletcher's support came directly from chief executive James Weinstein, who was moving the organization toward value-based contracts as aggressively as he could. Telehealth makes a lot more financial sense under such arrangements; Weinstein intends for DHMC to be ahead of the game.

Tensions between the Dedicated Team and the Shared Staff are sharpest when the success of the innovation initiative implies reduced demand (or some other direct threat) to the Shared Staff. In such circumstances, innovation leaders have little hope of thriving without vigorous support from the top.

Consider an effort to build a new model for primary care at AtlantiCare, the largest health system in Atlantic City, New Jersey. The basic hypothesis was straightforward: If you spend more on primary care, you improve outcomes and save money downstream. The clinic served working adults with substantial care needs, typically those with multiple chronic conditions. It provided additional services focused on engaging patients in taking better care of themselves.

The Dedicated Team in this case was the primary care clinic; the Shared Staff, the specialists who served the same patients. The conflict is obvious. The very reason for the existence of the Dedicated Team was to minimize the need for the services that the Shared Staff provided.

The head of the primary care clinic took a positive and collaborative approach to dealing with the Shared Staff, but little progress would have been possible without the direct support of the CEO. He gave that support because he and his board of directors saw that the future of AtlantiCare was tightly coupled to the future of Atlantic City's economy — its casinos. The city had long since ceased to be the sole destination for gaming on the East Coast. To thrive, the city had to be competitive, and that meant reining in health care costs. If AtlantiCare did not bring costs down, both the city and the health system would spiral downward together.

One could view Atlantic City as the canary in the coal mine for the entire nation. Every developed nation on the planet has substantially lower health care costs than the United States. If we don't fix health care, we won't be competitive. Our economy and our health care system will stagnate in parallel.

Manage the Partnership: Summary of Observations and Recommendations

1. Tension between the Dedicated Team and the Shared Staff is inevitable.

2. The most important measure you can take to reduce tension is to ensure that the Shared Staff has adequate resources for both of its jobs — innovation plus ongoing operations.

3. One way to increase the resources available to the Shared Staff is to set up an internal transfer payment that compensates the Shared Staff from your budget. Doing so may not be necessary. If you are fortunate, the Shared Staff is composed of providers who will be reimbursed by payers for services that are part of

your innovation effort.

4. Never antagonize the Shared Staff.

5. Even the best innovation leaders sometimes need help from senior leaders, most often in the form of direct adjudication of conflicts between innovation and ongoing operations.

Action Steps for Senior Executives

1. Emphasize dual excellence — simultaneous success in both innovation and ongoing operations.

2. Allocate time and energy to resolving conflicts where innovation and ongoing operations rub shoulders. Your investment here will dramatically improve the odds of success for innovation leaders.

3. Ensure that every Shared Staff is adequately resourced for both of their jobs — ongoing operations and innovation.

4. Consider altering expectations or incentives for the Shared Staff, with an eye toward improving the partnership with the Dedicated Team.

Chapter 9
Recognize the Boundary Between
Innovation and Change

A MULTIPLICITY OF PROVIDERS CARE FOR BREAST CANCER PATIENTS, including surgeons, medical oncologists, radiation oncologists, radiologists, plastic and reconstructive surgeons, pathologists, geneticists, primary care physicians, physical therapists and social workers. Because so many providers are involved, the fragmented nature of the U.S. health care system is particularly ill suited for breast cancer care. Sadly, many patients around the country bounce from specialist to specialist and institution to institution receiving conflicting advice that they must sort out on their own. The treatment steps that a patient follows, far from being the result of a well-coordinated plan grounded in the collective wisdom of a team, may depend most heavily on which doctor the patient happened to see first.

Breast cancer patients spoke clearly and forcefully when Dartmouth-Hitchcock Medical Center conducted a focus group discussion in the late 1990s. After initial diagnosis, patients felt they were on their own, needing to schedule separate appointments with as many as five specialists, each on separate days and in different parts of the hospital. Some were driving long distances for each appointment. Patients also highlighted what they saw as poorly coordinated care, limited communication among doctors and a lack of a clear overall care plan.

It was a wakeup call for change. In response, DHMC created the Comprehensive Breast Program and soon named Dale Vidal, a plastic and reconstructive surgeon, its leader. Ideally, Vidal would have designed a new care model from scratch. She faced sharp constraints, however. In particular, the volume of breast cancer patients at DHMC was not large enough to build a new and experimental model for breast cancer care alongside what already existed. (In theory, Vidal could have enabled such a move by trying to attract a greater volume of breast cancer patients to DHMC, perhaps by recruiting specialists from competing health systems. This would have been an uphill climb, however, as DHMC's market share in its service area was already high.)

Thus, the only way Vidal could rebuild breast cancer care from scratch would be to simultaneously dismantle the care model that already existed. The mandate for so bold an approach simply wasn't there.

§

The partnership between the Dedicated Team and the Shared Staff is a design that enables moving forward with innovation initiatives while simultaneously sustaining excellence in ongoing operations. Most innovations in health care delivery can proceed in exactly this way.

As Vidal's story illustrates, however, there are some situations in which this approach proves impractical. Although innovations in health care delivery tend to be tiny relative to the industry and small relative to a health system or hospital, even a minimum-scale innovation effort can be quite large compared with the related clinic or department. Sometimes there simply aren't enough patients to make it possible to run two operations side by side — one traditional and the other new and experimental.

In such cases, progress requires a difficult choice, one that dramatically raises the stakes: a choice to build the new model of care while dismantling the existing one. This leap can be difficult to make. It requires senior leaders make a much higher level of commitment, one that may only gel in the face of a clear crisis.

I'm going to label such projects change efforts rather than innovation initiatives. This distinction, I acknowledge, appears academic, because innovation efforts can involve dramatic change for the people most directly involved. However, there are stark managerial differences between the two endeavors. Consider:

- Change efforts break down and rebuild what exists; innovation efforts build something experimental while minimizing impact on what exists.

- Change demands a much higher degree of confidence in the

new care model. You know it is a better place to be and you are determined to get there.

* Change tends to be messier than innovation, because change efforts tend to impinge much more heavily on non-volunteers, often people who are quite happy with the status quo. A combination of patience and heavy persuasion may win the day, but in many cases, and especially if time is tight, change efforts involve force, up to moving people involuntarily into new positions or out of the organization. One CEO put it to me quite concisely: "If you want to change the culture, you have to change the people."

* Change efforts tend to create pain points distributed throughout the affected organization. Innovation efforts, by contrast, require particular attention to managerial "hot spots" — specific and discrete moments of pain — especially conflicts over resources at overlaps between innovation and ongoing operations.

If you are involved in a change effort, you are moving outside of my area of expertise, and, thus, outside of the scope of this book. There are many terrific books about change management. You can't go wrong by starting with John Kotter's *Leading Change*.

My purpose in this chapter is to explore what is possible when you are faced with constraints like Vidal's. That is, you see an ideal model of care that is far different from what exists, but the only way to build it is to first break down what exists, and you simply don't have the ability for so aggressive a plan.

If you are determined to move forward, your best option, frankly, might be to move to an urban medical center with a much larger patient volume and a senior leadership team that wants to support your work. Short of that, you might try patience. You might want to consider just how much positive change you can bring about with a less aggressive approach, while awaiting a bigger opportunity.

§

Vidal's story illustrates just how much positive impact is possible, even with limited authority and a limited mandate. Indeed, Vidal had no budget.

She could create change only by persuading colleagues to alter their behavior and by trying to squeeze new activity into slack time.

Before I describe the details, it is worthwhile to take a step back and contemplate how you would design the ideal care model for breast cancer, if you could build it from scratch.

There is certainly room for disagreement, but you might concur that the principles for a better approach to care include:

- After initial diagnosis, the patient is referred to a team, not a specialist.

- The patient can look to a single provider who is responsible for coordinating care from beginning to end.

- When patients require multiple specialist visits, they are scheduled on the same day.

- Specialists suggest treatment options and explain the benefits, risks and side effects of each.

- Specialists collaborate when they have differing points of view on the merits of a treatment option. They bring the best available evidence to bear and speak with one voice.

- The patient works with the coordinating provider to determine which treatment option is best. Breast cancer patients frequently have choices among multiple reasonable treatment options.

You might also agree that the most likely team model is one in which the specialists are on a Dedicated Team. They work full time, subspecializing in breast cancer care; they work side by side each day. Some specialists whose involvement in breast cancer care is less intensive might remain on the Shared Staff, but having at least the surgeons, medical oncologists and radiation oncologists together would be crucial to achieving the potential benefits of a team approach to care.

The providers who took on the care coordination role would also be on the Dedicated Team. Reasonable health professionals could certainly differ on the qualifications necessary for this role. It could be a primary care physician who chooses to specialize in breast cancer care and would continue to manage each patient's care in the post-acute phase. It might be a health coach, nurse, nurse practitioner or physician assistant who was adept at guiding patients through their options. It could be one of the specialties, provided that the collaborating specialists were confident that patients would receive unbiased advice.

§

Committed to DHMC and her patients, Vidal was determined to get as far as possible within the existing structure. Her accomplishments might inspire optimism about the potential for positive change even when the organizational structure must be accepted as a given rather than reinvented. Vidal:

- *Specified additional roles for social workers who worked with breast cancer patients.* Social workers contacted patients immediately after initial diagnosis. These first days tended to be extremely stressful for patients. Now, at least, they immediately felt that they had support. The social worker could also provide some measure of continuity in the care experience from beginning to end.

- *Created a single administrative point of contact.* Patients could make one call and, if possible, have multiple appointments scheduled on the same day.

- *Moved patient appointments to a single location.* Vidal claimed for the Comprehensive Breast Program an occasionally used clinic space two half-days per week. Almost all patient appointments were scheduled on these days. Patients could easily move from one appointment to the next. Furthermore, specialists shared a central bullpen, making it far easier to discuss treatment plans or even to see the same patient simultaneously. Vidal even managed to convert an adjacent closet into a small computer room where radiologists could discuss images with other specialists.

- *Used weekly tumor board meetings to hammer out standard care protocols.* Specialists met to discuss selected cases. Standard treatment plans for common situations were formalized. Over time, the tumor board narrowed its agenda to focus on complex cases or those in which the various specialties disagreed on the best approach to care.

• ***Expanded the meetings to include all providers.*** Nurses and social workers were able to bring a fuller picture of patients' lives, wants and needs into the discussion. Making the meeting more inclusive also enhanced the cohesion of the team as a whole.

• ***Ensured that patients fully understood their choices.*** To help ensure that patients received the treatment plan that best fit their goals and values, Vidal worked closely with DHMC's Center for Shared Decision Making, where patients could draw upon a number of resources and decision aids to learn about their options. She also created a Decision Quality Report that, on quick review, highlighted situations in which patients made a treatment choice that seemed inconsistent with their goals. In such cases, physicians revisited the decision with patients to ensure that they were making a choice with which they were comfortable.

By 2009, patient satisfaction scores had risen dramatically and were openly posted online. In that year, Vidal stepped down to become the chief of plastic surgery. A breast cancer surgeon, Kari Rosenkranz, replaced Vidal as the head of the breast program.

§

After taking on her leadership role, Rosenkranz spoke with several others around the country about their approach to breast cancer care. She believed that DHMC, while imperfect, was ahead of the pack. Simply sustaining her predecessor's gains was a priority.

For her to do so will require vigilance. Improvements that are "squeezed in" to slack time can be fragile. What is squeezed in can easily be squeezed out by other initiatives. The Comprehensive Breast Program clinic space, for example, could become unavailable in various building expansion and improvement plans.

Furthermore, the program is not its own organizational unit; it is an overlay on top of the more dominant and traditional organizational structure. The physicians at DHMC who treat breast cancer still identify far more strongly with their individual specialties than with breast cancer care. Rosenkranz herself is on a hallway with other surgeons, not other breast cancer doctors, and she spends more time managing the surgical residency program than managing the breast program.

While contemplating her replacement, Vidal had been quite cognizant of the possibilities of backsliding and loss of team cohesion. She viewed Rosenkranz's collaborative and persuasive leadership style as a crucial asset. The entire team responded well to her.

Rosenkranz's ambitions for the Comprehensive Breast Program extended beyond simply sustaining the status quo. Opportunities for improvement certainly remained. Nonetheless, it is hard for me to see how significant further gains can be made without taking the major step of dismantling and rebuilding breast cancer care from scratch.

Rosenkranz had seen what could be possible if such a step were taken. She had done her residency at MD Anderson in Houston, which treated a much heavier volume of breast cancer patients. High volume made full-time subspecialization in the disease more practical and more common. No tumor board was necessary; the specialists collaborated every day. Also, a primary care team — an internist and a group of nurses — guided and coordinated care after the initial treatment.

In 2014, the possibility of a major reorganization at DHMC, one in which physicians are organized by medical conditions and service lines rather than by specialty, was under active discussion. Though such a reorganization would be good for the coordination of care, there are opposing considerations. Organizing by specialty works well for training new physicians and for advancing the field, for example. Furthermore, not all physicians *want* to subspecialize in a single medical condition. A middle road between doing nothing and reorganizing the entire medical center would be to selectively reorganize, creating Dedicated Teams for the disease states that stood to benefit most from improved coordination of care.

Recognize the Boundary Between Innovation and Change: Summary of Observations and Recommendations

1. Change efforts break down and rebuild what exists simultane-

ously; innovation efforts build something experimental while minimizing the impact on what exists.

2. When an effort to build a new care model is of similar size to the existing clinical operation, moving forward may require a change effort — breaking down and rebuilding.

3. If your institution is willing to take this step, you'll find principles from a related management discipline, change management, to be a helpful adjunct to the principles in this book.

4. If your institution will not take this step, you are left with a tough choice. You might need to leave your institution to pursue your innovation initiative elsewhere. Alternatively, you might choose to create as much positive change as you can within the existing organizational structure.

Action Steps for Senior Executives

1. Develop clear and distinct agendas for change and innovation.

2. Be cautious about encouraging the development and pursuit of ideas that would require a change effort unless you stand ready to forcefully support the endeavor.

Chapter 10
Prove It Works

....

IN 2010, PARTNERS HEALTHCARE IN BOSTON LAUNCHED AN EXPERI-
ment in primary care. The Ambulatory Practice of the Future, or APF, led
by primary care physician David Judge had grown to 14 employees by 2014.
It served 3,500 patients, all of whom were employees of Partners or their
dependents. As Partners self-insured its employees, the organization was at
financial risk for APF's entire panel.

The explicit purpose of the APF was to test a hypothesis that we have
already seen twice in this book. If you spend more on primary care, it will
be more than offset by the savings that you generate downstream. The APF,
however, was pushing this hypothesis to the limits. It did not focus on the
highest-cost pediatric patients, as Clark and Murphy did at Intermountain,
or on working-age adults with multiple chronic conditions, as AtlantiCare
did. Instead, it served a population that was not too different from the commercially insured population at large.

Furthermore, the APF was being aggressive, planning to nearly double
the typical per-patient spending on primary care. Was there enough potential
downstream savings in this population to justify the extra investment? Judge
and his team needed to prove that their initiative worked. They needed to
assess the APF's impact on both outcomes and costs.

.... § ....

The overarching purpose of this chapter is to explore best practices and
pitfalls for proving the effectiveness of an innovation in health care delivery.
In this particular dimension of innovation management, your scientific train-

ing is an enormous advantage over most businesspeople. You know and are comfortable with a process for disciplined experimentation — the scientific method. This advantage, however, is paired with a major disadvantage. The specific model for experimentation that you are most accustomed to, and instinctively gravitate toward, is the randomized controlled trial.

I will open the chapter with a discussion of the strengths and weaknesses of randomized controlled trials, and then outline an alternative framework for disciplined experimentation in health care delivery. I'll close the chapter by describing several potential pitfalls when running experiments inside established organizations.

·· § ··

The randomized controlled trial is often referred to as the gold standard of clinical evidence. As many of you are well aware (though a quick review seems worthwhile), the trial earns this label because of its single overwhelming advantage. Far better than any other experimental design, it eliminates confounding variables that could distort or confuse the cause-and-effect relationship of interest, as shown in Figure 9.

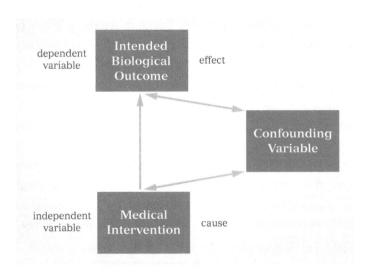

Figure 9

For example, consider a study with an experimental design far weaker than a randomized controlled trial — say, an observational study that compares the weight loss of a group that takes a diet pill to that of a group that does not. Perhaps the data clearly show that those who take the diet pill also lose more weight.

It would, of course, be a tremendous mistake to jump to the conclusion that the diet pill caused the weight loss. Correlation does not imply cause and effect. In this case, it seems far more likely that the people who chose to take the pill were also more disciplined about diet and exercise. These additional activities, diet and exercise, are the confounding variables that make it difficult to say anything conclusive about the diet pill. (A confounding variable is one that is correlated with both the independent variable and the dependent variable in a study.)

Now imagine that a more rigorous experiment randomized the test subjects before the trial to ensure that there were equivalent diet and exercise habits in both test and control groups and to neutralize any other potential confounding variables as well. And, imagine that the results of the randomized controlled trial indicated that this diet pill did indeed cause weight loss. Assuming there are no major flaws in the study and similar studies achieved the same result, the case is closed. The diet pill caused the weight loss.

§

But wait. The trial has weaknesses as well. Most critically, gold standard levels of evidence are generally only achievable by narrowing the scope of the trial. There is, generally speaking, a tradeoff between experimental rigor and experimental breadth.

Yes, the study proves a cause-effect relationship between diet pills and weight loss. But what other cause-effect relationships might also exist? The human body is incomprehensibly complex. There could be biological pathways that designers of the trial did not anticipate and did not look for. Perhaps the diet pill, over the long term, increases the risk of certain forms of cancer. It might take many years longer than the duration of the trial to prove it.

Randomized controlled trials often prove cause-effect relationships that are even narrower — between an intervention and some biological measure such as blood pressure, tumor size, urinary flow or even measures at the microscopic or molecular level. But what about the subsequent cause-effect relationship between these biological measures and the stuff that patients really care about?

Recall the five D's of patient-centered outcomes: death, disease, dysfunction, discomfort and dissatisfaction. As illustrated in Figure 10, even drugs that are supported by the most unassailable clinical trials may be premised on a nothing more than an informed guess about the cause-effect connection between the dependent variable in the trial and an outcome that is actually meaningful.

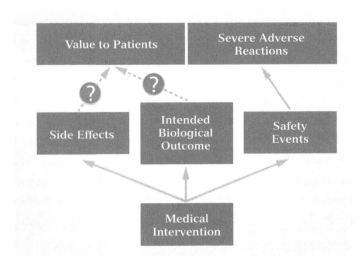

Figure 10

As such, a randomized control trial can omit an ocean of relevance. Indeed, the "gold standard" moniker may be more than a bit too grand. Drug approval requires extensive evidence of safety but is not typically contingent on any scientific effort to validate effects downstream of the dependent variable. It is also not contingent on any scientific effort to assess how patients value those outcomes or trade them off against side effects. Indeed, a new drug, proven by unassailable clinical trials, could have negative net value for patients. That is, the value that patients place on the intended effects of the drug could be overwhelmed by the negative value that patients place on the side effects.

Once a new drug is approved, the trial's narrow scope is quickly forgotten. Physicians have enormous discretion to move usage of a drug well outside the zone in which it was tested. For example, because patients with co-morbidities are typically excluded from trials, physicians will inevitably prescribe to an older and sicker population. And physicians have the discretion to prescribe "off label" for diagnoses that were not even contemplated in the trial.

My desire is not to take a position on the adequacy or inadequacy of the government requirements for approving medical innovations. Perhaps a narrow focus makes sense for innovation in the biosciences. Maybe the success or failure of most such innovations boils down to the existence or non-existence of one crucially important cause-effect relationship, sometimes one at the molecular level.

My intent is simply to show that there is a flip side to the strengths of the gold standard test. Implicit in the design of the typical clinical trial is a value judgment: A high standard of proof is prized; a narrow scope is accepted. Randomized controlled trials can leave a great deal unproven.

§

As leader of an innovation effort in health care delivery, you will need to prove your case in a difficult context. You will face those who oppose your efforts, especially those who perceive that they are competing with you for resources. Your opposition's weapon of choice may very well be pointing out that the evidence supporting your innovation falls short of a randomized controlled trial.

How will you answer?

Depending on the context, you might start by pointing out that your opponents are themselves advocates for treatments that are supported by flimsy evidence that falls far short of any gold standard. Often "current standard of care" is confused with "fully supported by rigorous science." You should not be pressured into producing gold standard evidence to displace a status quo that has little going for it other than the fact that it is the status quo.

Beyond this, however, I'd like you to argue that the typical trial makes little sense for innovations in health care delivery. There are two main thrusts to the argument.

The first is that sacrificing breadth for rigor is undesirable for innovations in health care delivery. The intended impact of innovations in health care delivery is not seen at the molecular level, or at the microscopic level or at the organ system level. The impact is seen at the population level. These are system innovations. They typically have an impact on multiple system and population variables, through several interrelated cause-and-effect chains, ultimately with an impact on outcomes and costs. Why would it make sense to give overwhelming attention to a single cause-effect relationship while leaving everything else to experienced judgment?

For example, imagine where the logic of clinical trials as practiced in the biosciences might lead for Partners Healthcare's primary care initiative, the Ambulatory Practice of the Future. Judge could conceivably run a randomized controlled trial that indicated that his practice was successful in bringing down the HgA1c levels of a carefully selected subset of the diabetics that he serves (those with no co-morbidities). On the basis of a successful study, the APF's approach to high-intensity primary care would be "approved."

Judge could then proceed to deliver higher-intensity primary care to all diabetics. Indeed, he would be able to "prescribe" the APF "off label" to any patient who he thought might benefit. All of this could be done without any convincing evidence that the APF achieved its overarching goals — better outcomes and lower costs at the population level.

You may have noticed that I have been careful to address the weaknesses of *typical* randomized controlled trials. The experimental method is not the enemy. As one researcher pointed out to me, randomization is just a coin flip, one that puts patients in either an experimental group or a control group. In some cases, it may be possible to run a randomized controlled trial *without* sacrificing breadth for rigor — one that captures not just a single biological cause-effect relationship but much broader phenomena, especially outcomes relevant to patients and costs. If so, and especially if the trial can be conducted quickly and cheaply, then of course the randomized controlled trial is better than a less rigorous study.

The typical randomized control trial, however, is lengthy and expensive. It may not be worth the effort.

This brings me to the second thrust of the argument: Innovations in health care delivery involve substantially different risks than innovation in the biosciences. Intuitively, the standard of proof in an experimental design ought to be tied to the consequences of drawing a false conclusion. This logic is familiar; it underpins the U.S. legal system. In a criminal trial, with wrongful incarceration at risk, the standard of proof is beyond a reasonable doubt, but when mere money is at risk in a civil suit, the standard is a preponderance of the evidence. The same principle guides the regulation of health care innovation. To paint an extreme, nobody would expect the same level of testing for a new design for crutches as for a new cancer pill.

Biosciences-driven innovations — new treatments — carry substantial risk. They are invasive. Cutting, irradiating, ingesting chemicals and implanting devices all risk bodily harm to patients. Before we allow physicians to go forth with such actions, we certainly should demand a high level of proof that the risk is small and there is at least some benefit to offset the risk.

By stark contrast, innovations in health care delivery involve little risk to patients. Consider the four innovation categories in this book — standardize and delegate, coordinate, prevent, improve decisions. It is hard to identify any risks to patients in the latter three. What is the intervention? It is more talking and thinking — engaging patients, encouraging patients to take better care of themselves, giving more information to patients, creating better care plans, and having more frequent conversations among providers serving the same patients. Can we agree that there is nothing here that is on par with cutting, irradiating, ingesting chemicals, and implanting devices?

The first of the four categories is the exception. There is a meaningful risk of harm in standardizing and delegating. Overly aggressive delegation of tasks can increase medical errors. Nonetheless, decades of experience in industry suggests that standardization of processes tends to push errors down, not up. Leaders of initiatives in this category must be ready to demonstrate continuous improvement in their rate of medical errors.

This exception aside, the main risk associated with innovations in health care delivery is not bodily harm, it is financial loss. Some innovations in health care delivery will fail. The consequence? Instead of saving the system money, these innovations will cost it even more.

Given this, does it make sense to demand that an innovation in health care delivery show data from a randomized controlled trial to *prove* that it saves money? Many who are steeped in the habits of the biosciences seem to expect such proof, but it is hard to reconcile the sudden anxiety over financial loss. After all, in what fraction of trials for new biosciences-driven innovations was saving money even a consideration? And how many drugs with marginal benefits (albeit randomized controlled trial-proven marginal benefits) have been approved *despite* their enormous cost?

Where the biggest risk is financial loss, the sensible standard of proof is: Proven to the satisfaction of those who write the checks. With value-based contracts in place, the people writing the checks (and bearing the risks) are typically leaders of provider organizations. They are the ones who are held accountable for the bottom line and are well compensated to take on that responsibility. They do not need a government oversight body to protect them from innovations in health care delivery that could fail to deliver savings, the way patients deserve such oversight before they swallow a novel chemical concoction.

Consider what other industries might look like if the burden of proof inherent in clinical trials was applied to business decisions. One of the key conjectures underpinning Amazon's business model, for example, is that if the company sells bestselling books at a loss, it will more than make up for that loss with sales of more profitable products. Can you imagine Jeff Bezos presenting RCT data to shareholders or government regulators to get approval to expand this practice beyond a small trial?

Innovations in health care delivery look much more like business bets than medical bets. Scientific proof is not the standard by which business bets tend to be judged. Instead, they are evaluated based on data, trends and experienced judgment.

Of course, "experienced judgment" is fallible. Skilled business executives know that their judgments are far more likely to be sound when there is a clear hypothesis and a disciplined hypothesis-testing process in place. Even then, interpreting trends can be challenging.

Business leaders are constantly faced with difficult judgment calls in response to *worse-before-better* trends. An investment is made today in hopes of a return tomorrow. The tough call takes the form of a decision on whether to continue to invest in the face of a downward trend. Will the trend reverse itself? Will it do so quickly enough and robustly enough to earn an acceptable return on continued investment? As a physician innovator, you will have to make similar calls.

There is, however, another trend type to be particularly aware of, one that is unusually common in health care delivery. It takes exactly the opposite form — not worse-before-better, but better-before-worse. I think of it as the trend that lies.

You are well advised to be wary of the better-before-worse possibility. This is what can go wrong: You innovate. You see improvement. You're excited. You choose to scale up. Unfortunately, it turns out that the improvement that you saw was temporary, unsustainable or altogether unrelated to your innovation.

Thus, as you scale up, all you are doing is adding cost. Better-before-worse.

The best advice I can offer when trying to discern whether a quick success is really a mirage is to engage the best scientists in your organization. Ask them: Why might this trend be misleading? They will ask terrific questions, rooted in their knowledge of the weaknesses of observational studies.

Keep in mind that your scientist-partner's job is to remain skeptical until presented with definitive proof. Your job is different; it is to make a sound business decision in the face of uncertainty. The discussion with the scientist will be useful in this regard. It will help you see and consider all of the possibilities before making a decision.

§

Now, by suggesting that the typical randomized controlled trial is of questionable relevance to innovation in health care delivery, I hardly mean to suggest that you as an innovation leader are under no obligation to prove the merit of your work. You are, and you will want to prove your case — to yourself, to your supporters and to your funders.

Indeed, you'll want to learn from your experiment even as it is in progress, and as quickly as you can. You'll want to be able to identify the weak spots, make changes, adapt and zero in quickly on success. Your innovation likely involves multiple discrete interventions. You'll need to know: Which of the cause-effect relationships that you hypothesized seem to be real? Which are stronger or weaker? And for which patients? Which trends are promising? Which are ephemeral?

You'll be more successful on all of the above if you treat your innovation initiative as much like a *disciplined experiment* as possible. Disciplined, however, does not imply just like an randomized controlled trial. You do not want to be needlessly obsessive about high standards of proof; nor do you want to sacrifice the big picture for a misguided overemphasis on a single cause-effect relationship.

You'll want, instead, to accept the realities. An innovation in health care delivery is a messy experiment executed in a messy context. It is not run over a fixed timeframe; instead, it is ongoing, continuous, living and adaptable. It is not focused on a single cause-effect relationship; instead, it tests a theory that connects multiple overlapping cause-effect chains.

This is not bench science.

Nonetheless, the process of the scientific method is still your friend. Follow it as best possible and you'll learn fast. When you launch, there will no doubt be a number of conjectures built into your plan, and perhaps even a wild guess or two. Through disciplined experimentation, you'll systematically gather evidence to support or challenge your conjectures and you'll hasten your arrival at your desired destination — better outcomes, lower costs.

§

Before we can dive into the details of disciplined experimentation, we must address a fundamental complication. In innovation in health care delivery, there is no true bottom line. Instead, there are two goals — improving outcomes and reducing costs. Which is more important? Is an innovation that improves one but not the other valid? What tradeoff between outcomes and costs ought we be willing to make if we cannot win on both counts?

I will use Figure 11 to frame this discussion. The dual impact of any innovation in health care delivery could be plotted on this graph — change in outcomes on the vertical axis, change in costs on the horizontal axis. Initiatives that improved both costs and outcomes would land in the upper left quadrant of the graph. Clear failures would be plotted in the bottom right quadrant. The tougher cases, those that improved one objective but not the other, would land in either the upper right or lower left quadrants.

Where could the dividing line between success and failure be drawn? There is at least a theoretical answer. Some policymakers and cost-benefit analysts have attempted to address the tradeoff between outcomes and costs by measuring — quality-adjusted-life-years — and trying to establish a dollar value per QALY. Practically speaking, setting public policy based on an explicit dollar-per-QALY threshold is a dicey proposition. Still, there must be some cost-benefit tradeoff that is implicit in the health care decisions that we are already making. I have read figures ranging from $50,000 to $250,000 per QALY; the exact figure is unneeded for our purposes.

Included on Figure 11 is a diagonal line that represents this dollar-per-QALY threshold. In principle, one could judge any innovation (either biosciences or health care delivery) a failure if it fell below this diagonal line. The wedge that is in the upper right quadrant but below the diagonal line — an area where biosciences-driven innovations can easily fall — might be described as a low value innovation. Outcomes improve, but at great cost.

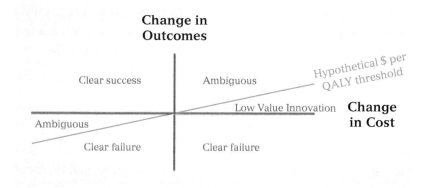

Figure 11

Now, only the two zones marked Ambiguous in Figure 11 need further discussion. Let's start in the lower left. Here, for results to be above the line, a marginal decline in outcomes would have to be accompanied by a substantial reduction in costs. While a cold-hearted analysis of QALY math might suggest that such an outcome might be acceptable, no respectable physician would occupy this space. Furthermore, the last thing we need in the movement to fix health care is to deal with charges that costs are being cut by stinting on care.

The upper right quadrant is far more interesting. Note that almost all biosciences-driven innovations will fall in this quadrant of the graph. Indeed, if we demanded that new medical interventions improved outcomes *and* reduced costs, the entire biotech, pharmaceutical and medical device industries would have screeched to a halt decades ago.

Should innovations in health care delivery be held to more stringent standards than innovation in the biosciences? An innovation in health care delivery might improve outcomes while *raising* costs and still produce an overall impact that outpaced even the best biosciences-driven innovation of the year! Why shouldn't we delighted to support such an initiative?

Unfortunately, even in a world of value-based contracting, it is not at all clear how such an initiative would be paid for. One option would be to limit the new services to patients who paid more for a premium health plan. Beyond this potential exception, however, I don't think there is a need to encourage innovations in health care delivery in the upper right quadrant. Even though these initiatives may outperform biosciences-driven innovations, the system imperative to reduce costs is inescapable, and there are plentiful opportunities to do so while improving outcomes. Thus, upper right quadrant innovations ought to be deprioritized as long as there are plentiful upper-left alternatives — and I suspect this will remain the case for many years to come.

························ § ························

So we have set the bar. We are looking for innovations that land in the upper-left quadrant. They both improve outcomes and reduce costs.

There is a caveat, however. Innovation initiatives are dynamic. They do not have a discrete stopping point. They do not suddenly land on the graph at their final destination. They are ongoing, living experiments. They follow a trajectory, and often that trajectory is well described as worse-before-better.

This raises a crucial question: What results are acceptable temporarily, on the way to the upper left quadrant, as a new care model is refined? I suspect that the lower left quadrant (lower costs but worse outcomes) is best to avoid even temporarily. The upper right quadrant (higher costs but better outcomes) ought to be viewed as comfortable, if only temporarily. For complex initiatives, funders need be patient, even waiting in the upper right quadrant for as long as a few years. Initiatives need to have time and space to test their cause-effect conjectures and zero in on the best care model.

For example, the APF effort started and will remain in the upper right quadrant for years. By late 2013, Judge had analyzed the first two years of data and the first 1,000 patients. The analysis showed clear savings, but not enough savings to offset the additional expenditures on primary care. Judge perceived, however, that he'd have enough time to prove his case.[11]

In general, a process of disciplined experimentation will help you extend your grace period. By continuously assembling the evidence that your experiment is generating, you'll build the best possible argument for sustaining your initiative through its worse-before-better trough.

························ § ························

We have now set the bar a bit more precisely. Your objective is to prove that your initiative can deliver results in the upper left quadrant (better outcomes and reduced costs) with some leeway to operate temporarily in the upper right quadrant (better outcomes but higher costs) as you refine your care model. Onward, then, to the specific practices for structuring an innovation in health care delivery as a disciplined experiment.

It should come as no surprise that step one is to express a clear hypothesis. The form of the hypothesis is critical. As discussed, innovations in health care delivery are rarely focused tests of single cause-effect relationships. Instead, the underlying hypothesis is generally well described as multiple overlapping cause-effect chains.

Figures 9 and 10, earlier in this chapter, illustrate a useful approach to diagramming such a hypothesis. The elemental building block of such diagrams is a single cause-effect relationship illustrated by an arrow connecting two variables. Here is a suggested process for sketching a complete diagram:

1. Create a short list of the major activities that will consume the bulk of your team's time. Most of these will be service delivery activities — coaching patients, engaging patients, seeing patients by videoconference and planning their care. Ideally, there are one to five activities on your list that collectively account for the bulk of your budget.

2. Consider these activities one at a time. Think through how each impacts patient-centered health outcomes (the five D's) and costs. Make a list of intermediate outcomes — variables that might show a change before an impact on outcomes or costs was detected. For example, if you are investing in health coaching for diabetics, blood sugar levels (HgA1c) and incidence of diabetes-related acute care events might be two such variables.

3. Draw a separate cause-effect chain for each activity. Connect each activity through its intermediate outcomes to its ultimate impact on costs and/or outcomes. (See, for example, Figure 12.)

4. Consider each outcome. For each that may be difficult to measure, consider substituting a proxy measure that is likely to be closely correlated.

5. Noting the overlaps (outcomes that appear on more than one chain), combine your collection of separate cause-effect chains into a single cause-and-effect diagram.

6. Highlight the critical unknowns. These are the cause-effect relationships that you feel are both highly uncertain and highly

consequential. You will want to test these cause-effect hypotheses quickly and cheaply, if possible.

Over the years, I have seen that the primary risk when creating these diagrams in managerial settings is that once people start to get comfortable with them, they can get carried away. The diagrams can easily become overly complex. One team I worked with literally covered a wall with cause-effect chains.

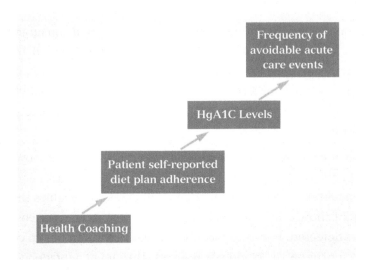

Figure 12

Keep in mind that what you are creating is a framework for ongoing review, analysis and learning. If the hypothesis is tremendously complex, collecting and making sense of the data will become a Herculean task that you simply do not have time for.

Furthermore, keep in mind that you will want to involve your team in the learning process. To engage, each team member must understand the hypothesis. As a rule of thumb, your hypothesis should be simple enough that team members could study the diagram to refresh their memory in the five minutes before a meeting. If the diagram fills more than one page, it is probably too complex. If there are more than about five activities at the bottom of the diagram or more than 20 variables on the diagram, then it is probably too complex. Note that the error of making the hypothesis too complex can be seen as the opposite of the error of hewing too closely to the randomized controlled trial model for experimentation and giving overwhelming focus to just one cause-effect relationship.

§

To fully express your hypothesis, you will want to add predictions to your cause-effect diagrams. These predictions need not be precise. Your goal is to estimate the *trends* that you expect to see from launch to success. For each dependent variable on your cause-effect map, you want to address questions like: How much improvement can we achieve? Ten percent, or 10X? Over what time period? When will most of the improvement happen? When will the improvement trajectory be flat? When will it be steep? Might results get worse before they get better? Might they get better before worse?

You can express answers to these questions by sketching an expected trend over time. Unless you have reason to believe that you can predict more accurately, rough estimates are appropriate. You might show only approximate labels on the axes of these trend graphs, such as "weeks" or "years" on the horizontal axis, and 10 percent or 10X on the vertical axis. Your goal, with both the cause-effect map and the trend graphs, is to express a qualitative hypothesis. You want *shared understanding* within your team about how, how much and when, roughly, the initiative is expected to succeed so that you can have clear and productive conversations about progress to date and next steps.

One way or another, somewhere along the way, spreadsheets are going to work their way into the picture. Perhaps you will be asked to develop a spreadsheet model of your initiative in order to get approval and funding.

A spreadsheet is an alternative means for expressing your hypothesis, and it can give a more formal and rigorous appearance. I advise that you first develop the qualitative hypothesis in the form of a cause-effect map plus trend graphs and only later build a spreadsheet. It is possible that the spreadsheet-building exercise will help you improve your qualitative hypothesis, but do not let the spreadsheet *replace* the qualitative hypothesis. The spreadsheet gives a false sense of precision. It also obscures the cause-effect logic underlying your hypothesis in favor of showing a lot of digits. Few people on your team will have the time or the inclination to try to dig into the spreadsheet to understand how it works. The cause-effect map and the trend graphs are far superior tools for generating shared understanding and structuring clear and productive conversations when it is time to review progress and discuss next steps.

§

Now, let's move forward in time, from clarifying your hypothesis to gathering the evidence to support or challenge it.

Disciplined experimentation implies carefully and rigorously comparing what you thought was going to happen to what actually happened. The right tool for the comparison can be enormously helpful: a trend graph. Innovations in health care delivery are ongoing, living experiments; thus, you will be comparing predicted trends and actual trends. As time passes, the actual trends become clearer and your judgment about the accuracy of your prediction firmer.

You guide your initiative by systematically gathering data to test each cause-effect conjecture in your hypothesis. You'll want to engage your team in conversations about the latest evidence frequently — as often as new data is available that may lead to a substantive redirection of your initiative. Monthly is often the appropriate interval.

The overarching question in these review meetings is: Are we on a trajectory to success? Given that the upper right corner of Figure 11 (better outcomes but higher costs) is a far more comfortable place to be than the lower left corner (worse outcomes but lower costs), you should be patient in demonstrating reduced costs but quite impatient in showing improved outcomes.

Depending on the nature of your initiative, it might be several months or quarters before you can detect any change in the bottom lines of outcomes and costs. In such cases, the upstream variables on your cause-effect chains become particularly relevant. These are your leading indicators — your early warnings that signal when your initiative is headed toward unexpected terrain.

As you gather evidence, you may conclude that some of your hypothesized cause-effect relationships are quite strong; others weak or non-existent. As you draw such conclusions, you update your qualitative hypothesis. By systematically evaluating each cause-effect chain, you may find that you naturally surface more specific cause-effect conjectures that you can subsequently test to further refine or redirect your initiative.

For example, the APF is experimenting with a wide range of interventions, including lengthy face-to-face interactions between patients and health coaches, greater email and video conference access to primary care physicians, mental health screenings, outreach to patients prioritized by their chronic disease status, and a team-based approach to care that requires frequent meetings to discuss the status of each patient. Some of these interventions may have high costs but little impact. The APF can accelerate its trajectory to success by systematically testing the cause-effect conjectures associated with each intervention.

Keep in mind that in many cases it will be helpful to discern not just whether the cause-effect relationships exist or do not, but *for which patients* the cause-effect relationships are most powerful. This suggests a significant multiplication of the analytical work in assessing your hypothesis, as every data point must be broken down by patient type. It is worth the effort. For example, it seems likely to me that the APF will prove successful for some subsets of patients but not all. The venture's sustainability may hinge on discovering which ones.

To fully test your hypothesis, you will likely be mining data from multiple sources, including electronic medical records and the financial systems of your health institution or practice. Much of the work may be challenging. Sometimes the data that you need will be hard to come by. Sometimes you will be frustrated by financial systems that make it easy to see charges to payers for various services but much harder to estimate actual costs. (Anyone who has looked closely knows that one has bizarrely little to do with the other.)

However, you might be pleasantly surprised by the richness of the cost information available within health care delivery institutions. One hospital cost analyst I spoke with had the ability to break down care delivery activities by episode, patient and department. As it happened, however, the capability was rarely used. Also, health insurers can be important partners here. They are increasingly adopting the mind-set that the biggest contribution to fixing the system that they are in a position to make is offering useful data and analytics that help physicians improve care on the front lines. Old barriers to sharing information, such as competitive concerns and privacy issues, are coming down.

.... § ....

The gathering and analysis of data that will guide the direction of your initiative will not happen in a vacuum. Rather, it will occur within the context of a larger organization that is gathering and analyzing data to guide its future.

Let me be clear: Your organization's planning process is not your friend. Its purpose is not to run a disciplined experiment but to administer a proven operation. There are many dangers here. I will conclude this chapter with several related recommendations.

Create a separate and stand-alone plan for your initiative, starting with a blank page. Established organizations are guided by a simple hypothesis — that this year will look an awful lot like last year. This is convenient. It makes planning quite efficient — merely a matter of refreshing key planning templates with updated data.

The danger lies in the habit of thinking only in terms of small variations from the past. When leading innovation in health care delivery, you want to rethink everything from scratch. Just as you take a "clean slate" mind-set to designing the Dedicated Team, you take a clean slate mind-set to developing plans.

Of course, you will have to engage in your organization's planning process for certain limited purposes. You will need to win the necessary resources for your initiative. Also, not to be overlooked, you will need to coordinate resource planning for the Shared Staff. Otherwise, however, you want to keep your distance.

Create a separate forum for discussing results. In general, you do not want to be in meetings discussing the performance of your initiative alongside colleagues who are discussing the performance of proven operations. Instead, you want to convene a different meeting at a different time, one that is focused only on your initiative (or perhaps yours plus other innovation initiatives that your organization is running in parallel.)

The two meetings must be separated because they revolve around questions that are diametrically opposed to each other, and it is nearly impossible to go back and forth between the two. When the topic is ongoing operations, the key questions are: How far off plan are we? How do we get back on plan? The validity of the plan is rarely questioned, only the discipline in executing it. For innovation, by stark contrast, the validity of the plan (the hypothesis) is the central question.

I have seen that when innovation leaders have their efforts reviewed in the wrong meetings they inevitably end up being put on the defensive. They are blamed for falling short of a speculative plan or they are blamed for underperforming against metrics that are commonly used by the organization but have little relevance for the innovation initiative. Sadly, once defensiveness rises, learning stops.

Meet frequently. The frequency with which you meet to discuss the results of your innovation initiative sets a "speed limit" on learning. Pay no attention to the frequency of your organization's review meetings. You want to meet as often as new information is available that might cause you to rethink your hypothesis.

Try to influence which senior leader evaluates your individual performance. You want someone who is comfortable taking a long-term perspective, someone who wants to take a risk with you on your innovation initiative, and if at all possible, someone who has experience evaluating the progress of innovation efforts.

Negotiate sensible accountabilities. Most of your peers in your organization will be expected to deliver on plan. Individual physicians may be expected to generate a certain level of billings, and departments as a whole will be expected to be on budget.

It makes little sense, however, for innovation leaders to be held strictly accountable to plan. The plan is too uncertain. Still, you should not expect to be left unaccountable or to simply have the "freedom to fail." You, too, must be held accountable. But for what? You can volunteer possibilities in each of three categories of accountability:

1. **For results.** Though it is rarely productive to hold someone accountable for a speculative result, some of the intermediate outcomes on your cause-effect map may be relatively certain. If so, you ought to volunteer to be held accountable for these outcomes. In established organizations, hitting your numbers gives you credibility. At the APF, Judge reported patient satisfaction (APF scored very highly) plus a wide range of well-established metrics from chronic care guidelines, for example.

2. **For doing the work.** As an alternative to being held accountable for outcomes, you can volunteer to be held accountable for

inputs. Where available, quantitative measures are preferable. For example, at the APF, Judge agreed to deliver at least 70 percent of the normally expected number of reimbursable patient visits (leaving 30 percent for non-traditional and non-reimbursable interactions with patients, such as phone calls and emails.) A less satisfying alternative: You can ask to be assessed based on qualitative observations of how hard you are working and how hard the team you are leading is working.

3. *For disciplined experimentation.* To evaluate whether you a running a disciplined experiment, the person evaluating your performance must observe you closely — especially your thought process and the way you run meetings with your team to discuss progress and next steps. In the best-case scenario, the person evaluating you engages with you in frequent collaborative discussions about the latest data and lessons learned. The evaluation of whether you are running a disciplined experiment can be broken down into many more specific questions, including: Do you have a clear hypothesis? Does your team understand the hypothesis? Are you investing adequate time and energy in gathering data, analyzing data, and assessing and updating your hypothesis? Are you evaluating results at an adequate frequency? Are you focused on the most critical unknowns? Are you learning as quickly as possible? Do you only change direction when there is clear evidence suggesting the need to do so? Are you reacting quickly to new information? Are you willing to confront the truth, including, possibly, a failed initiative? Are your predictions getting more accurate over time? You should welcome this form of accountability. Doing so will help you and your team zero in on a successful care model as quickly as possible.

Prove It Works: Summary of Observations and Recommendations

1. As a model for experimentation, the typical RCT is of dubious relevance to innovation in health care delivery. RCTs tend to focus narrowly, on a single cause-effect relationship, and to presume risks to patients that are not present for innovations in health care delivery.

2. Innovations in health care delivery must meet two goals: They should improve outcomes and reduce costs. There should be substantial leeway, however, to increase costs temporarily as the new care model is developed and refined.

3. Develop a hypothesis for your initiative in the form of multiple overlapping cause-effect chains. Make a prediction for each outcome on your cause-effect map in the form of an approximate trend over time.

4. Disciplined experimentation implies carefully, rigorously and regularly comparing predictions and outcomes. Guide your initiative by systematically gathering evidence to test each cause-effect relationship in your hypothesis.

5. Your organization's planning process is not your friend. Develop a separate, stand-alone plan for your initiative and discuss results in a separate forum.

6. Negotiate your individual accountabilities in three categories: for outcomes, for doing the work and for running a disciplined experiment.

Action Steps for Senior Executives

1. Evaluate innovations in health care delivery as business bets, not medical interventions.

2. Support innovators in health care delivery by making it easier to gather the data necessary to prove the success of their endeavors.

3. Demand that innovation leaders follow a process of disciplined experimentation. Be patient for results.

PART III
ACTION

....

Chapter 11
Standardize and Delegate

Y OU'VE DONE IT! YOU'VE MADE IT THROUGH THE "MEAT AND POTA-
toes" of this book, the detailed managerial recommendations for leading
innovation in health care delivery.

My goal in Part III is to introduce you to as many additional examples of
innovation in health care delivery as I can. Some will naturally be of greater
interest to you than others. Feel free to skim. While Part II was sequential
and cumulative, you can jump in and out of Part III however you desire.

In four chapters, I will present four categories of innovation initiatives
named by the primary means through which value is created: standardize
and delegate, coordinate, prevent, and improve treatment decisions. Each
chapter will open with a few new principles that are specific to the innovation
type, then continue with the examples.

Keep in mind that the examples do not all fit neatly into just one of the
four categories. Some combine two, three or even all four strategies for improving value in a single effort. In sharing each story, I have focused on the
primary strategy.

Recall that each chapter in Part II ended with a summary of observations
and recommendations. These are highlighted as they arise in each example,
as are the new principles introduced at the beginning of each chapter. While
the recommendations from Part II are broadly applicable, those from Part III
are specific to the innovation type.

§

At the Connecticut Joint Replacement Institute, efficiency was a point of pride. The surgeons at this high-volume knee and hip replacement center, a part of the Saint Francis Hospital and Medical Center in Hartford, Connecticut, routinely completed eight operations a day. The team honed its processes, standardized each step and eliminated sources of complications and errors.

We will take a closer look at CJRI and two other examples of innovations in health care delivery that seek to radically improve value using the same principles that long ago guided the industrial revolution. Standardization of work steps, specialization of labor and delegation of tasks to capable but lesser trained and lesser salaried workers — these are, and always have been, reliable pathways to increased quality and productivity.

An exploration of the value that can be created through standardization and delegation is an ideal first stop in Part III. These are the most elemental initiatives. The same principles described in this chapter will continue to appear in the remaining chapters in more of a support role.

Note that standardization and delegation are two separate and distinct pathways to improved value. I have chosen to present them in the same chapter because they so often travel together. When a work process is standardized — that is, scripted and predictable — it becomes much easier to train capable workers with less education and lower salaries to do the work. Note that there is no law that dictates that they *must* travel together. Indeed, one of the three examples in this chapter illustrates a focus on delegation only.

§

The following principles guide innovations in health care delivery that create value primarily through standardization and delegation.

Principle 1: Focus on mainstream patients; serve unusual patients through other means. Many physicians instinctively react negatively to the notion of standardizing health care. After all, health care is not manufacturing. Patients are not standard. They are highly variable in their wants, needs and medical conditions.

All true. Many segments of the health care industry are simply not amenable to standardization. Indeed, throughout the industry, one of the dangers inherent in efforts to standardize care is that patient preferences might be trampled upon in favor of one-size-fits-all treatment plans. Standardization efforts ought to be limited to processes for delivering care, excluding decisions about treatment plans.

Standardization efforts are most powerful where large numbers of patients have predictable and similar needs. A focus on such patients is a bit against the grain. The culture of medicine emphasizes not the mainstream patient but the most unusual one. The most heroic doctor is the one who can make the unusual diagnosis, perform the most daring procedure or lead the most precarious rescue. Stories of such episodes are the lifeblood of physician war stories.

Does it make sense, however, to build a health system around these most unusual patients? Many parts of the health system we have today look as though this was the mind-set. We build, for example, high-tech general hospitals that are, in theory, ready for anything. For patients with mainstream needs, however, such hospitals are needlessly costly and burdensome to navigate.

I have touched on many differences between the MD and MBA mind-sets through this book. Here is another. The typical MBA does not embrace the exceptionally complicated customer. To the contrary, the typical focus is on identifying a set of customers with consistent, well-defined needs, and meeting those needs with repeatable and predictable processes. Customers with unusual or complicated needs may be avoided altogether, or, if not, will be served through separate processes. They will be treated as exceptions to the rule, not as the base case for system design.

Of course, the health system as a whole does not have the option to simply avoid complicated patients. It certainly can, however, design separate processes for standard patients and complex ones. And it is for the mainstream patients where there is enormous opportunity to achieve higher levels of value through standardization and delegation.

Principle 2: Take pride in system performance. The notion of routinizing work, such that one does the same job the same way every day, will not appeal to every physician. Many of you value independence and prize flexibility. You take pride in customizing your approach to each individual patient as circumstances demand.

But not all of you. I have interviewed several doctors who take great satisfaction from their efforts to build safe, reliable and efficient clinical operations. These conversations have occasionally given me the impression that I'm speaking not with a physician but with a manufacturing engineer.

Make no mistake, however. These physicians have lost neither their humanity nor their desire to serve their patients as well as possible. They simply see a different means to the same end — by shifting their primary focus from the patient in front of them at any given moment to the overall performance of a clinical team. They have seen that standardization and delegation can be a pathway to reducing errors, improving outcomes and making health care more affordable for all.

Principle 3: There is value in standardization at any level of specificity. In some industries, standardized process can become extremely specific. An auto manufacturing process might go so far as to lay out the precise order in which bolts are tightened.

Such specificity would be extreme in health care. Sometimes enormous gains in value are achieved through only modest, high-level standardization of workflows, as we will see shortly in a primary care example. And, in even the most heavily scripted processes I have seen, it is well understood that each patient is different. There is plenty of room for physicians to make judgment calls about when exceptions are called for.

Principle 4: Delegate a bit at a time to gradually discover the limits of what is possible and safe. There is a clear risk with delegating tasks in health care — that providers will be pushed to operate beyond their training and will deliver low-quality or even unsafe care.

Two of my colleagues at Dartmouth, Al Mulley and Pat Lee, use a simple graph, Figure 13, to describe optimal delegation. On the horizontal axis is preparation for a task, and on the vertical axis is the complexity of the task. On the 45-degree line that bisects the axes, people work at the so-called top of their license. Beneath the line, the problem is inefficiency. Above the line, the problem is poor quality or unsafe care.

There may be a few examples of operations above the line in health care today. As an example, one physician described to me his first few unsupervised nights in the ICU caring for the 20 sickest patients in a hospital.

Still, I think you'll probably agree that there are a far greater number of examples below the line. Most physicians spend a substantial fraction of their time with tasks that require far less than their many years of medical training. Today's norms about who takes on which tasks developed in an age when the forces of cost containment were weak and there was little push for delegation.

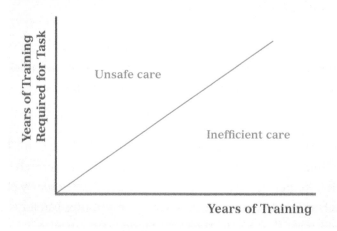

Figure 13

The 45-degree line is an abstraction. You find it by pushing toward it a bit at a time, watching closely what happens and adjusting.

Principle 5: Start by redesigning the team from scratch. Clean-slate team design has been a theme throughout this book. It is worth restating here because efforts to standardize and improve processes seem like fodder for traditional process improvement tools. The problem with the process improvement mind-set is that it takes the existing team structure for granted and works to make that structure move faster, better and cheaper. Much larger gains are possible when team redesign is treated as step one, saving traditional process improvement techniques for subsequent fine-tuning.

Principle 6: Team compositions for standardization and delegation initiatives are familiar but bottom-heavy. The emphasis on delegation naturally leads to greater staff-to-physician ratios. We will not otherwise see significant departures from traditional health care organizational structures in this chapter.

Principle 7: Do not wait for a value-based contract to move forward with an innovation focused on standardization and delegation. You may add more providers that deliver non-reimbursable services, but at the same time, you may make physicians more productive and thus increase reimbursements.

Three examples follow. The first two focus on primary care; the first in a mainstream setting, the second in a particularly challenging one. The final example is CJRI's high-volume joint replacement operation.

In 2004, University of Utah Health Care opened a new clinic based on a model for primary care that had been redesigned from scratch. The "Utah Model," as it is sometimes called, is not suitable for particularly challenging patients, such as adults with multiple chronic conditions. But for the majority of patients with routine needs, the Utah Model works well.

Focus on mainstream patients; serve unusual patients through other means. (Ch. 11)

This may be straightforward primary care, but it is high-value straightforward primary care. In fact, after shadowing in one of the university's primary care clinics, I felt dumbstruck by the elegant simplicity of it all. It was hard for me to believe that the model hadn't already gained much wider adoption.

The core idea: The clinic employed only two categories of health workers, primary care physicians and medical assistants. The medical assistants were cross-trained to take on a wide range of common tasks in primary care. Michael McGill, the leadership force behind the endeavor, had studied a variety of innovative models for primary care around the country and humbly and readily admitted that he had "proudly stolen" the idea.

> In team design, first ask: What skills are needed? Then: Who
> possesses those skills or can readily be trained? (Ch. 6)

> Originality is not required. (Chs. 5 & 6)

The clinic's basic processes are perhaps best described through the eyes of a patient. Upon entering the clinic, the patient is greeted by a medical assistant who remains with the patient throughout the visit. After escorting the patient to the exam room, the MA, following symptom-specific questionnaires, takes what might be considered a "rough draft" of the patient's history and enters it into the medical record. The MA also orders routine tests and reconciles prescriptions, all before the doctor enters the room.

While the physician is in the room, the MA assists. For example, he or she makes the appropriate notations in the medical record, following oral instructions from the physician. And, in parallel with the ongoing conversation between doctor and patient, the MA might email a prescription to a pharmacy, make a referral or order further tests. Because the physician is freed from administrative burdens, he or she can focus all attention on the patient.

Once the physician leaves the room, the MA prints a summary of the visit and reviews it with the patient, reinforcing the patient's understanding of the conversation with the physician. The MA also enters the necessary billing information and completes the patient's chart, later to be reviewed, amended and approved by the physician.

This is the model in its essence. It is an effort to standardize the sequence of events in a typical primary care visit, albeit only at a high level, and an effort to delegate as much of the physician's work as possible to the MA.

> There is value in standardization at any level of specificity. (Ch. 11)

Over time, the team at the University of Utah refined its understanding of just how much work the MAs could take on. With adequate additional training, the MAs gradually added drawing blood, taking EKGs and X-rays, and giving injections to their responsibilities. Between visits, MAs also took on greater responsibilities for the overall function of the team. They helped manage information flow, for example, by routing electronic messages within the office and informing patients of normal lab results. (Appointment scheduling and routine referrals were handled offsite at a call center that managed several University of Utah clinics.)

> Delegate a bit at a time, so as to gradually discover the limits of what is possible and safe. (Ch. 11)

Defining a new team is first a matter of establishing roles, responsibilities and hierarchy — in this case, just for the two roles. Once the basic structure is in place, however, there is the task of zeroing in on exactly how many staffers are needed in each role. In due time, the University of Utah settled on a ratio of five medical assistants for every two physicians.

> Over time, refine your team design, including who does what and your estimates of the needed panel sizes and staffing ratios. (Ch. 6)
>
> Team compositions for standardization and delegation initiatives are familiar but bottom-heavy. (Ch. 11)

The growth of the university's health system afforded a valuable opportunity to further improve the model, reimagining the physical layout of the clinic as part of building a new facility. After careful debate and study, the team chose a floorplan with exam rooms that could be entered from either of two corridors, one on each side. The corridor for patients was designed with an eye toward patient experience, the one for providers with an eye toward productivity.

Beyond lowering system costs, the University of Utah found that the model had a strong positive impact on patient satisfaction. With constant attention from the medical assistants, patients felt well cared for. To further elevate this positive result, Magill and his colleagues began emphasizing hospitality skills in their evaluation of potential new MA hires.

Another benefit was a reduction in physician burnout. Because the MAs did most of the administrative work of handling EMRs, physicians were able to discharge the abhorred practice of putting in a couple of hours of additional work at night to catch up on the EMR notes they had been unable to complete during the work day. A final benefit, Magill said, was an across-the-board improvement in clinical outcomes.

Although the university was unable to profit from its efforts as handsomely as might be possible in the emerging world of value-based contracting, this was an instance in which the fee-for-service model was at least not in direct opposition. Because the physicians were more productive when they delegated more tasks to MAs, they were able to bill for more visits, offsetting the additional cost of the MAs.

> Do not wait for a value-based contract to move forward with an innovation focused on specialization and delegation. (Ch. 11)

Once Magill and his colleagues had proven the model to their satisfaction in one clinic, they naturally wished to spread it to more clinics in the University of Utah health system. At that point, an innovation effort became a change effort. As challenging as innovation can be, change can be harder.

In this particular case, the degree of resistance that Magill faced might seem surprising. After all, the model not only worked for patients and for the system, it made physicians' lives easier. Yet many physicians did indeed resist. Magill said, "I found that there was an enormous capacity for people to be simultaneously jealous (of the newly redesigned model) and resistant to change."

Rather than fighting the resistance head on, Magill pursued a more patient and measured strategy. He watched physicians closely as they listened to presentations about the new model. He then focused on those who seemed most interested, ignoring the rest until momentum began to build behind the new model.

Innovation initiatives typically draw in volunteers, but change
initiatives impact non-volunteers. Patience and heavy persuasion
may be adequate. (Ch. 9)

The Boston Health Care for the Homeless Program, like many programs for the disadvantaged, survived for decades on a tight budget. It is exactly this reality that makes the program a likely hotbed of innovation.

Want to find out what the future of high-value care might look like? Wouldn't it make sense to examine programs that have had no choice, for years, but to learn to do more with less? Studying health care innovations that have taken hold in the developing world is one obvious option, but there are also plenty of programs in the United States that are interesting in exactly the same way.

The bulk of services at the Boston Health Care for the Homeless Program can be described as primary care, but that is about where the similarities with the prior example end. While the Utah Model is designed for mainstream patients, the needs of the patients served by the Boston program were complex and variable. There was little opportunity for the program to pursue standardization of care. As we will see, however, there was plenty of opportunity for delegation. Indeed, the program's financial viability has depended on its willingness to push as close to the 45-degree line (Figure 13) as possible.

For patients who worry each day about finding food and shelter, routine health care is not a priority. So, by the time patients seek care, their medical conditions are typically more advanced. Treatment is complicated by what the staff refers to as tri-morbidities — concurrent medical, psychological and substance abuse challenges.

Furthermore, the patients tended to be mistrustful of institutions. Treating them required building a comfortable relationship, and that demanded a deep understanding of their lives, right down to street-level economics. Patients did not necessarily come to the clinic, so the program reached them on the street and in shelters. It also treated patients in a respite center, to aid with transitions out of the hospital.

> Define which patients you will serve. (Ch. 5)

As we have discussed, innovation in team design lies at the very core of innovation in health care delivery. Jim O'Connell, who has led the Boston homeless program since the 1980s, saw just how critical teamwork would become for the organization almost immediately upon his arrival.

Though medical school had offered O'Connell no substantive introduction to how to work with nurse practitioners, a profession then just emerging, he developed what became an extraordinarily valuable partnership with one. O'Connell had the deeper medical expertise, but his partner, Yoshiko Vann, was more familiar with the lives of disadvantaged populations and more able to build trust. O'Connell quickly saw that without trust there could be no effective treatment. He needed Vann. Over time, O'Connell also learned just how much of the work of primary care Vann could handle on her own.

> In team design, first ask: What skills are needed? Then: Who possesses those skills or can readily be trained? (Ch. 6)

> Delegate a bit at a time, so as to gradually discover the limits of what is possible and safe. (Ch. 11)

The experience led O'Connell to implant teamwork as the program's cultural pillar. He believed that if the organization were to succeed as a whole, there would need to be deep partnerships across all professions — including physicians, psychiatrists, physician assistants, nurse practitioners, nurses and social workers. Formal hierarchy was far less important than rich collaboration. The mind-set was reflected in the leadership structure, which included two equal partners who led operations under O'Connell — a physician medical director and a chief operating officer who was a nurse by training.

What the Boston Health Care for the Homeless Program could not do, however, was create a single team design that would be effective across its many and diverse settings. The staff, nearly 300-strong, served patients in more than 70 locations, and patient needs differed dramatically from one location to the next. O'Connell encouraged frontline leaders to develop the right team structure for their particular context. There were, in effect, leaders of innovation in health care delivery throughout the program.

A variety of team designs emerged. O'Connell himself led a "street team" that included a psychiatrist, two nurse practitioners or physician assistants, a caseworker and a clerical worker. Another team was composed of a physician and three nurse practitioners or physician assistants. There were also a few teams for particular medical conditions, including an HIV team. If you are contemplating innovation in primary care, you could find plenty of food for thought by getting to know the Boston program.

Although the program did not try to establish any hard rules about teams, a handful of principles about the assignment of tasks and responsibilities did emerge. For example:

- Though they rarely had to assert it, physicians retained final authority of medical decision-making, and as required by law, had to formally approve and periodically review prescriptions of certain medicines.

- Physicians and psychiatrists cultivated close partnerships.

- Physician assistants and nurse practitioners were treated as nearly interchangeable. Either could be assigned the role of primary caregiver for a particular patient. Physicians tended to take on the most complex patients.

- Physician assistants and nurse practitioners had tremendous latitude to take on as much responsibility as they and their physician partners were comfortable with. The best took on more complex patients and worked more independently.

- Nurses took on unusually heavy managerial and operational responsibilities.

- Nurses and case managers acted as advocates for patients.

These relatively loose guidelines left plenty of room for interpretation. This was occasionally stressful on the front lines, as service providers of various stripes worked among themselves to determine who tackled which tasks and who accepted which responsibilities. Some degree of conflict was inevitable, but this was healthy conflict. Working out who does what is a normal aspect of innovation in health care delivery.

The Boston homeless program has yet to make the effort to formalize its knowledge about team structures — what works, what does not and where. This would be the natural next step, if it were to expand its programs or choose to systematically share its expertise with other programs.

> Once you know what works, consider growth. Also, consider sharing your knowledge with other leaders of innovation in health care delivery. (Ch. 6)

Although there is no single team structure at BHCHP, O'Connell's heavy emphasis on delegation is apparent from the overall staff makeup. On average, in primary care in the United States, physicians outnumber physician assistants plus nurse practitioners nearly 4-to-1. At the Boston program, the physicians were in the minority by a ratio of 2-1. That results in a tremendous cost savings, given the difference in salaries between physicians and PAs/NPs.

> Over time, refine your team design, including who does what and your estimates of the needed panel sizes and staffing ratios. (Ch. 6)
>
> Team compositions for standardization and delegation initiatives are familiar but bottom-heavy. (Ch. 11)

Despite the challenging conditions under which the Boston Health Care for the Homeless Program worked and despite its lean staffing model, it delivered safe, conscientious and effective care, improving the lives of those served. What we cannot be sure of is how the outcomes would compare to those that might hypothetically be achieved if the same patients had access to more expensive, traditionally staffed, primary care.

This begs the question: What is the optimal staffing model for primary care in general? Could staffing ratios like Boston's outperform? The primary reasons that outcomes might be better is that physicians would focus their time and energy on the patients who most need their expertise. As a result, they would sharpen and sustain their unique knowledge and skills. Morale might also improve as a result. In addition, the skill level of physician assistants and nurse practitioners would rise as they were challenged to take on more work. On the flip side, the primary reason to believe that the quality of care might be worse is that non-physicians would be pushed to operate beyond their skill level.

Wherever your own intuition lies, the program's experience forces us all to wrestle with the possibility that we may not know, even to the nearest order of magnitude, where the optimal staffing model for primary care lies. I suggest that this ought to provide plenty of inspiration for potential innovators in primary care.

§

Can a surgeon operate on two patients simultaneously? Crazy as this may sound to the uninitiated, it is exactly the idea that underpins the high-volume surgical operation at the Connecticut Joint Replacement Institute. The pathway to this apparent magic? Heavy-duty standardization and delegation.

Though surgeons have traditionally stayed in the operating room from the beginning to the end of a knee or hip replacement operation, their expertise is really only required for a subset of the steps involved. Delegate the rest to other personnel in the operating room and suddenly it becomes possible for surgeons to move back and forth between operating rooms, staying with patients only from the moment of incision until time to close.

At a high level, the typical sequence of steps for a Connecticut Joint Replacement Institute surgeon is:

1. Finish work on the first patient of the day in Operating Room 1, leaving a physician assistant in charge to close the wound, awaken the patient and move the patient to the recovery room.

2. Go to the waiting room to speak with the third patient of the

day and answer any final questions before surgery. This patient
is then moved into Operating Room 1 and prepped.

3. Go to Operating Room 2, where the second patient of the day is
    already fully prepped and ready for surgery.

Anesthesiologists run through a similar sequence to cover two patients
simultaneously. Nurse anesthetists are in charge of anesthesia when the
anesthesiologist is out of the room.

Steve Schutzer, the physician innovator who led the Connecticut Joint
Replacement Institute from founding, had heard of the development of similar
operations elsewhere in the country. He had even seen a video that showed
that surgeons had their hands on the patients for only about one-third of
their time in the operating room. He was determined to build a high-volume
surgical center of his own. Chris Dadlez, CEO at Saint Francis Hospital and
Medical Center, opened the door to make it possible.

Originality is not required. (Chs. 5 & 6)

From the beginning, the joint replacement center was envisioned as a
hospital within a hospital, one with its own Dedicated Team. Following an
overall staffing plan that he and Dadlez had agreed to, Schutzer had the authority that he needed to recruit, hire and compensate staffers. He also had
the flexibility to design the team from scratch.

Unlike the Shared Staff, the Dedicated Team has enormous
flexibility. It can accommodate entirely new roles and team
designs. (Ch. 7)

Joint replacement surgeons hardly worked alone. Schutzer's team included anesthesiologists, hospitalists, physician assistants, physical therapists,
nurses and surgical technicians. Because he had never run a high-volume
surgical center, it was not obvious how many personnel would be needed

in each category. Schutzer found valuable the advice of consultants from a free-standing orthopedic hospital in another part of the country, one that had already built a high-volume center. They assisted with staffing estimates and operating processes.

> In team design, first ask: What skills are needed? Then: Who possesses those skills or can readily be trained? (Ch. 6)

The most striking characteristic of the team: fewer surgeons but more support staff per procedure. With experience, the Connecticut Joint Replacement Institute was able to refine its understanding of how many people were needed and when, even adopting a practice of bringing on additional hourly staff during predictably busy times each week. Over time, the institute also further refined roles on the team. For example, one nurse in the operating room monitored safety. A critical care nurse took responsibility for interviewing patients before surgery to identify any potential issues that might lead to a last-second delay and disrupt operations.

> Team compositions for standardization and delegation initiatives are familiar but bottom-heavy. (Ch. 11)
>
> Over time, refine your team design, including who does what and your estimates of the needed panel sizes and staffing ratios.(Ch. 6)

Schutzer went out of his way to create a distinct team identity for the staff, even creating distinct uniforms. The foundational elements of the culture Schutzer created were a focus on overall team performance — safety, efficiency, patient satisfaction, process excellence and cost — and a continuous drive to learn and improve. Surgeons did not sit on a pedestal.

> Guard against the reemergence of traditional norms, behaviors,
> habits and assumptions. (Ch. 6)
>
> Take pride in system performance. (Ch. 11)

Schutzer recruited both inside and outside Saint Francis. Some were very comfortable in the high-speed, high-efficiency and hard-work culture. But a few, including one surgeon, realized that they preferred their prior jobs and departed. You might naturally wonder if burnout was an issue given the repetitive nature of the work. It wasn't. In Schutzer's view, the culture was self-selecting. Those who were attracted to the challenge of maximizing system performance enjoyed the work.

> Be equally open to hiring insiders and outsiders. Avoid teams composed
> entirely of insiders, especially those that have a history of working
> closely together or otherwise already know each other well. (Ch. 6)

To desire to improve was not enough. The institute also needed data. From the first day, Schutzer was committed to research to drive continuous improvement. His colleague Courtland Lewis, both an orthopedic surgeon and a researcher, led the effort to build a patient registry that, by 2013, included data from more than 10,000 joint operations.

Improvement work is most effective when there is a steady baseline from which to measure the impact of changes. Early on in the joint replacement institute's life, all of the surgeons and anesthesiologists gathered to develop a basic protocol. All agreed to conduct their standard knee and hip replacement operations in the same way almost every time, with the latitude to make exceptions for unusual patients. Proposals for permanent alterations to the protocol were discussed when all of the surgeons and anesthesiologists gathered for periodic meetings. The arguments over which protocols were best could be sharp; the team resolved to settle disputes through data and analysis.

> Disciplined experimentation implies carefully, rigorously and
> regularly comparing predictions and outcomes. Guide your
> initiative by systematically gathering evidence to test each
> cause-effect relationship in your hypothesis. (Ch. 10)

Lewis and his staff of four analysts tackled a range of data-driven inquiries from which patients were most likely to suffer complications (those suffering from anxiety or depression, as it turned out) to how much antibiotic cement should be used (not much if any, the data showed). Improving safety was Lewis' top priority, with a goal of eliminating all complications. He personally reviewed every complication, and he and his team analyzed the data to assess root causes. Over time, the protocols became ever more specific and operations more consistent. The institute even adopted the practice of ensuring, as best possible, that the same teams worked together in the operating room every day.

The standard approach to surgeries worked well because the institute, in general, took on straightforward cases. It did take on some less-common surgeries, such as revisions, in which an artificial joint replaces an older artificial joint. In these cases, surgeons only worked on one patient at a time. To minimize the potential of these non-routine cases to disrupt operations, they were always scheduled at the end of the week, after all of the routine surgeries were complete.

> Focus on mainstream patients; serve unusual patients through
> other means. (Ch. 11)

The Connecticut Joint Replacement Institute looked to streamline the entire process, not just what happened in the operating room. The staff was particularly attentive to recovery time. Patients wanted to go home. Furthermore, discharging patients in a timely manner was critical to prevent congestion in the recovery ward from impacting the operating room. Research showed that the keys to a quick recovery were getting patients moving quickly and starting their physical therapy early. This, in turn, depended on controlling pain. The anesthesiologists used the institute's pain and patient satisfaction data to perfect methods of administering multi-modal anesthesia.

Though movement was critical, pushing too aggressively could be dangerous. Falls were the most common impediment to recovery. The nursing team on the recovery floor collected data to discover who was falling and when. As it turned out, it was not the elderly who fell the most but the youngest patients, who tended to think they were ready for more than they really were. The institute's Call Don't Fall program asked patients to sign an informal agreement that they would not get up without calling for help. It slashed the number of falls each month.

The journey was not one without any bumps in the road. As Schutzer built his team, for example, he experienced firsthand just how different working with the Dedicated Team could feel compared to working with the Shared Staff. For example, early on, post-surgical patients were seen by St. Francis' staff hospitalists. Schutzer found, however, that the hospitalists were strained by the additional volume. They were not giving joint institute patients the attention that Schutzer believed they needed; nor were they adequately oriented to the particular needs of joint replacement patients. So Schutzer contracted with an outside practice, adding to the team hospitalists who could be more fully dedicated to his program and more adaptable to its processes.

> The Shared Staff has nearly no flexibility. It can do more work, but not different work. (Ch. 8)

> Ensure that the Shared Staff has adequate resources for both of its jobs — innovation plus ongoing operations. (Ch. 8)

Schutzer also discovered that the relationship between a Dedicated Team and the rest of the organization can be a delicate one. The double reporting lines that are inevitable for the Shared Staff are a common source of strain. Some members of his Dedicated Team reported both to Schutzer and a hospital leader.

In addition, some of the Saint Francis staff thought that the joint replacement institute was getting "special treatment." Others, even in absence of any financial data, suspected it was losing money. Some were angry about the physical moves within the hospital that were required to accommodate the institute; others grumbled that it was only able to do what it was doing by picking the "low hanging fruit," meaning the most straightforward surgeries. Some were put off by the way the institute aggressively demanded that its needs be met in its early days.

> Tension between the Dedicated Team and the Shared Staff is
> inevitable. (Ch. 8)

A more collaborative approach spearheaded by a new executive director calmed the waters. It also helped that one of the surgeons on the joint institute team, Bob McAllister, had a long history within St. Francis and could help build a bridge to the rest of the organization. (Schutzer and the other institute surgeons were new to St. Francis.)

> Even the best innovation leaders need help. (Ch. 8)

Schutzer was also aided enormously by Dadlez. His commitment to the Connecticut Joint Replacement Institute was unwavering. He frequently spoke about why the institute was important for the hospital, emphasizing how it could be a powerful reputation builder for St. Francis.

The hospital's investment paid off financially. The hospital, surgeons and anesthesiologists all made money. Indeed, the spread of high-volume surgical operations will be aided enormously by the reality that service providers usually win financially when these initiatives succeed, even in a fee-for-service environment.

> Do not wait for a value-based contract to move forward with an
> innovation focused on specialization and delegation. (Ch. 11)

Nonetheless, Schutzer anticipated a future of value-based contracts and was eager to move forward. In 2012, he signed the first bundled payment agreement with a payer, ConnectiCare. Negotiations were complicated because four parties were involved — payer, hospital and two medical groups, the surgeons and the anesthesiologists — all of which had to settle on terms for dividing compensation and risk. Ultimately, all three providers protected themselves by signing a stop-loss contract with a reinsurer.

Schutzer was pleased with the overall result and confident that the institute had the foundation for managing risks. He was eager to grow operations. He and his team went about systematically spreading the word about outcomes, even studying the strength of the institute's reputation by ZIP code. He also planned to add facilities to make the institute more accommodating to patients and families who needed to travel to get to Hartford.

Schutzer needed to recruit new surgeons to support growth. It was easier in some cases than others. Potential hires tended to fear loss of independence and loss of flexibility in the operating room. In Schutzer's view, however, it was only a matter of time before joint replacement surgeons had no choice. High volume would become the norm, not the exception. To hasten that day's arrival and to maximize the positive impact of his work on the system as a whole, Schutzer openly shared what he was doing with anyone who was interested.

> Once you know what works, consider growth. Also, consider sharing your knowledge with other leaders of innovation in health care delivery. (Ch. 6)

§

A Few Questions to Consider

1. Who are the mainstream patients in your practice? Do they have predictable, and predictably similar, needs?

2. How closely are the non-physicians in your practice working to the top of their license?

3. What fraction of your day is consumed by tasks that do not require your many years of medical training and experience and could probably be delegated?

4. Could you transfer some of the pride that you take in serving individual patients to pride in overall clinical performance?

Chapter 12
Coordinate

....

THE AMBITION OF THE BONE AND JOINT INSTITUTE, LAUNCHED IN 2012 at Hartford Hospital, is high-value end-to-end care for all acute orthopedic conditions. Courtland Lewis, an orthopedic surgeon, led the effort to build the institute. It was his next career step after heading up the research team at the Connecticut Joint Replacement Institute, described in the previous chapter.

Wherever appropriate, the Bone and Joint Institute planned to increase value by emphasizing non-surgical options for treatment of musculoskeletal injuries. But in Lewis' view, the more critical driver of value in orthopedics was longitudinal coordination of full episodes of care, from presurgical preparations to post-surgical physical therapy and rehabilitation.

The team, to be more than 100-strong, would include orthopedic surgeons, neurosurgeons, rheumatologists, endocrinologists, metabolic bone disease specialists, anesthesiologists, radiologists, physiatrists and a rehabilitation team. Most would be housed in a stand-alone musculoskeletal hospital, within which would sit a high-volume surgery center. Lewis anticipated that physical proximity of all of the providers would enhance communication and coordination of care.

Another improvement was the creation of what Lewis referred to as the PREPARE Clinic. The facility would simplify the patient's experience by consolidating all of the necessary presurgical steps in one location. Lewis also planned investments that would ease patients' post-surgical recoveries, enabling more to go home sooner.

While facilities were still under construction, the Bone and Joint Institute team, including a quality expert, was already hard at work defining care processes and treatment protocols. It also was designing and building the sophisticated IT infrastructure that would help Lewis measure performance and guide operations. Early on, Lewis also focused on establishing a culture of performance excellence and labored to resist the natural organizational instinct to gravitate toward existing habits and routines.

§

Poor coordination of care burdens patients with unrealistic expectations. They are asked to execute vague, complicated or conflicting care plans on their own, or to somehow follow through on impossibly inconvenient or time-consuming care steps. Most patients fall short. Some may feel like they are failing, or even that they are incapable of taking care of themselves. A few may disengage from health care altogether and suffer the predictable consequences.

Because the health system is so fragmented, coordination failures are widespread. The opportunities for improvement are omnipresent. Lewis' plans for the Bone and Joint Institute illustrate an all-out effort to capture these opportunities.

In this chapter we will explore coordination efforts more fully. There are some similarities between coordination efforts and the standardization and delegation efforts discussed in the prior chapter. Both try to make collections of related service activities move faster and with fewer errors. The difference in this chapter is the willingness to look broader and longer. It's not just the service activities under your own roof, but those across the system. It's not just the services delivered yesterday, today and tomorrow, but those delivered over the past few months and the next few.

§

The following principles guide the pursuit of initiatives focused on better coordination of care.

Principle 1: Be clear about how better coordination will improve care in your particular context. Better coordination of care can create value by defining and executing clear care plans, by eliminating useless duplication of services and by avoiding missed or unnecessarily delayed care steps. The goal is to execute all necessary care steps — on time and just once.

There is one more source of value, though an indirect one. Efforts to improve coordination often result in providers specializing more narrowly, in just one medical condition or few similar ones, achieving deeper knowledge and skill in those areas. The process of gaining deeper knowledge is catalyzed by frequent opportunities to interact with others on the team who are doing the same.

Principle 2: Feel free to start small, bringing together just two services commonly used by the same patients. At the Bone and Joint Institute, Lewis is building clinical teams that deliver coordinated services across an entire episode of care. That's an aspiration that can seem overwhelming at first. Do not let the potential size of the objective dissuade you from getting started. There is value to be created in every step along the way.

Principle 3: Team structures for coordination initiatives are horizontal, not vertical. In traditional medicine, a provider's closest work relationships tend to be with peers with similar training and specialization. By emphasizing relationships that cross specialties or provider types, initiatives focused on coordination turn this structure inside out.

The logic for doing so is simple. When multiple providers who treat the same patients are on the same team and under the same roof, conversations about patients and their care plans are more likely and coordination is improved.

Be sure to consider exactly when these conversations can happen. It may be unrealistic to expect to squeeze these conversations into the minimal slack time on most providers' schedules, even when the providers all work on the same team and in the same location. Regular meetings such as daily huddles may be necessary.

Principle 4: Coordination initiatives generally benefit from an explicit "team captain" who is responsible for longitudinal execution of care plans. A coordination approach that relies strictly on informal communications among providers — even those on the same team and under the same roof — may be inadequate for the task. Coordination of care will improve when a single provider or small team accepts full accountability for patients. This provider or team invests time in carefully developing the care plan with the patient, communicating the care plan to other providers and then following up to ensure that all care steps are executed. The job often demands proactive contact with patients to check on their health status and keep them motivated and on track.

Principle 5: Build shared goals, a clear structure and mutual respect. This recommendation from Chapter 6 is worth repeating here because initiatives focused on coordination of care often involve more than one Shared Staff, and they may be spread across multiple departments or even multiple institutions. As a result, building shared goals, a clear structure and mutual respect is harder and takes longer.

Principle 6: Invest in measurement. Initiatives focused on co-ordination can present particularly steep measurement challenges because they often span multiple organizational boundaries and multiple information systems. Collecting the data to prove that your initiative improves outcomes while reducing costs might involve a lot of frustrating legwork. One simplifying option is to start by serving patients who are all insured by the same payer, one that is willing to share data.

Principle 7: You may not need to wait for a value-based contract. Under fee-for-service contracts, care coordination initiatives usually involve unreimbursed work. That makes these initiatives challenging to fund, but not impossible. Sometimes care coordination efforts pay for themselves through reductions in hospital variable costs attached to a DRG-reimbursed hospital stay or through reduction in readmission penalties. Also, because efforts to standardize and delegate often dovetail nicely with those focused on coordination, it is possible that savings from the first will pay for the second.

The four examples in this chapter range in ambition from coordinating just two providers to efforts like the Bone and Joint Institute, which aim to coordinate entire episodes of care. The examples will take us into diverse settings: dermatology, physical therapy in the ICU, home health and oncology.

§

Telehealth technologies have the potential to bring providers together and make coordination of care easier, and technology lies at the heart of this first story. Carrie Kovarik, a dermatologist, first put telehealth technology to work in Malawi, where she worked to bring dermatological services to a rural and poor population.

It did not take long for her to see the potential to do the same within the confines of her own institution, the University of Pennsylvania Health System. She knew that disadvantaged populations in the United States, especially those in rural areas, had little access to dermatologists. Waits as long as one year were common; many community hospitals did not have a dermatologist on staff.

Kovarik saw that the value of telehealth extended beyond expanding access. It could also improve the coordination of care between dermatologists and primary care physicians. Specifically, it could eliminate heavy duplication of the social aspects of medicine. Did the dermatologist really need to build a relationship with a patient with a routine dermatology problem, given that the primary care physician had already done so?

> Feel free to start small, bringing together just two services
> commonly used by the same patients. (Ch. 12)

Like radiology, dermatology is a visual medical specialty that lends itself naturally to remote consultation. In roughly 90 percent of cases, Kovarik said, a dermatologist is able to make a diagnosis on the basis of visual information alone, without even reading a patient's clinical history. Dermatologists train by studying thousands of photographs. Having a patient travel to an office, shuffling all of the necessary paperwork and investing time in building a relationship between doctor and patient, was almost always wasted motion.

> Be clear about how better coordination will improve care in your
> particular context. (Ch. 12)

Instead, Kovarik and her team trained primary care doctors to take photographs of "good-enough" quality in their own offices and submit them, along with a brief clinical history, to University of Pennsylvania Health System via a cellphone app. (The app was developed by the American Academy of Dermatology. Kovarik had been directly involved; the effort was partly grant funded.) By 2014, roughly 85 physicians were doing so from 14 primary care clinics, including 10 Philadelphia health centers.

When a primary care doctor submitted a request via the cellphone app, the UPHS department of dermatology received an email or text alert. Within 48 hours, a dermatologist responded directly. Many responses required the dermatologists less than one minute to handle. (Quite a time savings compared to an office visit!) Less than 10 percent of remote consults led to a subsequent in-person appointment with a dermatologist.

Kovarik was eager to see the practice expand. Unfortunately, under 2014 rules in Pennsylvania, teledermatology consults were not reimbursable. The dermatologists who were participating in the program were doing it strictly on a volunteer basis and viewing it as charitable work. Because so many of the consultations could be completed in just a minute or two, the volunteering dermatologists had little difficulty squeezing the work into slack time in their schedules. It was not disruptive to their workflow.

It was equally important that the use of the cellphone app did not disrupt the workflow in the primary care office. Early revisions to the design of the app focused on ease of use. By 2014, Kovarik no longer saw any significant barriers to adoption. Physicians needed to download the app, read a short document, learn how best to light and frame pictures, and for certain conditions, learn which additional photos might be needed for diagnosis. All of this preparation work took about one hour.

At this point, note that this is an effort without a Dedicated Team. It was all Shared Staff using only slivers of slack time. Some coordination of care work is possible in this manner, and might even be possible within traditional continuous improvement programs, except that such programs generally do not span multiple organizations.

> The Shared Staff has nearly no flexibility. It can only take on tasks that: a) are already familiar or can be quickly learned, b) fit naturally into existing roles, and c) fit naturally into existing work processes. (Ch. 7)

If value-based contracting continues to spread, however, Kovarik does anticipate Dedicated Teams of dermatologists who do nothing other than teledermatology consults. Such a Dedicated Team would spend most of its time in a room full of computers, quite possibly *not* located within a larger clinical facility. The Dedicated Team would have the flexibility to completely reinvent roles, responsibilities and work processes, perhaps creating value in unanticipated ways.

> The Dedicated Team has enormous flexibility. It can accommodate entirely new roles and team designs. (Ch. 7)

The specialty was gradually becoming more comfortable with teledermatology, as evidence accumulated that quality of care remained high even when dermatologists worked remotely.

§

Like Kovarik's efforts, the following story illustrates the substantial value that can be unlocked by coordinating just two services. This time, the value created lies not in eliminating useless duplication but in avoiding costly delays. The institution is Johns Hopkins, and the coordination effort focuses on ICU nurses and physical therapists.

> Feel free to start small, bringing together just two services commonly used by the same patients. (Ch. 12)
>
> Be clear about how better coordination will improve care in your particular context. (Ch. 12)

Dale Needham, who had long studied quality improvement in critical care, led the initiative alongside Mike Friedman, a physical therapist and MBA, and Michael Palmer, head of physical medicine and rehabilitation. The leadership trio set up a formal trial to test a simple hypothesis: If physical and occupational therapists were stationed in the ICU full time, alongside ICU nurses, outcomes would improve.

> Team structures for coordination initiatives are horizontal, not
> vertical. (Ch. 12)

All three were familiar with the struggles that patients, particularly elderly ones, tended to face after a stay in intensive care. While in the ICU, patients tend to eat poorly, sleep poorly, receive medications that adversely affect cognition and become deconditioned by bed rest and inactivity. All can lead to complications and readmissions that have more to do with the hospital stay than the medical condition that brought the patient to the hospital in the first place. Needham, Palmer and Friedman all believed that getting patients mobilized sooner could reduce the incidence of these complications. The sooner the patients were up and moving the better.

Putting the physical therapists in the ICU full time proved to be an important choice. The effort needed the degree of flexibility that is only available to full-time teams. To make the initiative work, both the physical therapists and ICU staff needed to deeply redefine their routine processes and workflows.

> The Dedicated Team has enormous flexibility. It can accommodate
> entirely new roles and team designs. (Ch. 7)

Getting patients moving meant, first of all, reducing a patient's level of sedation and at least approximating a normal sleep-wake cycle. When patients were awake more often, they asked for and needed attention more often, increasing the total workload on the nursing staff and disrupting normal routines. Further adaptation on both sides was needed to accommodate collaboration for making decisions about who, when and how to mobilize. ICU nurses, appropriately, were worried about safety — falls in particular. Each group needed to learn from the other and each needed to adapt their normal routines to give the experiment the best possible chance at success.

The work stressed both sides; the direct involvement of the physicians and the commitment of the nurse leaders in the ICU was critical for keeping the endeavor on track.

Even the best innovation leaders need help. (Ch. 8)

The outcomes were impressive. Extensive analysis of data derived from a four-month trial revealed improved clinical outcomes and reduced harms, delirium and lengths of stay — all clear wins for the patients and for the system.

The analysis also showed that Johns Hopkins could win financially by continuing the program, even under current reimbursement rules. Although an annual budget increase of approximately $350,000 was required for the additional staff in the ICU, this would be offset by an annual savings of approximately $800,000 from avoided variable costs associated with fixed-fee hospitalizations. There would be fewer days in the hospital, fewer imaging tests and fewer labs.

You may not need to wait for a value-based contract. (Ch. 12)

There are a couple of important points to note about this figure. First, from the perspective of any other industry, this is a jaw-dropping return on investment — more than 100 percent — with the savings coming in the same year the investment is made. Second, the $800,000 figure is conservative insofar as it only includes direct variable costs. It excludes any allocation of fixed costs or any avoided penalties for readmissions.

On the strength of the analysis, Johns Hopkins expanded the program. The team developed and refined training modules to prepare physical therapists for working in the ICU environment. Physical therapists developed familiarity with the wide range of medical conditions typically encountered in the ICU and learned how to work with the ICU staff to assess which patients were safe to mobilize and to discern how far a given patient could be pushed.

Meanwhile, ICU nurses studied the routines of physical therapists and even picked up some of their skills so that they could also mobilize some patients. Thus, the learning needed to go both ways. The team would only thrive if there was mutual respect and if each side understood the roles and capabilities of the other.

Build shared goals, a clear structure and mutual respect. (Ch. 12)

Once you know what works, consider growth. Also, consider sharing your knowledge with other leaders of innovation in health care delivery. (Ch. 6)

Needham, who was well known nationwide, routinely engaged with other institutions to try to catalyze the innovation's spread.

§

It was a shocking discovery. As part of an in-depth research effort to pinpoint the root causes of her hospital's readmissions, Pamela Duncan, an experienced epidemiologist and the director of innovations and transitional outcomes at Wake Forest Baptist Hospital, found a single home health agency whose patients had nearly a 50 percent likelihood of a return trip to the hospital within 30 days. A few other agencies had similarly alarming rates.

It would have been easy to blame the agencies with the high readmission rates, but Duncan chose instead to partner with them and work to coordinate the efforts of the hospital team with that of the home health agency. She suspected that the most challenging patients — those with the most difficult medical conditions, the most challenging living conditions, the fewest resources and the least support from family and friends — were concentrated in just a few neighborhoods and served by just a few agencies. A safety net hospital, Wake Forest Baptist served a high proportion of poor patients.

Feel free to start small, bringing together just two services commonly used by the same patients. (Ch. 12)

Hospitals, in Duncan's experience, paid little attention to patients once they were stabilized and discharged. The quality of turnover to post-acute service providers was variable. There was no accountability for a good handoff.

As a result, many patients experienced declining health after discharge. Unfortunately, they tended to delay care, avoid care or simply did not know when they should seek care. An unplanned emergency room visit was often the result. Duncan believed this was an area in which Wake Forest Baptist Hospital could improve.

> Be clear about how better coordination will improve care in your
> particular context. (Ch. 12)

With Duncan's support, the hospital created a small team to support and more closely coordinate with the home health agency. The hospital team, led by a geriatrician named Franklin Watkins, assumed continuing accountability for patient care. The team included two nurse practitioners, a social worker, a pharmacist, a hospice care representative and a pastoral counselor.

> Coordination initiatives benefit from an explicit "team captain"
> who is responsible for longitudinal execution of care plans. (Ch. 12)

The bulk of the coordination work took place during one-hour calls, twice per week. The calls included both the hospital team and the home health agency's team of nurses, therapists and social workers. The partners discussed which patients were at greatest risk of decline, they updated care plans and they acted quickly to prevent declines in health status when needed. Absences from the conference calls were extremely rare.

In addition to the conference calls, the hospital team stayed in touch with the home health agency throughout each week. During work hours, a nurse practitioner was always available to the home health agency by phone. The hospital team also made periodic house calls. In Watkins' view, the home visits were absolutely crucial. They were the only way to develop an adequate understanding of what it was going to take for the patient to heal. Watkins made house calls a weekly routine in his own schedule, as did the two nurse practitioners.

The two nurse practitioners on the team executed the bulk of the team's work. In essence, the pair shared a full-time job. Because their other part-time responsibilities were isolated from and independent of the home health partnership, they had the flexibility inherent in Dedicated Teams. Their commitments to the home health partnership were firmly scheduled and took first priority. The remaining members of the hospital team were Shared Staff and worked on the initiative only part time.

> The Dedicated Team has enormous flexibility. It can accommodate entirely new roles and team designs. (Ch. 7)

It took more than a year to negotiate the structure of the partnership with the home health agency. Then it took ongoing training and discussion on both sides to ensure that everyone was operating by the same playbook and understood their respective roles. The partners collaborated on standard treatment protocols and developed clear instructions for patients on whom to call and under what circumstances. They established guidelines that allowed patients to self-assess their health status as in the green zone, yellow zone or red zone, and even agreed on particulars such as exactly what to tell CHF patients about how much weight gain over what time period is cause for a phone call to a care provider.

> Build shared goals, a clear structure and mutual respect. (Ch. 12)

Both Duncan and Watkins were pleased with the results of their efforts by 2014. In their judgment from direct observation, the program was having an impact. Readmission rates were trending in the right direction, declining from 27 to 18 percent. (The duo made two cautions about the figure. First, a more accurate analysis would require payer data that showed admissions to other area hospitals. Second, there were concurrent hospital initiatives that likely contributed to the decline.) Beyond readmissions rates, Duncan and Watkins were gathering data on patients' function to further show that their work was improving outcomes.

Invest in measurement. (Ch. 12)

Both were interested in cautiously growing the program while sustaining its quality. They estimated that they could be serving at least five times as many patients in the local area. Growth would require elevating the profile of the program internally, sufficiently that the hospitalists understood its value and referred patients more frequently.

Careful attention to the program's economic sustainability would also be required to support growth. As Wake Forest Baptist Hospital was not by 2014 significantly engaged in value-based contracting, the business case for the initiative weighed the gains of reduced readmission penalties against the losses of the hospital team's unreimbursed work plus forgone profits from reduced admissions. The economic case would look better as readmissions penalties increased. It also could be improved with refined predictive analytics that helped the team focus its energies on the patients most likely to be readmitted. As it stood, Wake Forest was "risk-stratifying" patients based on a few data points such as history of admissions and socioeconomic status based on ZIP codes.

Once you know what works, consider growth. Also, consider sharing your knowledge with other leaders of innovation in health care delivery. (Ch. 6)

§

Cancer care is complicated. Many specialists are involved, and new treatment options continue to emerge. Patients can easily be overwhelmed and confused. Cancer care may be Exhibit A in illustrating the opportunity for improved care coordination.

Medical oncologist John Sprandio and his nine-physician, three-office practice, Consultants in Medical Oncology and Hematology outside Philadelphia, have chosen to step up to the plate and accept end-to-end accountability for coordination of patient care. In Sprandio's view, medical oncologists are in the best position for care coordination. They tend to have the longest-term relationships with cancer patients, and they are almost always directly involved at junctures in which coordination of care could be a problem.

Coordination initiatives benefit from an explicit "team captain" who is responsible for longitudinal execution of care plans. (Ch. 12)

Consultants in Medical Oncology and Hematology had applied for and received designation as a patient-centered medical home. The practice handled all aspects of cancer care for its patients, partnering with primary care physicians for other medical conditions.

In cancer treatment, well-coordinated care starts with careful care planning and monitoring. To make this easier, Sprandio's practice invested heavily, through an in-house IT team, in the development of a software overlay that was integrated with a commonly used oncology EMR package. The physicians documented the care plan in the overlay.

Be clear about how better coordination will improve care in your particular context. (Ch. 12)

In addition, the overlay presented the most relevant clinical data in a standard physician-centric format. Longitudinal information was presented in formats that enabled the physician to quickly assess trends in the patient's health status. Physicians could make notes and annotations directly in the overlay. The overall design reinforced the key steps in the care process. Sprandio estimated that he spent 90 percent of his clinical computer time working within this overlay.

Close coordination of care required that Consultants in Medical Oncology and Hematology maintain close relationships with neighboring institutions that delivered related services. Physicians in the practice had privileges at two nearby hospital systems and rounded on hospitalized patients daily. EMR links enabled the physicians to review lab reports and radiological reports. The organization's main offices were adjacent to surgical and radiation oncology practices, making care more convenient for patients and enabling routine face-to-face conversations among the three disciplines. The physicians of Consultants in Medical Oncology and Hematology routinely participated in biweekly tumor board meetings in the hospitals where they practiced and connected by phone with other specialists following a structured progress note format.

> Team structures for coordination initiatives are horizontal, not
> vertical. (Ch. 12)

A final critical piece of the effort to coordinate care was its work to keep patients engaged. Cancer care is an endurance event for patients. Treatments last for months or years and patients can easily feel overburdened or overwhelmed. Better coordination of care requires that patients participate as part of the team.

Consultants in Medical Oncology and Hematology defined a new role within its practice, patient navigators. (Other institutions used the same title, but Sprandio saw significant differences between the way the health systems he was familiar with defined the role and the way his practice did.) These navigators maintained frequent contact with patients over the full length of their treatments. They guided patients, kept them feeling upbeat and supported, ensured they remained engaged in their care and understood their care plan, and they encouraged them to complete every step in their care process.

> Invent new titles and job descriptions for your team members.
> (Ch. 6)

To reduce the overall burden on patients, the navigators coordinated required insurance approvals and scheduled all tests and appointments and follow-up visits. They also followed up to make sure that outside appointments and tests actually had occurred as scheduled. If a patient was due for a CT scan and no image showed up in the patient's EMR on that date, the navigator called the patient to find out what happened. If the patient responded that he was feeling too sick to leave home, the navigator would direct the patient to triage staff to see if an unscheduled visit was appropriate. If the patient simply forgot, the navigator rescheduled.

There were four patient navigators on staff, two in the larger office and one in each of two smaller offices. Though the job required comfort with multitasking, it was not difficult to learn. Most of the navigators had been administrative assistants or receptionists, and it only took a few days of shadowing plus some study of a handbook of written instructions to get up to speed. The navigators found the job rewarding, especially the close contact with patients. Consultants in Medical Oncology and Hematology had seen little turnover in the position.

The practice also invested in a substantial education program for new patients, helping them anticipate what was to come and become aware of situations in which they should ask for help rather than toughing it out. The practice was particularly proud that it had slashed hospitalizations and ER visits for its chemotherapy patients by 50 percent and 70 percent, respectively, between 2007 and 2011.

Although much of this effort to coordinate care is not reimbursable, the practice is healthy financially. Sprandio and his team have pursued continuous process efforts for years to run as leanly as possible. Indeed, the practice operates at well below the staff-per-physician average.

You may not need to wait for a value-based contract. (Ch. 12)

The practice has extensively standardized work flows and information flows within the office and pushed to delegate as much as possible. To make it possible for nurses to interview patients before a physician appointment, for example, the team developed standard interview protocols and symptom scores for common side effects of cancer treatments such as nausea, vomiting, insomnia and depression.

Each patient's health status and current trends could be consistently graded so the physician already had all of the relevant information in a useful format before the patient walked through the door. The team also agreed on automatic standard responses to more than 25 specific symptom reports. If a patient reported mouth sores, for example, the first step was a particular mouthwash. The triage nurse only needed to involve the physician when the symptom reports fell outside of clear standards.

Sprandio was eager to see the spread of value-based contracts. He had signed a few such contracts by 2014. He was optimistic that the contracts would soon be much more commonplace and he anticipated that they would probably combine patient-centered medical home-style standards with bundled payments for full episodes of care.

In Sprandio's view, oncology as a whole needed more impetus to improve. Value-based contracts would provide it. To spur the growth and development of high-value oncology practices, Sprandio had launched a consulting practice, Oncology Management Services, which worked with both payers and providers.

> Once you know what works, consider growth. Also, consider sharing your knowledge with other leaders of innovation in health care delivery. (Ch. 6)

A Few Questions to Consider

1. Consider your patients with the most complex needs. Do they have clear care plans? How able are they to execute their care plans without help? Is there a provider who knows that he or she is accountable for helping patients execute their care plans?

2. Who are the providers outside your clinic or practice that see your patients most frequently? How effective are the handoffs to these providers? What tends to fall through the cracks?

3. How frequently do your patients endure duplicated care steps? Why? What are the consequences?

4. What care steps do your patients most frequently delay or avoid altogether? Why? What are the consequences?

Chapter 13
Prevent

....

IS IT POSSIBLE THAT ONE LINCHPIN IN THE BATTLE TO FIX HEALTH care might be a new type of health worker, one who needs little medical education? That was the bet made by AtlantiCare, the dominant provider of health care services in Atlantic City, New Jersey.

In its Special Care Center, health coaches, many with no more than a high school education, engaged low-income adults with multiple chronic conditions in taking better care of their health. The coaches proved effective, able to slash emergency room visits and hospitalizations for these patients.

The underlying hypothesis here is by now quite familiar: Spend upstream, save downstream. All of the initiatives in this chapter share this same basic bet and all serve narrowly defined patient populations — those who appear likely to have heavy but avoidable health needs in the near future.

.... § ....

Collectively, these programs illustrate the following principles for initiatives focused on prevention:

Principle 1: Prevention generally requires educating, counseling and engaging patients in improving their health behaviors. Medical interventions may also play a role, but most effective preventive medical interventions (and some that may not be so effective) are already in mainstream use. The interventions that remain take the form of service delivery, not pills.

Principle 2: Long-term patient engagement requires a strong and trusting human bond between patient and coach. There is still a great deal to learn about how such bonds are most reliably created and how to identify and recruit health care providers who are up to the task. The examples in this chapter suggest that frequent and sometimes lengthy interactions are needed, and that conversations initiated by the provider rather than the patient are particularly potent. They also indicate that learnable techniques such as motivational interviewing can be catalytic, and that success may be most likely when the patient and the provider share life experiences, socioeconomic class, heritage, culture, language or values. Relative to these considerations, depth of medical knowledge appears to be of distant relevance.

There are some promising technologies for patient engagement such as smartphone apps that help patients track exercise and food consumption. At this point, however, these tools appear more likely to be a complement to, rather than a substitute for, a close patient-provider relationship.

Principle 3: Predict who will benefit most from the intervention that you design. Prevention-focused initiatives are expensive. You are unlikely to achieve an overall reduction in costs if you spend money on low-risk patients. Because of their deep access to data, payers may be particularly valuable partners for these initiatives. Sophisticated computer models will also play a role.

Principle 4: Team designs for prevention initiatives put non-physician providers and providers with little or no prior medical training on the front lines. The physician's role, then, is to supervise, clarify roles and establish policies that guide when a patient's case should be escalated to a team member with more in-depth medical expertise.

Principle 5: Anticipate and mitigate conflicts with downstream providers. Your initiative, if successful, may very well reduce demand for the services of your colleagues, either in your organization or your community. How will this conflict be resolved? Is there any way that you can co-opt these providers and bring them onto your team in a meaningful way? If not, do you have the senior organizational support you may need to neutralize these conflicts?

Principle 6: Anticipate a worse-before-better cost trajectory. With these initiatives, you will be spending now to save money later. Ensure you negotiate an adequate time period to prove that the initiative is both improving outcomes and reducing costs. To sustain your funders' confidence, collect clinical data that show that you are on a trajectory to success, such as blood sugar levels for diabetics.

Principle 7: Be wary of the possibility of a better-before-worse trend for measures of utilization. To make good judgment calls about the power of these initiatives, it is important to recognize that there is a great deal of churn in the population of high-utilizing patients. It is not the same patients, year in and year out. For example, a patient might be categorized as a high-user the year that he first acquires a chronic disease, but drift out of the high-use category the following year because he has become proficient at self-managing his condition.

Therefore, early indications of success can be misleading. It is possible that the innovation had no causal connection to the decrease in utilization. It was going to happen anyway. This phenomenon is described by some researchers as a "reversion to the mean." If you pursue a prevention initiative, you must be careful not to take credit for it. If you do, you may add a great deal of cost with little impact over the long term. Better-before-worse.

Principle 8: If funding is tight, start with the patients whom you are most confident that you can help. Then expand. If possible, sort (or "risk stratify") the patient population into several groups based on your assessment of the likelihood of intense downstream health care needs. Start by serving the highest-risk group. Once you demonstrate that you are both improving outcomes and reducing costs, expand to serve more patients.

We will look at six initiatives with a prevention focus in the following contexts: bone health, metabolic syndrome, congestive heart failure, AIDS and two distinct approaches to serving low-income adults with chronic disease.

§

If you serve elderly patients, then you already know that hip fractures and the lengthy hospitalizations that follow can be dangerous. They can lead to loss of function, loss of independence, the introduction of new medical problems and the exacerbation of existing ones. Astonishingly, nearly 30 percent of hip fracture patients die within one year.[12] For many people, the risk of bone fractures rises with age, as osteoporosis decreases bone density.

In 1998, Richard Dell, an orthopedist at Kaiser Permanente Southern California's Downey Medical Center, initiated an effort later dubbed Healthy Bones to treat osteoporosis. The intervention was straightforward. Patients came to the medical center for education on bone disease and lifestyle counseling, including avoidance of trip hazards in the home. Depending on the results of a bone density scan, some patients also received a prescription for bisphosphonates.

> Prevention requires educating, counseling and engaging patients
> in improving their health behaviors. (Ch. 13)

The trick was getting the right patients to the clinic. Using EMR data, the organization's IT systems calculated a risk score based on more than a dozen parameters, including age, gender and prior fractures. Then, the IT system sent automated alerts to high-risk patients. A care manager called or emailed patients who did not respond to encourage them to make an appointment. The care manager also followed up when patients did not pick up their prescriptions. Dell was proud that his team succeeded in getting more than 90 percent of at-risk patients into the medical center for a bone scan, much higher than the nationwide average.

> Predict who will benefit most from the intervention that you
> design. (Ch. 13)

After demonstrating the program's effectiveness in reducing both costs and fractures, Dell led the expansion of Healthy Bones across all of Kaiser's medical centers. The team structure varied somewhat, but in general orthopedists served as the leaders and primary champions.

Most of the work, however, was in the hands of care managers. In the Downey Medical Center, Dell worked alongside a nurse practitioner who worked full time in the program seeing patients in person and managing their care. Dell's Shared Staff included a radiology technician and an endocrinologist who worked with about 5 percent of patients, those whose medical conditions were particularly complex.

The organization's various medical centers competed with one another for the most effective Healthy Bones programs, pushing each other toward a goal of 100 percent screening rates. The company's analysis showed that the savings far outweighed the cost of the program. Fracture rates were down nearly 40 percent on average and nearly 50 percent in some locations.

Dell was frustrated by the lack of progress nationwide. In his view, there was nothing particularly complicated about implementing the program. He had seen that many physicians regarded osteoporosis as a natural part of aging, something that little could be done about. To the contrary, the progression of osteoporosis could be slowed and the risk of hip fracture slashed. The necessary steps: regular exercise, smoking cessation, adequate calcium and Vitamin D, and in some cases, medications such as bisphosphonates that suppress cells that remove bone.

Fee-for-service reimbursement was the obvious barrier. Nonetheless, according to Dell, more than 100 similar programs around the country were making progress anyway. The proliferation of value-based contracts ought to provide a strong tailwind for continued progress.

§

Payers have a powerful asset in their hands — a treasure trove of data that can be mined for deep insight into patients and costs. For example, nearly one in three adults in the United States has metabolic syndrome, and their health care costs are 60 percent higher than the population average.

Metabolic syndrome, a precursor to both heart disease and diabetes, is the co-occurrence of three of five symptoms above gender-specific thresholds — abdominal girth, blood pressure, fasting blood glucose, triglycerides and cholesterol. The more symptoms you have above a threshold, the higher your health costs. Of the five symptoms, abdominal girth is most strongly predictive of future health care consumption.

These insights are from Aetna, one of the nation's largest health insurers. There, Greg Steinberg, MD, and his colleague Adam Scott, an MBA, led an innovation team that was launched in 2011 and focused on developing datasets and analytics that could guide both employers and health care providers in improving care.

Using claims data, lab data and biometric data from employer health screenings, the team developed a quantitative model that did more than simply identify the most expensive patients retrospectively; it predicted the progression of metabolic syndrome. Specifically, it estimated what a patient's status on each of the five metrics would likely be in 12 months, assuming no intervention.

> Predict who will benefit most from the intervention that you design. (Ch. 13)

By 2014, the company offered reports from the model to several of its national accounts and worked with these clients to develop a custom menu of wellness and disease management programs. It might, for example, suggest wellness programs for patients who were predicted to have two or three symptoms and more aggressive disease management programs for those predicted to have four or five. These programs always had, at their core, close one-on-one relationships among patients and health coaches or nurses to educate patients on metabolic syndrome and its risks and to engage them in efforts to improve their health behaviors.

Aetna also forged a partnership with a small company to develop and pilot a new intervention program. Patients were segmented into one of eight groups based on a saliva-based genetic test that helped estimate, for example, a patient's ability to control appetite and helped indicate the type of intervention to which that the patient was most likely to respond. Patients were further evaluated on their readiness to engage in programs focused on improving health behaviors. Each patient met weekly with a health coach by video conference, logged food consumption and monitored physical activity on a wearable device.

Prevention requires educating, counseling and engaging patients
in improving their health behaviors. (Ch. 13)

Aetna characterized the preliminary results as very promising. Of particular note, roughly 60 percent of patients remained engaged in the program after nearly a full year.

§

Might the future of health care belong to the computer scientists? Perhaps not, but what about physicians who also have deep interest in computers?

One such physician is Ruben Amarasingham. Though he ultimately chose medicine as his career, he had nurtured a deep interest in computer programming in his youth. His research during his fellowship at Johns Hopkins explored the use of predictive modeling to improve care at the bedside. Ultimately, his unusual combination of skills put him in a perfect position to launch a new prevention initiative.

Parkland Hospital, in Dallas, Texas, is a safety net health system. Resources are scarce; demand, overwhelming. In the late 2000s, Amarasingham, serving as medical director of care management, reviewed patient records looking for patterns that might lead to improved allocation of resources.

Noticing a high rate of readmissions for patients with congestive heart failure, Amarasingham followed an instinct that data in electronic medical records could be used to estimate the likelihood of readmission. He'd soon built a predictive model that relied on both medical and socioeconomic inputs. The model sorted patients into five quintiles, the top two of which he labeled high risk.

Predict who will benefit most from the intervention that you
design. (Ch. 13)

The hospital already had a staff of case managers who worked to help patients stay healthy after discharge. The value of the predictive model was that it allowed the case managers to set priorities and use their time more effectively. All patients received the standard of care; those in the high-risk categories received extra attention. (Note that this is one of the few initiatives covered in the book that did *not* involve the creation of a new kind of team.)

Care managers checked in quickly post-discharge, for example, to ensure that high-risk patients understood the nature of CHF and understood their treatment plan, including the importance of staying on the prescribed medications. They also asked whether patients were receiving the support they needed at home, and explored other options such as nursing homes, if needed. Further, they helped arrange follow-up care, such as pharmacy or dietary consults or appointments with primary care physicians or specialists, and even ensured that the patients showed up as scheduled for tests and appointments. The care managers found that most of their work could be handled by telephone, reserving home visits for only the most difficult cases.

> Prevention requires educating, counseling and engaging patients
> in improving their health behaviors. (Ch. 13)

Because of the use of the predictive model, Parkland cut the 30-day readmissions rate (including readmissions to all Dallas hospitals) by 19 percent (from 26.2 to 21.2 percent) for all CHF patients.[13] After some recognition for his work at Parkland, Amarasingham began to receive calls from other hospitals around the country about what he was doing. Interest was heightened when the Affordable Care Act instituted increased penalties based on hospital readmission rates.

In 2012, Amarasingham founded the Parkland Center for Clinical Innovation, a nonprofit spin-out from Parkland Hospital, to commercialize predictive modeling software and related services. By 2014, the Parkland Center team included more than 64 clinicians, mathematicians, computer scientists and consultants, and was developing predictive models for a range of medical conditions, including diabetes, acute myocardial infarction, chronic kidney disease, asthma, pneumonia and sepsis.

§

By 2014, thanks to remarkable advances in the biosciences, it was quite possible for HIV/AIDS patients to live a normal life. Doing so, however, required meticulous attention to medications. When the disease was poorly controlled, the body's viral load steadily increased, the immune system weakened, the virus became more drug resistant and the required medical therapies more complicated and more expensive. Eventually, there were no more options. The stakes were high, both for patients and for public health. Sicker patients were more infectious and tended to pass along the disease in a more severe and more challenging-to-treat state.

Getting patients to stay on proper medications may sound simple in theory. The reality, however, is that even highly educated people often fail to finish a short course of antibiotics. HIV/AIDS therapy is far more demanding. It requires a lifelong commitment and it demands near perfect adherence. Newer therapies made it possible for some HIV patients to take just one pill daily, but it needed to be taken at roughly the same time each day.

Many patients faced significant barriers, including poverty, substance abuse, mental health issues, difficult housing and living conditions, and widespread misinformation about the disease. Also challenging were the side effects of the medication, which could include sleep disturbances, nausea, vomiting, diarrhea and a worsening of diabetes. Perhaps the steepest barrier was the social stigma that many patients still associated with their diagnosis. Many patients struggled to accept the diagnosis and were reluctant to seek proper care. Some hesitated to carry their medications with them, for fear that their diagnosis might accidentally be revealed to friends, family members or co-workers.

Launched in 1990, the HIV/AIDS program at Christiana Care Health System in Delaware worked to overcome these hurdles and deliver the best possible care to its patients. In 2014, the program was led by full-time medical director Susan Szabo and program manager and nurse Arlene Bincsik. The program served roughly two-thirds of the state's HIV/AIDS population, most of whom lived in and around the city of Wilmington.

Though the medical decision-making for AIDS patients could become complicated, the bulk of the staff's work lay in engaging, educating and supporting patients in sticking with their treatment plans. In Bincsik's view, a trusting relationship between a patient and a specific member of the program's staff, typically a nurse, was critical. When a patient missed an appointment, it had to be the trusted nurse that called to follow up, not a stranger or a machine. Nurses often contacted patients frequently to ensure that they were staying on track.

> Prevention requires educating, counseling and engaging patients in improving their health behaviors. (Ch. 13)

Rather than simply prod, the nurses attempted to diagnose the underlying reason for any missed care steps. If, for example, the patient lacked transportation, the team would work to find a solution. In some cases, patients found helpful the habit of stopping by the clinic weekly to have their pillboxes filled. When the nurses struggled to locate a patient, they sometimes collaborated with state service agencies that were in touch with the same patient.

The need to build trust with patients guided the program's recruiting and hiring. In particular, Bincsik looked for candidates with a strong desire to bring care to the underserved. New hires also had to have strong social interaction skills. Health care experience and technical skills were less relevant by comparison.

> Team designs for prevention initiatives put non-physician providers, and sometimes providers with little or no prior medical training, on the front lines. (Ch. 13)

The program strove to deliver comprehensive primary care. Indeed, for about 80 percent of patients, it was their sole source of care. Many patients preferred this arrangement as it allowed them to minimize disclosure of their disease. Because it was possible to live indefinitely with HIV/AIDS, patients faced many of the same health challenges as the population at large. The treatment of the common chronic diseases, such as diabetes, heart disease and hypertension, could not be separated from the treatment for HIV/AIDS.

Indeed, the medications for these common chronic ailments often worked in opposition to those for HIV/AIDS.

In 2014, the program's staff numbered just more than 50, spread across several locations. Beyond Szabo and Bincsik, the management team included Robin Bidwell, who headed quality improvement. Most of the staff worked full time within the program, including approximately five nurse practitioners, 14 nurses, four pharmacists, seven social workers/mental health providers and eight secretaries. Roughly 12 additional physicians contributed to the program part time, from multiple disciplines including internal medicine, OB/GYN, psychiatry and infectious disease.

> The Dedicated Team has enormous flexibility. It can accommodate
> entirely new roles and team designs. (Ch. 7)

Bincsik described the program's approach as multidisciplinary team medicine. No physician or nurse practitioner kept a set panel of patients that they were individually responsible for. Each afternoon, the team met for rounds, discussing the care plan for every patient seen in the clinic that day. In the main clinic, Szabo attended along with nurse practitioners, nurses, pharmacists and social workers.

Because many of the program's services were not reimbursable and many patients were uninsured or poorly insured, only about 10 percent of the program's funding came from insurance reimbursements. Federal and state government programs provided another 40 percent, and the balance came from Christiana Care itself. Christiana had an economic rationale for doing so — keeping uninsured patients healthy and out of the hospital and emergency room — but Bincsik believed Christiana was also motivated by a desire to serve the community. She did not know of many other medical institutions that were funding similar programs at the same level.

To keep the program financially sustainable, its leaders had long emphasized the importance of every team member working at the top of his or her license and capability. Bidwell's quality improvement work helped prioritize their efforts. She had created a performance dashboard that allowed her to monitor the health and care of the program's patient population in aggregate.

Both Bincsik and Bidwell credited Szabo's orientation toward teamwork and quality improvement as critical elements of the program's overall performance.

Take pride in system performance. (Ch. 11)

Like so many innovators in health care delivery, Bincsik and her colleagues were eager to see the creation and expansion of similar programs elsewhere. They spent a significant amount of time mentoring other programs, both in Delaware and out of state. Proliferation of value-based contracts will further catalyze expansion of similar programs.

Once you know what works, consider growth. Also, consider sharing your knowledge with other leaders of innovation in health care delivery. (Ch. 6)

Though the payoff can be much greater than the cost, it is expensive to employ new types of health workers to engage patients in improving their health behaviors. What if the same result could be achieved not by employees but by friends? Was it possible to bring together small groups of patients with similar health conditions and for those patients to become friends who support each other and push each other toward success? Could patients coach patients?

This is the aspiration behind an emerging model of care, group medical appointments. Cooper Hospital, in Camden, New Jersey, is one of several around the country that is experimenting with the approach.

Cooper serves a population that is both poor in the economic sense and in health. These patients generally faced difficult barriers to improved health, such as a limited social support network, sharp financial constraints and poor living conditions. They often felt alienated from the health system.

Long-term patient engagement requires a strong and trusting
human bond between patient and coach. (Ch. 13)

Because Cooper consistently faced financial losses from treating the poor, the management team turned to Jeffrey Brenner for help. Brenner was well known in the Camden health care community for his work at the Camden Coalition, the nonprofit organization he directed. An article in *The New Yorker* magazine, "The Hot Spotters," had raised the profile of Brenner's work. Brenner agreed to develop a proposal for a new Advance Care Center at Cooper.

Around that time, Brenner was introduced to Kathy Stillo, a Columbia MBA who had led an internal consulting team at Bristol-Myers Squibb for more than eight years but had become avidly interested in health care delivery. The two agreed to join forces, and they soon established the Advance Care Center's goal: Help patients achieve better health and stay out of the hospital primarily by helping them become more successful in self-managing their conditions. Their first initiative was group medical appointments.

Each group appointment was dedicated to a single medical condition. By 2013, the list of conditions included asthma, COPD, diabetes, headache, hypertension and metabolic kidney disease.

Appointments lasted 90 minutes. The time was divided equally into an individual clinical evaluation and a group conversation. Upon arrival, patients entered into a queue and proceeded, production-line style, through a series of clinical stations. A medical assistant took the patient's vitals, a nurse recorded the medical history and a pharmacist reconciled the medications. At the last station, a physician spent five to seven minutes with each patient in a semiprivate consultation.

After the clinical evaluations were complete, the patients congregated for the group discussion. A behaviorist explained the group norms, solicited questions and facilitated comfortable conversation until all the patients had arrived. A physician then led the discussion. Some of the discussion time was typically dedicated to a preselected topic of relevance to the group such as medication usage or eating strategies during the holiday season. Remaining time was dedicated to questions or topics suggested by patients.

Though it was too early to draw conclusions about the impact of group visits on outcomes or system costs, patients clearly liked group visits. One indicator: In the population that the Cooper Advance Care Center serves, no-show rates were as high as 30 to 40 percent, but patients were showing up for their center appointments, and in many cases showing up early.

Although patients spent less time in one-on-one interaction with physicians, they spent more time in the same room as the physician, and this enabled a more thorough discussion. Also, because more appointments were available, patients had improved access to care. Patients also had the opportunity to learn from the concerns and experiences of others, make social bonds and form informal support networks.

Two Cooper physicians worked part time at the Advance Care Center in leadership and advisory roles. Steve Kaufman was the assistant medical director, and Ed Viner served as a liaison to the Cooper physician community and the medical school, helping to build support among physicians, negotiate the politics and manage stakeholders. For example, some accounting challenges arose, such as properly allocating revenues and costs associated with a patient's care when that care spanned both the center and other internal entities.

Even the best innovation leaders need help. (Ch. 8)

To increase patient volume, Stillo needed Cooper's physicians to willingly, if not eagerly, refer patients to the center. Some physicians referred right away, but the overall response from physicians was mixed. Some did not wish to depart from the traditional care model, in which the one-on-one physician-patient relationship was the fundamental building block. Some were uncomfortable putting a new and unfamiliar clinic with an unproven care model in between themselves and their patients. Although the Advance Care Center, in principal, intended to augment rather than replace the care patients received one-on-one at Cooper, some physicians may have been worried about the possibility that the center would reduce demand for their services. Others physicians, particularly the younger generation, were interested in reforming health care and eager to explore new possibilities for simultaneously improving care and remaining financially sustainable. The center minimized the tension by working first with those who were most willing.

Anticipate and mitigate conflicts with downstream providers. (Ch. 8)

Physicians who embraced the center appreciated how it enabled more productive use of their time. Freed from routine examinations and the need to educate each patient on disease basics, physicians were able to focus their time on unique and unusual patient needs and on higher-order cognitive work and complex decision-making. Stillo worked to make the center as attractive as possible to physicians so they could simply show up, engage in patient consultations, lead the education session and depart.

The Shared Staff has nearly no flexibility. It can only take on tasks that: a) are already familiar or can be quickly learned b) fit naturally into existing roles, and c) fit naturally into existing work processes. (Ch. 7)

By September 2013, the center employed nearly 30 full-time staffers, mostly receptionists, medical assistants and LPNs. Stillo was the full-time executive director; her role was running the daily operations of the team and managing relationships with grantors and other institute partners. Brenner devoted 30 percent of his time to the center as medical director, spending the remainder of his time continuing to guide the Camden Coalition.

Brenner and Stillo had the flexibility to design the Dedicated Team from scratch. The group visit model was a big change for the center's full-time staff, changing both their roles and the rhythm of daily work. Indeed, more than half of the original staff chose to leave. Stillo wasn't entirely surprised. She supported those who chose to depart.

The Dedicated Team has enormous flexibility. It can accommodate entirely new roles and team designs. (Ch. 7)

Brenner and Stillo chose to start the center's work with group visits because they saw that almost every stakeholder could emerge a winner. Critically, group visits could be a financial win for the hospital. Even though the facilitated group conversation was not reimbursable, the clinical evaluation met insurance requirements for an individual patient visit, so the center could be paid as though it had seen each patient separately.

Despite the favorable fee-for-service financial model, the Advance Care Center still needed funding to get started. Cooper Hospital raised the capital needed to redesign and attractively refresh the physical space for group visits and to recruit and hire staffers.

§

In Atlantic City, New Jersey, it seems that every institution has had an interest in containing health care costs. The city's economy relied on casinos, and there were far more of them up and down the East Coast than there had once been. Casino executives were anxious about remaining competitive, and about escalating health care costs in particular. Atlantic City's biggest union, one that served low-wage service employees such as pit bosses, maids and cooks, was equally vexed. For years, most every penny of increased compensation that it had negotiated for its members was gobbled up by the rising cost of health insurance. AtlantiCare, the city's dominant health system, was as concerned too. If the casinos struggled, the city struggled, and if the city struggled, the health system struggled.

This confluence of interests created the conditions for an unusual partnership between a health system and one of its biggest customers — the union. The tangible product of the partnership was the Special Care Center, described briefly in this chapter's opening.

Betsy Gilbertson, who managed the trust that provided health benefits for the union, had been galvanized by a 2005 consulting report that laid out a model for intensive primary care that promised improved outcomes and dramatically reduced costs. She had been sent the report by one of its authors, Arnold Milstein. Health coaching, a concept under development at the time, lay at the model's foundation.

Gilbertson and Margaret Belfield, a senior executive at AtlantiCare, partnered to implement the clinic. They were advised by Rushika Fernandopulle, a physician who had long been interested in and involved with efforts to redefine primary care. Fernandopulle had experience working with Milstein implementing intensive primary care models in other settings.

In building the Special Care Center, the team hired health coaches not based on medical qualifications, but on their social skills. The job of the health coach was to build close and trusting relationships with patients and engage them in a disciplined effort to modify their health behaviors — more exercise and better diet in particular. Over time, the team concluded that a crucial qualification for health coaches was that they shared culture and language with their patients. Health coaches needed to understand their day-to-day lives.

> Long-term patient engagement requires a strong and trusting
> human bond between patient and coach. (Ch. 13)

The Special Care Center nurse practitioners and physician assistants trained and supervised health coaches, stepping in directly when patients were actively sick. The medical director supervised the entire operation, managing relationships with specialists who saw center patients, delegating as much work as possible to the center team and spending only a fraction of time seeing patients.

> The Dedicated Team has enormous flexibility. It can accommodate
> entirely new roles and team designs. (Ch. 7)

The initiative stumbled when choosing the first medical director, however. He simply couldn't break out of old habits, Fernandopulle said. He did not naturally gravitate to a focus on team performance. Furthermore, instead of challenging specialists to work with the center to reduce what Fernandopulle saw as an egregiously excessive frequency of visits, he instinctively developed close and cordial relationships with them.

> Before you create you must forget. Team design is a blank slate
> endeavor. (Ch. 6)

Fernandopulle stepped in for roughly a year to run the center himself during the search for a permanent medical director. The eventual permanent new hire, Ines Digenio, had experience working in resource-poor settings in Africa and South America. As a result, she had a natural inclination to delegate to get the most out of the health coaches. She also had an instinct for identifying and eliminating wasteful practices. In her effort to cut back on unnecessary specialist visits, she had AtlantiCare's explicit support.

Team designs for prevention initiatives put non-physician providers, and sometimes providers with little or no prior medical training, on the front lines. (Ch. 13)

Anticipate and mitigate conflicts with downstream providers. (Ch. 13)

Even the best innovation leaders need help. (Ch. 8)

Digenio and her operations manager, Sandy Festa, gathered the entire team for a huddle each morning to review the cases that had been active during the prior 24 hours and to update care plans. Each team member had an opportunity to offer input. Digenio had the final say on medical matters, and Festa on operations. The team reported any hospitalizations and ER visits during the huddle and cheered each day when there were none to report.

Take pride in system performance. (Ch. 11)

Overall, the Special Care Center monitored five categories of performance indicators — health status indicators, hospital admissions and readmissions, finances, staff satisfaction and patient satisfaction. Although the center needed to be alert to the possibility of better-before-worse trends, the highlights reported in 2012, based on internal analysis, were promising: outpatient procedures down 23 percent, hospitalizations down 41 percent, emergency room visits down 48 percent and endocrinologist visits down nearly 90 percent. The center estimated a savings of more than $2,000 per member per year for the union.

> Be wary of the possibility of a better-before-worse trend for measures of utilization. (Ch. 13)

The Special Care Center benefited from the lessons of an earlier project at Boeing that Milstein and Fernandopulle collaborated on. Milstein had persuaded Boeing to give the intensive primary care model a try for the 10 percent of its employees who had the highest health care expenses. Fernandopulle led the effort full time on the ground.

Boeing's unwillingness to ask its employees to change primary care physicians presented a complication. It forced Fernandopulle to try to weld health coaching services on to existing primary care practices. The practices were happy to accept the funding for additional services, but were not inclined to rethink the way they organized or operated. Primary care physicians continued to operate as usual; they simply prescribed additional attention for the high-utilizing patients. (Some felt ethically conflicted because they could not make the offer to all patients.) Nurses stayed in their existing roles but were expected to learn how to coach patients on the fly, with no particular training.

> The Shared Staff has nearly no flexibility. It can only take on tasks that: are already familiar or can be quickly learned, fit naturally into existing roles and fit naturally into existing work processes. (Ch. 7)

Though the Boeing effort improved outcomes and lowered costs, Fernandopulle felt certain that it had fallen far short of full potential. He noted, for example, that nurses varied widely in their ability to coach patients, a task that called little on their training and expertise. The experience left Fernandopulle with a deep desire to see what could be achieved with a clean-slate effort. He wanted a Dedicated Team; he got one in Atlantic City.

.... § ....

A Few Questions to Consider

1. Of the patients you serve, who are the most likely to be high users of care next year? Why?

2. Do you have access to any data or analytics that might help you make good predictions about who these patients are?

3. What do these patients need to do to stay healthy and out of the hospital?

4. What type of intervention might keep these patients on track?

Chapter 14
Improve Treatment Decisions

A T THE UNIVERSITY OF CALIFORNIA SAN FRANCISCO MEDICAL
Center in 2014, many newly diagnosed cancer patients were matched
with an assistant, a premed undergraduate college student, who helped them
consider their treatment options. The students were members of UCSF's
Patient Support Corps.

The students did not have medical training and did not give advice. Instead,
the students listened, took notes, asked open-ended questions of patients to
explore what was most important to them, helped patients develop lists of
questions for the doctor, asked doctors for clarifications during appointments,
and recorded and summarized the conversation with the doctor. Patients
and physicians alike were generally satisfied with the program, believing it
improved communication and decision-making.

§

It is not hard to describe an ideal medical decision. All reasonable treatment
options are considered, including the option of doing nothing. The available
evidence on the benefits, risks and side effects of each option is reviewed.
The value that the patient places on each potential outcome is weighed, as
is the patient's tolerance for risk. Finally, the optimal decision is calculated.

Reality intrudes. Physicians face relentless time pressures. Patients, for
their part, are sometimes so stricken in the immediate aftermath of a serious
diagnosis that they hardly recall a word that the physician says about treatment options. Actual decisions fall far short of ideal.

This would all be of little concern were the medical decision-making process well described as physician experts putting science to work to identify the unambiguously superior and rational treatment option. The black-and-white medical decision, however, is the exception, not the rule. Medical decision-making is awash in shades of gray — and ever more so as advances in the biosciences deliver ever more treatment options.

Studies of variations in care show just how little consistency there is in medical decision-making around the country.[14] Total consumption of care varies nearly three times from the highest-spending regions to the lowest; variations in the use of specific procedures can be much higher. There is no appealing explanation for such enormous variation. These studies render a clear, if uncomfortable, verdict. The state of medical decision-making is poor.

Of all of the initiatives discussed in the book, those focused on improving decision quality feel intuitively to me like the most powerful. I suppose that I have saved the best for last.

That is certainly not to say that achieving perfect decision-making is easy. Hardly so. In fact, this is an area in which perfect can easily become the enemy of good. The goal is not perfection, it is simply moving closer to the ideal. Doing so is a relatively straightforward matter of giving patients additional support and resources so that they can slow down, consider the options, reflect on what matters and, in collaboration with their physician, make a choice.

Historically, advocates for improving medical decision-making have seen their work as a cause — as the right and ethical thing to do. To date, however, the moral force of the cause has funded only a handful of initiatives. It has been a cause in need of a business case.

Under value-based contracting, the business case just could not be any simpler. Here is a simple hypothesis for you: Informed patients choose less care. There is already substantial evidence to support this hypothesis (along with a few counterexamples).[15] And then there is common sense. Medical decision-making is gray. It is, therefore, subject to biases and incentives, nearly every one of which points in the same direction — toward more care.

The antidote? More information and more careful decision-making.

Might it make sense to spend a few hundred dollars on improved decision-making before choosing a procedure that costs tens of thousands of dollars? If informed patients choose to have the procedure just 1 or 2 percent less often, the initiative breaks even financially. Studies of some treatment decisions have shown reductions in use an order of magnitude higher.[16] It hardly takes precision mathematics to see that these initiatives make economic sense, in addition to being the right thing to do.

Given physicians' deep commitment to patient care, I'd like to think that most everyone reading this book would be excited to find ways to guide patients through the dense fog of major medical decisions. These initiatives are also potent at a macro level. One of the reasons we have runaway health costs — perhaps the biggest reason — is a particular form of market failure. Market theory assumes perfectly informed "consumers," but patients suddenly confronted with a frightening and unfamiliar diagnosis are a long way from perfectly informed. Initiatives that focus on improving medical decisions take direct aim at this market failure.

§

The initiatives described in this chapter, plus conversations with experts in the field, suggest the following principles:

Principle 1: Better decisions are achieved by giving patients facing consequential treatment decisions more time, more support and better information. Patients are understandably emotional and stressed after receiving serious diagnoses. It takes time for patients to learn about the options and reflect on what matters most to them. Having a detached partner with whom to discuss the options can be very helpful.

Principle 2: With a bit of training, many people can succeed in the role of supporting patients in considering their treatment decisions. Social workers, chaplains, undergraduate students interested in medical careers and nurses are just a few examples of the backgrounds of people involved in decision support. Important attributes include a genuine desire to partner with patients, skill in listening, the ability to build trust and ease in the use of open-ended questions to help patients think about what they value. Medical expertise is not necessary beyond an ability to understand the same information about treatment options that the patient receives.

Principle 3: Isolate people in patient support roles from any and all financial considerations. The minute that patients think the information they are receiving is biased by financial considerations, the game is over.

Formal separation of decision support staff from business staff could help cement trust. Other industries have developed "Chinese Wall" organizational partitions that isolate certain personnel from business considerations. In the media industry, for example, newsroom personnel are strictly isolated from business staffers to eliminate any possibility of advertisers influencing the way the news is reported.

Another possibility is the development of nonprofit organizations with patient decision support as a core mission. These organizations would hire, train and deploy decision support staff. Health systems that operate under value-based contracts would happily fund these nonprofits, even while remaining at arm's length. (The nonprofits would maintain independent boards of directors.)

Principle 4: Measure decision quality. Assessing the progress of an initiative focused on improving medical decisions is a bit tricky. Which outcomes should be measured? Imagine for a moment that an initiative like UCSF's later showed increasing mortality in cancer patients. Would such a result indicate that the initiative had failed? Or would it indicate that informed patients had prioritized quality of life over length of life?

Ambiguity in the value that patients place on specific outcomes is, of course, part of the justification for these initiatives. In general, these initiatives might very well lead to a worsening trend in a specific medical outcome, but such a result might very well be offset by other improvements that patients value more.

Clinical data are of limited relevance in assessing these initiatives. An alternative is near at hand. Several researchers have been working on methods for evaluating the quality of the decision-making process based on feedback from patients.[17]

Principle 5: When asked, physicians should offer a treatment recommendation. Initiatives of this type are often surrounded by a perception that the fundamental goal is to shift power and control from the physician to the patient. I do not see this as a particularly helpful framing. Physicians and patients ought to be viewed as partners, not rivals. Furthermore, an objective of placing the full burden of decision-making on the patient's shoulders seems unrealistic. Consequential treatment decisions are difficult. Even with time, information and support, patients may lack the confidence to choose a treatment option on their own.

A better goal is simply getting all of the relevant information on the table. This requires a two-way exchange of information. Patients need to learn about their options; physicians need to learn about what patients value (and patients may need time and assistance in reflecting on what they value before they can articulate it). Once this exchange of information has happened, the quality of the decision will almost certainly be higher, regardless of how exactly the patient and the physician interact on their pathway to a final choice.

Principle 6: Consider bringing all providers who treat the same medical condition together on the same team. You may be doing this anyway, to achieve better coordination of care, as described in Chapter 12. A similar team design pays off when improving medical decision-making. Richer and deeper interaction among providers who treat the same medical condition improves each provider's ability to engage patients in thoughtful consideration of all of their choices.

Principle 7: Generally speaking, decision support staffers are on the Dedicated Team, treating providers on the Shared Staff. Team designs for these initiatives are relatively free of complications, especially because decision support staffers are likely to meet the exceptions outlined in Chapter 7 that enable them to be on the Dedicated Team even if decision support is not a full-time role.

In this chapter we explore a three initiatives, starting with UCSF's, that endeavor to bring treatment decisions closer to the ideal.

The Patient Support Corps at UCSF was inspired by the Stanford doctoral research of Jeff Belkora that showed that newly diagnosed cancer patients often faced consequential decisions with a poor understanding of their options. Conversations between patients and providers were impeded by any number of barriers; limited time was just one of them. Newly diagnosed patients were naturally anxious and confused. They were overloaded with information. Some patients froze under the stress of the serious diagnosis and were simply unable to actively engage in deliberate thinking about their options. Others diligently prepared questions for their doctors, only to shy away from actually asking them. Some patients brought family members to help with note taking, but family members were also emotionally involved. A few patients contemplated recording the conversation with the physician, but were nervous about asking permission to do so. Patients commonly struggled to express their priorities and preferences to clinicians.

Patients needed more help, but how could it be provided? Belkora found that the use of decision aids — publications or videos that explained the treatment options in a concise and layperson-friendly form — were helpful and far better than Internet searches. However, decision aids alone were inadequate.

Ultimately, Belkora partnered with the director of the Breast Care Center at UCSF, Laura Esserman, who suggested that the undergraduate students who worked at the center as interns could be trained to support patients. With little budget for the program, Belkora couldn't pay stipends to the interns, and payers did not typically reimburse for decision support service.

So Belkora devised a clever workaround. He struck an arrangement with the University of California, Berkeley, whereby college students could earn course credit by joining the Patient Support Corps. The program quickly drew a heavy volume of applications. The approach had the advantage that students were naturally distant from the influence of financial incentives within the health system.

> Isolate people in patient support roles from any and all financial
> considerations. (Ch. 14)

Belkora and his team suspected that many college students had the necessary skills — listening closely, asking good questions, taking outstanding notes, producing concise summaries and researching the available decision aids. The team used several techniques to screen applicants, including asking students to write essays that included anecdotes that demonstrated these skills. Some students had to be eliminated because their course schedules did not mesh with that of the clinic, and some were eliminated because they did not show a genuine interest in improved medical decision-making. Those who passed these screens were called in for an interview, and about half of those were offered positions.

> With a bit of training, many people can succeed in the role of
> supporting patients in considering their treatment decisions.
> (Ch. 14)

Given that the students had no other jobs in the health system, Belkora had tremendous flexibility to shape their roles. The students could be considered the Dedicated Team. Those who accepted their positions prepared for their roles by practicing specific interview protocols and participating in five increasingly challenging role-plays.

In general, Belkora viewed the students' relative ignorance of medicine as an advantage. It made it easier for students to genuinely walk in the shoes of the patients and made them more likely to ask good clarifying questions of doctors.

Newly diagnosed patients received a phone call from Belkora's team. They were advised to do quite a bit of homework before their initial appointment at the medical center, review educational materials and decision aids and formulate a list of questions for the physician. They were also advised to bring a note taker and a recording device to the appointment, so that they could later review and reflect upon all that was said.

For many patients, this felt like a lot to do. As a result, they were welcoming when asked if they would like to be matched with a student in the Patient Support Corps who could help them. By 2014, almost all patients who wanted help received it.

Once the student and patient were matched, the student phoned the patient, developed a preliminary list of questions and pointed patients to the relevant decision aids. After an adequate period of time for the patient to review the educational materials and reflect, the student and patient finalized a previsit plan, including a question list. During the appointment, the student listened, took notes and occasionally asked questions not on the list. After the visit, the student created a written summary of the visit, including the physician's answers to each of the questions on the prepared list. Patients were often grateful for the help; some even stayed in touch with the student who assisted them.

One of the pleasant surprises for Belkora was seeing how readily both patient and physician accepted the involvement of a student. Interestingly, some patients agreed to accept help without really believing that a student could provide any value. These patients simply liked young people in general and wanted to help them along on their journey to becoming doctors. Physicians often viewed the students similarly, at least at first. They were accustomed to having physicians-in-training around and enjoyed the mentoring aspects of their work. Both patient and provider were then pleasantly surprised by just how much value the students delivered. Both reported that conversations about medical decisions felt more focused and productive.

Measure decision quality. (Ch. 14)

Based on his early experience with the program, Belkora developed a standard staffing model. Students could reasonably work with about 30 patients per year, one per week while school was in session. A program coordinator could oversee a cadre of 30 students serving about 900 newly diagnosed patients per year. Roughly half of all newly diagnosed patients agreed to be matched with a student, and about half of those were actually matched, though Belkora expected these rates to increase over time as clinics and patients became accustomed to the program. Thus, a program with a single coordinator was suitable for a cancer center seeing a few thousand newly diagnosed patients per year.

With some grant funding from the Informed Medical Decisions Foundation, Belkora had developed job descriptions, policies, procedures, training manuals and operating manuals for the program. Belkora hoped to see similar programs arise at other institutions around the country and wanted UCSF to serve as a technical support center as new programs launched.

§

It is worthwhile to briefly reprise the advance care planning initiative at Gundersen Lutheran, previously described in Chapter 7. As you may recall, Bud Hammes created a small team of facilitators who met with patients to discuss their care plans at end of life.

Although UCSF's Patient Support Corps discussed imminent treatment decisions with patients while Hammes' team engaged in discussions of future decisions, the similarities between the two programs are otherwise impossible to miss.

Both are simple but powerful. Both put people with relatively little medical knowledge into positions in which they help patients consider their treatment options. Both recruit with an emphasis on listening skills, ability to gain trust and ask good questions. Both train recruits on specific interviewing techniques. And, in both cases, the decision support staff worked only part time on the initiative, but, following the guidance in Chapter 7, could be considered part of the Dedicated Team.

> With a bit of training, many people can succeed in the role of supporting patients in considering their treatment decisions. (Ch. 14)

> The Dedicated Team is generally composed of people working full time on the innovation initiative. There are limited exceptions. A part time team member can be considered part of the Dedicated Team only if that team member's two jobs are completely independent of and isolated from one another. (Ch. 7)

§

Most adults at some point in their lives will be challenged by an episode of acute low back pain. The first time it happens, the pain can be so severe, unfamiliar and shocking that the natural instinct is to race to the doctor looking for an immediate fix. Many patients expect what feels like the biggest "hammer" to address the problem: surgery.

The instinct has fueled a rapid explosion in surgical procedures for addressing back pain. This growth has been far from uniformly distributed, however. Because the evidence base is thin, there is little consensus about which treatments for back pain are effective and when. If you find interesting the variation in overall consumption of care — nearly three times the difference between the highest- and lowest-spending regions — then you'll certainly find the variation in spine procedures jaw-dropping — as much as 20 times from the highest-use to lowest-use regions.[18]

Add to this picture one more stark fact: For many patients, the only solution needed is patience. Most low back pain resolves on its own within a few weeks.

Jim Weinstein, orthopedic surgeon and by 2014 president and CEO of the Dartmouth-Hitchcock Medical Center, had a particular point of view about how to improve medical decision-making in this context: Put all of the providers who treat back pain in the same place and on the same team. That way, a surgeon, for example, is never faced with the stark choice between telling a patient "I can operate" or "You'll need to seek help elsewhere." Furthermore, if the orthopedic surgeons, neurosurgeons, pain doctors, physician assistants, physical therapists, occupational therapists, radiologists, nurses and nurse practitioners all routinely interact, then they become more familiar with all that is possible in treating back pain.

> Consider bringing all providers who treat the same medical
> condition together on the same team. (Ch. 14)

Weinstein founded the Dartmouth-Hitchcock Spine Center in 1997. Simply by virtue of being close by, the providers in the spine center have frequent conversations about particular patients and their treatment plans. (The ideal would be for all providers to be working in the spine center full time, but some worked part time because either volume was inadequate to keep them occupied full time or to be able to treat patients with conditions other than back pain.)

The team also met weekly to discuss particularly interesting cases. To ensure that all treatment options had an equal footing, Weinstein worked to create an egalitarian culture within the spine center. Over time, team members become more proficient at explaining all treatment options and the benefits, risks and side effects of each.

> Better treatment decisions are achieved by giving patients facing consequential treatment decisions more time, more support and better information. (Ch. 14)

The staff of the spine center made a strong recommendation for surgery only when there was clear evidence to support it. To deepen the evidence base, Weinstein initiated and led a grant-funded $27 million study of more than 3,500 patients. The study confirmed that surgery was appropriate for patients with three specific conditions — spinal stenosis, degenerative spondylolisthesis and herniated discs — provided that the patients met certain specific clinical criteria.

For other patients, there remained tremendous uncertainty in spine care. Simply by virtue of having all options available in the same place, however, physicians and patients alike tended to gravitate toward less risky, less aggressive and less expensive interventions such as physical therapy. That patients choose this route once fully informed seems unsurprising. Why not try something simple first, especially given that it does not foreclose the option of pursuing a riskier intervention later?

Dartmouth-Hitchcock Medical Center has one of the lowest rates of spine surgery in the United States.

§

A Few Questions to Consider

1. Which are the most difficult and consequential medical decisions that your patients face each year?

2. How would you evaluate the quality of the two-way information exchange before these decisions? Does the patient fully understand the benefits, risks and side effects of each treatment option? Do you fully understand what the patient values?

3. Would these patients benefit from an assistant who could guide

them through their treatment options, someone who could help them reflect on what they value most?

4. How much do you think the treatment decisions would change if all of the providers involved in delivering all of the treatment options worked on the same team and discussed patient treatment plans with each other every day?

Conclusion

....

YOU'RE STILL HERE? THAT'S TERRIFIC, BUT IT IS TIME TO STOP READ-
ing and start innovating.

Seriously, after contemplating several possible options for a chapter-length
conclusion to this book, my only conclusion was that I'd be repeating myself.
Doing so would be a waste of your time and mine. If you want to reinforce the
key ideas in what you have just read, I recommend you reread the summaries
at the end of each chapter in Part II, and perhaps the principles listed at the
beginning of each chapter in Part III.

I look forward to continuing to learn about the terrific work that all of you
are doing to fix health care. I suspect that I will be hearing much more about
information technology and how it can play an even more catalytic role in
innovation in health care delivery than the illustrations in this book suggest.
I also suspect that I will be hearing much more frequently about initiatives
that take the form of partnerships with organizations *outside* the health system. These initiatives will take direct aim at what are often called the social
determinants of health — housing, food, education, public safety, employment
and so much more. As many of you know, these variables, collectively, are far
more strongly correlated with good health than the total of all that happens
within health care. Initiatives of this type will be more challenging, but, as
is often said, nothing worth doing is ever easy.

I will be continuing to work hard to catalyze innovation in health care
delivery. For example, I will be teaching in Dartmouth's Master of Health Care
Delivery Science program, where I have the privilege of working directly
with a great number of inspired physician innovators. As I have been doing

for many years, I will also be traveling to deliver speeches, presentations and workshops.

I will also be developing as many resources as I can to supplement this book, including toolkits and discussion guides that will help you take action tomorrow to move your ideas forward. I will also be preparing videos, short online courses, interviews with innovators and much more. They will be available online. Please keep in touch by visiting the book's website, www. physicianinnovators.net, or by emailing me at chris.trimble@dartmouth.edu.

I hope that we cross paths soon.

Chris Trimble
May 2015
Dartmouth College
Hanover, New Hampshire

Acknowledgements

···

IN A VERY REAL SENSE, THIS BOOK WAS WRITTEN BY INNOVATORS and for innovators. My greatest debt is to those of you who are already doing the critical work of building new teams that can deliver better outcomes at lower costs. You are not only fixing health care, you are lighting the path for many more innovators who will follow in your footsteps.

I could not have learned so much about the innovations currently in progress without the help of many. Thank you to all of you who spent time with me, shared your experiences with me, allowed me to shadow you as you served your patients, and tolerated my many follow-up questions: Bill Abdu, Jonathan Abrams, Kyle Allen, Ruben Amarasingham, Kim Beekman, Jeff Belkora, Margaret Bellfield, Robin Bidwell, Arlene Bincsik, Barry Bock, Ed Clark, Andrea Cronin, Chris Dadlez, Richard Dell, Art DeTore, Ines Digenio, Lollie Dubiel, Pam Duncan, Elizabeth Dwelle, Cliff Eskey, Rushika Fernandopulle, Sandy Festa, Bert Fichman, Michael Friedman, Jessie Gaeta, Maureen Geary, Elizabeth Gilbertson, Bud Hammes, Eric Hartmann, Rowley Hazard, AJ Horvath, David Judge, Shreya Kangovi, John Keggi, Stephanie Kelly, Donna Kostak, Carrie Kovarik, Steve LeBlanc, Vivian Lee, Court Lewis, Maureen Lowry, Robin Mackey, Mike Magill, Greg Makoul, Robin Marcus, Arnold Milstein, Sohail Mirza, Dawne Mortensen, Nancy Murphy, Jim O'Connell, Adam Pearson, Scott Pingree, Sarah Pletcher, Steven Poplack, Kate Roche, Kari Rosenkranz, Larider Ruffin, Katherine Schneider, Steve Schutzer, Adam Scott, Nate Simmons, Sanjay Sinha, Deb Smith, Kim Soboleski, John Sprandio, Elizabeth Stedina, Greg Steinberg, Kathy Stillo, Bob Taube, Dale Vidal, Franklin Watkins, Jim Weinstein, Linda Wick and Gabe Wishik.

I found many examples of innovation only because of help from others. Thank you for the introductions and connections: Jeff Brenner, Todd Dunn, Rushika Fernandopulle, Eric Isselbacher, Brent James, Greg Makoul, Roy Rosin and Mike Zubkoff.

There were also several generous physicians who allowed me to shadow them during the early days of my effort to write this book, to help me get my feet wet in the world of medicine. They deserve particular thanks for fielding my most naïve questions. Thank you, Steve Liu, Sean Uiterwyk, Tim Gardner, Kenny Rudd and Joe Perras.

I received tremendous support from my friends at the American Association for Physician Leadership. In particular I'd like to acknowledge my editors Bill Steiger and Gina Bowden Pierce, as well as Peter Angood, Hans Zetterstrom, Dawn McKnight, Karen Scheidt Kaminskas, and Tricia Barnes.

I am grateful for the opportunity to work alongside so many outstanding faculty colleagues here at Dartmouth. Of greatest importance, Al Mulley provided the institutional support for making this book possible. Also, through collaboration on several projects, Al quickly deepened my understanding of the most fundamental flaws in health care, both in the United States and abroad. I am also grateful for more than a decade of collaboration with Vijay Govindarajan, my co-author on five prior books about innovation. The ideas in these prior books formed the foundation for this work.

My understanding of the industry has grown enormously thanks to opportunities to listen to and learn from many Dartmouth colleagues. Thank you, in particular, to Paul Barr, Paul Batalden, Ethan Berke, Nan Cochran, Don Conway, Bob Drake, Glyn Elwyn, Elliott Fisher, Paul Gardent, Bob Hansen, Adam Keller, Punam Keller, Michael Lewis, Don Likosky, Ellen Meara, Katy Milligan, Manish Mishra, Bill Nelson, Gene Nelson, Greg Ogrinc, Rob Shumsky, Jonathan Skinner, Elizabeth Teisberg, Rachel Thompson, Dale Vidal, Eric Wadsworth, Scott Wallace, Jim Weinstein, Gil Welch, Jack Wennberg and Mike Zubkoff.

The research that led to this book was fun and inspiring. It was also a great deal of work. I was thrilled to have help from Alana O'Brien, Michael Lewis, Thom Walsh, Eric Weinberger and Irene Wielawski. Bravest of all were those who volunteered to review the manuscript. The book is much better for the insightful feedback I received from Alice Andrews, Peter Angood, Patrick Brophy, Chris Blaski, Ed Clark, Addi Faerber, Rushika Fernandopulle,

Tim Foster, Robert Greene, Eric Isselbacher, Patrick Lee, Shreya Kangovi, Stephen Liu, Kedar Mate, Al Mulley, Gene Nelson, Mark Nunlist, Chal Nunn, Merritt Patridge, Alok Sharan, Jeff Thompson, Sean Uiterwyk, Dale Vidal, Eric Wadsworth, Thom Walsh and Gil Welch.

Finally, I could not have finished this book without the love and support of my family. Among so many gifts, my parents, Henry and Sally, taught me to write. My wife and children, Lisa, Hank and Mateo, extended an incredible reservoir of patience. Where's Dad? He's still in his office ...

Thank you all.

About the Research

This book is an adaptation of prior work to the specific context of health care. It was preceded by more than a decade of fieldwork on the subject of the best practices for executing innovation initiatives inside established organizations.

Innovation execution is a complex, dynamic and difficult-to-quantify phenomenon. Studying it effectively does not involve statistical analyses of large data sets. Instead, generating new theory requires assembling, examining and comparing multiyear case histories of innovation initiatives. All of my prior work used grounded theory as the methodology — a qualitative, clinical, longitudinal, field-study-based approach. If you are interested in a more thorough exploration of the formal academic underpinnings of this past work, I refer you in particular to two of my earlier books with Vijay Govindarajan, *The Other Side of Innovation* (2010) and *Ten Rules for Strategic Innovators* (2005), both published by Harvard Business School Press.

For this book, I continued to use the same methodology. I conducted more than 70 interviews with physician innovators and their teams, developing five lengthy case studies and more than 20 short ones. The bulk of my research time was consumed by immersing myself in writing, analyzing and comparing these case studies. As I did so, I tried to understand: What is working? What is getting in the way? What are the root causes of struggle? Are these struggles unusual or likely to be experienced by most physician innovators? I also compared these case studies to those I have developed in other industries. This work helped me address the crucial question: *What is unique about innovation in health care?* There was plenty, and that was what made this project so compelling to me as a researcher.

Endnotes

....

1   See, for example, Sinsky, Christine A., *et al,* "In Search of Joy in Practice: A Report of 23 High-Functioning Primary Care Practices," *Annals of Family Medicine,* 11:3, p. 272-8, May/June 2013.

2   This figure is frequently used. I'm not sure if there is any hard data behind it, but I suspect we can agree that the proportion of health care spending that is physician-directed is quite high.

3   Prasad, Vinay, *et al.,* "A Decade of Reversal: An Analysis of 146 Contradicted Medical Practices," *Mayo Clinic Proceedings,* August 2013: 790-798.

4   Wennberg, John E., *et al.,* *The Dartmouth Atlas of Health Care,* Chicago: American Hospital Publishing, 1996. For related and more recent publications, see dartmouthatlas. org/publications/reports.aspx.

5   My colleagues in the economics department will likely take issue with that statement. They will point out that healthy price competition is essential for a well-functioning economy, one that directs resources to where they are most needed. Okay, sure, perhaps the term *necessary* is better than *neutral.* Eating, breathing and sleeping are *necessary* for human life, but they fall short of activities anyone would describe as admirable or exciting.

6   My economist colleagues' voices are in my head again. The indirect impact is less determinate. It depends on what the winner does with the profits.

7   There are certainly a few examples, such as new clinics in drugstores like CVS or Walgreens. And, Iora Health is an entrepreneurial firm with enormous ambitions that is seeking to reshape primary care. nytimes.com/2015/03/29/upshot/small-company-has-plan-to-provide-primary-care-for-the-masses.html?_r=0

8   From Chapter 1 in Fletcher, Robert H., *et al.,* *Clinical Epidemiology: The Essentials,* 5th Edition, Philadelphia: Wolters Kluwer, 2014.

9   See chw.upenn.edu.

10   Rules for reimbursement of such consultations are under review and very well may have changed by publication date.

11   In 2014, Judge left Partners to join Iora Health, a startup that specializes in intensive primary care.

12   Jorma, Panula, *et al.,* "Mortality and cause of death in hip fracture patients aged 65 or older – a population-based study," *BMC Musculoskeletal Disorders,* 12:105, 20 May 2011.

13   For full details, see Amarasingham, *et al.*, "Allocating scarce resources in real-time to reduce heart failure readmissions: a prospective, controlled study," BMJ Quality and Safety, 2013, 0:1-8

14   Wennberg, John E., *et al.*, *The Dartmouth Atlas of Health Care*, Chicago: American Hospital Publishing, 1996. For related and more recent publications, see dartmouthatlas.org/publications/reports.aspx.

15   Al Mulley, Chris Trimble, Glyn Elwyn, "Patient Preferences Matter: Stop the Silent Misdiagnosis," The King's Fund, 2012.

16   Hawker, Gillian A., *et al.*, "Determining the need for hip and knee arthroplasty: the role of clinical severity and patients' preferences," *Medical Care,* 39:3, p 206-16, 2001.

17   Barr, P.J.; Thompson, R.; Walsh, T., Grande, S.; Ozanne, E.; Elwyn, G. "The psychometric properties of CollaboRATE. A fast and frugal patient-reported measure of the shared decision-making process," *J Med Internet Res.* 2014 Jan 3;16(1).

18   Wennberg, John E., *et al.*, *The Dartmouth Atlas of Health Care*, Chicago: American Hospital Publishing, 1996. For related and more recent publications, see dartmouthatlas.org/publications/reports.aspx.

Science and Technology Annual Reference Review

Science and Technology Annual Reference Review 1989

Edited by
H. Robert
Malinowsky

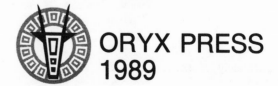

ORYX PRESS
1989

The rare Arabian Oryx is believed to have inspired the myth of the unicorn. This desert antelope became virtually extinct in the early 1960s. At that time several groups of international conservationists arranged to have 9 animals sent to the Phoenix Zoo to be the nucleus of a captive breeding herd. Today the Oryx population is nearly 800, and over 400 have been returned to reserves in the Middle East.

ISBN 0-89774-487-X
ISSN 1041-2557

CONTENTS

INTRODUCTION

The idea for *Science and Technology Annual Reference Review* (*STARR*) came from the realization that, although there are books and journals that review science and technology reference publications as part of their coverage, and although librarians have expressed the need for a single source devoted *solely* to the review of science and technology references, no one had developed a publication of this type until now.

STARR will annually review as many science and technology reference books as possible, both those published in the current year, and, especially in this first issue, some older "standby" titles from earlier years. This represents a formidable task, given conservative estimates that some 1,200–2,000 science reference books are published in the United States and Canada each year.

There is a definite need for this annual. Librarians today face increasing difficulties both in evaluating and funding reference book purchases. Reference publications are often expensive, overlapping in coverage, and misleadingly titled. Accordingly, it is becoming more and more common for libraries to delay purchase of certain books until a review has been published; public and school libraries have long followed this policy, with academic libraries beginning to follow suit. It is hoped that *STARR* will be of special use to those collection development officers who seek a second opinion when considering the purchase of a reference book—especially a new, expensive title.

WHAT IS INCLUDED

Each volume of *STARR* will contain descriptive and, in most cases, evaluative reviews of reference materials at all levels, from those for school libraries to those for special and academic libraries. Reviews are arranged alphabetically within subject and type of publication. Books for which more than one subject category applies have been listed under the category that was determined to be most appropriate.

While most of the publications reviewed fall into predictable reference categories, three

types may not always appear to be of obvious reference value.

(1) Textbooks that are unique in their coverage (those in their fifth or later editions; especially medical textbooks)
(2) How-to books (included on a very selective basis)
(3) Exam reviews books (a neglected reference source)

ELEMENTS OF EACH ENTRY

Each entry contains some or all of the following elements:

title
edition
author(s)
editor(s)
compiler(s)
publication information, including publisher division and distributor, if applicable; publication year; indication of the number of pages; and inclusion of illustrations, bibliographies and indexes
price
series title
ISBN
ISSN
SuDoc number
review
reviewer name

THE REVIEWERS

The reviewers are librarians from all types of libraries, from school and public to academic and special/research, though most of them are librarians employed at special libraries, many of whom belong to Special Libraries Association (SLA). They range from general reference librarians to specialists to retired professionals. Each has a strong background in the discipline that he or she reviews and his or her own style of writing, which makes each review unique.

Ideally, each review includes the following elements: a description of the book and a comparison with similar titles; a detailing of

its good and bad points; and a recommendation for purchase.

This first volume contains over 600 reviews. Future editions of *STARR* will continue to review as many science and technology reference titles as possible, with the intention that at least 1,000 can be covered each year.

INDEXES

There are four indexes: (1) title, which, due to the unusual length of many subtitles, includes main titles only; series titles as well as former titles are included; (2) name, which includes authors, editors, compilers, original authors of earlier volumes, et al.; (3) subject; and (4) type of library, including college, public, school, special/research, and university, for which the publication is recommended.

COMMENTS

The editor welcomes comments and suggestions from users and publishers. Write to him at the following address:

H. Robert Malinowsky
1250 W. Grace, 1st Floor
Chicago, IL 60613

ACKNOWLEDGMENTS

I would like to thank all publishers who submitted books for review in this first edition, as well as to all reviewers, who worked hard to give a fair review to the books that were received. Thanks also to Al Covert of Oryx Press in Phoenix, who received and mailed out the review copies; to Jan Krygier and Jean Bann at Oryx Press, who advised me and edited the manuscript; and to all of the individuals behind the scenes who helped make *STARR* a "star" publication.

CONTRIBUTORS

The following individuals contributed one or more reviews to *Science and Technology Annual Reference Review 1989.*

Charlene M. Baldwin
Science and Engineering Librarian
University of Arizona
Tucson, AZ

Julie Baldwin
Research Assistant
School of Library and Information Science
University of Wisconsin at Milwaukee
Milwaukee, WI

Judith B. Barnett
Associate Professor
University of Rhode Island
Wakefield, RI

Anne Best
Media Specialist
East Middle School
Aurora, CO

Julie Bichteler
Associate Professor
Graduate School of Library and Information
   Science
University of Texas at Austin
Austin, TX

James E. Bird
Reference Librarian
University of Maryland
College Park, MD

Raymond Bohling
Reference/Bibliographer
University of Minnesota
Minneapolis, MN

Craig S. Booher
Librarian
Appleton, WI

John T. Butler
Assistant Librarian
University of Minnesota
Minneapolis, MN

Jo Butterworth
Librarian
Mathematical Sciences Research Institute
Berkeley, CA

Janet K. Chisman
Database Coordinator
Washington State University
Pullman, WA

Bill Coons
Information Literacy Specialist
Cornell University
Ithaca, NY

James L. Craig
Biological Sciences Librarian
University of Massachusetts at Amherst
Deerfield, MA

Jean E. Crampon
Head Librarian
University of Southern California at Los
   Angeles
Los Angeles, CA

George M. A. Cumming, Jr.
Reference Librarian
Boston Public Library
Boston, MA

Juanita A. Cutler
Research Assistant
School of Library and Information Science
University of Wisconsin at Milwaukee
Milwaukee, WI

Sandra K. Dawson
Mathematics Department Chairperson
University of Illinois at Champaign–Urbana
Urbana, IL

Jimmy Dickerson
Chemistry Librarian
University of North Carolina
Chapel Hill, NC

Carolyn Dodson
Life Sciences Reference Librarian
University of New Mexico
Albuquerque, NM

Richard Eimas
Curator of John Martin Rare Books
University of Iowa
Iowa City, IA

Rita C. Fisher
Reference Librarian
Washington State University
Pullman, WA

Marilyn Kay Gage
Librarian
Oklahoma State University
Stillwater, OK

R. Guy Gattis
Assistant Coordinator of Research Information
University of Michigan
Ann Arbor, MI

Sharon Giovenale
Reference Librarian
University of Rhode Island
Kingston, RI

Joanne M. Goode
Head Mathematics Librarian
University of Kentucky
Lexington, KY

Kimberley M. Granath
Assistant Biological Sciences Librarian
Oklahoma State University
Stillwater, OK

David Hines
Research Assistant
School of Library and Information Science
University of Wisconsin at Milwaukee
Milwaukee, WI

Linda Jacobs
Head of Circulation
Thomas Jefferson Library System
Jefferson, MO

Pat Jameson
Research Assistant
School of Library and Information Science
University of Wisconsin at Milwaukee
Milwaukee, WI

Sue Ann M. Johnson
Assistant Physical Sciences and Engineering
   Librarian
Oklahoma State University
Stillwater, OK

Berta Keizur
Technical Information Specialist
University of California at Livermore
Livermore, CA

John Laurence Kelland
Reference Bibliographer for Life Sciences
University of Rhode Island
Kingston, RI

Frank R. Kellerman
Biomedical Reference Librarian
Brown University
Providence, RI

Jody Kempf
Chemistry Bibliographer/Reference Librarian
University of Minnesota
Minneapolis, MN

Yves Khawam
Research Assistant
School of Library and Information Science
University of Wisconsin at Milwaukee
Milwaukee, WI

James S. Koga
Coordinator of Online Reference Service
California State Polytech University
Pomona, CA

Joan Latta Konecky
Reference Librarian
University of Nebraska Medical Center
Lincoln, NE

Ahmed Kraima
Research Assistant
School of Library and Information Science
University of Wisconsin at Milwaukee
Milwaukee, WI

James Krebs-Smith
Head, Information Centers Branch
National Agricultural Library
Beltsville, MD

Barbara Kruser
Research Assistant
School of Library and Information Science
University of Wisconsin at Milwaukee
Milwaukee, WI

Ruth Lewis
Biology Librarian
Washington University
St. Louis, MO

Charles R. Lord
Engineering Librarian
University of Washington
Seattle, WA

H. Robert Malinowsky
Professor/Bibliographer for Science and En-
   gineering
University of Illinois at Chicago
Chicago, IL

Loren D. Mendelsohn
Public Services Librarian
Wayne State University
Detroit, MI

Lillian R. Mesner
Cataloger/Reference Librarian
University of Kentucky
Lexington, KY

Earl Mounts
Reference Information Specialist
Alcoa Labs
Alcoa Center, PA

John Nisha
Librarian
College of Engineering
West Hartford, CT

Ze-hui Niu
Research Assistant
School of Library and Information Science
University of Wisconsin at Milwaukee
Milwaukee, WI

Fred O'Bryant
Science Bibliographer
University of Virginia
Charlottesville, VA

Gerald J. Perry
Reference Librarian
Rush University
Chicago, IL

Mary Peyovich
Research Assistant
School of Library and Information Science
University of Wisconsin at Milwaukee
Milwaukee, WI

J. L. Pfander
Science and Engineering Librarian
University of Arizona
Tucson, AZ

Anne Pierce
Director of Engineering Student Services
College of Engineering
West Hartford, CT

Kenneth Quinn
Engineering and Physical Sciences Librarian
California State Polytech University
Pomona, CA

Jelena Radicevic
Research Assistant
School of Library and Information Science
University of Wisconsin at Milwaukee
Milwaukee, WI

Aziz Rhazaoui
Research Assistant
School of Library and Information Science
University of Wisconsin at Milwaukee
Milwaukee, WI

Robert B. Marks Ridinger
Subject Librarian
Northern Illinois University
DeKalb, IL

Robert G. Sabin
Science and Technology Bibliographer
Auburn University
Auburn, AL

Marsha L. Selmer
Map Librarian
University of Illinois at Chicago
Chicago, IL

Janice Sieburth
Head Librarian
Pell Marine Science Library
West Kingston, RI

Steven L. Sowell
Associate Librarian
Indiana University
Bloomington, IN

Angela Marie Thor
Marine Studies Librarian
University of Delaware
Lewes, DE

Debra A. Timmers
Assistant Librarian
Oshkosh, WI

Michael Troudt
Research Assistant
School of Library and Information Science
University of Wisconsin at Milwaukee
Milwaukee, WI

Linda Vincent
Research Assistant
School of Library and Information Science
University of Wisconsin at Milwaukee
Milwaukee, WI

Marilyn Von Seggern
Head, Owen Science and Engineering Reference Library
Washington State University
Pullman, WA

Camille Wanat
Head, Engineering Library
University of California at Berkeley
Berkeley, CA

Lucille M. Wert
Professor Emeritus
University of Illinois at Champaign-Urbana
Urbana, IL

William H. Wiese
General Science Librarian
Oklahoma State University
Stillwater, OK

Stanley Wilder
Research Assistant
University of Illinois at Chicago
Chicago, IL

Mitsuko Williams
Associate Professor
College of Veterinary Medicine
Urbana, IL

Daniel Y. H. Wong
Research Assistant
School of Library and Information Science
University of Wisconsin at Milwaukee
Milwaukee, WI

Linda R. Zellmer
Head, Geology and Maps Library
University of Wyoming
Laramie, WY

REVIEWS

AGRICULTURE

★

★
GENERAL SOURCES
Biographical Sources

1. Agricultural and Veterinary Sciences International Who's Who. 3rd. Harlow, Great Britain: Longman Group UK Ltd., Dist. by Gale Research Co., Detroit, MI, 1987. 1195p. in 2v. $450.00. Reference on Research. 0-582-90159-6.

This two-volume set in the series *Reference on Research* takes the first step in filling an information void concerning prominent agricultural and veterinary scientists and their activities. First published in 1979 as *Who's Who in World Agriculture*, this third edition provides biographical details of 7,500 senior scientists from over 100 countries. The second edition, brought out in 1985, contained biographical sketches for 12,000 scientists. The publishers have planned to issue a new edition every three or four years. The editors sent invitations to be included to officers in scientific societies, directors in agricultural laboratories and consultancies, heads of academic departments, and members of the editorial boards of relevant journals. The listings exclude retired scientists and emeritus professors. Biographical profiles organized in alphabetical order include education, current position, previous experience, appointments, membership in societies, major publications, interests, address, and telephone number. A country-subject index lists the scientists in each country by general subject, such as agricultural economics, agricultural engineering, anatomy and physiology, animal production, biochemistry, biophysics, botany, fisheries, food science, forestry, horticulture, microbiology, oceanography, plant production, soil science, veterinary medicine, and zoology. No other points of access are provided. The international breadth and the number of scientists listed prevent a thorough treatment. Only a small percentage of United States scientists are represented. Furthermore, the listings appear to emphasize scientists of the United Kingdom. For example, the subject index for scientists of the United Kingdom fills six pages, whereas the listing for the United States fills only three, about the same number as Australia. Honduras, Gabon, Belize, and Reunion receive directory attention with one scientist each. Agricultural administrators, librarians, conference organizers, recruitment officers, market researchers, businesspeople, government officials, and ag-

ricultural consultants will find this set useful. A companion volume available separately is *Agricultural Research Centres: A World Directory of Organizations and Programmes.*   LINDA JACOBS

Directories

2. Food, Hunger, Agribusiness: A Directory of Resources. Thomas P. Fenton; Mary J. Heffron, eds. Maryknoll, NY: Orbis Books, 1987. 131p., illus., bibliog., index. $9.95pbk. 0-88344-531-Xpbk.

The authors compiled this resource guide on food and hunger issues "to direct concerned citizens in First World countries to the books, periodicals, audiovisuals, and other resources they need to study in order to take informed and effective action to correct injustices in the ways their governments and businesses treat Third World nations and peoples." This volume expands the resources presented in the *Third World Resource Directory* (Orbis, 1984) and is part of a series of annotated guides to Third World resources. Other titles in the series include *Women in the Third World, Latin America and Caribbean,* and *Asia and Pacific.* Fenton and Heffron are codirectors of the Third World Resources project at the Data Center, a public-interest information library in Oakland, CA. The publisher and supporter of the book, the Catholic Foreign Mission Society of America (Maryknoll), recruits and trains people for overseas missionary service. This small volume cannot begin to include all organizations and other resources concerned with world hunger, but presents those "resources that stress the relationship of hunger [in the Third World] to its political and economic origins." The book is aimed at "citizen advocates, teachers, librarians, nutritionists, clergy, economists, volunteers in soup kitchens," and people who have experience in the Third World. The authors admit bias in favor of a "radical analysis" of Third World affairs, believing that there is "a crying need" for radical change that will come about "through struggles (though not necessarily violent) between the 'powerful' and 'powerless'." The scope of the work includes organizations, books, periodicals, pamphlets, articles, audiovisuals, and curricula. The chapter arrangement is by form of the resource. In general, each chapter contains annotated entries, combined with supplemental lists and further notations of other sources of information. Organizational annotations include the religious affiliations, focus, activities, re-

sources, and publications of each organization. Books and articles date for the most part from 1980 to 1986 and are presented along with summaries of their essential message. An excellent source of thought-provoking perspectives on food and hunger issues, this book underlines the extent to which religious organizations have become involved with world hunger. Separate indexes provide access by organization, individuals, titles, geographical areas, and subjects. "Bibliographical data on all of the resources are stored in a computerized database" that can be accessed "through the PeaceNet electronic communications network." *LINDA JACOBS*

Guides to the Literature

3. Selective Guide to Literature on Agricultural Engineering. Gayla Staples Cloud, comp. Washington, DC: Engineering Libraries Division, American Society for Engineering Education, 1985. 23p. $5.00pbk. Engineering Literature Guides, no. 4. 0-87823-104-8pbk.

Students use literature guides almost as much as the librarians do. The American Society for Engineering Education (ASEE) has produced guides to the various engineering disciplines, and number four in the series deals with agricultural engineering covering a variety of topics including food engineering, machinery, and buildings, plus some general engineering sources. The guide is arranged by publication type, and listings within each type are alphabetical by author. Included are yearbooks and surveys, standards and specifications, dictionaries, and encyclopedias. Database entries include lesser-known services such as *Agricultural Research Review*, available through NewsNet Inc. Ms. Cloud has provided a handy tool for those dealing with the agricultural engineering discipline. *ANGELA MARIE THOR*

Handbooks

4. CRC Handbook of Agricultural Energy Potential of Developing Countries. James A. Duke; Alan A. Atchley; Kenneth T. Ackerson; Peggy K. Duke. Boca Raton, FL: CRC Press, 1987. 4v., illus., bibliog., tables, index. $449.00. 0-8493-3640-6 set.

This four-volume set is primarily a handbook, as the title implies, with emphasis on aspects relevant for identifying the biomass energy for many of the developing countries of the world. It is arranged alphabetically by country. Each chapter contains summary information on (1) "agriculture" (compiled mostly from FAO data), covering land/animal distribution, farm animals, manure production per animal per year and agricultural production (2) "Soils" (3) "Productivity, including Net Primary Productivity (NPP) and chaff factors" (4) "Brief economic demography"

(5) "Energetics summaries" and (6) "Climate, including the Holdridge Life Zone," a means to identify in a list of 1,000 economic plants that have been reported from the same or similar life zones. L. R. Holdridge constructed a chart (published in 1947) which differentiates the vegetation of dry land areas of the world into 100 closely equivalent formations separated by temperature, precipitation, and evaporation lines of equal value. Approximate elevation is an additional influencing factor when using the chart. The introduction explains what is generally included in each category, a table of the Holdridge Life Zones, and a detailed table of economic plants and their ecological distribution.

Following the summary information are vegetation and weather station maps and, when available, soil maps. Tables of climatological data for the country by locale complete the individual country entries. Some of the data were generated by the authors, but many were compiled from many different sources. The list of 278 references is listed at the back of each volume, along with a volume index by crop and farm animal. There are two appendices: one includes the conventional and energetic yields of specific crops, and the other lists the phytomass files by crop. The set is similar to other CRC publications. Organizations which have an extensive interest in the developing world will find it a useful addition to their collections. *RITA C. FISHER*

5. Midwest Plan Service Structures and Environment Handbook. 11th rev. Ames, IA: Midwest Plan Service, 1987. 1v. various paging, illus. $25.00pbk. "MWPS-1." 0-89373-057-2pbk.

This is an established basic reference book for planning farmstead systems and facilities. It covers 12 separate topics, each with an application to some area of farm management of buildings, crops, or animals. These topics include (1) loads—dead, snow, wind, floor, suspended, stored materials; (2) materials and selection—wood, fasteners, concrete, caulks, paints, plastics; (3) construction—foundations, framing, pavements, tanks, fencing; (4) design—strength, indeterminate structures, wood, concrete, steel, granular materials, beam and frame formulas; (5) planning—farmstead, economics, livestock housing, grain centers and drying, crop storage, shops; (6) environmental fundamentals—theory, animal heat and moisture; (7) control systems—insulation, heat and moisture, systems; (8) ventilation applications—livestock housing; (9) fruit and vegetable storage; (10) energy—solar energy, energy saving, cost estimating; (11) waste management—properties, handling, treatment, utilization, domestic sewage; and (12) utilities—water supply, electric wiring. The text is well written and detailed in its description with many illustrations, charts, and diagrams. For example, the section on fencing covers types of fencing needed for a specific animal, posts, installation, and gates. Safety is stressed in every area by recommending the wearing of protective gloves or clothing when working with certain kinds of materials. There is a certain amount of mathematics that may cause some

people some concern, but, by and large, the text is easily understood by the layperson. The appendix section includes information on climatic data with maps showing winds, rainfall, and lake evaporation; metric units; and weights and measures. This is an excellent handbook covering a wealth of information that would be useful to any farm or ranch manager, student, layperson, and libraries serving agricultural communities.

H. ROBERT MALINOWSKY

★
ANIMAL CULTURE
Handbooks

6. Beef Housing and Equipment Handbook. 4th. Ames, IA: Midwest Plan Service, 1987. 1v. various paging, illus., bibliog., index. $7.00pbk. "MWPS-6." 0-89373-068-8pbk.

Beef producers have special problems in facilities for raising beef. This handbook presents recommendations for beef producers concerning the design and operation of buildings and equipment. It is written in a very readable style with a minimum of jargon but supplemented with many excellent illustrations, line drawings, and charts. The information includes planning data, farmstead planning, building construction and materials, ventilation, cow-calf facilities, cattle feeding facilities, cattle handling facilities, feed handling, water and waterers, utilities, manure management, and fences and gates. Ranchers and students will find this a useful handbook, and libraries in rural areas will want a copy for reference. *H. ROBERT MALINOWSKY*

7. Dairy Housing and Equipment Handbook. 4th. Ames, IA: Midwest Plan Service, 1985. 1v. various paging, illus., bibliog., index. $6.00pbk. "MWPS-7." 0-89373-062-9pbk.

This is a comprehensive and practical handbook on housing dairy cattle and the equipment that is used in the dairy industry. The text is well written and supplemented with numerous line drawings and charts. It is divided into 13 sections that cover data summary, total dairy facility, replacement animal housing, milking herd facilities, milking center, special handling and treatment, building environment, manure management, feeding facilities, silo capacities and feed data, utilities, equipment plans, and concrete.

H. ROBERT MALINOWSKY

8. Handbook of Aviculture. Frank Woolham. Poole, Great Britain: Blandford Press, Dist. by Sterling Publishing Co., New York, NY, 1987. 368p., illus. (part col.), bibliog. $65.00. 0-7137-1428-X.

This concise volume is designed to provide the amateur aviculturist with the basic information on keeping birds in captivity and has a recurrent theme of conservation and the role played by bird breeders in preserving endangered species. Recipes for 20 diet mixtures are given, as are designs and dimensions for cages and enclosures. One chapter is devoted to breeding, and another, by a veterinarian, covers diseases. The main part of the book, however, is an illustrated listing of 300 avian species along with description, range, status, avicultural characteristics, feeding, breeding, and general management. A novice aviculturist or a person considering the hobby of keeping birds will find this a readable, anecdotal account, but the more serious person will need a more comprehensive manual. *CAROLYN DODSON*

9. Jensen and Swift's Diseases of Sheep. 3rd. Cleon V. Kimberling. Philadelphia, PA: Lea and Febiger, 1988. 394p., illus., bibliog., index. $52.00. 0-8121-1099-4.

Prepared by a well-published authority in sheep pathology, C. V. Kimberling, this is the third revision to a long-valued veterinary textbook and reference source published under the same name in 1974 and 1982. Nearly all chapters have been revised slightly, but particularly prominent revisions are the chapters dealing with the respiratory system diseases of lambs, the addition of several new syndromes and pathologic conditions, and updated information in the treatment and prevention of diseases. Illustrations, however, are almost entirely the same as in the second edition and are, unfortunately, shown only in black and white.

Maintaining the same format as the previous editions, the content of the third edition is in three parts: (1) "Diseases of Breeding Sheep and Nursing Lambs," (2) "Diseases of Feedlot Lambs," and (3) "Diseases of Adult Sheep." In each part, the disease conditions are organized under affected organ systems such as digestive system, central nervous system, kin, etc. For each disease name, discussions follow under uniform headings of definition, occurrence, etiology and pathogenesis, clinical signs and post-mortem lesions, diagnosis, and prevention and treatment. An average of about 15 references follow each disease discussion and bring the readers up to date on diagnostic methods, treatment, and preventive measures of diseases affecting these economically unstable farm animals.

The overall physical characteristics of this book are very attractive. The print is sufficiently large, very dark, and on crisp, good quality paper. The currentness as well as the authoritativeness of information makes this book a must for students in veterinary medical or animal science curricula, as well as for veterinary practitioners. Sheep producers and serious sheep business investors would also benefit from this book, and libraries serving veterinary schools will want it for their reference collections. *MITSUKO WILLIAMS*

10. Raising Turkeys, Ducks, Geese, Pigeons, and Guineas. Cynthia Haynes. Blue Ridge Summit, PA: TAB Books, Inc., 1987. 354p., illus., index. $24.95, $16.95pbk. 0-8306-0803-6, 0-8306-2803-7pbk.

Keeping in mind both beginners and experienced poultry handlers, Cynthia Haynes devotes 50 percent of her book to turkey production and the rest to similar treatments of ducks, geese, pigeons, and guineas. The author derives much of her knowledge of poultry production from eight years' experience operating a small country hatchery and breeder farm. Packed with practical advice, the book provides detailed information on the selection of breeds, hatchery procedures, space requirements, handling considerations, housing, preparation for show, equipment, feeding, butchering, and diseases. The audience that will find this book useful includes those raising birds for pleasure or profit, 4-H projects, show, or food production. It is especially recommended for any public library serving a rural population. One shortcoming is the occasionally choppy writing that sometimes makes the text difficult to follow. Well-illustrated with photographs, drawings, and charts, the book includes building plans for range shelters, brooders, feeders, incinerators, etc., suitable for the small-scale poultry handler. An appendix provides sources for equipment and pharmaceuticals. The book does not list hatcheries that will ship the young poultry; the author suggests contacting other handlers, poultry slaughter houses, feed stores, etc. It does contain a limited index. Recommended for school and public libraries, particularly in rural areas, and for academic agricultural science collections.
LINDA JACOBS

11. Sheep Housing and Equipment Handbook. 3rd. Ames, IA: Midwest Plan Service, 1982. 116p., illus., glossary, index. $6.00pbk. "MWPS-3." 0-89373-052-1pbk.

Raising and maintaining sheep is a complicated industry, and housing them is just as complicated. This handbook "summarizes the design, selection, and operation of buildings and equipment for sheep production." With clear and concise text supplemented with excellent line drawings, the book covers planning sheep facilities, barns, barn and lot layouts, barn environment, manure management, feed storage and handling, treating and handling facilities, equipment for raising orphan lambs, utilities, equipment, and construction materials. Sheep ranchers and students in high school will find this a very useful handbook. Libraries in rural areas will need it for reference. *H. ROBERT MALINOWSKY*

12. Swine Housing and Equipment Handbook 4th. Ames, IA: Midwest Plan Service, 1983. 112p., illus. $6.00pbk. "MWPS-8." 0-89373-054-8pbk.

This handbook outlines the agricultural engineering recommendations for swine producers. It includes discussions on the design and operation of buildings and the equipment necessary for efficient swine operation. The text is concise and supplemented with many line drawings and tables. The sections covered in this handbook are site selection, scheduling, farrowing, nursery, growing and finishing facilities, gestation and breeding facilities, combining buildings, pasture production, insulation, mechanical ventilation, manure pit ventilation, cooling systems, natural ventilation, manure management, slotted floors, utilities, swine handling, existing buildings, grain-feed centers, equipment, and building and equipment materials. *H. ROBERT MALINOWSKY*

Textbooks

13. Fundamentals of Dairy Chemistry. 3rd. Noble P. Wong; Robert Jenness; Mark Keeney; Elmer H. Marth, eds. New York, NY: Van Nostrand Reinhold Co., 1988. 779p., illus., bibliog., index. $79.95. "An AVI Book." 0-442-20489-2.

Fundamentals of Dairy Chemistry consists of various chapters, each written by different authorities, covering the various aspects of milk composition and dairy chemistry. Of particular interest in a reference collection are the numerous tables, charts, and bibliographies that have been difficult to locate in one volume. The editors have streamlined this edition by eliminating the early, less significant bibliographic references. They have also added two new chapters pertaining to new developments in the field of dairy science.

This is a textbook and as such is written in textbook style with each chapter building on the previous chapter. It is well written and comprehensive in its coverage. The detailed index facilitates locating information within the text and, thus, makes it a useful reference source. For libraries with collections in dairy science, food science, and nutrition this is a highly recommended reference source on a topic that has few reference books. *JAMES S. KOGA*

★ FOOD SCIENCE
Dictionaries

14. Knight's Foodservice Dictionary. John B. Knight. Charles A. Salter, ed. New York, NY: Van Nostrand Reinhold Co., 1987. 393p. $31.95. "A CBI Book." 0-442-24666-8.

This special dictionary "brings together for the first time definitions from all aspects of the food-service industry: basic terms, basic ingredients, cost controls, culinary arts, foodservice equipment, management information systems, menu analysis and development, nutrition, sanitation and safety, service and merchandising, and bar management." It is arranged alphabetically letter-by-letter, with each entry in full capitals.

The definitions are generally brief but complete. Words within in a definition that are defined elsewhere in the dictionary are in boldface. This is an interesting dictionary to thumb through and read at random to discover such interesting terms as "Charm Price—Price ending with an odd cents value. It is thought to attract customers," "Dipsa—Any food that causes thirst," and "Japanese Roulette—Game sometimes played in Japan of eating puffer fish, which is safe only if it is prepared correctly—otherwise it is deadly poisonous." Some entries are composite entries such as "cheese," which lists 70 kinds of cheeses with a very brief description of each. Under liquors and liqueurs are listed 68 kinds. This is an excellent dictionary for many types of special libraries, bringing together numerous terms that are scattered in several sources.

H. ROBERT MALINOWSKY

Guides to the Literature

15. Food and Nutrition Information Guide. Paula Szilard. Littleton, CO: Libraries Unlimited, Inc., 1987. 358p., bibliog., index. $45.00. 0-87287-457-5.

This comprehensive and well-organized guide should be of considerable interest to anyone seeking information in the broad areas of food and human nutrition, particularly college students and faculty and research librarians. Coverage of this guide is primarily English-language materials published in the last 10 years, with most citations annotated. The first of the six major sections are devoted to describing the scope, evolution, and published literature of the field. The following section gives constructive guidance on undertaking a research paper and fully utilizing a research library. Also included are helpful descriptions of relevant indexing/abstracting services and online databases. The four remaining sections detail resources which are specific to the areas of human nutrition, dietetics, food science and technology, and related areas (i.e., food service and the social sciences).

Though a few minor inaccuracies were noted, they were generally the result of information becoming out-of-date and do not detract from the overall usefulness of this excellent guide. Timeliness is always a challenge with books of this nature—it can only be adequately addressed by regular updates. The author states that an important goal of this book "...is to convey a more structured, systematic approach to information seeking so that the less practiced library user can approach each problem with a plan for locating information...." Indeed, even more seasoned library users could benefit considerably if this guide were kept close at hand. *JAMES KREBS-SMITH*

Handbooks

16. Elements of Food Engineering. 2nd. Ernest L. Watson; John C. Harper. New York, NY: Van Nostrand Reinhold Co.,

1988. 308p., illus., bibliog., index. $34.95. 0-442-22677-2.

This second edition of *Elements of Food Engineering* (first edition published in 1976) provides a concise review of major engineering fundamentals and their applications to food processing. Written for students studying food science, the first chapter covers an explanation of dimensions and units with emphasis on conversion into the SI system of measurement. Each of the other 11 chapters is on a specific topic such as the first law of thermodynamics, fluid flow, heat transfer, mixtures of gases and vapors, refrigeration, and dehydration. The chapters are consistently organized with a beginning paragraph that states an overview of the topic and the use of the information in food processing. A brief discussion of the basic engineering elements includes formulas to calculate the processes involved and explanations of needed numerical values along with their derivation. Simple graphs and diagrams illustrate key concepts and relationships under varying conditions. A number of examples follow which cover a practical problem that could be encountered in food manufacturing and the method of calculating the conditions needed for a satisfactory product. Another section asks the reader to solve a food processing problem where it may be necessary to find the rate of a reaction, a quantity of ingredients needed, or the necessary heat or time for warming a substance. No answers are given, but hints are sometimes provided. A concluding section lists symbols used in the chapter with a brief definition.

The use of practical examples (such as calculating the thermal conductivity of an avocado, the rate of heat transfer to a vegetable puree in a simple heat exchanger, or the time necessary to cool a box of butter to a consistent temperature) gives meaning to the basic concepts that are covered more thoroughly in textbooks. A bibliography of approximately 50 books published between 1953 and 1983 is provided, and the subject index includes the names of mathematical constants and chemical reactions.

The authors are university engineering faculty with many years of teaching experience. Upper-level chemical engineering students will find this book useful to reinforce the fundamentals learned in courses, to review formulas and calculations, and to utilize as a reference source. It is appropriate for academic engineering libraries and special libraries in the food industries.

JANICE SIEBURTH

17. A Guide to Food Additives and Contaminants. K. T. H. Farrer. Park Ridge, NJ: Parthenon Publishing Group, Inc., 1987. 157p. bibliog., glossary, index. $38.00. 0-940813-11-4.

Originally published in Australia by Melbourne University Press as *Fancy Eating That* (1983), this is a detailed discussion of food chemistry in general and of three specific groups of chemicals—additives, naturally occurring contaminants, and artificially added compounds. Beginning with a brief overview of the history of food preserva-

tion, three chapters cover those toxins which may occur within foods (including fish and shellfish) as a part of their growth cycles, such as goitrogens, carcinogens, caffeine, and related alkaloids; the contamination of food by chemicals such as the Minamata (Japan) mercury disaster of 1956; and food additives. Emphasis throughout the text is on the abysmal state of popular education regarding food processing and the substances involved in it and what can be done to alleviate this state of affairs. Microbial contamination is treated in a separate chapter. Readers should be aware that the information in the volume is somewhat dated. Chapter 7, on regulation and control, covers Australia, Canada, New Zealand, and the United States, as well as the creation of the Codex Alimentarius by the World Health Organization and Food and Agriculture Organization of the United Nations in 1962. A glossary and reference bibliography are provided. Suitable as an introductory work on the subject for classroom use and for college and university libraries. Research libraries and laboratories will find the concise summary useful.

ROBERT B. MARKS RIDINGER

Manuals

18. **The Commercial Food Equipment Repair and Maintenance Manual.** Roland Greaves. New York, NY: Van Nostrand Reinhold Co., 1987. 189p., illus., index. $29.95pbk. "A CBI Book." 0-442-22755-8pbk.

No manual can eliminate the need for service calls when it comes to correcting some difficulty, but this book is designed "to provide general information on maintenance, that will enable you to perform some minor repairs and to avoid unnecessary service calls." The emphasis is on general maintenance, preventive maintenance and troubleshooting. Part 1 gives general information on gas and controls for gas-fired systems, electricity, refrigeration, steam, testing instruments, and warranties. The following six parts cover specific types of food processing equipment: food preparation, cooking, sanitation, serving, beverage, and auxiliary equipment. Specific types of equipment covered include cutters, mixers, dough dividers, dough rollers, food cutters, food processors, peelers, proof boxes, saws, slicers, broilers, convection ovens, deck ovens, fryers, griddles, kettles, microwave ovens, ranges, steamers, dishwashers, disposals, pot and pan washers, warmers, ice makers, ice dispensers, soft ice cream machines, toasters, coffee urns and makers, dispensers, and conveyors. Each chapter explains what the machine is and does then gives information on maintenance, including a preventive maintenance schedule. The checklist of troubleshooting problems is listed with possible solutions. Finally, it gives a recommended list of spare parts to have on hand in case of emergencies.

Anyone in the food-service industry should have this manual full of practical tips. It is especially useful for technical libraries that serve students who are preparing to enter the food-service industry.

H. ROBERT MALINOWSKY

PETS
Encyclopedias

19. **The Practical Encyclopedia of Pet Birds for Home and Garden.** Don Harper. New York, NY: Harmony Books, 1987. 208p., illus. (col.), bibliog., index. $17.95. 0-517-56543-3.

This colorful encyclopedia was first published in 1986 in Great Britain as *Pet Birds for Home and Garden*. Birds have been popular pets around the world for thousands of years. This is a hobby that can take as little time as you want or a great deal of time, depending on the kind or kinds of birds you decide to have.

Don Harper shows just how easy it is to raise birds for fun, sport, breeding, or exhibition. His book is divided into two parts, the first being the practical section covering the basics of having birds as pets, in the home, or in the garden. Useful information is given on choosing a bird, catching and handling the bird, feeding, health care, breeding, cages, taming, aviaries, flights, henhouses, and enclosures for waterfowl. The color illustrations and excellent diagrams make the very readable text authoritative and interesting. The second section is an encyclopedia of species suitable for pets, including finches, parrots, softbills, pigeons, doves, bantams, pheasants, quail, and waterfowl. Each species is fully described with a color photograph. In addition, distribution, size, diet, sexing, compatibility, and pet appeal are outlined at the beginning of each description. The book ends with a list of books for further reading, a general index, and a species index.

For the seasoned bird hobbyist, this would be an excellent general encyclopedia. For someone interested in birds as pets, this book would give a great deal of information to help decide on a pet. The section on species makes this a very good reference book for school and public libraries.

H. ROBERT MALINOWSKY

Handbooks

20. **A Complete Introduction to Finches: Completely Illustrated in Full Color.** Jurgen Nicolai. Neptune, NJ: TFH Publications, Inc., 1987. 93p., illus. (col.), bibliog. $7.95, $4.95pbk. 0-86622-255-3, 0-86622-293-6pbk.

Based on *Breeding Birds at Home* by the same author and publisher, this book fails to live up to its current title. Although part of a series of books for the beginning pet keeper, this book is more suited to the experienced bird fancier interested in breeding birds in aviaries. The range of birds discussed includes not only finches but also thrushes, warblers, flycatchers, and doves. The first half of the book is somewhat advanced for the beginner because it discusses the nesting habits of various species without a complete introduction to the raising of finches. The second half of the book devotes itself to descriptions of a variety of birds that are imported for sale to bird fanciers. Each discussion includes the native origin of the bird, its natural habitat, and a few brief comments on its habits, hardiness, food needs, and breeding requirements. About half of the species are represented by good quality color photographs. The page layout has narrow inner margins that would prevent rebinding. A good reference source for school and public libraries.

LINDA JACOBS

21. A Complete Introduction to Labrador Retrievers: All the Information You Need about Selecting and Keeping a Labrador Retriever—Featuring Sections on the Breed's History, Training, Health Care, Breeding, and Showing. Anna Katherine Nicholas. Neptune, NJ: TFH Publications, Inc., 1987. 126p., illus. (col.), bibliog. $7.95, $4.95pbk. 0-86622-361-4, 0-86622-379-7pbk.

Intended for distribution through pet stores, this book provides practical advice on selecting, raising, training, and showing the labrador retriever. The author presents the history and development of the breed in a lackluster style, with a heavy emphasis on the major breeders who have brought the labrador to its present state of popularity in England and North America. Notably missing from this abundantly illustrated volume are photographs or illustrations of the champion dogs mentioned in the text. In fact, the book's major shortcoming is the lack of coordination between the text and the illustrations, which appear to be a haphazard selection of pet products, such as bags of dog food, chew toys, and pet carriers, interspersed with anatomical illustrations more appropriate to a veterinary guide. The narrow inner margins of the pages make the book somewhat difficult to read and would prevent rebinding. The text itself provides useful information, such as feeding guidelines, housing requirements, obedience training, major disease concerns, breeding, puppy care, and an introduction to the show world. School and public libraries will find this an appropriate reference source on labrador retrievers.

LINDA JACOBS

22. A Complete Introduction to Snakes: Completely Illustrated in Full Color. Mervin F. Roberts. Neptune, NJ: TFH Publications, Inc., 1987. 93p., illus. (col.),

bibliog. $7.95, $4.95pbk. 0-86622-272-3, 0-86622-352-5pbk.

Intended for distribution through pet stores and addressed to the pet keeper who has "just begun to take an interest in snakes," this book touches on many fascinating aspects of snake lore. With a breezy style, the author advises you to "start slowly; shun venomous species; be observant; gain experience; handle and house your animals humanely; and if you want to make money, real money, stay out of the snake business." Other topics include transportation of snakes, anatomy, reproduction, diseases, and a brief introduction to the scientific classification of snakes. Popular native pet species, such as racers, water snakes, garter snakes, and bull snakes, are discussed, with information on habits and food requirements. (The author's style leads him to make one misleading statement that leaves the uninformed reader guessing, "Alaskans seem to be immune" to poisonous snake bites.) Plentiful photographs and illustrations of a wide range of colorful snakes add to the text. This book will appeal especially to children in the upper elementary grades, but the quality of the binding and narrow inner margins of the pages may limit its lifespan.

LINDA JACOBS

Histories

23. Dogs of the American Indians. William Pferd, III. William W. Denlinger; R. Annabel Rathman, eds. Fairfax, VA: Denlinger's Publishers, Ltd., 1987. 192p., illus., bibliog., index. $19.95. 0-87714-126-6.

William Pferd III presents a much more ordered view of American Indian dogs than the one given in *The Natural History of Dogs*, where it is stated that, "The more primitive inhabitants of the American continent, comprising both Eskimos and Amerindians, domesticated and kept a motley crew of dogs derived from any wild canine species available and suitable for the purpose." From Pferd's research emerges a work that "attempts to provide...a single source for all known references to the varieties of Indian dogs" and "a history of dogs of the American Indian from prehistoric times to the late 1800s." Sources are drawn from anthropology, archaeology, and history; the records of early explorers, missionaries, and pioneers; plus the works of artists and photographers. This wide variety of resource material is melded into an interesting collage of history, art, and science to provide insight into the relationship of the dog and the Indian.

The first third of the book contains several general chapters covering the historical and artistic record and the groups of Old and New World dogs. They introduce the remaining chapters dealing with specific breeds of Indian dogs. These include chapters on dogs of the Great Plains, Arctic, sub-Arctic, Northwest Coast, Southwest, Inca, and southern South America. About 18 breeds are described. For each, there is a discussion of its distribution, use and general appearance, plus accounts of the qualities of the breed

and its relationship to the Indian. Black-and-white reproductions of art works serve to illustrate the social bond between dog and man as well as the physical characteristics of the various breeds.

There is a bibliography of over 80 references to the very scattered literature. There is some inconsistency in the format and completeness of the journal references. For example, a reference to an article in the journal *Evolutionary Biology* gives no volume, issue, or page numbers. The index is adequate, although it consists entirely of the names of people, places, breeds, and tribes. The addition of entries for some general concepts might have made it more useful.

Overall, this sturdy, well-bound book is a welcome addition to other dog books. It presents a great deal of information from scientific research to popular accounts in a readable format. The chapters on the specific dog breeds are especially noteworthy. They can be read straight through or enjoyed in whatever order suits the reader's fancy. School, public, and academic libraries would find this a valuable source of information on American Indian dogs. *JANET K. CHISMAN*

Manuals

24. Beagles: Everything about Purchase, Care, Nutrition, Breeding, Behavior, and Training. Lucia Vriends-Parent. New York, NY: Barron's Educational Series, Inc., 1987. 78p., illus. (part col.), index. $3.95pbk. 0-8120-3829-0pbk.

Vriends-Parent, a biologist and beagle owner, has based this book on "many years of experience with beagles and on a careful study of beagle biology and genetics." Written in a concise, easy-to-read style, this manual provides a complete guide to beagle ownership and care. The book is divided into eight main sections. "Considerations before you Buy" stresses the responsibility involved with pet ownership and the importance of dealing with a reputable breeder when purchasing a pet; "Basic Rules of Beagle Care" covers housebreaking, the dog's living quarters, and necessary equipment the owner must purchase to provide adequate care for the pet; "Grooming" discusses all aspects of proper pet hygiene, including the prevention and treatment of fleas, lice, and ticks; "The Proper Diet" explains the components that constitute a dog's balanced diet and the differences in the various types of dog food on the market; "Raising Beagles" deals with the breeding aspect of pets; "If Your Beagle Gets Sick" describes various diseases, their symptoms, and proper procedures to follow for treatment, as well as first aid procedures for other health problems that arise; "Simple Obedience Training" instructs the owner on how to train the pet to obey basic commands; and "Understanding Beagles" provides a history of the breed, as well as a breed description and standard that has been adopted by the American Beagle Club in rating these dogs at beagle shows.

Discussions are supplemented with helpful drawings. Eight pages of color photographs are also scattered throughout. A useful list of important kennel clubs and their addresses is included. The book is very readable; boldface type for individual topics adds clarity and makes subject location quite easy. However, although this paperback edition will probably suffice for the individual owner, it may not be durable enough for the wear and tear of a library setting.

Beagles is recommended for its pleasing format and its clear, informative style. Its thoroughness in discussing most aspects of beagle care makes it worthwhile for both established beagle owners and those still considering the possibility of pet ownership, as well as for a reference source for school and public libraries.

MARY PEYOVICH

25. Doberman Pinschers: Everything about Purchase, Care, Nutrition, Diseases, Breeding, Behavior, and Training. Raymond Gudas. New York, NY: Barron's Educational Series, Inc., 1987. 79p., illus. (part col.), index. $3.95pbk. 0-8120-2999-2pbk.

This book is one in a series of pet owners' manuals that are published by Barron's. It is compact, easy to understand, and includes, or at least touches upon, virtually everything a reader might need to know about the history, breeding, and care of doberman pinschers. Raymond Gudas, with the help of consulting editor Matthew Vriends, has written an informative manual for the owner of this rather misunderstood pet. Using both black-and-white drawings and color photographs, Gudas has divided the manual into nine main chapters, taking the reader from "Considerations Before You Buy" to "History of the Breed." Each chapter is further subdivided by specific topic discussions, ranging from "Taking Your Pet Home" to "Dogs and Children."

This is not a technical or scholarly manual. It is intended for the general public and, as such, tends to cover a lot of material in a little space. The text is written for the layperson, with a minimum of scientific description. (In his discussion of arranging transportation for breeding, Gudas writes, "If the stud you have selected is near enough for you to drive your pet to the location, you'll save a few bucks and spare the animal the inevitable of relatively minor discomfort of air travel.")

However, the material is comprehensive and very informative, and the two longest chapters—"Training" and "Ailments and Illnesses"—are well written and could be useful for any dog owner, because the text discusses many aspects which are common to other breeds of dog as well. Of particular interest might be the chapter entitled "History of the Breed," which offers a highly informative discussion of the doberman's roots from 1880 to the present. This chapter concludes with the American Standard for doberman pinschers of the American Kennel Club, giv-

ing the AKC qualifications for distinguishing the doberman breed. The book also includes a listing of addresses of kennel clubs from the U. S. and the around the world.

Doberman Pinschers is a useful book, especially if one is considering this breed as a pet or is just curious to learn more about the species in general. School and public libraries will find it a quick reference source. *MICHAEL TROUDT*

26. Labrador Retrievers: Everything about Purchase, Care, Nutrition, Breeding, Behavior, and Training.
Kerry V. Kern. New York, NY: Barron's Educational Series, Inc., 1987. 71p., illus. (part col.), index. $3.95pbk. 0-8120-3792-8pbk.

The labrador retriever had its beginnings in the 1800s in England and came to the United States in 1917. It is a hunting dog retriever but has other attributes such as being a companion or therapy dog, guide dog, military or police dog, and disaster work dog. It is an excellent, lovable pet. This owner's manual covers how to buy, caring, feeding, grooming, dealing with the dog's illnesses, history, training, and breeding. A list of retriever clubs, kennel clubs, and publications is included. Good black-and-white sketches and beautiful color photographs supplement the informative text. This is an excellent manual for school, public, and private libraries as well as for veterinary science libraries.

H. ROBERT MALINOWSKY

★
PLANT CULTURE
Encyclopedias

27. Encyclopaedia of Cultivated Orchids: An Illustrated Descriptive Manual.
Alex D. Hawkes. Boston, MA: Faber and Faber, 1987 (c1965). 602p., illus. (part col.), glossary, index. $80.00. 0-571-06502-3.

Noted orchidologist Alex D. Hawkes produced this comprehensive encyclopedia more than 20 years ago. Although new species and hybrids are available, the volume remains a useful reference book, "designed for easy use by the orchid collector, whether he is botanically trained or not." Almost 700 genera in Orchidaceae are arranged alphabetically, with discussions of their general features, genetics and hybrids, and cultural directions. Each species in general cultivation is described under its genus with information about flowering season, ideal temperature requirements, and distribution. Correct pronunciations are noted in the glossary and in the index. There is also a list of multigeneric orchid hybrids with a cross-index and a checklist of genera with citations to authoritative descriptions.

RUTH LEWIS

28. Taylor's Guide to Ground Covers, Vines and Grasses: A Complete Guide to Gardening with Ground Covers, Fully Illustrated with 454 Color Photographs.
Boston, MA: Houghton Mifflin Co., 1987. 495p., illus. (part col.), glossary, index. $14.95pbk. 0-395-43094-1pbk.

This is another encyclopedia based on the fifth edition of *Taylor's Encyclopedia of Gardening*. Section 1 goes over the basic botany of the plants and provides some explanation of botanical terms, advice on how to get started in gardening, weather zones, garden design basics, lawn care, record-keeping techniques, buying advice, and a list of nurseries. Section 2 contains 454 color plates demonstrating the four groups of plants: foliage ground covers, flowering ground covers, vines, and grasses. Section 3 includes the text explaining the plants by the same groups and arranged alphabetically within the groups. The fourth section contains the appendices which include plant charts, a section on pests and diseases, and a good glossary.

The main part of this work is contained in sections 2 and 3. The photographs are of very high quality, showing the shapes and growth habits of the plants. Unfortunately, in some instances, photos are not close enough to demonstrate flowering structures or leaf structures. The text entries make up for this, however, because they contain line drawings of the small flowers. The captions accompanying the color plates contain essential information such as common and scientific names (genus and species), height, effective seasons, best growing zone, and sun and soil conditions. Then a page notation sends the reader to the encyclopedia entry which enlarges on all of the above elements and also indicates the family of the plant. Many of the encyclopedia entries are updated and revised from the basic Taylor's encyclopedia, and all are arranged alphabetically by genus. Sections 2 and 3 should be used together.

The book has a good index that contains both the common and scientific names of the plants. The boldface numbers refer the reader to the plates, and those not in boldface refer the reader to the text. This adds to the book's usability.

This encyclopedia is a good beginner's guide to ground cover, vines and groves and would also provide a great deal of information to a slightly more experienced gardener. The contributors to the volume are professionals in the field and have designed a document that would add much to public and college libraries. For the professional horticulturist, however, it would seem a little elementary. *LILLIAN R. MESNER*

29. Taylor's Guide to Houseplants: A Complete Guide to Gardening with Houseplants, Fully Illustrated with 427 Color Photographs.
Boston, MA: Houghton Mifflin Co., 1987. 463p., illust. (part col.), glossary, index. $14.95pbk. 0-395-43091-7pbk.

This little encyclopedia is based on the fifth edition of *Taylor's Encyclopedia of Gardening*. It is arranged in four sections, each section giving a considerable amount of basic information. Section 1 goes over basic botany, with explanations of terms, how to choose plants, plant care, potting techniques, propagation, and plant charts. Section 2 contains 427 color plates of the plants by major groups: hanging plants, small foliage plants, orchids, bromeliads and flowers, showy foliage plants, lacy leaves, and succulents and others. Section 3 contains the text entries of the plants arranged alphabetically by genus. Section 4 contains the appendices on orchids, decorating with plants, buying plants, and a glossary.

Sections 2 and 3 are the heart of the book. The photographs are, in the main, very high quality and give a good demonstration of what the actual plants should look like. If there were a complaint, it might be that some photos do not demonstrate the small plant structure of the flowers. The text entries make up for this, however, because they contain line drawings of the smaller structures. The captions that accompany the plates contain essential information such as the size of the plant, how much light it needs, and the proper temperature and humidity. Then a page notation sends the reader to the encyclopedia entry which enlarges on all of the above elements and also indicates the family of the plant. Many of the encyclopedia entries are updated and revised from the basic Taylor's encyclopedia. Sections 2 and 3 should be used together.

Completing the book is an index that contains the common and scientific names of the plants, with both plates and text indexed.

Beginners, as well as experienced gardeners, will find this a very useful encyclopedia. Public and college libraries will find it a useful reference source for information on house plants.

LILLIAN R. MESNER

30. Taylor's Guide to Shrubs: A Complete Guide to Gardening with Shrubs, Fully Illustrated with 486 Color Photographs. Boston, MA:

Houghton Mifflin Co., 1987. 479p., illus. (part col.), glossary, index. $14.95pbk. 0-395-43093-3pbk.

Based on the fifth edition of *Taylor's Encyclopedia of Gardening*, this encyclopedia is arranged in four sections, with each giving basic information. Section 1 goes over basic botany with explanations of botanical terms, shrub shapes, how to get started, planting, pruning, propagation, keeping records, and a calendar of maintenance activities. Section 2 has 486 colored plates in four groups: flowering shrubs, colorful foliage shrubs, fruiting shrubs, and green foliage shrubs. Section 3 is an encyclopedia with entries alphabetically arranged by genus. Section 4 contains the appendices with shrub charts, pests and diseases, a list of nurseries, and a glossary.

The color plates and text in sections 2 and 3 are the heart of the encyclopedia. The photographs are, for the most part, excellent. Some, however, do not show the small plant structures. The text entries correct this problem through line drawings of the flowers and smaller structures. The captions accompanying the plates contain essential information such as how tall the shrubs grow, the kind of soil needed, and when flowering shrubs bloom. Then a page notation sends the reader to the encyclopedia entry which enlarges on all of the above elements and also indicates the family of the shrub. Many of the encyclopedia entries are updated and revised from the basic Taylor's encyclopedia. Sections 2 and 3 should be used together. The index contains both the common and scientific names of the shrubs.

Beginning horticulturists will find this a very useful guide and public and college libraries will need it for their reference materials pertaining to shrubs. *LILLIAN R. MESNER*

Handbooks

31. A Book of Herbs and Spices: Recipes, Remedies and Lore. Gail Duff.

Topsfield, MA: Salem House Publishers, 1987. 120p., illus. (col.). $16.95. 0-88162-281-8.

Decorative illustrations touched by humor spice up this otherwise bland herbal compendium with a British flavor. Dragons, plump pigs, and medieval gentlefolk share the beribboned pages with recipes for Spiced Beef, Woodruff Yoghurt Ice, and Deep Fried Comfrey Leaves with Parsley Mayonnaise. Along with a thumbnail sketch of the plant, the two-page entry for each of 30 herbs and 17 spices presents history, the method of cultivation, and legend and lore. For example, "the legendary bird called the phoenix was said to collect cinnamon, spikenard and myrrh with which to make the magic fire from which it would be reborn." The Greeks and Romans believed that the more you cursed basil at the time of planting, the better it would grow.

The author suggests culinary, medicinal, and household uses for each herb and spice. For example, savory added to cabbage adds flavor and reduces cooking odors, and woodruff tea eases nervous headaches. "Old herbalists recommended rubbing bee and wasp stings with mint leaves." The author also entertains us with short quotations from earlier authors, such as Nicholas Culpeper, who in 1649 wrote that "Carraway confects...is a most admirable remedy for those that are troubled with wind." Preliminary notes discuss growing and preserving herbs and the general use of herbs and spices in the kitchen and for medicine. The sample recipes, one for each herb and spice, may send one scurrying to find haricot beans, sorrel, blackcurrants, bergamot, candied angelica, duck giblets, and a scale that measures in ounces. The book lacks a bibliography and ignores the almost inevitable demand for sources of the herb plants listed. Gardening advice also appears to reflect the milder British

climate. Previous works by the author include *Food from the Country, Fresh All the Year, Good Healthy Food, Oriental Vegetarian Cooking,* and *A Book of Pot-Pourri.* LINDA JACOBS

32. Chemicals for Crop Improvement and Pest Management. 3rd. M. B. Green; G. S. Hartley; T. F. West. New York, NY: Pergamon Press, 1987. 370p., illus., glossary, index. $53.00, $30.00pbk. 0-08-030250-5, 0-08-030249-1pbk.

This third edition was published in 1987 and represents a significant contribution to the literature. The current title reflects the state of the art of chemical pest control, and it supersedes the previous two editions. The first edition, completed in 1969, was entitled *Chemicals for Pest Control*; the second, published in 1977, was entitled *Chemicals for Crop Protection and Pest Control.* The current title reflects the ongoing improvements in and evolution of this text.

To quote the authors, "...Times change, and chemicals are being increasingly used not just to protect crops from competition from other plant species and from infection by plant pathogens but also systemically to cure such infections after they have occurred and to increase the yields and qualities of those parts of the plants that constitute the crop and make them easier to harvest....The practice of pest control in which the aim was to kill as many offending species [as possible] has been replaced by the concept of pest management in which the infestation is kept merely to the economic threshold of crop damage and thus an ecological balance of pests, predators, and parasites is maintained."

Although the authors are from Commonwealth nations, the text is global in its intent, and most of the various chemicals/treatments can be transferred to the United States. The information is presented in a surprisingly interesting writing style (an exception among scientific texts); and the authors clearly state in the preface the intent, scope, and limitations of this particular text. At various points, the authors remind the reader that this book is not "a practical handbook of crop protection and pest control," that it is not a pesticide applicator's manual, that it does not cover the details of machinery design or chemical application, and that it is not "a general text on control of agricultural pests." What it is about is chemicals.

The book is divided into 26 chapters, with a preface, glossary, general index, and index of active chemicals. The first six chapters introduce the detailed information to follow. The initial two chapters, "Toxic Chemicals in Nature" and "Pesticides in Agriculture," provide an excellent overview and summary/rationale for the judicious use of crop protection chemicals. The next three chapters—"The Shape of the Industry," "Technological Economics of Pesticides," and "Pesticides and Energy"—relate the production and usage of various chemicals in America and Britain and list the energy inputs for producing and using the chemicals. Chapter 6 gives a clear summary of the types of pesticides and their method of transport within plants. Chapters 7 through 20 cover specific types of chemicals, situations which call for particular applications, and desired outcomes (such as plant growth modification). Specific types of chemicals covered include oils, synthetic and organic insecticides, heavy metals, and fungicides. Each chemical description is accompanied by tables, ring structures where appropriate, and suggestions for further reading. The remaining chapters discuss the application of pesticides, formulation, fumigants, pest resistance, pesticide safety, and the future of pesticides (here again, remarkably good writing and an apt summary of the social, economic, and biological questions affecting our food supply).

The glossary is only four pages long, but it is thorough and easily understood. The 11-page index of active chemicals and a list of properties, one of the most useful features of this text, delineates the melting points, saturation concentrations, water solubility, and lethal doses of over 500 different chemicals.

This is a book of valuable and not-to-be-overlooked information. It should be in the general stacks of every academic or special library with collections that cover either agriculture or the life sciences or chemistry. BILL COONS

33. Flowers for All Seasons: A Guide to Colorful Trees, Shrubs and Vines. Jeff Cox; Marilyn Cox. Emmaus, PA: Rodale Press, 1987. 312p., illus. (part col.), index. $24.95. 0-87857-726-2.

This book concentrates on woody perennial flowering trees, shrubs, and vines that may be evergreen or deciduous but, unlike herbaceous perennials that die back to the roots each year, keep their wood stems, branches, and trunks alive all year. Five chapters guide the reader through the creation of a flowering garden for any size yard and for any taste. The first chapter, "Creating Combinations of Flowering Plants," shows how the plants can be arranged for the best effect. Chapter 2, "Flowers Month by Month," describes the flowers according to the month in which they bloom, and chapter 3, "A Color Tour of Flowers Through the Seasons," is a color-photographic atlas of the flowers grouped by season. Chapter 4, "Guide to Flowering Trees, Shrubs, and Vines," is a detailed listing of the flowering trees, shrubs, and vines arranged in alphabetical order by the scientific name giving pronunciation, common name, zone where grown, plant type, mature size, shape of mature plant, culture, flowering aspects, cultivar or specialty of the plant, and special comments. Finally, chapter 5, "Maintaining the Garden's Beauty," is a brief discussion on pruning. There are three appendices: "Common Name Index"; "Mail-Order Sources for Ornamental Trees, Shrubs, and Vines"; and "Hardiness Zone Map."

This is an excellent book, easy to read, and full of useful information on flowering trees, shrubs, and vines. No homeowner with an interest in having a beautiful yard should be without it. Libraries will need it for its detailed listing of the trees, shrubs, and vines commonly used in the United States. H. ROBERT MALINOWSKY

34. Grain Drying, Handling and Storage Handbook. 2nd. Ames, IA: Midwest Plan Service, 1988. 1v. various paging, illus., bibliog., index. $5.00pbk. "MWPS-13." 0-89373-071-8pbk.

This is a revised edition of *Planning Grain-Feed Handling for Livestock and Cash-Grain Farms.* "This book can help you make good decisions while planning farm grain handling, drying, and storage systems. Planning involves selecting specific equipment and their capacities, organizing them into a system, and properly locating the system." An initial chapter covers the guidelines in planning the drying, handling, and storing of grains. This is followed with a section each on handling, drying, and storage. All aspects are covered with examples and excellent illustrations. The final section covers locating and developing a grain center.

Any farmer in the grain business will find this a useful handbook, as will students studying agriculture. The numerous charts and graphs with the excellent text make this a good choice for reference shelves in libraries serving agricultural areas. *H. ROBERT MALINOWSKY*

35. Low Temperature and Solar Grain Drying Handbook. Ames, IA: Midwest Plan Service, 1983 (c1980). 86p., illus., glossary, index. $5.00pbk. "MWPS-22." 0-89373-048-3pbk.

"This book gives the principles of low temperature grain drying and recommends practices for the North Central region. When and how supplemental heat can be useful is presented, with emphasis on applying heat from solar energy." Low-temperature drying includes the use of air flow, fans, and low temperature dryers. Each is discussed in detail and includes numerous charts, diagrams, and examples. The use of solar energy is covered in detail, including the principles of collecting, storing, and using solar energy. Plans for several types of collectors are included. Energy-efficient farms will find this a very useful handbook. *H. ROBERT MALINOWSKY*

36. Native Trees, Shrubs, and Vines for Urban and Rural America: A Planting Design Manual for Environmental Designers. Gary L. Hightshoe. New York, NY: Van Nostrand Reinhold Co., 1988. 819p., illus., bibliog. $79.95. 0-422-23274-8.

There was a time when a trip to the nursery provided mainly imported ornamental plants. Although this is often still the case, there is a trend toward using native plants to landscape everything from homes to highways. Most guides to landscape plants deal mainly with the exotics. Hightshoe's *Native Trees, Shrubs, and Vines for Urban and Rural America* is a welcome addition to the literature, offering much needed information on native species. Professor Hightshoe, a member of the landscape architecture department, Iowa State University, knows the advantages of using native plantings, their hardiness and ease of care. To aid in the selection of these materials, he has divided the book into two parts: Part 1 covers trees, and part 2 covers shrubs and vines. Each part has an "Elimination Key" followed by a "Master Plates" section. The "Elimination Key" helps one narrow selections based on specific design criteria such as bark color, branching texture, or flood tolerance. Each criteria lists suitable plantings by both scientific and common name. The next step is to check selections in the "Master Plates" section. The plates are arranged alphabetically by scientific name and include primary characteristics such as form of fruit, secondary characteristics such as size or bark, and specific design data such as large canopy or dense mass. A range map shows the geographic distribution by county. (Note: "The region of application and use of this manual includes southeastern and south central Canada; north central, northeastern, central and east central U.S.") Drawings depict the overall image of the plant along with important details such as fruits or flowers. The volume covers over 700 species and will prove quite useful to any landscape architecture collection. *ANGELA MARIE THOR*

Manuals

37. The European Garden Flora: A Manual for the Identification of Plants Cultivated in Europe, Both Out-of-Doors and Under Glass, Volume I: Pteridophyta, Gymnospermae, Angiospermae—Monocotyledons (Part I). J. Cullen; P. S. Green; J. Lewis; V. A. Matthews; D. A. Webb; P. F. Yeo; J. C. M. Alexander. New York, NY: Cambridge University Press, 1986. 430p., illus., bibliog., glossary, index. $99.50. 0-521-24859-0.

This is volume I of a two-volume set that is intended to serve as an identification manual for cultivated plants used in gardening or landscaping in Europe. It states that it has been compiled to meet the needs of the "informed amateur gardener" as well as the professional plant taxonomist. The manual is arranged in taxonomic order under the Latin names (useful for "floras such as this which cover an area in which several languages are spoken"). Full-length descriptions are given for families, genera, and species. Brief information is given on cultural requirements and methods of propagation. There are references to books, articles, and registration lists for each family and genus. Citations to illustrations (up to four per species) are also given. Additional information includes geographic distribution of the plant "in the wild," the hardiness code—"a tentative indication of the lowest temperature a species can withstand"—and flowering time.

Volume II was published two years prior to volume I. Both have the same informative introduction and useful glossary, and there is an index in each volume. One difference is the map

in the introduction showing the mean January isotherms for Europe and hardiness codes has been updated for volume II. The map in volume I has been corrected from volume II (1984).

This work, although limited geographically in scope, does provide a "scientifically accurate and up-to-date" source of information on "plants cultivated for amenity in Europe." Recommended for university and specialized collections.

J. L. PFANDER

38. The European Garden Flora: A Manual for the Identification of Plants Cultivated in Europe, Both Out-of-Doors and Under Glass, Volume II: Monocotyledons (Part II). S. M. Walters; A. Brady; C. D. Brickell; J. Cullen; P. S. Green; J. Lewis; V. A. Matthews; D. A. Webb; P. F. Yeo; J. C. M. Alexander, eds. New York, NY: Cambridge University Press, 1984. 318p., illus., bibliog., glossary, index. $65.00. 0-521-25864-2.

See entry for volume I of this title.

39. How to Grow Vegetables Organically. Jeff Cox; Organic Gardening Magazine. Emmaus, PA: Rodale Press, 1988. 312p., illus., bibliog., index. $21.95. 0-87857-683-5.

This is an informative, one-volume compendium for the home gardener. The first half of the book consists of general chapters on planning, soil preparation, planting, care, and container gardening. The section on garden planning is excellent, with different ways to organize the garden presented in a clear and easy-to-understand format. Soil improvement is the lesson of the soil chapter with the emphasis on composting and natural fertilizers. The chapter on planting covers starting seeds indoors and subsequent care of seedlings, direct seeding into the garden, succession planting, and interplanting. Watering, fertilizing, mulching, weeding and cultivation, insects, and diseases are dealt with in the chapter on care. There are charts on common insect pests and diseases. The chapter on container gardening gives instructions on growth media to use, container sizes, light and temperature requirements, planting, watering, and fertilizing, plus suitable varieties.

The second half of the book is an encyclopedia of about 40 commonly grown vegetables with entries arranged by the name of the vegetable. Each entry contains brief sections on soil preparation, how much to plant, how to plant, handling seedlings, growing guidelines, keys to top yields, harvesting and storing, pests and diseases, nutritional value, and choosing varieties. Entries average four to five pages with one of those pages often containing line drawings of the various vegetables being discussed. There is also a list of 40 sources of seeds which can be used as a beginners' guide for obtaining catalogs. A 15-item bibliography (eight of them published by Rodale) introduces the gardener to a very select list of other gardening books. This bibliography should be supplemented with the books listed under the "organic gardening" heading in the subject section of *Books in Print* to obtain a more balanced view. The index is complete and easy to use with entries to information in the encyclopedia as well as in general chapters.

Although this book is intended mainly for the beginner, its charts and encyclopedia can serve as a basic reference source for the seasoned gardener as well. The authority of the Rodale Press in this area assures one of the accuracy and usefulness of the information presented. This is an attractive book. The print is very readable, the illustrations pleasing, and the paper heavy with a feeling of substance. The placement of the drawings and charts plus the use of different type sizes for emphasis almost makes up for the lack of color.

JANET K. CHISMAN

40. Methods for the Diagnosis of Bacterial Diseases of Plants. R. A. Lelliott; D. E. Stead. Oxford, Great Britain: Blackwell Scientific Publications for the British Society for Plant Pathology, 1987. 216p., illus., bibliog., index. $32.50. Methods in Plant Pathology, v. 2. 0-632-01233-1.

This manual contains clear descriptions and methodologies for persons in agricultural extension services who need to identify bacterial plant diseases for farmers and horticulturists. It could be used as a laboratory text for university students because the material is well laid out, from general descriptions of bacterial characteristics to specific pathogen details. The reference value is in the guide to the collection and preservation of cultures including diagnosis and technique. It is written in a very readable style, but more diagrams are needed for the novice. *ANNE PIERCE*

Textbooks

41. Agricultural Insect Pests of Temperate Regions and Their Control. Dennis S. Hill. New York, NY: Cambridge University Press, 1987. 659p., illus., bibliog., glossary, index. $95.00. 0-521-24013-1.

Those who garden know the damage insect pests can cause. Problems, however, come with identifying the pest and choosing the proper treatment. As consultant in entomology and plant protection in the U.K., Dennis Hill is more than familiar with the difficulties involved with pest infestations. He saw a need for a textbook that could be used by students in areas such as applied entomology and crop protection. What started as a single textbook for agricultural entomology in the tropics has become two books, this one and one covering the U.K., Europe, and North America.

Following the introduction, the basics in pest ecology and pest control are covered, discussing various methods from the chemical to the biological. Those pesticides in current use are discussed in general terms in chapter 8 giving properties,

use, and formulations. Rate of application is not given because recommendations vary from country to country. Chapter 9 is arranged by the insects' scientific names and gives common names, host plants, damage description, life history, control, distribution maps, drawings of the insect, and, in some cases, a depiction of the type of damage inflicted. The final chapter contains 73 major temperate crops along with their major and minor pests. The list can be used for initial pest identification, because it does give a brief damage description. From here, one needs to go to the index to find the proper page since the insects are arranged according to *Imms' General Textbook of Entomology* (Richards and Davies, 1977) in taxonomical order. Because the volume is a textbook, the arrangement of information does not provide for easy reference. The index, however, is quite good giving both scientific and common names of all species included. The glossary lists many of the common terms used in applied entomology. Though most of the information can be found in greater detail within other specific sources, *Agricultural Insect Pests* provides a good starting point and is recommended for academic and special libraries.   *ANGELA MARIE THOR*

★
VETERINARY SCIENCE
Handbooks

42. **Bolton's Handbook of Canine and Feline Electrocardiography.** 2nd. N. Joel Edwards. Philadelphia, PA: W. B. Saunders Co., 1987. 381p., illus., bibliog., glossary. $34.95. 0-7216-1847-2.

This second edition is an updated and expanded version of Bolton's *Handbook of Canine Electrocardiography* (1975). It includes both the canine and feline species. According to the author, this edition has an "increased emphasis on clinical signs, diagnosis, and management." The text is arranged in a workbook format to assist and teach the clinician and student to read and interpret an electrocardiogram (ECG). Abnormal beats and rhythms of the heart (arrhythmias) are identified with many sample ECG illustrations. Clinical signs and methods of management for each arrhythmia are discussed. Also, various changes on an ECG as a result of diseases and drugs are given. One chapter is devoted entirely to examples of normal electrocardiograms. The last and largest chapter is a self-assessment section in which the reader is asked questions to help him or her interpret an ECG. A list of common abbreviations, a glossary of terms used in veterinary cardiology, and a table of cardiac drugs and their approximate dosages are included. After reading this book and proceeding through the programmed text, the reader should have a good practical understanding of the recognition and management of disease conditions as detected by the electrocardiogram. Small animal clinicians and veterinary students will want a copy for ready reference.   *ROBERT G. SABIN*

43. **Current Therapy in Equine Medicine—2.** N. Edward Robinson. Philadelphia, PA: W. B. Saunders Co., 1987. 761p., illus., bibliog., index. $75.00. 0-7216-1491-4.

This is a collection of articles by veterinary and medical experts on the therapy of common medical problems of the horse. Each article contains a description of a disease or condition, the appropriate procedures for diagnosing it, the recommended therapy for treating it, and measures to take for preventing it from happening. The editor states in the preface that this second edition is approximately 80 percent rewritten to reflect more up-to-date information and that each article represents the author's opinion of the best current therapy known for treating a particular disease. New sections have been added on medical problems of the foot, behavioral disorders, and the problems of endurance horses. Each article concludes with supplemental readings. The appendix gives normal clinical pathology data and a table of common drugs and their approximate doses. This is an essential volume for veterinary clinicians and veterinary students engaged in the practice or study of equine medicine.
          ROBERT G. SABIN

44. **Diseases of the Cat: Medicine and Surgery, Volume 1.** Jean Holzworth. Philadelphia, PA: W. B. Saunders Co., 1987. 971p., illus. (part col.), bibliog., index. $75.00. 0-7216-4763-4.

This is a massive volume devoted exclusively to the diseases and surgery of the cat. Written by a group of recognized experts in their respective fields, chapters cover such topics as nutrition and nutritional disorders, sedation and anesthesia, soft tissue surgery, orthopedic surgery, basic and clinical immunology, viral diseases, bacterial diseases, mycotic diseases, protozoan diseases, and metabolic diseases. An opening chapter on the sick cat discusses such general aspects as keeping medical records, recognizing signs of illness, conducting the examination, performing diagnostic procedures, and handling and housing the sick cat. The anatomy and physiology of the cat are discussed in chapters dealing with the cardiovascular and hematopoietic systems. Also, the eyes, ears, and skin of the cat are covered in separate chapters. Latest developments in various fields such as feline leukemia and methods of detecting cardiovascular diseases are included. Lengthy bibliographies follow each chapter, and many excellent black-and-white photographs accompany the text. Overall, this is an outstanding text that will long be considered the definitive work in the field of feline medicine.   *ROBERT G. SABIN*

Manuals

45. **Barron's First Aid for Your Dog.** Fredric L. Frye. Hauppauge, NY: Barron's Educational Series, Inc., 1987. 19p., illus. (col.). $9.95. 0-8120-5828-3.

Imagine you are home on a quiet afternoon. You are relaxing, your dog, Arf, is playing with his toys. Suddenly, you notice Arf coughing, hacking, pawing at his mouth. What do you do? Well, if you had a copy of *Barron's First Aid for Your Dog*, you could quickly find out what is wrong and how to treat it. The tabbed index allows rapid reference to the section you need. Each topic provides information on the problem signs and the action you need to take. Color drawings provide visual aid for symptom diagnosis as well as for treatment procedures. Dr. Frye, professor at the School of Veterinary Medicine, University of California, Davis, has created a concise, easy-to-use emergency manual for dog owners, and a very useful quick reference source for school and public libraries.

ANGELA MARIE THOR

46. The Equine Veterinary Manual.
Tony Pavord; Rod Fisher. New York, NY: Howell Book House, Inc., 1987. 172p., illus., index. $19.95. 0-87605-863-2.

The Equine Veterinary Manual is an ambitious title for this book. It conjures up a picture of a precise reference document covering veterinary concerns with the horse. The book is, however, a general text for a layperson who probably has had no experience with a horse. The only sense in which this could be considered a manual is that it covers a number of areas instead of developing one area about the horse. Its 172 pages go over 14 different areas including 10 anatomical and physiological systems plus sections on "the healthy horse," "you and your vet," "veterinary examinations," and "current and future techniques." The extent of coverage for these areas is very brief and rather superficial. For example, the paragraph on respiration under "the healthy horse" is less than 180 words long; the anatomy and physiology of the eye is covered in less than 1,000 words. Each section on a system covers a general explanation of the structure and function of the system, what conditions can be associated with the system, and what preventative or therapeutic care is necessary concerning that system. The book suffers from poor photography, or perhaps the reproduction was poor, resulting in information lost from the pictures. One picture of an ear is so dark that the condition being demonstrated (a case of plaque) is barely visible. Because the contents do not indicate what is included in the sections, a good index would help in locating the needed information. Unfortunately, the index is rather brief and leaves out some general terms. The terms "limp," "swelling," "labour" (this is a British book with British spellings), "lice," "lactation," and "milk" were not found. The section on poisoning by plants listed a number of plants, but not all of them were indexed. It is a bit difficult to pinpoint the audience for this book, but a teenager would have

no difficulty with it because it is not scientific. It would be a good addition for a school or public library as a circulating general reference book.

LILLIAN R. MESNER

Textbooks

47. Fundamental Techniques in Veterinary Surgery. 3rd. Charles D. Knecht; Algernon R. Allen; David J. Williams; Jerry H. Johnson. Philadelphia, PA: W. B. Saunders Co., 1987. 349p., illus., bibliog., index. $34.95. 0-7216-1397-7.

The art of surgery may take a great deal of time and experience to develop, but students must begin somewhere. The third edition of Knecht's *Fundamental Techniques in Veterinary Surgery* proves once again that it is a worthy beginning point for the veterinary student with its presentation of fundamental surgical techniques for both small and large animals This book covers all aspects of veterinary surgery from instrumentation to wound dressing to resuscitation procedures. New material in the third edition includes a chapter on surgical laboratory procedures which contains alternative techniques for common small animal surgical procedures. Additionally this work includes chapters on instrumentation, suture materials and patterns, operating room conduct, wound dressing for small animals and horses, application of techniques, casts and splints for small animals and horses, and cardiopulmonary resuscitation. Most chapters include references along with text, drawings, and photographs.

The illustrations are one of this book's major strengths; both photographs and drawings are used to show application of various surgery techniques. A few photographs include artistic enhancements, and the majority are clear, well labelled and captioned, and arranged in a step-by-step format for following the various steps used in the techniques. Another strength of *Fundamental Techniques in Veterinary Surgery* is that it shows techniques for both small and large animals. Small-animal techniques are illustrated on either cats or dogs, whereas the large animal techniques are always performed on the horse.

One feature that users may appreciate is that, for all specific brand-name materials mentioned in the text, the authors include the name and location of the particular manufacturer. The authors have attempted to limit their materials to those that are currently available, but the chapter on casts and splints for small animals does include the application of yucca board splints, even though this product is nearly absent from the market. The primary users of this book will be veterinary medical students and faculty, but it should not be overlooked for inclusion in the reference collections of larger academic libraries. Users of the earlier editions of this work should not be disappointed with this latest effort.

KIMBERLEY M. GRANATH

ASTRONOMY

★
GENERAL SOURCES
Catalogs

48. A Catalogue of Southern Peculiar Galaxies and Associations, v. 1: Positions and Descriptions, v. 2: Selected Photographs. Halton C. Arp; Barry F. Madore. New York, NY: Cambridge University Press, 1987. 2v., bibliog. $125.00 set, $60.00 v. 1, $75.00 v. 2. 0-521-12579-X set, 0-521-33086-6 v. 1, 0-521-33087-4 v. 2.

In 1973 Halton C. Arp, compiler of the *Atlas of Peculiar Galaxies* for the northern hemisphere, was invited to do a similar survey for the southern skies by the planner and builder of a new United Kingdom Schmidt telescope in Australia. Because of recent advances in telescope technology which made possible fainter-surface-brightness detection and side-field registration, Arp knew it would be possible to penetrate to fainter magnitudes than had previously been possible in either hemisphere. He accepted the challenge, beginning the project which was to involve many astronomers and photographers and result 14 years later in this two-volume set.

The purpose of this catalog is to "highlight the most spectacular phases of galactic evolution, and perhaps formation," in the southern hemisphere, and also to "serve the practical needs of optical astronomers." The galaxies chosen for inclusion were identified on Schmidt telescope survey plates (down to slightly less than 0.2mm) as departing from the norm. Twenty-four categories were later developed to group the galaxies by such "peculiarities" as galaxies with interacting companions, interacting doubles, triples, quartets, and quintets, ring galaxies, three-armed and multiple-armed spirals, dwarf galaxies, and so on. This highly technical and detailed galaxy catalog contains tabular descriptions and positions of approximately 7,000 objects, with accompanying photographs in volume 2 of about one-sixth of these. The list of galaxies in volume 1 is organized by alphanumeric catalog name and includes the relative data such as right ascension, annual precession, and galactic and supergalactic latitude and longitude, as well as a brief description, numerical coding of the description, and cross-identification to other catalogs. The care with which the data was gathered, double-checked, rechecked, analyzed, and photographed leads one to view the *Catalogue* as a highly reliable, accurate, and valuable reference for researchers and advanced students in the field of astronomy. *MARILYN VON SEGGERN*

49. The Classification of Stars. Carlos Jaschek; Mercedes Jaschek. New York, NY: Cambridge University Press, 1987. 413p., illus., bibliog., index. $79.50. 0-521-26773-0.

Beginning with a brief review of the necessity for understanding historical origins of taxonomic systems in observational astronomy, this volume proceeds to analyze and compare in detail the methods of spectroscopic and photometric classification. The first six chapters explore these areas, and the second section treats each group of established stellar types. Specific classes of stars covered include O, B, A, F, G, K, and M, with material on white dwarfs and ZZ Ceti stars listed under "degenerates." Frequent reference is made throughout the text to the SIMBAD (Set of Identifications, Measurements and Bibliography for Astronomical Data) database of the Centre De Données Stellaires at the Observatoire De Strasbourg, where both authors are staff members.

Each individual section is followed by a listing of all cited references, which are keyed to a group of abbreviations provided in the introduction. Subjects are presented in a lucid manner, with illustrations, tables, and spectral diagrams clearly labelled and identified. Contemporary satellite data, such as that from the IRAS infrared station, is incorporated effectively.

Although suitable for a large public library with a well-developed science and technology collection, this work is most clearly applicable to university research libraries and private institutes with extensive holdings in astronomy, spectroscopy, and the philosophies of science. Researchers should be aware that the authors have published and edited extensively in the field of stellar spectroscopy. *ROBERT B. MARKS RIDINGER*

Dictionaries

50. The Facts on File Dictionary of Astronomy. 2nd. Valerie Illingworth, ed. New York, NY: Facts on File Publications, 1987 (c1985). 437p., illus. $19.95, $12.95pbk. 0-8160-1357-8, 0-8160-1892-8pbk.

This title was first issued in 1979, a revised second edition was published five years later, and it is the paperback version of this second edition which is considered here. The book contains over 2,300 entries, along with more than 85 diagrams and illustrations and 12 tables as appendices. All in all, this dictionary gathers into one convenient spot most of the astronomical data and terminology a nonspecialist is likely to require. The definitions are clear and usually succinct, ranging from a line or two up to a page or so in length for the more complex or important topics.

The definitions are quite good and contain current information, although some will require more than a passing knowledge of mathematics to understand fully. However, this level of sophistication makes the work suitable for ready reference by professional astronomers, as well as the educated layperson. Related terms are clearly marked for cross-reference to their own definitions. With a good typeface and layout, this dictionary remains a premiere acquisition for both individual astronomers and libraries not wishing to acquire more detailed and expensive reference works such as the *McGraw-Hill Encyclopedia of Astronomy*, one of this dictionary's few real competitors.   *FRED O'BRYANT*

Field Books/Guides

51. Leslie Peltier's Guide to the Stars. Leslie C. Peltier. New York, NY: Cambridge University Press, 1986. 185p., illus. $11.95pbk. 0-521-33595-7pbk.

Guide to the Stars is an inviting introduction to astronomical observation using only binoculars and the unaided eye. Here the author, the late Leslie Peltier, one of America's most accomplished amateur astronomers, continues his mission of recruiting people to the joys of stargazing. Written as a trail guide, the work is usefully organized around seasonal observing with individual chapters on the solar system bodies, novae and nova hunting, comets, and meteors. Special insights are provided in chapters on observing the lunar surface and on variable stars, the author's expertise. The locating of celestial objects through basic knowledge of the constellations is emphasized, with a variety of small star maps supporting the text. Suiting the intended audience, the chapters are short and content basic. The result is appealing though, as Peltier weaves anecdote, analogy, legend, history, and poetry with the personal tips and knowledge of an experienced observer. Quality of the illustrations varies. A recommended addition to school and public libraries.   *JOHN T. BUTLER*

Histories

52. History of the Royal Astronomical Society, 1820–1920 J. L. E. Dreyer; H. H. Turner, eds. Boston, MA: Blackwell Scientific Publications reprinted for the Royal Astronomical Society, 1987 (c1923). 258p., illus., index. $78.95, $32.95pbk. 0-632-02173-X, 0-632-02175-6pbk.

Originally issued in 1923, this reprinted history was planned as a collection of chapters written by the fellows of the society, with each being responsible for one decade of the first century's activities. Because several of the group did not complete their assignments, others of the society were induced to replace them. Twelve portraits of famous members of the Royal Astronomical Society, such as John Herschel, illustrate a text that, although technical, is well written and readable. Appendices list officers of the society for its first century and recipients of the society's medals. An author and subject index is provided.

The volume should be consulted with its companion covering the period 1920–1980 for a comprehensive picture of the Royal Astronomical Society's involvement in both practical instrumentation and the history of the discipline of astronomy. Suitable for college and university libraries and for research institutions with collections emphasizing the history of science.   *ROBERT B. MARKS RIDINGER*

53. History of the Royal Astronomical Society, Volume 2, 1920–1980. R. J. Tayler, ed. Boston, MA: Blackwell Scientific Publications for the Royal Astronomical Society, 1987. 262p., illus., bibliog., index. $70.00, $35.00pbk. 0-632-01791-0, 0-632-01792-9pbk.

Intended as a complementary volume to the *History of the Royal Astronomical Society, 1820–1920*, this volume begins with the Annual General Meeting in 1920. As with its predecessor, each chapter was written by a fellow of the society, although "the chapters were written independently and no serious attempt has been made to provide cross-referencing between them." Despite this, the text has a depth of detail clearly presented.

Of interest is the shift in the views of the universe and observational instrumentation and techniques between the beginning of the twentieth century and the 1980s. Appendices include listings of the officers and senior members of the society, together with bibliographic data on their publications; recipients of the society's various medals and awards; those individuals who presented lectures in behalf of the society, together with references to the published texts of these addresses; and statistical data on the organization. This volume is suitable for college and university library collections and for research institutions with an interest in the philosophy and history of astronomy and scientific investigation.   *ROBERT B. MARKS RIDINGER*

Tables

54. A Supplement to the Tuckerman Tables. Michael A. Houlden; F. Richard Stephenson. Philadelphia, PA: American

Philosophical Society, 1986. 564p., bibliog., tables. $35.00. American Philosophical Society Memoirs Series, v. 170. 0-87169-170-1.

This work, issued as *Memoirs of the American Philosophical Society*, volume 170, supplements two of the society's earlier volumes. The *Tuckerman Tables: Planetary, Lunar, and Solar Positions at Five-Day and Ten-Day Intervals* were published in volume 56, 1962 (data for 601 B.C. to A.D. 1) and volume 59, 1964 (data for A.D. 2 to A.D. 1649). These volumes contain tabulations of the positions of the five bright planets (Mercury, Venus, Mars, Jupiter, and Saturn), the Moon, and the Sun. They are used for the dating of ancient and medieval astronomical observations, and for determining the accuracy of early measurements and calculations.

Finding errors in Tuckerman's tabular positions of the planet Mars, the authors produced this supplementary volume "to make available revised positions for Mars throughout the period 601 B.C. to A.D. 1649 and to enable the apparent magnitude of each planet to be estimated at any time during this period." In a four-page introduction, the authors discuss the *Tuckerman Tables* and the content and use of their supplement. There is a short bibliography at the end of the introduction which contains references to other tables and related papers. Following the introduction are 564, well-produced, easy-to-read, columnar pages with information for each of the years covered: (1) the eliptical longitude, latitude, and apparent magnitude of Mars; (2) the longitude of the sun; (3) the Julian Calendar date; (4) the apparent magnitudes of Mercury, Venus, Jupiter, and Saturn. The apparent magnitudes of the bright planets is unique to this supplement. Anyone who has the *Tuckerman Tables* would want to acquire this work. It is of interest to academic libraries with archaeoastronomy or history of science collections. *SUE ANN M. JOHNSON*

BIOLOGY

★
GENERAL SOURCES
Abstracts

55. Ecological Abstracts. Norwich, Great Britain: Geo Abstracts, Ltd., 1974-. v. 1- , index. Bimonthly. £155.00 per year. 0305-196X.

Geo Abstracts publishes this abstract journal six times a year, arranging the abstracts by broad subject areas. The selections encompass global, marine, tidal, estuarine, freshwater, and terrestrial ecology. They also include applied ecology—pests and pollution, conservation, and agricultural, forestry, and fisheries ecology. Also represented are evolution, palaeoecology, and historical ecology. This journal is similar in scope to *Ecology Abstracts*, produced by Cambridge Scientific Abstracts, a monthly abstracting journal. The editors do not clearly define their reasons for inclusion or exclusion of items. Selections come from such journals as *Oikos, Canadian Journal of Zoology, Ecology, Journal of Ecology, Annales Botanici Fennici, Canadian Journal of Fisheries and Aquatic Science* and *Environmental Monitoring and Assessment*. In addition to journal articles, citations may refer to books, monographs, conference proceedings, or reports.

Each issue contains approximately 1,300 abstracts, over 90 percent of which cite articles written in English with abstracts written by their authors. Each issue contains a species index, but for additional access points, one must wait for the annual index. This index is similar to a keyword-in-context index except the indexed word is located in the conventional left-hand position. Each item is indexed an average of seven times by subject. In addition, a regional index refers to specific items by abstract number. In comparison, Cambridge's *Ecology Abstracts* provides 1,000 abstracts each month with author and subject indexes. A time lag of over six months occurs between journal publication and the appearance of the abstract. Geo Abstracts also publishes *Geographical Abstracts* and *International Development Abstracts* and also offers its abstracts in an online database, GEOBASE, available through DIALOG. GEOBASE dates from 1980 and covers selections from the world literature on geography, geology, ecology, and related disciplines. Recommended for all libraries with a strong research in ecology. *LINDA JACOBS*

Bibliographies

56. Discovering Nature with Young People: An Annotated Bibliography and Selection Guide. Carolyn M. Johnson, comp. Westport, CT: Greenwood Press, 1987. 495p., index. $49.95. 0-313-23823-5.

The compiler of *Discovering Nature with Young People*, Carolyn M. Johnson, is a librarian and writer. She states in the preface that the items in this bibliography are primarily for the use of children from ages eight to 16, and for those who work with these young people, such as teachers, librarians, youth-group workers, and parents. The book should be of special interest to science teachers whose students are in this age range. All of the items listed are intended to help children or young adults relate to and better understand nature through the use of written and audiovisual materials.

The book's contents are grouped into five major parts: written works for young people, audiovisual items for young people, items for educators and parents, selection sources, and activity supplies and aids. Included are fiction and nonfiction works on animals, plants, earth science, astronomy, wilderness areas and national parks, biographies of persons with backgrounds in these fields, and works relating to the enjoyment and appreciation of nature. There are over 2,400 numbered items, followed by several hundred additional unnumbered ones. The majority of the items are annotated, and, for many, the precise age range of intended audience is indicated. Johnson includes a wide variety of formats: books, magazines and magazine articles, films, videotapes, records, cassettes, computer programs, games, kits, posters, puzzles, and miscellaneous supplies. The appendices mention many additional items, and also contain a list of publishers, producers and distributors. There are four separate indexes: author, title, subject, and media type. This is a recommended book for school and public libraries and highly recommended for schoolteachers. *WILLIAM H. WIESE*

Dictionaries

57. Dictionary of Microbiology and Molecular Biology. 2nd. Paul Singleton; Diana Sainsbury. New York, NY: John Wiley and Sons, 1987. 1,019p., bibliog.

$149.00. A Wiley Interscience Publication. 0-471-91114-3.

Dictionary of Microbiology and Molecular Biology is a revised edition of *Dictionary of Microbiology* published in 1978. In compiling the second edition, the authors have tried to provide the most current definition available reflected in the actual usage in current journals and texts. They have also, where appropriate, given former meanings, alternative meanings, and synonyms. The scope has been expanded from the first edition to include the classical aspects of microbiology to the current developments in related areas of bioenergetics, biochemistry, and molecular biology. The terms are arranged in alphabetical order, with acronyms treated as words. Various other practices are used for hyphenated words, Greek prefixes, numerals, subscripts, and the like. These practices are explained in detail in the "Notes for the User." Cross-references are abundant. Many of the definitions include external references to papers, articles in books, or journal articles. These are numbered and keyed to a list of references at the end of the volume. If a term has more than one meaning in different contexts, the definitions are numbered; however, the order of the definition does not reflect in any way the appropriateness or frequency of usage.

There are five appendices that cover molecular diagrams for pathways, cycles, fermentations, and biosynthesis. A key to the journal abbreviations and the book references are included at the end. This is an outstanding dictionary. Its binding, however, may not hold up to heavy use. The review copy has already split with little use. It does lay flat when open, which is extremely important for someone doing research in reading an article and identifying a term in the dictionary. There are no illustrations to aid in the definitions, but this does not cause any great problems. This is a highly recommended dictionary for academic and special libraries.

H. ROBERT MALINOWSKY

Directories

58. The Biotechnology Directory: Products, Companies, Research and Organizations. 4th. J. Coombs; Y. R. Alston. New York, NY: Stockton Press, 1987. 500p., bibliog., index. $150.00. 0-935859-13-6.

The Biotechnology Directory, currently in its fourth edition, is a listing of a variety of products, services, and organizations, covering all aspects of biotechnology, including genetic engineering, waste treatment and recovery, and a variety of industrial and biomedical applications. The *Directory* is divided into three parts covering "International Organizations and Information Services," "Government Organizations and Professional Societies or Associations," and "Companies, Research Institutes, and University Departments." There are also several indexes at the end of the volume.

Part one has two sections: an alphabetical listing of international organizations, and a listing of information services grouped into three categories. These categories are databases and abstracting services, journals, and newsletters. Each section is followed by a subject classification index. Part two lists national organizations by country, and each national listing separately lists government organizations and professional organizations. This part has no separate indexes. Part three lists companies and research centers by country. The national listings are not broken down into subsections. The indexes cover part three and include (1) alphabetical index of companies and research centers, (2) buyers guide alphabetical index (this is an alphabetical list of products and services primarily for use with the third index), (3) products, research and services buyers' guide (this is a list of companies and organizations by product or service), (4) buyers' guide classification index (this is a classified list of products and services, primarily for use with the third index), and (5) list of advertisers. Most entries include telephone, telex, or fax numbers, and many include a brief description, although the minimum information provided for each entry is the address. This directory, which could be significantly improved by the inclusion of telephone numbers and description for all entries, attempts to collect together all biotechnology organizations and services; information professionals should find the listing of information services particularly useful. It is possible, however, to find much of the same information by using other directories. The *Life Science Organizations and Agencies Directory*, published by Gale Research, duplicates most of the entries, and the *Thomas Register* provides similar access to services and products.

The primary advantage of *The Biotechnology Directory* is that it provides subject-specific indexing by product or service and that its limited subject scope facilitates the process of locating the organization, service, or company of interest. Although the indexes in this directory are very useful, they also have many difficulties. Except for the list of advertisers, none of the indexes gives page numbers. The absence of a comprehensive indexing system covering all parts of this directory also creates problems: There is no index covering part two; there is no immediate indication that the indexes at the end of the volume only cover the listings in part three, and there are no instructions on how to use the buyers' guide alphabetical index. If these difficulties could be corrected, this directory would be much easier to use. These problems notwithstanding, individuals or organizations seeking information on biotechnology products or services will still find this directory to be one of their most valuable tools. University and research libraries will find this a useful directory as will large public libraries.

LOREN D. MENDELSOHN

Exam Reviews

59. Barron's How to Prepare for the Advanced Placement Examination AP Biology. 3rd. Gabrielle Edwards; Marion Cimmino. New York, NY: Barron's Educational Series, Inc., 1987. 435p. $9.95pbk. 0-8120-3875-4pbk.

See review for *How to Prepare for the College Board Acheievement Test Biology: Including Modern Biology in Review.*

60. How to Prepare for the College Board Achievement Test Biology: Including Modern Biology in Review. 9th. Maurice Bleifeld. New York, NY: Barron's Educational Series, Inc., 1987. 336p. $8.95pbk. 0-8120-2986-0pbk.

For some students, study and preparation for competency in biology begins early in high school and continues into the highly competitive medical school and scientific research milieus. Two new guides, *Barron's How to Prepare for the Advanced Placement Examination AP Biology* and *How to Prepare for the College Board Achievement Test Biology: Including Modern Biology in Review* (also known as *Barron's How to Prepare for the College Board Achievement Test CBAT Biology*), have been published to assist students preparing for either advanced placement and/or college credit (AP) or admissions-related college board examinations (CBAT) in biology.

The AP guide, in its third edition, was prepared by Gabrielle I. Edwards and Marion Cimmino, both of the Franklin D. Roosevelt High School, Brooklyn, NY. The CBAT guide, now in its ninth edition, was prepared by Maurice Bleifeld, principal emeritus of the Martin Van Buren High School, Queens Village, NY. Central features of both guides are sample or model tests and recommended test-preparation and test-taking strategies. The CBAT is administered to high school students in anticipation of college attendance. Students applying to college may need to take one or more college board achievement examinations, of which biology is one. Students who have completed advanced placement college-level course work while still in high school may elect to take the AP exam, and high scorers may be exempted from first-year college biology and/or may receive college credit. Similar in format, style, and tone, both guides are highly recommended to the student audiences they serve and to public and high school library reference collections. *GERALD J. PERRY*

Handbooks

61. ASTM Standards on Materials and Environmental Microbiology. ASTM Subcommittee E35.15 on Antimicrobial Agents. Roberta A. Storer, ed. Philadelphia, PA: American Society for Testing and Materials, 1987. 307p., illus., bibliog., index. $39.00. "PCN: 03-535087-48." 0-8031-0982-2.

An academic branch library may purchase this reprint volume without duplicating the seven volumes of the *Annual Book of ASTM Standards* from which it was compiled. This would also be a good choice for the appropriate subject collection. The index seems thorough, but one is not able to go directly from it into the volume. Reference had to be made back to the section table of contents to look for the citation number to determine which section contained the correct pages. Each reference may include subsections covering such things as scope, applicable documents (e.g., standards), significance and use, terminology, apparatus, materials, safety-precautions procedures, and interpretation of results. *JAMES L. CRAIG*

Indexes

62. European Handbook of Plant Diseases. I. M. Smith; J. Dunez; D. H. Phillips; R. A. Lelliott; S. A. Archer, eds. Boston, MA: Blackwell Scientific Publications, 1988. 583p., illus. (part col.), bibliog., index. $125.00. 0-632-01222-6.

Aimed at the practicing plant pathologist, the *European Handbook of Plant Diseases* is an outstanding and thorough work that will become a standard reference in the field. Supported by contributions from over 170 authors, the goal of this handbook is "to provide a comprehensive treatment of the diseases of cultivated plants and major forest and amenity trees in Europe."

Sponsored by the British Society for Plant Pathology and the European and Mediterranean Plant Protection Organization, the volume is divided into 17 chapters with pathogen and host indexes covering viral, bacterial, and fungal pathogens. A large number of black-and-white photographs, tables, and 32 color plates accompany the text. Pathogen entries are arranged into chapters by major systematic groups. Genera are in alphabetical order in each chapter, as are species within each genus. Each entry typically includes the following: scientific name, synonyms, and anamorph names; basic description of pathogen; host plants; host specialization; geographical distribution; disease symptoms; epidemiology; economic importance; control; and any special research interests. Each chapter ends with an extensive bibliography covering all of the entries of the chapter.

The pathogen index covers all names and synonyms mentioned in the text with the accepted names printed in bold type. The host index is arranged by scientific name with cross-references from common English names to scientific names. Although many of the plant diseases discussed occur in geographic areas outside of Europe, this handbook will be most useful to libraries with strong interests in the region or comprehensive collections in plant pathology.

STEVEN L. SOWELL

Manuals

63. Biology Projects for Young Scientists. Salvatore Tocci. New York, NY: Franklin Watts, 1987. 127p., illus., bibliog., index. $11.90. Projects for Young Scientists. 0-531-10429-X.

"This book contains ideas for biology projects that can be done either in school or at home." Each chapter begins with some background information on the topic under discussion and is followed with some ideas for projects that are designed to be done without any highly sophisticated equipment. After a chapter on "Where Your Biology Project Can Lead," eight chapters cover "Origin of Living Things," "Chemicals in Living Things," "Photosynthesis and Respiration," "Cell Structure and Function," "Genetics," "Development of Animals and Plants," "Systems in Living Things," and "Populations and Evolution." Topics discussed include water balance in larger organisms, the dark side of photosynthesis, respiration without oxygen, cell division, mapping genes, regeneration, and natural selection. This is a very well-written book with good illustrations that will help students understand some of the complex areas of biology. It is recommended book for school and public libraries.

H. ROBERT MALINOWSKY

64. Gene Cloning and Analysis: A Laboratory Guide. G. J. Boulnois, ed. University of Leicester Gene Cloning Course, comp. Palo Alto, CA: Blackwell Scientific Publications, 1987. 136p., illus., bibliog., index. $32.50. 0-632-01927-1.

Gene Cloning and Analysis: A Laboratory Guide is "Based on the experimental protocols for an intensive, practical course...held in Leicester during September 1985 and July 1986." It is an update of R. H. Pritchard's *Basic Cloning Techniques* (Blackwell, 1985).

Not intended to be comprehensive, the manual touches on key methods "used on a day-to-day basis in the School of Biological Sciences at Leicester." Each of the eight methods described is covered in minimum of three sections: introduction, procedure(s), and references, the last containing three or more citations, which have been mentioned in the introduction or procedure(s) sections. The procedures are detailed. This volume is, therefore, useful in those collections where student users come looking for "experiments on cloning." *JAMES L. CRAIG*

★ BOTANY
Atlases—Science

65. River Plants of Western Europe: The Macrophytic Vegetation of Watercourses of the European Economic Community. S. M. Haslam. New York, NY: Cambridge University Press, 1987. 512p., illus., bibliog. $125.00. 0-521-26427-8.

This is a comprehensive atlas of the macrophytic communities of waterways of western Europe, with maps, charts, tables or diagrams on nearly every page. The introductory chapters present a detailed overview of the aquatic environments of eleven countries. The geological history and the use of rivers from the Paleolithic period to the present are described along with the geographic, geologic, and botanic classification of rivers. This is followed by detailed discussions on the importance of climate in determining species distribution, the patterns of river discharges for each country, and the distribution of water plants in relation to physical factors.

The author presents his system of ranking the trophic status of habitats, the relation of nutrient status of main rock types, the geographic distribution of the aquatic macrophytes of the political area, and the recent changes in river management that impact river vegetation.

CAROLYN DODSON

Dictionaries

66. The Plant-Book: A Portable Dictionary of the Higher Plants, Utilising Cronquist's An Integrated System of Classification of Flowering Plants (1981) and Current Botanical Literature Arranged Largely on the Principles of Editions 1–6 (1896/ 97–1931) of Willis's A Dictionary of the Flowering Plants and Ferns. D. J. Mabberley. New York, NY: Cambridge University Press, 1987. 706p., bibliog. $34.50. 0-521-34060-8.

The Plant-Book, as the subtitle implies, is a dictionary of all generic and family names of vascular plants, as well as English common names. Entries under families provide classification, distribution, botanical details, and main uses. Generic entries include family, number of species, distribution, and English common name as well as appropriate botanical, horticultural, agricultural, and medical aspects. The appendix lists Cronquist's system for the arrangement of angiosperms and also provides a bibliography.

This is the latest edition of that old friend, J. C. Willis's *Dictionary of the Flowering Plants and Ferns*, first published at the turn of the century. The *Dictionary*, a basic component of every botanical library, came out in eight editions. The last was edited by H. K. Airy Shaw in 1966. D. J. Mabberley's *The Plant-Book* is a direct descendent, with a few modifications. The outdated Englerian phylogenetic system is replaced by the modern Cronquistian system. An invaluable addition is the inclusion of references to recent generic monographs. Then, to conserve space, generic synonyms given in the earlier edition are for the most part omitted.

This update of a classic handbook should be on the shelf of every botanical, horticultural, or agricultural library, along with the previous editions which it complements.

CAROLYN DODSON

Encyclopedias

67. Encyclopaedia of Ferns: An Introduction to Ferns, Their Structure, Biology, Economic Importance, Cultivation and Propagation. David L. Jones. Melbourne, Australia: Lothian Publishing Co.. Pty., Ltd., 1987. 433p., illus. (part col.), bibliog., glossary. $44.95. 0-85091-179-6.

This comprehensive guide for the fern cultivator, by an Australian horticultural researcher, covers fern biology and classification, cultural requirements, pests, propagation, specialized fern culture, and descriptions of 700 species of ferns amenable to cultivation. All parts are authoritative and profusely illustrated.

The general sections on economic importance, structure, reproduction, and classification of ferns and fern-allies are extensive enough to serve as a source of information on nonhorticultural aspects of ferns. The sections dedicated to cultivation are excellent sources of practical information for the greenhouse or outdoor gardener. Here one can look up soils, nutrition, fertilizers, pests, diseases, propagation, hybridization, potting, containers, and housing.

The major portion of the work is dedicated to the illustrated list of fern species. Each account includes names, distribution, frond type, description, and growing requirements. Finally, there are appendices listing species according to cultural characteristics, a glossary, and four-page bibliography.

CAROLYN DODSON

68. The Encyclopedia of Cacti. Willy Cullmann; Erich Göetz; Gerhard Gröener. Portland, OR: Timber Press, 1986. 340p., illus. (col.), bibliog., index. $45.00. 0-88192-100-9.

The purpose of *The Encyclopedia of Cacti* (first published as *Kakteen*), according to the authors, is to "present a practical book, easy to read and with up-to-date horticultural and botanical information to suit enthusiasts at all levels." The book fulfills all of its aims in a superior format, beginning with the book jacket and ending with the last end paper. This encyclopedia is well worth the money for readers at all levels. In this revision of Willy Cullman's earlier edition, the text was rewritten within the guidelines of the purpose of the original book. Thus, the 1986 book continues to be readable and informative for both the beginner and advanced cactus admirer.

Erich Göetz of the Botanical Institute of the University of Hohenheim was responsible for the identification keys, cactus systematics, botanical descriptions, and drawings. Dr. Gerhand Gröener was in charge of photography, cultivation, and propagation. The book's greatest assets are the over 400 color photographs and a detailed dictionary of over 750 species and varieties found in the Southwest, U.S., Mexico, Central America, and South America. The first part of the book is a comprehensive description of cactus structure and its adaptation to survival in extremely harsh environments. Individual chapters are devoted to structure, classification, habitat, culture, grafting, collecting, diseases, geographical distribution, and propagation. An identification key of the genera and species of the cacti concentrates on fairly recognizable characteristics. Illustrations accompany the key to help the reader understand the dichotomous selections.

The alphabetical sequence of the genera and species of the cacti comprise almost two-thirds of the volume. All genera are included, and explanation is given for the selection of the names used. For each entry, the genus name and the Latin meaning are listed first. A physical description of the plant is included as well as a color photograph. The number of known species is taken either from Donald's list of 1974 or from Ritter. Native countries and cultural conditions are given. In certain difficult groups such as *Lobvia*, a key is inserted within the alphabetical sequence.

Additionally the encyclopedia contains various helpful sections. The appendices include a glossary of common specialized terms, the authors of cactus names, a list of associations and publications, and a list of plant suppliers. The Latin name index lists all generic and specific names found in the book. The end pages provide maps of the states within the countries which are habitats for the cacti discussed.

This encyclopedia is recommended for purchase. For cactus collectors, this purchase is a must.

MARILYN KAY GAGE

69. Shrubs of the Great Basin: A Natural History. Hugh Mozingo. Reno, NV: University of Nevada Press, 1987. 342p., illus., (part col.), index. Max C. Fleischmann Series in Great Basin Natural History. $27.95, $16.95pbk. 0-87417-111-3, 0-87417-112-1pbk.

Written for the layperson, this encyclopedic treatise covers 67 species of Great Basin shrubs, with details of the life histories, origin of name, familial characteristics, evolution, adaptation to environment, economic uses, and other interesting information. The color photography is outstanding, as are the line drawings of the shrubs that are native to this part of the country. It is a highly readable book for the nonscientist who wants more information than is given in the usual field guides. Although this excellent book is mainly of interest to westerners, it is of great interest to any botanical library where detailed histories and descriptions of plants in the United States are needed.

CAROLYN DODSON

Field Books/Guides

70. **Colorado Flora: Western Slope.** William A. Weber. Boulder, CO: Colorado Associated University Press, 1987. 530p., illus. (part col.), glossary, index. $19.50. 0-87081-167-3.

Colorado Flora: Western Slope is a guide to the flowering plants, ferns, and conifers of Colorado west of the Continental Divide. Professor Weber, a botanist and teacher in Colorado with wide field experience elsewhere and the curator of the University of Colorado Museum, is well qualified to produce such a manual. He is author of five editions of *Rocky Mountain Flora*. Several features, including size and price, make this a useful guide for readers at many levels including amateur vacationers and field botanists. For advanced users, plant families are arranged alphabetically, and there are indexes to common names and genera. Synonyms are included for all generic terms, including those that Weber has created. Because the species are limited to western Colorado, the keys are somewhat simpler than in books with a broader scope, making it useful for the amateur botanist and an important reference guide for Colorado public and school libraries.

RUTH LEWIS

71. **Conifers.** Keith D. Rushforth. New York, NY: Facts on File Publications, 1987. 232p., illus. (part col.), bibliog., glossary, index. $24.95. 0-8160-1735-2.

Organized alphabetically by genus and featuring a comprehensive gazetteer, this book covers taxonomy and physiology of conifers, their uses and propagation, and habitat and pests. The book is well illustrated with black-and-white line drawings and maps. There is a short bibliography, useful glossary, and indexes of botanical and English names and subjects. This is a useful book for college, academic, and special libraries with a botanical collection. Public libraries will find it a useful addition as a source of information on trees. *ANNE PIERCE*

72. **Fall Wildflowers of the Blue Ridge and Great Smoky Mountains.** Oscar W. Gupton; Fred C. Swope. Charlottesville, VA: University Press of Virginia, 1987. 208p., illus. (col.), index. $12.95. 0-8139-1123-0.

This is an identification guide to the fruits of the wildflowers occurring in the Blue Ridge and Great Smoky Mountains. It is color-coded to quickly help identify the plant by fruit color. Within each group, the plants are arranged according to the time of fruiting. Each plant has a full-page color photograph and one page of description, including common name of the species, common name of the family, scientific name of species, and scientific name of the family.

Persons with no training in botany will appreciate this attractive and sturdy book with its pronunciation key to the scientific names of the 224 species and an index to both scientific and common names. The quality of the photography is excellent; the color appears true-to-life; and the camera focus on the particular parts of the plant seems adequate for identification. Plants are described as to general appearance of stem, leaf, floral, and fruit character. The poisonous nature and medicinal value are noted when appropriate, and the habitat is discussed, although no indication is given of where the plant occurs specifically within the Blue Ridge and Great Smoky Mountains. If the plant is not native, however, the country of origin is given. Propagation techniques are suggested for planting in home gardens. This book is recommended for use in the secondary school biology class and for the general public as both a field guide for fall field trips and as an informative and interesting handbook.

MARILYN KAY GAGE

73. **A Guide to Animal Tracking and Behavior.** Donald Stokes; Lillian Stokes. Boston, MA: Little, Brown and Co., 1986. 418p., illus., bibliog. $18.95. Stokes Nature Guides. 0-316-81730-9.

A Guide to Animal Tracking and Behavior is the sixth volume of the *Stokes Nature Guides* and is designed to help the user develop the ability to become wildlife detectives. Secretive mammals are found by learning from clues they leave as presented in Part 1: "Tracks and Signs" and Part 2: "Mammal Lives and Behavior." Ample illustrations help make one a better sleuth. According to the authors, "It is the first guide to picture all tracks and scats at their actual size." Thirty mammals are included in this guide. The main attraction of the book is its ease of use. Both back and front end papers serve as a fast table of contents. Part 1 includes chapters for use in the field. The guide to tracks is divided by tracks with the number of toes, and the tracks are arranged from smallest in size to largest. Animals whose tracks share common characteristics are grouped together in the track learning chapter, with the most commonly occurring tracks appearing first. The guide to scats is arranged in two groups, first the herbivores, then the carnivores and omnivores. Cross-references are provided to locate tracks, track learning, or behavior.

The guide to signs is divided into the following categories: "Injury to Trees or Shrubs"; "Digging, Scraping, and Tunneling in the Ground"; "Constructed Nests, Homes, and Dams"; "Disturbed Vegetation—Trails, Runways, Tunnels, Beds, Rolling Places"; "Natural Cavities"; and "Food Remains and Caches." Part 2 discusses the lifestyles of the 30 mammals. Each section contains a range map of the United States, an illustration of the representative mammal (usually includes the young), scientific name, general characteristics, how and when the animal moves, clues the mammal leaves, feeding styles, and family life. At the conclusion, a quick reference for the animal lists habitat, home range, food, mating, gestation, time of birth, number of young, number of litters, and age when the young become independent. The bibliography is useful for future study in that general texts are listed

first, followed by the bibliographic information for each animal. The citings include information from as late as 1984. The guide is very useful and entertaining, with drawings that enhance the text and provide detail needed for a guide to field trips. It is recommended highly for secondary school and college libraries. Every biology instructor should have a personal copy.

MARILYN KAY GAGE

74. Mushrooms Demystified: A Comprehensive Guide to the Fleshy Fungi. 2nd. David Arora. Berkeley, CA: Ten Speed Press, 1986. 959p., illus. (part col.), bibliog., glossary. $39.95, $24.95pbk. 0-89815-567-8, 0-89815-123-4pbk.

This is a very complete field guide to identification of mushrooms which can be used by the novice as well as the experienced collector. To aid the novice, the author prefaces his extensive key with sections on mushroom definition, environment, classification, collection, identification, and using a key, and he answers a number of questions which a novice collector will probably want and need answered. These sections are well written and informative and do not get bogged down in boring narrative. Although using scientific terms and defining them properly, the book uses illustrations or examples that are nonscientific and often somewhat humorous. Recognizing that, for the novice, these sections are necessary reading, the author has lessened the temptation to skip them to get into the actual field guide section.

To the experienced collector, the first sections are not as important, but because of the author's style, they can be read and enjoyed. One of these areas is the "How to Use the Keys" section, which includes sections on why there might be a discrepancy between the specimen being keyed and the book's description: "The Mushrooms' Fault," "The Keyer's Fault," and "The Key's Fault." Also particularly useful is a section called "Habitats," which indicates what specific mushrooms grow in certain habitats, ranging from specifically treed areas to lawns and pastures to sand to snow. The most important section of the book is the key. Each level of taxonomy contains an introduction to that group with characteristics and hints about that group. Each family includes diagrams of spore structure, which is helpful if you have a microscope at your disposal. After finally keying to genus and species, each contains information on the mushroom's anatomy, habitat, edibility, and additional helpful comments. These sections are liberally augmented with black-and-white photographs. If a color photo is present (at the center of the guide), it is referenced to a plate number next to the name of the mushroom. The color plates, in particular, are excellent, and each refers back to a page in the key. For the novice collector or the serious, experienced collector, this guide is a volume worth adding to his or her library. Even for an individual who is just interested in learning more about the mushrooms, this book is useful. It might even make a mushroom collector out of a couch potato. *ANNE PIERCE*

75. Mushrooms of Idaho and the Pacific Northwest: Volume 2—Non-Gilled Hymenomycetes, Boletes, Chanterelles, Coral Fungi, Polypores and Spine Fungi (Agaricales and Aphyllophorales). Edmond E. Tylutki. Moscow, ID: University of Idaho Press, 1987. 232p., illus. (part col.), bibliog., glossary, index. $14.95. 0-89301-097-9 v. 2.

Designed as a field guide to be used with a specimen in hand, this volume has the standard dichotomous keys to guide the hunter into the descriptions of the fungi. The coverage includes 363 taxa of non-gilled Basidiomycetes, 71 species of Boletaceae, 111 of coral fungi, 14 of Chaterelles, 121 of Polypores, and 49 of Spine fungi. Each section begins with a general description of the group and a list of species, followed by a key to the group, if it is small, or keys to subgroups in the larger groups. Each particular mushroom has a section headed by its common name with the genus and species under it, followed by a detailed description of the mushroom and a black-and-white photograph. There are only 15 color pictures demonstrating the five groups of mushrooms, but they are very good and give a lot of detail. The rest of the book is made up of black-and-white pictures that are sometimes quite dark and nondetailed; this is a limitation because color is an important part of identification, and this oversight leaves the user dependent on the written description of the colors. In the section on lists of species and descriptions, the author mentions that he used the color designations suggested in the "ISCC-NBS color-name charts." The user would almost always have to have those charts at the time of identification to determine what colors they are seeing.

The publisher's notes say that this was written for "mushroom enthusiasts," with or without mycological backgrounds, as well as well as for mycologists. It appears, however, to be written more directly for the former. The enthusiasts will need a certain level of sophistication, because the keys need to be understood and the language and descriptions are very technical. The glossary which is in the back of the book is a great help, and the introduction contains some good information about mushroom collecting, distinguishing edible and nonedible varieties, and the characteristics and development of mushrooms. Apart from the keys, the other points of access for the guide are the common and scientific name indexes. These make the guide a good reference tool as well as a useful field guide. The publisher's note states that this is volume two of the set with three more volumes to come. The whole set would make a good addition to a biological collection. *LILLIAN R. MESNER*

76. A New Key to Wild Flowers. John Hayward. New York, NY: Cambridge University Press, 1987. 278p., illus., index. $49.50. 0-521-24268-1.

A New Key to Wild Flowers is a field guide for identifying major angiosperms of the British Isles. It includes special simplified keys to difficult groups, ferns, cow parsleys, dandelion-like flowers, deadnettles, and docks. John Hayward developed the keys for amateur courses. "Subsequent extensive testing through the AIDGAP (Aids to Identification in Difficult Groups of Animals and Plants) organisation has further refined the key as a working tool for field and bench use." There are helpful line drawings on nearly every page, even in the glossary. There is an index to families and an index to plant names, including common names. Although there are many other field guides, this one is quite useful in that it has been developed for amateur use and has been thoroughly tested. Academic and special libraries with extensive botanical collectors will find this a useful addition to their reference collections.　　　　　　　　　*RUTH LEWIS*

77. Plants and Flowers of Hawaii. S. H. Sohmer; R. Gustafson. Honolulu, HI: University of Hawaii Press, 1987. 160p., illus. (col.), bibliog., index. $14.95. 0-8248-1096-1.

Dr. S. H. Sohmer, a botanist who has specialized in Pacific plants, is chairman of the Department of Botany, Bishop Museum, Honolulu. R. Gustafson is the collections manager of the botany department, Natural History Museum, Los Angeles, and has spent considerable time photographing Hawaiian plants, including taking most of the photographs for this book.

The authors describe well over 100 native species of Hawaiian plants and also mention some introduced species, several of which have had harmful effects on the native flora. In addition to providing an interesting and enjoyable introduction to the native flora of the Hawaiian Islands, Sohmer and Gustafson make the reader aware of the beauty, uniqueness, and scarcity of so many of these plants.

The text is relatively brief and is amply illustrated. Approximately the last two-thirds of the volume is reserved for 151 gorgeous full-color photographs of Hawaiian plants.

The plants are grouped according to vegetation zone in which they occur: coastal, rain forest, bogs, alpine, and so forth. Each photograph is numbered and captioned to include basic information such as scientific name (family, genus and species), common-name, type of plant, brief description, and indication as to on which Hawaiian island or islands the plant can be found.

The text, which occupies the first third of the book, contains an introduction followed by short chapters on the history of Hawaiian botany, biology of the Hawaiian flora, evolution of plant species, and Hawaiian vegetation zones. An index at the back of the book lists scientific names of plants that are illustrated or mentioned in the text. An added common-name index would have been helpful. The index is followed by a one-page bibliography of Hawaiian flora. Most libraries in Hawaii will want a copy of this guide, and academic and special libraries on the mainland will want it as a reference source on Hawaiian flora.

This is a recommended book for public, academic, and special libraries on mainland U.S. and also for school libraries in Hawaii because of the outstanding photographs.　　　*WILLIAM H. WIESE*

78. A Practical Guide to Edible and Useful Plants: Including Recipes, Harmful Plants, Natural Dyes and Textile Fibers. Delana Tull. Austin, TX: Texas Monthly Press, 1987. 518p., illus. (part col.), bibliog., glossary, index. $19.95. 0-87719-022-4.

Naturalists will find *A Practical Guide to Edible and Useful Plants* helpful for identification of Texas plants as well as giving historical significance of plant uses in the past. Tull is an environmental science educator, naturalist, and freelance writer working on her doctorate in science education. She has taught at the University of Texas and Austin Community College. There are 77 illustrations throughout the book and over 50 excellent color photos. The book is intended to reach readers looking for an "invaluable reference...[on] native plants and our natural environment." Tull has covered the identification of hundreds of plants, including common name (both in English and Spanish) and the scientific name; warnings about toxicity of plants; naming of plant parts that are edible versus parts that are poisonous; recipes for edible plants, e.g., "Tumbleweeds and Bacon," "Hackberry Sauce," and "Acorn Oatmeal Cookies," to name a few; identity of plants used for dyes (120 plants, with recipes), basket weaving or paper making, with instructions for use; medicinal plant use; representation of teas, berries, and spices; and dermatitis- and hay-fever-causing plant identification. Historical commentary is included, adding pleasurable reading. For readers with a botany background, Donovan Correll's and Marshall Johnston's *Manual of the Vascular Plants of Texas* would be more useful.　　　*BERTA KEIZUR*

79. Trees: A Quick Reference Guide to Trees of North America. Robert H. Mohlenbrock; John W. Thieret. New York, NY: Macmillan Publishing Co../Collier Books, 1987. 155p., illus. (part col.), bibliog., index. $24.95. Macmillan Field Guides. 0-02-063430-7.

This guide covers the more common trees of North America north of Mexico, listing 232 species, or one-third of the total for this area. For each tree, the common name, scientific name, field marks, and habitat are given; and for some, variation within species and similar species are described. Range maps are included, and each tree species is illustrated in color, showing leaves, twig, flower, or fruit.

Two appendices provide considerable information on history, economics, and more detailed descriptions for each family of trees. A further appendix advises the novice on equipment and methods for studying and identifying trees. Leaf shapes and tree silhouettes appear inside front and back covers.

The novice wishing to identify the more common trees will be well served by this guide, partly because the descriptions are brief yet detailed and clear. The illustrations are excellent, and the range maps are good because state boundaries are shown, not always the case in field guides. There is one limitation, however: Some species may not be covered. For a more extensive field guide covering 730 tree species, consult Brockman's *Trees of North America* (Golden Press, 1973) and Petrides's *A Field Guide to Trees and Shrubs* (Houghton Mifflin, 1958) which covers all species in eastern North America.

JOHN LAURENCE KELLAND

80. **The Trees of North America.** Alan Mitchell. New York, NY: Facts on File Publications, 1987. 208 p., illus. (part col.), index. $24.95. 0-8160-1806-5.

This handsome guide is authored by an authoritative source and beautifully illustrated with line drawings, paintings, and photos. Very useful as a field guide, it details typical habitat, leaves, bark, flowers, fruit, and silhouette. For the layperson, it contains practical information on selecting, planting and cultivating, pruning, and measuring various species. The beginning taxonomist would find it very useful; however, valuable information on reproductive cycles and medicinal or practical use is missing. Range maps are too general. It would have been more valuable if more ecological information was supplied. The lack of a bibliography prevents any check in determining accuracy of identification information.

ANNE PIERCE

Handbooks

81. **Diseases of Trees and Shrubs.** Wayne A. Sinclair; Howard H. Lyon; Warren T. Johnson. Ithaca, NY: Comstock Publishing Assoc, Cornell University Press, 1987. 574p., illus. (col.), bibliog., index. $49.95. 0-8014-1517-9.

Diseases of Trees and Shrubs is a pictorial survey of diseases of and environmental damage to forest and shade trees and woody ornamental plants in the United States and Canada. According to the authors, it was designed to serve as a diagnostic aid, as an authoritative reference to the diseases and pathogens illustrated, and as a guide to further information. The text emphasizes description, biology, and ecology. Some host and geographic ranges of pathogens are included as well as some information about biological and cultural control.

Arrangement is first by diseases caused by biotic agents versus other environmental stimuli. Second, it is generally by the plant part affected, and, finally, by taxonomic relationships. In the introduction, there is a "Reader's Guide to Plant Diseases and Injuries" which serves as an index by plant part affected or type of disorder and the causal agent. The index includes only Latin plant names, not common names, with photographs

that illustrate symptoms and signs visible with the unaided eye or a hand lens. References from the extensive list (well over 2,000) must be used to secure microscopic details. This list is current as of early 1986.

Emphasizing fungal diseases, the volume is not complete, and some significant diseases are not included, but it is useful in a library or laboratory environment if it is recognized that the book is not a definitive guide and that it is not meant to stand alone. It may well be necessary to consult any of the many references for more detailed information or other guides for diseases not included. The volume is too large and cumbersome for use as a field guide and is, therefore, recommended for any major academic or public library collection. *RITA C. FISHER*

82. **A Handbook of Mexican Roadside Flora.** Charles T. Mason, Jr.; Patricia B. Mason. Tucson, AZ: The University of Arizona Press, 1987. 380p., illus., bibliog., index. $14.95pbk. 0-8165-0997-2pbk.

The amateur naturalist traveling anywhere in Mexico will want to pack this excellent field guide for the trip, and libraries with an interest in Mexican flora will need this reference work. Charles T. Mason, Jr., professor of plant taxonomy and herbarium curator at the University of Arizona, has written a handy guide to the eye-catching roadside plants throughout Mexico. "As it is impossible to present in any meaningful manner the vast number of plants growing in Mexico, an attempt has been made to select those which, because of their flowers, the shape of their fruit, their unusual leaves, or overall peculiarities, have caused us in our many trips to Mexico to say, 'Oh! What is that?'" Mason has used "the outstanding characteristics that attract attention" to construct a plant identification key to more than 200 species. The species descriptions are arranged alphabetically by family, by genus, and by species. In addition to the scientific name of the species, each entry includes one or two Mexican common names; a brief description of the plant that provides information about its leaves, flowers, and fruit; and notes about its origin and geographical range. The book includes an illustrated glossary for any unfamiliar botanical terminology used in the species descriptions. Each entry is accompanied by detailed line drawings by Patricia B. Mason to help one identify the species.

The introduction will be useful for those unfamiliar with Mexico or with taxonomic nomenclature. A succinct, but highly informative, discussion of the basic topography and climatology of Mexico and their impact on vegetation types is provided. The introduction also has a clear explanation of scientific naming conventions. The guide concludes with a short bibliography of important works on Mexican flora and an index to the scientific and common names of the species and families described in the volume. Although this field guide is primarily for the amateur naturalist, libraries with a strong interest in the flora of Mexico also will want to add it to their collections.

STEVEN L. SOWELL

83. **North American Terrestrial Vegetation.** Michael G. Barbour; William Dwight Billings, eds. New York, NY: Cambridge University Press, 1988. 434p., illus., bibliog., index. $49.50. 0-521-26198-8.

Intended for "knowledgeable laypeople, advanced undergraduates, graduate students, and professional ecologists in both basic and applied fields," *North American Terrestrial Vegetation* focuses on the major plant communities of North America. The goal of the editors, who are both eminent professors of botany, is to provide "an accurate and challenging summary of what is currently understood about North American vegetation." The 14 contributors have met this goal well. This book is a "must-buy" for libraries with interests in the topics covered. The volume is divided into 13 chapters as follows: "Arctic Tundra and Polar Desert Biome"; "Boreal Forest"; "Forests of the Rocky Mountains"; "Pacific Northwest Forests"; "California Upland Forests and Woodlands"; "Chaparral"; "Intermountain Deserts, Shrub Steppes, and Woodlands"; "Warm Deserts"; "Grasslands"; "Deciduous Forest"; "Vegetation of the Southeastern Coastal Plain"; "Tropical and Subtropical Vegetation of Meso-America"; and "Alpine Vegetation."

Although the arrangement varies from chapter to chapter, all chapters include a frontispiece map, an introduction, a discussion of areas for future research, and a lengthy list of references cited. Most chapters have sections on climate, soils, paleobotany, modern environment, human impact on vegetation, successional patterns, and quantitative descriptions of the major types of vegetation. Many chapters also include "previously unpublished data, analyses, or models." Throughout the text are numerous maps, tables, graphs, diagrams, and black-and-white photographs. The volume concludes with an extensive index, primarily to geographic locations and scientific names of species mentioned in the text.

The material presented in *North American Terrestrial Vegetation* will "stand as the best summaries for some years to come." It is recommended for any college, university, or special library that has an interest in botany and ecology.

STEVEN L. SOWELL

84. **Weeds of the United States and Their Control.** Harri J. Lorenzi; Larry S. Jeffery. New York, NY: Van Nostrand Reinhold Co., 1987. 355p., illus. (col.), bibliog., glossary, index. $29.50. "An Avi Book." 0-442-25884-4.

Weeds are plants that appear where they are not wanted. Each year weeds are responsible for destroying 10 to 15 percent of the total value of agricultural crops in the United States. *Weeds of the United States and Their Control* describes over 300 weeds in 69 families and includes 305 color photographs. The one-page entry for each weed includes a distribution map with a general indication of the weed's location; a plate number; scientific name; common name; family name; a full description of the physical characteristics of the weed; and habitat, including the weed's native habitat and suggested control, including mechanical, cultural, biological, and chemical methods. The plate number refers to the section of color photographs for weed identification. Each photograph clearly shows the important identifiable characteristics on a black background. Beneath each photograph is the plate number and both scientific and common names.

The introduction describes the various types of weed control and includes nine tables of data. The tabular information includes common mulches; methods of herbicide application; herbicides for agronomic crops, vegetable crops, fruit and nut crops, turf, pastures, fence rows, waste areas, and industrial sites; and common names, trade names, and manufacturers of common herbicides. Following the main body of weed descriptions is a glossary of over 150 botanical terms. The definitions are brief and somewhat circular, but accurate, and the glossary includes most of the botanical terms used in the weed descriptions. Following the glossary is a bibliography listing the sources consulted for the names and descriptions used in this volume. The common names used are from a list prepared by the Weed Science Society of America. The index includes both scientific and common names. Where applicable, the index entry gives both plate and page numbers. Overall, this book is well organized, accurate, and attractive; the photographs are of excellent quality. However, there are some shortcomings of this book. Although the title of this work implies coverage of weeds in the entire United States, only the contiguous 48 states are included in the distribution maps. In addition, when comparing family coverage with other works of this type, there are differences. For example, *Weeds of the North Central States* has a smaller geographic coverage than *Weeds of the United States and Their Control*, but it includes four families (Capparaceae, Hydrophyllaceae, Martyniaceae, and Nyctaginaceae) not found in Lorenzi's work. There are also some typographical errors in this work: in the glossary entry for berry and index entries for *Brunnichia cirrhosa*, Cucurbitaceae, Compositae, and *Datura stramonium*. Regardless of these minor shortcomings, this book is highly recommended for its excellent photographs and its concise one-page format for each weed description. Agriculture and botany students and professionals, as well as home gardeners, will find this reference work to be helpful and easy to use. Its combination of identification and control methods makes *Weeds of the United States and Their Control* a worthy addition to academic and larger public libraries.

KIMBERLEY M. GRANATH

Histories

85. **Herbals: Their Origin and Evolution: A Chapter in the History of Botany, 1470–1670.** 3rd. Agnes Arber. New York, NY: Cambridge University Press, 1988 (c1986). 358p., illus., bibliog., index.

$24.95pbk. Cambridge Science Classics Series. 0-521-33879-4pbk.

Arber's *Herbals* "stands as the major survey of the period 1470 to 1670, when botany evolved into a scientific discipline separate from herbalism, a development reflected in contemporary herbals. Every work on herbals since 1912 has been indebted to Arber's classic." The second edition (1938) has long been unavailable. Several useful additions to an unaltered main text create this third edition. William T. Stearns has written a mostly biographical introduction and several annotations where more recent work adds information or where modern botanical names can be inserted. He has also more than doubled the bibliography. Two papers by Agnes Arber from 1940 and 1953 are included. There are subject indexes to the text and to the bibliography. Most libraries in botany, pharmacy, history, or art will want to add this volume to their collections.

RUTH LEWIS

Manuals

86. Aquatic and Wetland Plants of Kentucky. Richard R. Hannan; Marc Evans, eds. Frankfort, KY: Kentucky Nature Preserves Commission, 1986. 315p., illus., bibliog., index. $20.00pbk. Scientific and Technical Series, no. 5.

Aquatic and Wetland Plants of Kentucky is the fifth in a series published by the Kentucky Nature Preserves Commission, which is seeking to gather and disseminate information on the natural history of the state. Work on the volume in hand began in the mid-1970s and concluded about ten years later. The authors interpret the terms "aquatic and wetland plants" fairly broadly as including those "herbaceous vascular plants...that grow in water or in soil saturated, at least much of the year, with water." Works of this nature dealing with Kentucky wetland flora are relatively few, the authors listing but three. This present volume contains a good bibliographic section, as well as historical notes on the major Kentucky floras published in earlier times. The taxonomic section includes a key to families and a descriptive flora. All keys in the book are dichotomous, being divided into successively numbered "couplets," each of which is composed of a pair of "leads." By following these leads backward or forward through the key, one eventually comes to a family name. The families are arranged alphabetically in the descriptive section of the text. This alphabetic arrangement is a departure from the traditional taxonomic format, but it seems to work well and is probably easier to use for persons not professionally versed in taxonomy. Accompanying illustrations are nicely done line drawings, also keyed alphabetically to the descriptive section. A series of county distribution maps concludes the main portion of the text. The book concludes with an index of alternate and common names. The work is printed on glossy paper of good weight and is easy to see and read. The paper binding is sturdy enough but

will probably require rebinding where heavy use is expected. Margins are sufficient for rebinding. This work is recommended for all Kentucky libraries with an interest in botany, as well as for individuals and libraries in adjacent states where there is interest in aquatic and wetland plants.

FRED O'BRYANT

87. Ferns and Fern Allies of Kentucky. Ray Cranfill. Frankfort, KY: Kentucky Nature Preserves Commission, 1987 (c1980). 284p., illus., bibliog., index. $4.50pbk. Scientific and Technical Series, no. 1.

Ferns and Fern Allies of Kentucky is the first in a series published by the Kentucky Nature Preserves Commission, which is seeking to gather and disseminate information on the natural history of the state. The volume is designed as an introduction to the pteridophytes found in Kentucky, synopsizing information not usually found in more descriptive texts. This book gives an account of the distribution, habitat, and economic value of these plants, plus references to the literature. The first section includes a brief historical synopsis of fern collecting in Kentucky, a geographic and ecological analysis, and a generalized life history of the state's species. Section 2 treats the identification and descriptive keys to the various genera. Detailed ecological and geographical notes are given. The final section contains notes and geographical distribution maps at the county level. A brief index of Latin and common names rounds out the volume. The line drawings and several photographs are clear and easy to interpret. The paper binding is sturdy enough but probably will require rebinding where heavy use is expected. This work is recommended for all Kentucky libraries with an interest in botany, as well as for personal purchase by those in Kentucky or adjacent states with an interest in ferns.

FRED O'BRYANT

88. Flora of the British Isles. 3rd. A. R. Clapham; T. G. Tutin; D. M. Moore. New York, NY: Cambridge University Press, 1987. 688p., illus. (part col.), bibliog., glossary, index. $150.00. 0-521-30985-9.

This is an authoritative manual for taxonomists covering the flora of the British Isles. It "provides general descriptions which include life-form and chromosome numbers and some notes on phenology and mechanisms of pollination and seed dispersal and also on variability, distribution within and outside the British Isles, preferred habitats and commonly associated species." The authors do not claim to include all of the known varieties of putative hybrids but do indicate that it is the most comprehensive manual of its kind.

It is divided into three taxonomically arranged sections: Pteridophyta, Gymnospermae, Angiospermae. Each entry gives full description including color, size, flowers, when introduced, common name, and other pertinent scientific data.

This book is a must for all libraries interested in British flora, especially for those interested in ecology, geographical distribution, evolutionary history, and agricultural significance. The second edition is available in paperback.

H. ROBERT MALINOWSKY

89. The Marshall Cavendish Science Project Book of Plants. Steve Parker. Freeport, NY: Marshall Cavendish, 1988 (c1986). 43p., illus. (col.). $12.49. The Marshall Cavendish Library of Science Projects, v. 2. 0-86307-626-2.

"Science is all about discovering more about your world, finding out why certain things happen and how we can use them to help us in our everyday lives." This encyclopedic book for young people covers the field of botany. Through a series of chapters that discuss the plant world, flowers, pollen, seeds, vegetables, roots, chlorophyll, wood, and life around a tree, the student learns about botany. Experiments and projects are used to exemplify the various concepts. The well-written text and good illustrations make this a useful reference book for school libraries and an excellent textbook for the classroom.

H. ROBERT MALINOWSKY

Treatises

90. Medicinal Plants in Tropical West Africa. Bep Oliver-Bever. New York, NY: Cambridge University Press, 1986. 375p., illus., bibliog., index. $77.50. 0-521-26815-X.

In tropical West Africa, locally available medicinal plants offer the possibility of "reducing reliance on expensive imported drugs." In 1960, the Nigerian College of Arts, Science, and Technology in Ibadan published *Medicinal Plants in Nigeria*, an earlier work by the same author. *Medicinal Plants in Nigeria* formed the basis for the current work but is now out of print. For inclusion, the author selected plants whose uses were confirmed "by primitive populations in other parts of the world," those most "likely to have real therapeutic value" given the "known chemical and pharmacological information." The current work expands the region under consideration to tropical West Africa. It includes both indigenous and nonindigenous plants, but with an emphasis placed on those plants whose pharmacological properties may be less well known.

The author has arranged the discussion of plants according to their known primary pharmacological actions, grouping plants which have a similar action on the cardiovascular system, on the nervous system, on infectious diseases, and on hormone secretions in man. "Within each therapeutic group," plants are "assembled by chemical constituents." Each entry provides a Latin name and English common name, followed by the local medicinal uses, the known chemical constituents, and the known pharmacological and clinical actions. For botanical descriptions of the plants, the author directs the reader to *Flora of*

West Tropical Africa (1972) by John Hutchinson and J. M. Dalziel. Descriptions of native uses of plants are brief and general and do not specify dosages or the exact method of preparation. The author emphasizes the pharmacology of each plant as documented in the scientific literature, providing an 86-page bibliography. Grey-toned watercolor illustrations by the author decorate the text. Tables of plant families allow comparisons of the part of the plant used, the active constituents, and the "recognized or possible medicinal action." A general index provides additional access by genus and species only. Recommended for academic, medical, and special libraries.

LINDA JACOBS

★
ZOOLOGY
Atlases—Medical

91. Maps and Guide to Microdissection of the Rat Brain. Miklós Palkovits; Michael J. Brownstein. New York, NY: Elsevier Science Publishing Co.., Inc., 1988. 223p., illus., bibliog., index. $49.95. 0-444-01256-7.

Palkovits and Brownstein's guide is an addition to the general literature on brain dissections (they offer a good list of maps for avian and mammalian brain microdissections, too) and specifically a guide and maps for rat brain microdissections. The work is authoritative, as both men are Ph.D.-M.D.s and work in the areas of anatomy and cell biology. The book consists of maps of 266 nuclei or brain areas as well as text that explains microdissection and the brain areas that are demonstrated.

After a brief introduction, there is a chapter that goes into the techniques of microdissection including discussions of removal techniques, sectioning techniques, tools for punching tissue, punch techniques, homogenization for biochemical assay, and validation of methods. The next chapter goes into detailed descriptions of the areas demonstrated on the maps as well as directions on the removal of discrete brain nuclei. References are made to the numbered punches on the maps. The last chapter contains the maps themselves.

Each plate contains a photomicrograph of the right hemisphere of the rat brain with a line drawing of the left hemisphere. The right side is labelled with the numbered punch sites, and the line drawing is labelled with abbreviations for the anatomical structures. Each plate also contains other illustrations, a macrophotograph of the whole rat brain with the location of the plane of microdissection, and a graphic longitudinal view of the brain with the plane of microdissection.

There are a number of points of access to the volume. A detailed table of contents directs the user to the general areas of interest. A very detailed index directs the user to tables, figures, and sections of the text. There are also an alphabeti-

cal list of the abbreviations used in labelling with both English and Latin nomenclature and punch numbers, and an index of structures (in English) with their punch numbers and plate numbers.

Using this set of rat brain maps assumes a sophisticated level of anatomical knowledge. It would be a very valuable addition to biology and medical collections as well as a laboratory tool for practitioners.   *LILLIAN R. MESNER*

Atlases—Science

92. A Distributional Atlas of Kentucky Fishes. Brooks M. Burr; Melvin L. Warren, Jr. Frankfort, KY: Kentucky Nature Preserves Commission, 1986. 398p., illus.,. price not reported. Scientific and Technical Series, no. 4.

"The primary purpose of this atlas is to present an accurate synopsis of the distribution, systematics, habitat, and conservation status of Kentucky fishes in a format useful to professional biologists, resource managers, and non-professionals." The atlas precedes and will supplement a new *Fishes of Kentucky*, now in preparation, that will summarize the identification, biology, and distribution of Kentucky fishes. Both of the authors are with the Department of Zoology of Southern Illinois University at Carbondale.

A well-organized presentation of distributional maps of Kentucky fishes comprises the bulk of this treatise. Arranged in phylogenetic order, each map shows the record of occurrence of a single species in the major river systems of the state. The distribution of data spans a period "from about 1818 through November 1986 with the bulk of the records post-dating 1950." An inset map of North America gives a broader view of the distribution of that particular species.

The authors comment on the distribution as shown, the systematics of the species, and the habitat. Where appropriate, they indicate the conservation status, such as endangered or threatened. They also include their recommendations for any changes in conservation status. Discussions of the physical features of Kentucky and the zoogeography of the Ohio-Mississippi Basin will aid in the interpretive analysis of the factors involved in distribution. A final table summarizes the distribution of Kentucky fishes by major drainages.   *LINDA JACOBS*

Dictionaries

93. It's Easy to Say Crepidula! (kreh PID'yu luh): A Phonetic Guide to Pronunciation of the Scientific Names of Sea Shells and a Glossary of Terms Frequently Used in Malacology. Jean M. Cate; Selma Raskin. Santa Monica, CA: Pretty Penny Press, Inc., 1986. 155p., illus., bibliog., glossary. $19.95pbk. 0-938509-00-4pbk.

Cate and Raskin have compiled a pronunciation guide to the scientific names of sea shells that is directed primarily to the lay collector. As Raskin explains in the preface, there are several good reasons a collector should learn and use the scientific names of sea shells. Cate, a malacologist and former associate editor of *Veliger*, and Raskin, a writer and collector, have produced this guide to assist one in learning the scientific names. The authors have "listed only the most popularly collected families and species." About 80 genera are included. They have omitted scientific names which are easy to pronounce. A simplified phonetic spelling is used with capital letters and single quote marks indicating stressed syllables. The guide is organized alphabetically by genus then alphabetically by species. A few of the larger genera have been grouped together under a broader generic name. Family names are also indicated. At the beginning of most of the general lists, there are small black-and-white drawings which illustrate a species from the genus. Some of the species names are accompanied by short notes about the source of the name. In the introduction, the authors provide a short summary of some of the rules used in naming species, particularly those used in constructing name endings and patronymics. There is a short index to some of the popular names of sea shells and a useful, 11-page "Glossary of Terms Frequently Used in Malacological Publications." The work concludes with a brief bibliography of books on shells, all published since 1962. Although sea shell collectors will find this pronunciation guide useful, it will be of marginal value for research collections. The pronunciation rules found in many Latin dictionaries will be of greater use.
                    STEVEN L. SOWELL

Directories

94. The Directory of North American Fisheries and Aquatic Scientists. 2nd. Beth D. McAleer, ed. Bethesda, MD: American Fisheries Society, 1987. 363p., index. $25.00pbk. 0-913235-40-7pbk.

In expanding its title to include aquatic scientists, the second edition of the *Directory of North American Fisheries Scientists* has increased its listings by 22 percent to include over 10,000 individuals employed in a broad range of activities in fisheries and aquatic sciences. Its editor, Beth D. McAleer, currently editor of *Fisheries*, a bimonthly journal also published by the American Fisheries Society (AFS), began her employment with AFS as its membership secretary. According to *Fisheries*, "AFS promotes scientific research and enlightened management of aquatic resources for optimum use and enjoyment by the public. It likewise encourages a comprehensive education for fisheries scientists and continuing on-the-job training." The directory helps AFS meet this goal by identifying persons engaged in research, management, and education in the United States, Canada, and Mexico. The major subject areas or "specialties" covered are biology, fish culture, eco-

nomics, limnology, fisheries technology, oceanography, social sciences, and habitat and water quality. The directory begins with a short preface, brief use instructions, abbreviations key, and a key to subject specialties and job activities, followed by its three major sections. The first, listing individuals alphabetically by last name, gives the individual's highest academic degree, job title, employer, mailing address, telephone number, and numeric codes for "fields of expertise and professional activities." The entries in this section are arranged in three columns with the names in bold type. The type style and size are comfortable. The second section is a geographic index to the entries in the first section arranged in order by state, Canadian province, and Mexico. Because many of the state listings are quite long, this section would have been more useful if it had been subdivided by city. The last section is split into two parts: an index to the sixteen professional activity codes used in the entries, for example, "administration," "fisheries management," and "research," and an index to the specialty codes used, arranged into eight broad categories with 66 subcategories. This latter index is useful for finding people whose interests are in, for example, "fisheries engineering" or "exotic species." As in the second section, the two indexes in section three have many long listings that could have benefited from subdivision by location.

The data for the directory were compiled by updating entries in the first edition and sending questionnaires to over 9,000 people not listed previously. The return rate is not noted. In sampling the directory for individuals from a particular institution, it was found to be complete. Because AFS plans to update the directory, individuals omitted from this edition are "invited to contact the American Fisheries Society for inclusion in the next Directory." Despite its minor flaws, overall, the *Directory of North American Fisheries and Aquatic Scientists* is an effective and inexpensive paperback that will be useful for individuals and libraries with interests in this area. *STEVEN L. SOWELL*

Encyclopedias

95. Album of Prehistoric Man. Rev. and updated. Tom McGowen. New York, NY: Checkerboard Press, 1987 (c1975). 60p., illus. (part col.), index. $4.95pbk. 0-02-688514-X.

Written for school children, *Album of Prehistoric Man* is a survey of human evolution. Through a very interestingly written text and excellent illustrations by Rod Ruth, this book covers in 13 chapters "Ashes, Bones, and Stone Tools;" "Unknown Ancestors;" "The First Hominids;" "Hominids and First Humans;" "The Upright People;" "The Neanderthal People;" "The Ice Age;" "People like Us;" "The Cave Artists;" "The Ice Age Americans;" "The Great Change;" "Tombs, Temples, and Standing Stones;" and "Metal and Marks in Clay." A pronunciation guide and subject index are included.

This is a recommended book for school and public libraries. Although coverage may be brief, it is accurate and well presented, making it a good encyclopedic source for students and also for adults with no scientific background. *H. ROBERT MALINOWSKY*

96. Album of Sharks. Tom McGowen. New York, NY: Checkerboard Press, 1987 (c1977). 60p., illus. (part col.), index. $4.95pbk. 0-02-688513-1pbk.

Tom McGowen with the help of illustrations by Rod Ruth has written an interesting account of 11 different sharks—great white, hammerhead, mako, blue, thresher, tiger, whale, bull, wobbegong, dogfish, and Greenland. Each account is done in story fashion that weaves the history, facts, and traditions into an interesting account that makes fascinating reading for children. It is not technical, but the facts are accurate. Throughout the book the terms and names are given parenthetical pronunciations.

This is a good account of the major species of sharks that should make interesting reading for school children and a starting point for doing a paper on sharks. It is not intended to replace good scientific encyclopedias but rather to supplement technical information with a readable story about the sharks. *H. ROBERT MALINOWSKY*

97. Album of Snakes and Other Reptiles. Tom McGowen. New York, NY: Checkerboard Press, 1987 (c1978). 61p., illus. (part col.), index. $4.95pbk. 0-02-688503-4pbk.

The purpose of *Album of Snakes and Other Reptiles* is to explain which characteristics all reptiles share and show some of the differences through descriptions of a few of the more common ones found in literature that children may encounter. A brief introductory chapter discusses what a reptile is. This is followed by chapters that describe specific types of reptiles in a story fashion that makes fascinating reading. The facts are presented in nontechnical terminology intertwined with some folklore.

The very good color illustrations by Rod Ruth add to the interesting accounts. Covered in the 11 chapters are the crocodile, constrictor, rattlesnake, cobra, freshwater turtle, tortoise, sea turtle, monitor, lizards, glass snake, and tuatara. A pronunciation guide and an index are included at the end of the book. This is a good resource book for school children and a useful reference tool for school and public libraries. *H. ROBERT MALINOWSKY*

98. Amphibians. Giuseppe Minelli. New York, NY: Facts on File Publications, 1987. 57p., illus. (col.). $12.95. History of Life on Earth. 0-8160-1557-0.

This is a translation and adaptation of *I Anfibi*. Amply illustrated with colored drawings and diagrams, this slim volume presents an impressive amount of information in an interesting and accessible manner. The physical characteris-

tics of living amphibians are described in the context of their evolutionary history. It is an excellent reference source for school and public libraries written in nonscientific terminology.

CAROLYN DODSON

99. Coral Snakes. Sherie Bargar; Linda Johnson. Vero Beach, FL: Rourke Enterprises, Inc., 1987. 24p., illus. (col.), glossary, index. $11.66. The Snake Discovery Library. 0-86592-246-2.

The Snake Discovery Library by Sherie Bargar and Linda Johnson is a science series that would be a useful addition to any school or public library children's collection. The series is composed of 12 titles divided into two sets. Set 1 includes *Boa Constrictors, Cobras, Copperheads, Cottonmouths, Mambas,* and *Rattlesnakes.* Set 2 comprises *Anacondas, Coral Snakes, Kingsnakes, Pythons, Rat Snakes,* and *Tree Vipers.* This review is based on an examination of three titles from set 2: *Coral Snakes, Kingsnakes,* and *Pythons.*

Each title in the series follows the same general structure: a short introductory section followed by sections on habitat, appearance, senses, head and mouth, babies, prey, defenses, and relationships to humans. Each book has a table of contents and an index, and words in boldface are briefly defined in a glossary. Each title reviewed has 11 pages of beautiful color photographs by George Van Horn. A child's attention will quickly be attracted by these high quality pictures with captions that provide the scientific name as well as the common name for the species illustrated. Each book is sturdy, and its 7 1/2-by-8-inch size seems ideal for small hands. Although the large-print text makes the book easy to read, the print size is somewhat at odds with some of the long, complex sentences and the vocabulary level.

The books in the series include many basic facts about snakes. The consistent use of tables of contents and indexes will help the reader find specific information. The series would be excellent reference material for the upper elementary grades. *STEVEN L. SOWELL*

100. The Encyclopedia of Animal Evolution. R. J. Berry; A. Hallam, eds. New York, NY: Facts on File, Inc., 1987. 144p.,illus. (part col.), bibliog., glossary, index. $24.95. "An Equinox Book"; The Encyclopedia of Animals. 0-8160-1819-7.

The Encyclopedia of Animal Evolution is a short, but informative survey that will be an important addition to library collections serving students from high school through college and the interested layperson. Under the editorship of R. J. Berry, professor of genetics at University College, London, and Anthony Hallam, professor of geology at the University of Birmingham, 21 authors have contributed material for this work. Authors' entries are identified by their initials.

The book is divided into six major sections, each with several entries. The first section, "The Prehistoric World," describes the findings of paleontologists on what animals existed in earlier geological time periods, and entries in "The Background of Evolution" discuss Charles Darwin, his life, and his theory of natural selection as well as the evolutionary theories of earlier writers. "The Course of Evolution" covers the evidence on the development of life from simple to complex forms; "The Consequences of Evolution" looks at the origin of life, homology, and adaptation; and genetic variation and the origin of species are discussed in "The Mechanisms of Evolution." In the final section, "Man and Evolution," human evolution is considered, and the major controversies surrounding the theory of evolution are briefly described.

A short selected bibliography covering important historical works and recent publications is provided as guidance to further reading. A three-page glossary defining major terms used in the text and a brief, but adequate index to the text and illustrations are included. Excellent illustrations, mostly in color, appear on almost every page, and the editors' liberal use of diagrams, drawings, and photographs has enhanced and supplemented the text very well. *The Encyclopedia of Animal Evolution* is part of a four-volume series, *The Encyclopedia of Animals.*

STEVEN L. SOWELL

101. Frogs and Toads of the World. Chris Mattison. New York, NY: Facts on File Publications, 1987, 191p., illus. (part col.), bibliog., index. $22.95. 0-8160-1602-X.

As familiar as frogs and toads are to everyone, the books about them are scarce, especially if readability, coupled with scientific accuracy, is expected. For this reason, this book by Mattison fills the very much needed gap. A world-renowned amphibian and reptile specialist, Chris Mattison does not spare the readers from essential scientific information. He presents it in a nonthreatening way explaining scientific terms in an understandable manner. The most impressive feature of this book is its generous use of photographic and line-drawn illustrations—120 to be exact, with 60 in full, brilliant color. The first chapter addresses the most frequently argued point—appropriate use of the terms, frog and toad. Over the following nine chapters, the discussion focuses on the physiological and behavioral aspects of this highly important group of animals. The tenth chapter introduces the organism's anthropological relationship to mankind, and the eleventh and final chapter gives concise overall descriptions of frogs and toads grouped together under each family name.

Another special feature of this book is its bibliography, which lists many recent publications, grouped together under the topics of biology, taxonomy, guides and keys, and recordings. The author has provided some background information on the level of illustrations and readability for several references. Particularly useful is the list of recordings of animal sounds which are extremely difficult to identify and lo-

cate. Because frogs and toads were the first land vertebrates to vocalize, it is fitting for this book to introduce the readers to the recordings of their voices. The index consists of both common and Latin names and a few topical terms. The shortcomings of this book include the lack of anatomical description and not enough discussion of geographical distribution. A few more references in these two areas would have made the scope of this book a little more complete. Nonetheless, with its clear printing on high-quality paper and easy-to-follow headings in boldface print, this book is highly recommedned for public and school libraries. *MITSUKO WILLIAMS*

102. **Kingsnakes.** Sherie Bargar; Linda Johnson. Vero Beach, FL: Rourke Enterprises, Inc., 1987. 24p., illus. (col.), glossary, index. $11.66. The Snake Discovery Library. 0-87592-248-9.
See review under Coral Snakes.

103. **Pythons.** Sherie Bargar; Linda Johnson. Vero Beach, FL: Rourke Enterprises, Inc., 1987. 24p., illus. (col.), glossary, index. $11.66. The Snake Discovery Library. 0-86592-244-6.
See review under Coral Snakes.

104. **Reptiles.** Giuseppe Minelli. New York, NY: Facts on File Publications, 1987. 58p., illus. (col.). $12.95. The History of Life on Earth. 0-8160-1558-9.
Excellent, lavish illustrations of early reptiles occupy more than half of this large-format book, part of a six-volume series on the evolution of the vertebrates called *The History of Life on Earth*. Giuseppe Minelli, professor of comparative anatomy at the University of Bologna in Italy, also authored the other volumes in the series, including *The Evolution of Life, Marine Life, Amphibians, Dinosaurs and Birds* and *Mammals*. This volume is a translation and adaptation of *I Rettili*. The amazingly condensed narration describes the physical characteristics of reptiles and traces their evolutionary history from the late Carboniferous period, 300 million years ago, to the present. This excludes the period of the dinosaurs, which is dealt with in another volume, but includes animals fascinating in their own right, such as the *Anteosaurus*, a theriodont therapsid, whose illustration reminds one of the opening scenes of *Star Wars*.
The text requires tenacity in the mastering of new terms. For example, one learns that the lepidosaurs, a superorder of the diapsids, are divided into three orders, the eosuchians, rhynchocephalians, and squamates. To alleviate stress, the author intersperses accounts of the ancient and mysterious with their modern-day descendants such as the tortoise, the tuatara, the gecko, the marine iguana, and the Gila monster. In juxtaposition with the discussion of now-vanished members of the reptile family, Minelli warns of the threats to the marine turtle, the Komodo dragon, and the Galapagos tortoise. Well

coordinated with the text, Lorenzo Orlandi's illustrations add a great deal to its understanding. For example, the inside front leaf shows the evolution of the major reptile lines through the Palaeozoic, Mesozoic, and Cenozoic eras. Besides numerous illustrations of early and modern reptiles, one observes the supercontient, Pangaea, the basic features of a cotylosaur's skull, a schematic picture of the earth during the Permian and Triassic periods, and the internal skeletal structure of the cryptodire turtle. Orlandi has also illustrated four of the other volumes in the series. Although the illustrations will appeal to all ages, the small typeface and the number of unfamiliar words may dissuade all but the more serious reptile afficionados. However, the content and presentation make this volume of value to students at all levels. *LINDA JACOBS*

105. **Sharks.** John D. Stevens, ed. New York, NY: Facts on File Publications, 1987. 240p., illus. (col.), bibliog., index. $29.95. 0-8160-1800-6.
Sharks represents the combined effort of 17 noted authors and illustrators, most with United States or Australian affiliations. The book is a visual delight with scores of fine color photographs, paintings, and diagrams which depict numerous species of sharks as well as aspects of shark biology and behavior. The mostly well-written text complements these illustrations, and the result is a volume which will be informative and appealing to its intended audience, "...the interested general reader..." as noted by consulting editor Dr. John D. Stevens in his brief introduction. Although this book will be a useful reference work for the public and perhaps undergraduate library, there are several weaknesses that reduce its value as an upper-level undergraduate or graduate library resource.
The book is divided into three main parts. The first part includes chapters on shark biology, behavior, ecology, distribution, evolution, senses, and types of sharks. The chapter on types of sharks is particularly useful, with a representative species from each family illustrated along with general information on the overall family size, habitat, distribution, reproduction, and diet. The number of species in each family is also given. Information that is presented in the chapter on shark biology such as distribution maps and shark size comparisons could have been integrated with the chapter on shark types. Many of the diagrams in these beginning chapters are taken from previously published works, and proper citation (author, date) is given in the diagram captions. However, many of the full citations are not included in the bibliography (or further reading list) at the end of the book. All chapters in this section are well done; however, the shark ecology chapter could have been expanded to include more detailed information on the shark's interaction with its environment.
The second part of the book concerns shark attacks, with chapters covering the United States, Australia, New Zealand, the tropical Pacific, and South Africa. Fully one-quarter of the book is devoted to shark attacks. Each of the five chap-

ters in this section provides much of the same type of information, including number and distribution of shark attacks and potentially dangerous sharks. Although shark attack is no doubt an interesting topic to many, the treatment of this subject could have been handled in a more concise and less sensational way. The third part of the book, "Myth and Reality," includes chapters on the use of sharks, controlled and uncontrolled human encounters with sharks, repelling sharks, and the shark in myth and history. Although the material presented is interesting and enhanced by fine illustrations, as with the section on shark attacks, it could have been covered in a more concise manner.

At the close of the book there is a useful checklist of living sharks that includes common as well as scientific names. The one-page bibliography that follows contains 89 references with 75 percent of them dated prior to 1982. Even though we are told in this bibliography's brief introduction that "Much of the information in this book derives from original research undertaken by the contributors," it would have been useful to have included a complete list of the authors' publications on sharks. The book does have a detailed index. The publishers have done a fine job reproducing the color photographs, however, the binding of the book is not adequate for the heavy paper stock used. The review copy was split along both the inside front and back covers. This book will prove to be a useful reference work for the public and perhaps undergraduate library. It is not, however, recommended for upper-level undergraduate or graduate collections because of its lack of citation to the literature and the excessive coverage of certain subject areas as noted.

JAMES E. BIRD

106. **Snakes and Other Reptiles.** Mary Elting. New York, NY: Simon and Schuster, Inc., 1987. 50p., illus. (col.), bibliog. $7.95. "A Little Simon Book." 0-671-63664-2.

Snakes and Other Reptiles is a children's large-print book which briefly touches on various interesting facets of the biology and life history of some of the world's reptiles. It is not a field guide nor does it attempt to cover the full spectrum of species or any one particular geographic area. The section on snakes is the largest in this 50-page book. Some of the unusual and common features of various garter, giant, poisonous, tree, burrowing, and water snakes are described. Lizards are the next major focus of attention and include monitor, frilled, chameleons, iguanas, skinks, geckos, race runners, and Gila monsters. The remaining third of the book covers amphilisbaenids, tuataras, turtles, tortoises, terrapins, crocodiles, gharials, alligators, and camans. *Snakes and Other Reptiles* concludes with four interesting pages on baby reptiles and reptile folklore. A brief reading list refers readers to three slightly out-of-date books for more information. The author completed the text with the assistance of the education department of the Denver Museum of Natural History. Profuse, well-done illustrations, by Christopher Santoro, complement the text. This book seems intended for eight- to 12-year-olds. It would be most useful in public, elementary, and environmental education libraries.

BILL COONS

Field Books/Guides

107. **Amphibians and Reptiles in West Virginia.** N. Bayard Green; Thomas K. Pauley. Pittsburgh, PA: University of Pittsburgh Press in cooperation with the West Virginia Department of Natural Resources Nongame Wildlife Program, 1987. 241p., illus. (part col.), illus., glossary. $29.95, $14.95pbk. 0-8229-3819-7, 0-8229-5802-3pbk.

The 86 species and subspecies of amphibians and reptiles that have been collected in West Virginia are described. The accounts, in phylogenetic order, include description, habitat and habits, breeding, and range, along with a range map of West Virginia. A colored photograph illustrates each species. In addition, orders and families are characterized, with descriptions, biology, and range. The extensive introductory material includes the history of herpetology in West Virginia, the ecology of the state, as well as photographing and caring for herpetofauna. The authors' personal experiences from their many years of field work are scattered throughout the volume. Also provided are a key to the adult amphibians and reptiles, glossary, and 38-page bibliography of scientific literature. Green and Pauley, who have collected and published extensively on the herpetofauna of West Virginia, have compiled a concise, authoritative guide to the identification and biology of reptiles and amphibians for the informed layperson. A recommended book for public and academic libraries.

CAROLYN DODSON

108. **The Butterflies of Indiana.** Ernest M. Shull. Bloomington, IN: Indiana Academy of Science, Dist. by Indiana University Press, Bloomington, IN, 1987. 262p., illus. (part col.), bibliog., index. $25.00. 0-253-31292-2.

The foreword of this book, published by the Indiana Academy of Science, states that "almost as numerous as the butterflies themselves are books about butterflies. Why another one?" The *Butterflies of Indiana* is different in several respects: it represents a thorough inventory of the Lepidoptera of that state; it includes field observations of larval food plants and adult mating habits; and—most important—it may be the only treatise on butterflies ever written by a retired sociologist. "Butterflies are flying gems of exquisite beauty and design," writes Shull, for whom they have been a lifelong avocation. He has studied and collected them, and he also has several professional articles to his credit. In addition, the author's extensive field observations

have been complemented by the records and works of other scholars (studies from 1892, 1931, and 1972 were reviewed and incorporated in this work).

The *Butterflies of Indiana* was designed as a manual (note, it is not a key or field guide) to aid in identifying the butterflies presently known to occur in Indiana. It pulls together in one place information about Indiana's butterflies but does not attempt to deal with all aspects of the biology of butterflies. The 262 pages of this book are organized in four parts. Section 1, "Biology of Butterflies," includes brief information on color, patterns, mimicry, migration, collection tips, classification and identification techniques, conservation, and sap feeding (an interesting recipe to attract moths and butterflies—combine and ferment beer, overripe bananas, Kayro syrup, and brown sugar—is included). Sections 2 and 3 cover 53 skippers (Hesperiodea) and 96 butterflies (Papilionoidae). The account for each species includes a brief description and several paragraphs on distribution, habitat, and life cycle. Where it was observed, notes on mating behavior and larval food plants are included. Color plates (photographs) with the date and location of collection accompany each species description, as do county geographic distribution maps of the state. Section 4 contains a checklist of the butterflies of Indiana.

The *Butterflies of Indiana* is slightly marred by several technical oversights and inconsistencies; not enough thought or attention has been given to the reader. The glossary is inadequate, incomplete, and insubstantial; the indexes, both food plant and text, could be improved; and the lack of plate pagination and cross-referencing with the text (either to a page or a species number) is grossly inconsiderate. Despite these shortcomings, this book is recommended to any library with comprehensive collections on Lepidoptera, academic or public libraries in Indiana or the surrounding region, and inidividuals with a great interest in, or knowledge of, the subject. Those interested in a more general text on butterflies, however, would do best to look elsewhere.

BILL COONS

109. Close Encounters with Insects and Spiders. James B. Nardi. Ames, IA: Iowa State University Press, 1988. 185p., illus., bibliog., index. $12.95. 0-8138-1978-4.

James B. Nardi is a research scientist in the Department of Entomology at the University of Illinois. Dr. Nardi's own detailed black-and-white drawings appear throughout the text of this interesting book. *Close Encounters with Insects and Spiders* provides an introduction to some of the more common small creatures that we are likely to run across in our homes, lawns, gardens, and the nearby countryside.

The text assumes no special insect knowledge on the part of the reader. Because the author provides numerous suggestions for capturing insects and observing them in glass jars, terrariums, and aquariums, in addition to watching them as they go about their business in their natural surroundings, this guide and project book

is also interesting to read. It would be an excellent selection for junior high and high school libraries. *Close Encounters* should also be welcomed by public libraries, because it is capable of holding a general reader's interest while imparting considerable information about insects and scientific principles at the same time. It is a good choice for an adult of any age who does not know very much about insects but would like to learn more.

The author has arranged his book according to the types of surroundings in which the insects and their relatives can be found. There are five chapters: "Home, School, and Garden"; "Ponds and Streams"; "Meadows and Fields"; "Trees and Logs"; and "Beneath Our Feet." An introduction provides some general information about arthropods (animals with jointed legs, a group that includes the insects) and their classification. At the end of each chapter the scientific names are provided and explained for the insects discussed in that chapter (Class, Order, and Family). A list of books for additional reading is included, as are the names of some scientific supply companies where many of the mentioned insects can be purchased. There is a 15-page index at the end. Although not a very large volume, this is an attractive and readable hardcover publication, printed on good paper. *Close Encounters* should appeal to a wide variety of readers.

WILLIAM H. WIESE

110. A Field Guide to Hawks, North America. William S. Clark. Boston, MA: Houghton Mifflin Co., 1987. 198p., illus. (part col.), bibliog., index. $19.95, $13.95pbk. The Peterson Field Guide Series, 35. 0-395-36001-3, 0-395-44112-9pbk.

This is book number 35 in *The Peterson Field Guide Series*, a new guide written by William S. Clark, a hawk researcher and past director of the National Wildlife Federations's Raptor Information Center. It is sponsored by the National Audubon Society and the National Wildlife Federation and illustrated by Brian K. Wheeler, a wildlife artist whose work has appeared in *American Birds* and *Birding*. In addition to his color illustrations, there is a section of black-and-white photographs, most of which were provided by Wheeler and Clark. The introduction indicates that the account (description) for each of the raptors was reviewed by several persons familiar with that species, which should help ensure accuracy.

Thirty-nine species of North American diurnal raptors are separately described and illustrated, including species of hawks, eagles, falcons, kites, vultures, and the osprey and northern harrier The stated purpose of the field guide is to help anyone accurately identify these species when seen perched or in flight. Although the words "North America" appear on the title page, geographical coverage is limited to the United States and Canada.

The guide's arrangement is clear and consistent. The introduction explains how the book is organized and how one can get the most out of it. The introduction also includes general informa-

tion on identifying types of raptors by size, shape, and flight silhouette. In addition, there is a brief glossary of terms, plus illustrations, relating to a raptor's discernible physical features. The precise definitions will be especially useful for beginners, because these terms are used repeatedly throughout the book. Each species account includes a concise description, with any differences noted between the sexes, or between adults and immature birds. Major observable features are indicated in italics. Differences between similar species are also noted, to aid in precise identification. Flight patterns and behavior are covered, as are status and distribution, with a map indicating summer, winter, and permanent distribution in the United States and Canada accompanying most species accounts. Also mentioned for each species are unusual plumage, subspecies, etymology, and measurements.

Twenty-six plates are included in the center of the book, all but two of which are in color. Each plate includes brief descriptions of the species illustrated, with major identifying features again italicized. There is a contradiction in the guide regarding the California condor. The species account indicates that only a few individuals remain in the wild, whereas the description accompanying the color plate correctly reports that all wild condors have been taken into captivity.

The black-and-white photographs appear together near the back, following the species accounts. Both in-flight and at-rest pictures are included, and distinguishing features are labelled. Some of the photos are blurry or too dark, although even in the poorer ones, the labelled identifying features can usually be seen.

A long list of books and journal articles follows the photographs. The list is made more useful with the addition of its own index, searchable by species name and then by several broad topics under each species. References are included from the early 1900s up to 1987. A final index at the end of the guide lists both common and scientific names for the raptors, with page numbers indicated for text coverage and the photographs, plus plate numbers. It is obivious that considerable time and effort went into preparing this field guide so it could fulfill its stated purpose as an aid in the accurate identification of diurnal raptors. The text, drawings, and photographs all support each other to this end. This book was intended to be a working tool and should be welcomed by both the beginning hawk watcher and the seasoned expert. It is recommended for public, academic, and special libraries.
WILLIAM H. WIESE

111. Guide to the Mammals of Pennsylvania. Joseph F. Merritt.
Pittsburgh, PA: University of Pittsburgh Press for the Carnegie Museum of Natural History, 1987. 408p., illus. (part col.), bibliog., glossary. $34.95, $14.95pbk. 0-8229-3563-5, 0-8229-5393-5pbk.

The 63 mammals of Pennsylvania are described in this guide following an introductory chapter on the environmental features of the state. The accounts are in phylogenetic order, and each includes scientific and common names, description, and range maps (PA and USA). The accompanying two- to four-page text on each species covers ecology, behavior, reproduction, and, where applicable, the status of each in Pennsylvania. The 61 photographs, some colored, are of good quality, and the ample appendix contains track identification, dental formulae, a glossary, and a 15-page bibliography of scientific publications. Joseph Merritt is director of the Carnegie Museum of Natural History Biological Station, and his scientific consultant, Dr. David Armstrong, is an authority on mammals. This is an informative and accurate guide for the layperson who is interested in mammalogy of Pennsylvania. Public libraries in Pennsylvania would need this guide. *CAROLYN DODSON*

112. Rocky Mountain Mammals: A Handbook of Mammals of Rocky Mountain National Park and Vicinity.
Rev. David M. Armstrong. Boulder, CO: Colorado Associated University Press in cooperation with Rocky Mountain Nature Association, 1987. 223p., illus. (part col.), bibliog., glossary. $16.95, $8.95pbk. 0-87081-168-1, 0-87081-169-Xpbk.

Armstrong has written an excellent guide to the mammals of the Rocky Mountain National Park area for the tourist. A professor of natural sciences at the University of Colorado, Armstrong's purpose is "to allow visitors to increase their understanding of the mammals of the area—their diversity, their habits, and their complex relationships with the natural environment." This guide is a revised edition of the work first published in 1975 by the Rocky Mountain Nature Association.

The core of the text is the 66 species accounts of mammals now residing or known to once have resided in the park. Four species, bison, pronghorn, grizzly bear, and gray wolf, "have been extirpated within historic time." Each account includes the scientific name; a description; information on distribution, habitat, and natural history; plus selected references to the technical literature. The accounts are arranged "in phylogenetic sequence from order through genera" with species arranged alphabetically within each genus. Each order is accompanied by drawings by Bill Border. Color photographs are included for 16 species and many black-and-white photographs are provided in the section on ecological distribution.

This guide has extensive introductory material divided into four chapters. Three of the chapters are informative essays: "What Is a Mammal?," "Mammalian Distribution," and "How to Observe Mammals." Appendices include a key to mammal identification, a glossary of six pages, and a dis-

cussion of the origins of scientific names. The volume concludes with a bibliography of the literature cited in the text and species accounts. There is no index.

Although this guide is intended for use in the Rocky Mountain National Park and vicinity, it will also be useful for travelers "to mountainous parts of north-central Colorado and adjacent Wyoming." Public and school libraries in Colorado and Wyoming will find this a useful reference tool. *STEVEN L. SOWELL*

113. Whales of Hawaii: Including All Species of Marine Mammals in Hawaiian and Adjacent Waters.

Kenneth C. Balcomb, III. Stanley M. Minasian, ed. San Francisco, CA: Marine Mammal Fund, 1987. 99p., illus. (part col.), bibliog., index. $8.95. 0-9617803-0-4.

The Whales of Hawaii, published by the Marine Mammal Fund as part of their public education program, was produced by Stanley M. Minasian, written by Kenneth C. Balcomb, III, and illustrated by Larry Foster. Both Minasian and Balcomb co-authored *The World's Whales*, a guide to whales, dolphins, and porpoises published by Smithsonian Books in 1984. This book was "designed to acquaint the reader with the broad spectrum of marine mammals which typically occur around the Hawaiian Islands and to provide a glimpse of their wondrous lives." It is not technical, but it is an exceptionally concise, well-written, and informative handbook to marine mammals in the waters adjacent to Hawaii. Five species of baleen whales, six large-toothed whales, 11 species of dolphins, and one seal are described. In addition, a suggested reading list of 14 up-to-date and relevant references and an appendix, which mentions three species of whale and one of dolphin that might be sighted in the Hawaii area, are included with the text.

Descriptions of each species average three pages in length and include information about the mammal's distribution, natural history, and prevalence; descriptive remarks to aid in identification and the current status of the species as a result of the negative impacts of human activities are also included. The text is complemented with many excellent color action photos throughout, and each of the 28 species described in the book is accompanied by a pen-and-ink sketch with either color photographs and/or paintings (there are only two black-and-white photographs in the entire book). Although this book was written primarily to introduce Hawaiian residents and visitors to the marine mammals of that area, it does more than that. The exciting color photographs and effective writing bring *The Whales of Hawaii* to the reader, be they at home or in the library. It would complement either the reference or stack collections of public, school, or academic/special libraries. *BILL COONS*

Handbooks

114. Amphibians and Reptiles of Texas: With Keys, Taxonomic Synopses, Bibliography, and Distribution Maps.

James R. Dixon. College Station, TX: Texas A and M Univesity Press, 1987. 434p., illus., bibliog., keys, maps, glossary, index. $32.50, $16.50pbk. 0-90096-293-6, 0-89096-358-4pbk.

James R. Dixon, professor of wildlife and fisheries sciences at Texas A & M University, specializes in neotropical reptiles and has written extensively on Texas herpetology. This book began as an updating of *Amphibians and Reptiles in Texas* by G. G. Raun and F. R. Gehlbach (Dallas Museum of Natural History Bulletin 2:1-61, 1972). It provides several assorted aids to the study of Texas "herps." Almost a third of the book devotes itself to a bibliography of "scholarly literature concerning Texas amphibians and reptiles" from 1852 to 1986. Another third consists of distributional maps for each of the 204 known species, showing the record of its occurrence on a county-by-county basis in Texas. In addition, species accounts provide an index to the bibliography, by numeric notations that correspond to the numbered articles. The few black-and-white photographs provide little help in the identification of amphibians and reptiles, but for this purpose, the author has provided taxonomic keys to the species and subspecies. An attempt has been made to make these keys usable by beginning students, both by using terms familiar to beginners within the key and by providing a glossary of terms. The introductory text contains an analysis of the Texas herpetology literature, with reference to the major publication sources and the main contributors, together with a brief history of Texas herpetology. Although the book is slanted toward the more serious student, the Texas distributional maps will make it useful for the amateur if used in conjunction with general field guides to snakes and amphibians. The index provides access by common and scientific name only. *LINDA JACOBS*

115. Auks: An Ornithologist's Guide.

Ron Freethy. New York, NY: Facts on File Publications, 1987. 208p., illus. (col.), bibliog., index. $24.95. 0-8160-1696-8.

This is a readable guide covering the life histories of the Alcidae, the family of auks, puffins, murrelets, and other birds of northern hemisphere sea coasts. Introductory chapters cover evolution, distribution, and ecological significance of the family in authoritative, yet easily understood language. Each genus is discussed in detail in a separate chapter, accompanied by excellent color photos, line drawings, and maps. This is a recommended book for school, public, and college libraries. *CAROLYN DODSON*

116. Birding in the San Juan Islands.
Mark G. Lewis; Fred A. Sharpe. Seattle,
WA: The Mountaineers, 1987. 219p., illus.,
bibliog., index. $9.95pbk. 0-89886-133-0pbk.

In their introduction to *Birding in the San
Juan Islands*, Mark G. Lewis and Fred A. Sharpe
state that their book "...is not intended to be a
general guide to the field identification of birds"
but a summary of information on the birdlife of
the San Juan Islands gathered by the authors
over many years. They are right. This book, written by an ornithologist and a naturalist from the
Pacific Northwest, is a fine example of a region-
specific natural history with a focus on bird communities. It is well-written and informative and
will be a valuable reference work for the novice
natural history student, as well as for the wildlife
manager, the experienced naturalist, or the researcher. The book is divided into six main sections. In a brief three-page section entitled "The
San Juan Islands," the reader is introduced to the
locale. We learn about the topography of the
islands as well as the climate and plant life. In
the next section, "Human Presence in the Archi-
pelago," the authors discuss the threat to the
animal and plant life imposed by increasing human activity both on land and in nearby waters.
This threat takes the form of recreational activities, debris, and petrochemical development. This
section, although very informative, would have
been enhanced by references to the large body of
literature on these subjects. In "Going Birding,"
the authors provide a wealth of information to
the novice bird-watcher on such subjects as
birding etiquette and equipment. References to
various field guides that cover the birds of the
San Juan Islands are provided.

In a section on "Bird Habitats," the authors
describe thirteen avian habitats present in the
San Juan Islands and provide a brief list of bird
species inhabiting each. The "Site Guide" gives
the reader detailed information on the types of
wildlife and habitats to be seen while touring the
four main islands serviced by the Washington
State Ferry System or traveling the waterways by
ferry or by private boat. A map of each of the
four islands is provided. The maps are well
drawn; however, many of the place names are
difficult to read because of the shading used at
the land-water interface. The "Species Accounts"
section, the book's largest, is arranged according
to the American Ornithologists' Union *Check-List
of North American Birds* (1983) and supplements.
As each taxonomic order is presented, a brief
overview of the order is given. Each species account opens with the common and scientific
name of the bird. The account places the bird in
the San Juan Islands ecosystem, with information
on the bird's feeding habitats, nesting activities,
vocalizations, and when the birds can be expected
to be seen or heard throughout the year. Numerous references to the literature are provided
to document individual bird sightings. The excellent line drawings of many of the species enhance the text. Some of the individual species
accounts do contain field identification notes
"...particularly for those species whose proper
identification poses a specific problem in this re-

gion." A recommended addition to this section
would be length measurement ranges for each
species. The appendices contain a 12-month
checklist presented in bar-graph format (common
to very rare), a list of agencies concerned with
San Juan Islands wildlife, references and notes
(mostly sighting records), and a very useful three-
page bibliography. The index includes both scientific and common names, as well as place
names mentioned in the text. The few
misspellings of scientific names in the index do
not detract from its usefulness. Lewis and Sharpe
have produced a well-written and informative volume. The fact that this book is region-specific
should not restrict its acceptance to the
northwestern United States and Canadian library
collections. For those libraries, both public and
college and university, which collect in the area of
natural history, this book is a must. It is a valuable addition to a reference and/or circulating
collection.   *JAMES E. BIRD*

117. Birds in Minnesota. Robert B.
Janssen. Minneapolis, MN: University of
Minnesota Press for the James Ford Bell
Museum of Natural History, 1987. 35p.,
illus. (part col.). $35.00, $14.95pbk.
0-8166-1568-3, 0-8166-1569-1pbk.

Birds in Minnesota is not a field guide. With
that said, let us look at what it is. Some might
use it as an aid to bird-watching. Here they can
discover the status of a species, whether it is an
accidental or a resident, what route it migrates
on, and when is the peak time to see it in its
greatest abundance. The book is also an atlas of
breeding birds, providing ranges for each species
that nests in Minnesota. For those interested in
population studies, it offers up-to-date information on the number of nests found in the state
and the colony size. Finally, *Birds in Minnesota*
is a labor of love. Robert Janssen, a self-pro-
claimed "provincial Minnesota birder," has been
pursuing this activity since the age of 12. At that
time, he purchased Thomas Sadler Roberts's *The
Birds of Minnesota*, and vowed to "keep it up to
date." Through the years, Mr. Janssen has been
active in the Minnesota Ornithologists Union
(MOU), editing *The Loon*, the MOU's newsletter,
since 1958. Through the MOU he met Jan Green,
and together they fulfilled the vow of a 12-year-
old, publishing *Minnesota Birds: Where, When
and How Many* in 1975. The present volume
updates the 1975 volume, listing an additional 26
species over the 374 originally reported. Species
are arranged sequentially according to the *American Ornithologists Union (AOU) Checklist of
North American Birds* (6th ed.). Both common
name and taxonomic name are given for each
entry, however, the index lists common name
only. A very few pages of color photos of Minnesota specialties highlight the book, but, because
the book's intent is not identificational, pictures
do not accompany each entry.

ANGELA MARIE THOR

118. **Mammals in Wyoming.** Tim W. Clark; Mark R. Stromberg. Lawrence, KS: University of Kansas Museum of Natural History, Dist. by University Press of Kansas, Lawrence, KS, 1987. 314p., illus., bibliog., glossary, index. $25.00, $12.95pbk. 0-89338-026-1, 0-89338-025-3pbk.

The individual species accounts of 117 Wyoming mammals include a photograph, description, and the mammal's range and habitat (including a map of its distribution within the state), plus brief comments on reproduction, habits, and food preferences. The authors sometimes back up their comments with citations from the literature (approximately two citations per species).

Mammals in Wyoming updates Charles A. Long's *Mammals of Wyoming* (1965), a monograph written for a more scholarly audience. The current authors have dropped detailed descriptions, specific records of occurrences, and measurements that are of interest mainly to professional mammalogists and have added valuable information on the natural history of each species. As a result, this is a text that will appeal to a much wider audience.

The information is relatively current. For example, the book refers to the current status of the endangered black-footed ferret with a 1986 citation. It is hoped that future editions will follow the lead of Charles Schwartz's *The Wild Mammals of Missouri*, with its excellent full-page illustrations of each mammal, accompanied by details of their paws and skull. Such useful visual information can aid in the identification of each species by sight, tracks, or skull fragments.

An introductory section provides an overview of the history of Wyoming mammalogy, a discussion of biotic regions within the state, and an explanation of the species accounts that follow. Additional aids for study include a key to the orders and families of Wyoming mammals, a glossary, an extensive bibliography, and a technical key useful for identifying collected specimens. The index is limited to references to scientific and common names.

Tim Clark is a member of the faculty in biological science at Idaho State University. He is also the author of *Ecology and Ethology of the White Tailed Prairie Dog.* Mark Stromberg, formerly with the Nature Conservancy in Wyoming, is resident director of the Appleton-Whittel Biological Research Sanctuary in Arizona and is also a research associate at the University of Colorado, Boulder.

The authors ascribe the impetus behind this compilation to the energy "boom" and the obvious need to mitigate its negative effects. Since 1965, there have been several near extirpations of Wyoming mammals; the black-footed ferret and the pigmy shrew may be extinct. Every school library in Wyoming that serves fifth grade and over should have this book.   *LINDA JACOBS*

119. **Naturalized Birds of the World.** Christopher Lever. Harlow, Great Britain: Longman Scientific and Technical, Dist. by John Wiley and Sons, New York, NY, 1987.

615p., illus., bibliog., index. $197.00. 0-470-20789-2.

The *Naturalized Birds of the World* by Sir Christopher Lever is a reference work of rare quality: informative, meticulously researched and referenced, and a wonderful work to read. Lever is the author of two previous books on naturalized species: *Naturalized Animals of the British Isles* and *Naturalized Mammals of the World.* The author uses the term "naturalized," as defined by the *Oxford English Dictionary*, to mean "The introduction (of animals and plants) to places where they are not indigenous, but in which they may flourish under the same conditions as those which are native." The preface of the volume provides the user with, among other things, a glossary of the terminology used, as well as information on the classification, nomenclature, and sequence of presentation and the system of geographic distribution used.

In the beginning, Lever provides a detailed overview of the introduction of birds into new areas, including the various reasons and consequences for these occurrences. The main body of the work is individual species accounts. In the preface, Lever refers to the infividual species account as "...a monograph on an individual bird...," and indeed each one is. Each account contains the following sections: the natural distribution of the species with map, the naturalized distribution of the species with map, detailed accounts of the history of the species' arrival in their specific naturalized location, and brief details of the artificial distribution of allied, related, or associated species. Brief citations are included in a notes section at the end of each species account; full citations can be found in the 52-page bibliography toward the end of the volume.

Two indexes at the end of the volume provide access to the individual species accounts by geographic location and by vertebrate species. The geographic index is an alphabetical list of locations mentioned in the individual species accounts. It does not, however, index geographic locations mentioned in the discussions of allied, related, and associated species. The index to vertebrate species is one alphabetical list that includes both common and scientific names of the birds and other vertebrates discussed in the species accounts. Sir Christopher Lever has authored an outstanding reference work that will be a valuable addition to public libraries as well as to college and university reference collections.   *JAMES E. BIRD*

120. **The Tanagers: Natural History, Distribution, and Identification.** Morton L. Isler; Phyllis R. Isler. Washington, DC: Smithsonian Institution, 1987. 404p., illus. (part col.), bibliog., index. $70.00, $49.95pbk. 0-87474-552-7, 0-87474-553-5pbk.

Tanagers are colorful New World birds primarily found in Central and South America. This "comprehensive review of existing knowledge" will be of great value to ornithologists and will serve as an outstanding example of what a reference book should be. It is based on numerous

personal observations by the Islers, data gleaned "from over a thousand references," the in-depth study of museum specimens, and the unpublished research data of several individuals and institutions. In producing this work, the Islers hope "to stimulate field study that will contribute to the understanding and ultimately the conservation of tanagers and the ecosystems of which they are a part" and "to provide an up-to-date base for scientists engaged in ecological, zoogeographic, and taxonomic research." The authors have accomplished both goals brilliantly.

In addition to the standard prefatory material, the book includes a statement of objectives, an abbreviations list, a glossary, an outline of a typical "species account," and an essay, "The Nature of Tanagers." The essay provides an overview to the entire subfamily, Thraupinae. It discusses tanager size, appearance, distribution, habitat, and vocalization as well as social, feeding, and breeding behavior.

The heart of the text is the 242 species accounts. Each account includes the known information on average length, weight, geographic range, elevational range, habitat and behavior, vocalizations, breeding, and a detailed listing of the sources of information for the species. Each species account is accompanied by a range map (all are drawn to the same scale), and each is keyed to a color plate. Thirty-two plates, which were reproduced from beautifully detailed, original paintings by Morton L. Isler, illustrate 551 plumages and 23 flight patterns. The species accounts are grouped by genus and are preceded by genus accounts that "provide generalizations about habitat, behavior, vocalizations, and nesting." The Tanagers concludes with 18 pages of references to the ornithological literature on Thraupinae and an index which lists both the English and the scientific names for the species. This outstanding handbook is highly recommended, especially for academic and special libraries, but also for large public libraries.

STEVEN L. SOWELL

Histories

121. Idaho Wildlife. Jan Wassink. Helena, MT: American Geographic Publishing, 1987. 103p., illus. (col.), index. $14.95pbk. Idaho Geographic Series, no. 2. 0-938314-37-8pbk.

Idaho Wildlife is one of a series of books published by American Geographic Publishing which describes Idaho's geography, natural history, and cultural heritage. If this book is any indication of the quality of this series, they have a winner. *Idaho Wildlife* is organized into nine sections and contains 103 pages of well-written text interspersed with scores of exceptionally fine color photographs. Beginning with a brief description of the history of natural resource development and conservation in Idaho, the book moves on to a description of Idaho's various physical environments, and a section of practical hints on observing wildlife and nature's cycles follows. Most of the book (60 pages) is devoted to mammals and birds; two short sections (14 pages)

provide token coverage of key reptiles and fish. Species' physical characteristics, natural habitats, behaviors, natural history, and conservation are covered.

The author's writing is clear, and the scientific explanations are tempered with a humanistic sensitivity to the interplay between man and nature. Technical assistance was provided by individuals with Idaho's Fish and Game Department, the U.S. Bureau of Land Management, and other research organizations. This is not a definitive scientific guide or text, such as *Walker's Mammals of the World*, but it is a quick and delightful book for the layperson. Its utility and appeal should extend well beyond the borders of Idaho and the Intermountain West; many of the species covered live throughout the United States, and the photography is excellent. *Idaho Wildlife* has been cataloged as juvenile literature, but it is appropriate, instead, for a much wider audience. Individuals and libraries of all sorts would benefit from including this book in their collection. It is recommended without reservations.

BILL COONS

Manuals

122. Basic Anatomy: A Laboratory Manual: The Human Skeleton—The Cat. 3rd. B. L. Allen. New York, NY: W. H. Freeman and Co., 1987. 204p., illus. (part col.), index. $16.95. 0-7167-1755-7.

This is a laboratory manual, revised and expanded from the earlier edition, for an undergraduate gross anatomy course in which the cat is dissected. Similarities and differences between cats and human anatomy are pointed out. The material is organized by system and includes excellent illustrations of clear line drawings that are well labelled. The nerves, veins, and arteries are in color. The text is well written, and there is a 16-page index. This is basically a laboratory-use manual, but academic and special libraries will find it a useful reference source on comparing the anatomy of the cat and humans.

CAROLYN DODSON

123. New Caribbean Molluscan Faunas. Edward J. Petuch. Charlottesville, VA: The Coastal Education and Research Foundation (CERF), 1987. 158p., illus., bibliog., index. $38.50. 0-938415-01-8.

New Caribbean Molluscan Faunas contains descriptions of 109 recently identified living mollusks found in several widely separated areas of the Caribbean. Written by Edward J. Petuch, a widely published author on the topic, the book is based largely on his Caribbean mollusk research. After two introductory chapters, the book is arranged by location. Chapters on mollusks from Florida, the Bahamas, Honduras and Central America, Northern South America, and Brazil follow. Descriptions of two additional species from Florida are included in a four-page addendum which follows the bibliography and a sys-

tematic index. Although location plays a large part in the book's arrangement, no maps are included to show the location of specimens collected or molluscan communities. Black-and-white photographs of each species described are included on 29 plates (28 of which are listed in the table of contents). Although the photographs are useful in their depiction of shape, lack of color photographs will make use of this book difficult for nonprofessionals. In addition, several photographs of individual shells are washed out (possibly because of the printing process) or show glare from porcelaneous surfaces. Only recommended for university, special, and research libraries with collection on marine mollusks.

LINDA R. ZELLMER

124. Wisconsin Birds: A Seasonal and Geographical Guide. Stanley A. Temple; John R. Cary. Madison, WI: The University of Wisconsin Press, 1987. 364p., bibliog., maps, index. $22.50, $9.95pbk. 0-299-11430-9, 0-299-11434-1pbk.

This book, sponsored by the Wisconsin Society for Ornithology, presents the results of the Wisconsin Checklist Project, "a five-year (1982–86) program which used simple checklist information provided by volunteer observers to produce information on bird populations in Wisconsin." The first part of the book presents accounts for the 265 bird species most commonly found in Wisconsin. Each account is on a page by itself and presents in graphic format information on the probability of seeing the species in Wisconsin, the geographical range and abundance of the species within the state, and seasonal changes in abundance that occur during the year. The second part lists reports by county and months of 98 rare birds that are never reported more than 10 times during a year. This is a must book for bird-watchers in Wisconsin in that it indicates on a map where the species can be observed and on graphs the times of the year that the species are most abundant. It is recommended for university and research libraries, particularly those in Wisconsin and adjacent states and provinces. It is not a field guide because there are no descriptions of the species. An appendix discusses the Wisconsin Checklist Project, and the index permits access to the species in the main part of the guide. The tightly bound paper copy may not withstand heavy use.

H. ROBERT MALINOWSKY

Treatises

125. Birds of the Great Basin: A Natural History. Fred A. Ryser, Jr. Reno, NV: University of Nevada Press, 1985. 604p., illus. (part), bibliog., index. $29.95, $19.95pbk. Max C. Fleischmann Series in Great Basin Natural History. 0-87417-079-6, 0-87417-080-Xpbk.

"To the uninitiated, the Great Basin does not appear to possess much bird life." Ryser disproves this statement because "Whether seen or not, numerous kinds of birds frequent these desert shrublands and woodlands." The area abounds with birds of all types, including eagles, hawks, falcons, sparrows, owls, finches, pelicans, ibises, cranes, plovers, and dippers. "The Great Basin forms the northern part of the Basin and Range Province. To the west of the Basin lies the Sierra-Cascade Province, to the north the Columbia Plateau Province, and to the east the Colorado Plateau Province." It includes deserts, mountains, and wetlands.

The first 58 pages of the book discuss the environment of the area. This is followed by a chapter devoted to each of the types of birds: "Divers," "Specialized Fishermen: Pouch Users and Plungers," "Marshbirds," "Waterfowl," "Shoreline Birds," "Diurnal Birds of Prey," "Nocturnal Birds of Prey," "Upland Game Birds," "Cuckoos," "Hibernators," "Woodpeckers," and "Perching Birds." Each chapter begins with a brief paragraph indicating which orders live in the Great Basin and then offering a detailed description of the families. The descriptions are not meant to be a "field-guide type" of description but rather an interesting account of the bird including anatomical peculiarities, behavior, and natural history. The style of writing is excellent, enticing one to read more and more. For example, "A loon is certainly not designed for land life. Its legs cannot hold it off the ground, since they are positioned at the end of, instead of underneath the body. On land a loon moves by pushing itself along on its belly or by employing its wings as crutches. In recognition of their helplessness on land, loons build their nests immediately adjacent to water." There are beautiful black-and-white sketches of selected birds. In the middle of the book are 60 full-color photographs of birds. Unfortunately they are not keyed to any of the text. The author is a faculty member in the biology department at the University of Nevada and curator of the UNR Museum of Biology. He has studied the birds and mammals of the Great Basin for over 30 years. His excellent book should be of great interest to anyone who studies the birds in the Great Basin, whether scientifically or as an amateur. This is a highly recommended book.

H. ROBERT MALINOWSKY

126. Fishes of the Great Basin: A Natural History. William F. Sigler; John W. Sigler. Reno, NV: University of Nevada Press, 1987. 425p., illus. (part col.), bibliog., glossary, index. $32.50. Max C. Fleischmann Series in Great Basin Natural History. 0-87417-116-4.

Fishes of the Great Basin: A Natural History has been written for "fishermen, naturalists, ichthyologists, fishery biologists, pet owners, and people who just want to know about fish." The authors, both fishery biologists, have produced a comprehensive text that meets the information needs of this broad audience very well. Libraries with collections on ichthyology or the Great Ba-

sin will want to purchase this book. (The Great Basin covers most of Nevada and Utah and portions of Wyoming, Idaho, California, and Oregon.)

The text is divided into two sections. The first and shorter section is general with chapters on the Great Basin drainages, the history of fishing in the area, the Endangered Species Act, the evolution and classification of fishes, and a key to native and introduced fishes of the Great Basin (by Gerald R. Smith).

The second section provides life histories of over 90 fish species found in the Great Basin. These are arranged in phylogenetic order by families. Within each family, the species are listed alphabetically by genera. Common names are included with the scientific name of each species. Each life history is subdivided into nine topics: importance, range, description, size and longevity, limiting factors, food and feeding, breeding habits, habitat, and preservation of the species. A black-and-white line drawing accompanies each life history. Eleven colored line drawings are included to help differentiate some species of trout, sunfish, and perch. An "Annotated Checklist of Fishes of the Great Basin" follows the life histories. The volume concludes with an extensive glossary, a lengthy bibliography of literature cited, an index to fishes by scientific and common names, and a general index.

STEVEN L. SOWELL

127. Frogfishes of the World: Systematics, Zoogeography, and Behavioral Ecology. Theodore W. Pietsch; David B. Grobecker. Stanford, CA: Stanford University Press, 1987. 420p., illus. (part col.), bibliog., index. $67.50. 0-8047-1263-8.

Strange packages often disguise unexpected delights. The jacket cover of *Frogfishes of the World* promises to deliver more than anyone would ever want to know about a family of incredibly ugly fish, but inside is a fascinating exhaustive study of a very unusual group of organisms. This monograph is intended for the "professional ichthyologist" but is a model of clear exposition. The illustrations, color photographs, and overall book design are visually stimulating. From a historical literature survey to an analysis of locomotion, from a lengthy examination of taxonomic relationships to the demolition of cherished myths, the authors have produced a wonderful example of what good science and publishing can do for an obscure corner of knowledge.

The Antennariidae are a family of bizarrely shaped and colored predatory fish. Some of these fish have a unique form of locomotion. Some sport a rod, line, and bait to lure their prey close. The frogfishes are able to attack and swallow their victims in one gulp, even prey of their own size. The frogfishes are a weirdly intriguing group of creatures, and it is easy to understand the

interest that these fish might provoke. Having a longstanding interest and record of publications in this speciality. Pietsch's commanding grasp of the subject is very evident.

It is unfortunate that, because the focus of this book is very narrow and the target audience so specialized, this excellent book will appeal to so few. This is a recommended addition to any research collection with an interest in the life sciences and ichthyology in particular.

GEORGE M. A. CUMMING, JR.

128. Poisonous and Venomous Marine Animals of the World. 2nd rev. Bruce W. Halstead. Princeton, NJ: The Darwin Press, Inc., 1988. 1168p, 288p, illus. (part col.), bibliog., index. $250.00. "An International Oceanographic Foundation Selection." 0-87850-050-2.

Poisonous and Venomous Marine Animals of the World "incorporates the original three-volume work (published by the United States Government Printing Office in 1965, 1967, and 1970) and the first revised edition (published by the Darwin Press, Inc. in 1978) in a single, update volume." The primary purpose of the original three-volume set was to "provide a systematic organized source of technical data on marine biotoxicology, covering the total world literature from antiquity to modern times." The first revised edition was intended to provide an abridged, updated work within a single volume. This second edition brings all of this together in a very impressive volume that is well documented. With few exceptions, all the literature that has been referenced has been examined in its original form, with foreign works translated into English.

The book is phylogenetically arranged. Each species was selected on the basis of its commonality or uniqueness within a geographical area, its toxicity, or its medical importance. This excellent treatise is divided into 26 chapters under three sections: "History of Marine Zootoxicology," "Invertebrates," and "Vertebrates." The historical section is a chronology of marine zootoxicology with a rather extensive bibliography at the end.

The real meat of this treatise is in the sections on invertebrates and vertebrates. Each *phylum* is treated in the same manner. First there is an introduction that describes the *phylum* in general terms using, in most cases, common names. This is followed by a representative list of the animals that are venomous in this *phylum*. This list gives family, species, distribution, and the literature source. The biology is discussed in detail as is the morphology of the venom apparatus and mechanism of intoxication. Medical aspects cover clinical characteristics, pathology, diagnosis of stings or bites, treatment, and prevention. Public health aspects are then covered. The sections on toxicology, pharmacology, and chemistry are extremely detailed and important in the understanding of that particular venomous animal and how it can do harm.

There are some illustrations throughout the text, but the key illustrations are in a separately paged section to the back of the book. The plates depict a selection of the animals both in color and black-and-white photographs and also photographs of the actual bites and reactions to the human body. There is a glossary, personal name index, and a general index preceding the section of separately paged illustrations.

This is a highly recommended book for anyone who frequents the oceans and seas where venomous and dangerous animals may lurk. It is also an excellent research treatise for those doing research into dangerous animals. This particular field does not have as much research directed toward it as it should have, so a book of this type and uniqueness is very important.

H. ROBERT MALINOWSKY

CHEMISTRY

★
GENERAL SOURCES
Dictionaries

129. Compendium of Chemical Terminology: IUPAC Recommendations. Victor Gold; Kurt L. Lening; Alan D. McNaught; Pamil Sehmi, comps. Oxford, Great Britain: Blackwell Scientific, 1987. 456p., bibliog. $69.50, $48.45pbk. 0-632-01765-1, 0-632-01767-8pbk.

This compendium was initiated by the Interdivisional Committee on Nomenclature and Symbols of the International Union of Pure and Applied Chemistry (IUPAC) under the leadership of Victor Gold, professor of chemistry, King's College, London. Upon Professor Gold's death, this dictionary was continued by Kurt L. Lening (Chemical Abstracts Service) and Alan D. McNaught and Pamil Sehmi (Royal Society of Chemistry). It is the first step in the production of a "comprehensive and authorative chemical dictionary," being a dictionary of terms recommended by IUPAC for use by chemists in reporting research results.

The terms included in the dictionary were taken from the publications of the IUPAC divisions in the areas of analytical, inorganic, macromolecular, organic, and physical chemistry published before 1986. The dictionary does not include material from all chemistry-related divisions of IUPAC, e.g., Applied and Clinical Chemistry Divisions, or from the official documents with chemical content of other international organizations. It does not include names of specific compounds and includes only a few terms representing categories of compounds. The terms included represent chemical processes, chemical reactions, research procedures, observed phenomenon, methods of analyzing research results, statistical analysis, etc. The estimated 2,700 entries are arranged alphabetically by the first word in the term. Terms that begin with a Greek letter are alphabetized as though the letter were spelled out. Terms are printed in boldface, upper-case letters and are underlined so they stand out on the page. Some include the area of chemistry in which the term is used. In addition to the terms, each entry includes a definition (or definitions) of one line to one page and references to the literature and source documents where the term and the definitions were found. Three methods are used to indicate cross-references. Italicized terms

within a definition refer to other entries where relevant information can be found. "See also" references are used in cases where one definition supersedes another or when a second definition contains additional information. "See" references are used to refer from a term whose use is not recommended to the correct term. A "+" before an entry indicates IUPAC recommends that the use of that term be discontinued.

At the end of the dictionary is a list of literature references mentioned in the definitions and a list of IUPAC publications which were used as source documents. Although this promises to be a monumental reference tool in the future, its present use must be supplemented by other chemical and scientific dictionaries. It is, however, recommended for academic and research libraries as a working draft. *LUCILLE M. WERT*

130. Hawley's Condensed Chemical Dictionary. 11th. N. Irving Sax; Richard J. Lewis, Sr. New York, NY: Van Nostrand Reinhold Co., 1987. 1,288p., illus. $52.95. 0-442-28097-1.

This dictionary is the eleventh edition of a long-established reference source which describes approximately 20,000 chemical topics of interest. "Three distinct types of information are presented: 1) descriptions of chemicals, raw materials, processes and equipment; 2) expanded definitions of chemical entities, phenomena, and terminology; and 3) description or identification of a wide range of trademarked products used in the chemical industries." Also included are commonly accepted chemical abbreviations, short biographies of chemists of historic importance (including all Nobel Prize winners in chemistry), and brief descriptions of many American technical societies and trade associations. The entries are arranged alphabetically, letter by letter, e.g., "whitener" precedes "white oil," and many of the prefixes used in chemical names (o-, m-, p-, etc.) are disregarded for filing purposes. The reader is referred to the Introduction for a more complete discussion of the filing rules.

For each chemical entry, the following information is given, if available: the commonly accepted name (which is the filing element); synonyms, including IUPAC names; Chemical Abstracts Service Registry Number; molecular formula (structures are given in approximately 300 of the entries); properties, source, or occurrence; derivation, in general terms; grades available; hazard information; use; and shipping regulations. Synonyms are cross-referenced to the correct

heading. The coverage of chemical terms and the information given in the entries has not changed much from the tenth edition, although the type in the eleventh edition is larger and easier to read. In cases where significant changes in knowledge or technology have taken place, the entries have been updated. A good example of this is seen in the entry for "Superconductivity," which states: "Depending upon the substance, the maximum temperature [transition temperature] for the phenomenon is now [1985] 0-5-28K...." As in the previous editions, there are useful appendices. Appendix I, "Origin of Some Chemical Terms," is an interesting section, giving etymologies of approximately 225 chemical terms. Appendix II, "Highlights in the History of Chemistry," is new in this edition and, in five sections, gives important events and discoveries in the history of chemistry. The first section, "Chronology of Notable Achievements," lists 66 important people and their contributions to chemistry, beginning with Democritus (465 B.C.) and ending with Robert W. Woodward (who won the 1965 Nobel Prize in chemistry). The next three sections are brief essays on the histories of the American Chemical Society, *Chemical Abstracts*, and Center for History of Chemistry. The last section, "History of Five Major Industries," presents essays on the development of five major chemical industries: drug and pharmaceutical, paper, plastics, petroleum, and rubber. Although this information may be easily found elsewhere, it is also appropriate in this reference source. In only seven pages, this section does present important achievements and developments in the history of chemistry in an easily read, concise format. With the industry's emphasis on currentness and keeping up with the new technologies, it is easy to lose sight of chemistry's important beginnings.

Appendices three and four are to be used in conjunction with the trademarked names in the dictionary. Each entry for a trademarked product gives a superscripted number which refers to the manufacturer of that product, listed in numerical order in Appendix III. Addresses for the 587 manufacturers listed in Appendix III are given in Appendix IV. There are a number of very good dictionaries of chemistry, each having many of the same attributes but also suiting a slightly different purpose. Two other dictionaries which list chemicals and chemical terms of interest are Bennett's *Concise Chemical and Technical Dictionary* (4th ed., 1986) and *Hackh's Chemical Dictionary* (4th ed., 1972). These two sources include more entries (85,000 and 55,000, respectively) than Hawley's (20,000 entries), but Hawley's entries are more expanded (although only slightly more expanded than Hackh's). Even though Hawley's coverage is somewhat limited, it does include the most important terms in chemistry. Information on chemicals in this source is also similar to the information given in the *Merck Index* (which covers more chemicals and gives bibliographic information), but this dictionary also lists chemical terms and so would be useful even if a library already has the *Merck Index*. This dictionary is recommended to any library involved in chemical reference work and

should be used in conjunction with the aforementioned sources as well as with the more specialized dictionaries, when necessary.

DEBRA A. TIMMERS

Encyclopedias

131. The Biochemistry of Human Nutrition: A Desk Reference. Eva May Nunnelley Hamilton; Sareen Annora Stepnick Gropper. St. Paul, MN: West Publishing Co., 1987. 324p., illus., bibliog. $27.25. 0-314-29520-8.

The Biochemistry of Human Nutrition: A Desk Reference succeeds in its attempt to explain concepts of biochemistry associated with human nutrition in a language that is easily accessible to the general science reader. The entries are arranged alphabetically and each entry is more or less complete in itself. For those desiring information on related topics, cross-references are given to other entries in the volume. Also, for those in need of more in-depth treatment of these topics, there are frequent suggestions for further readings at the end of an entry.

This book will perhaps be most useful to those with some college-level biochemistry background who need to quickly reacquaint themselves with common biochemical terminology. The book is generally well written, and it does a good job of retaining its focus on those aspects of biochemistry which deal with human nutrition. It is recommended as a useful addition to any chemistry reference collection. *JIMMY DICKERSON*

Handbooks

132. Hazardous Chemicals Desk Reference. N. Irving Sax; Richard J. Lewis, Sr. New York, NY: Van Nostrand Reinhold Co., 1987. 1,084p., illus., bibliog., index. $69.95. 0-442-28208-7.

Information from this handbook, presenting toxicity information on approximately 5,000 common industrial and laboratory materials. The information is largely extracted from the more comprehensive handbook, *Dangerous Properties of Industrial Materials,*, 6th ed. (1984), by N. Irving Sax. This new book's scope is narrower, containing only 25 percent (or 4,700) of the 18,000 entries in the larger parent handbook. Selection is based upon the materials' "importance in the industry, their toxicity or fire and explosion hazard, or upon widespread interest in the material." The types of materials included are "basic chemicals, pesticides, dyes, detergents, lubricants, plastics, drugs, food additives, preservatives, ores, soaps, extracts from plant and animal sources, and industrial intermediates and waste products from production processes."

Because this source contains only 4,700 entries, it is not meant to be comprehensive in scope. Extensive spot-checking indicated that the most important materials are included. However,

there are some notable exceptions. Methyl triflate (Registry Number 333-27-7), an extremely toxic methylating agent, is not included. Neither are there entries under the more general headings Methylating Agents or Alkylating Agents. Two chemicals of recent interest, Aspartame (Nutrasweet) and Ibuprofen (Motrin), are not listed even though other artificial sweeteners (Saccharin and Sodium Cyclamate) and other analgesics (Acetol 2, or Asprin, and 4'-Hydroxyacetanilide, or Acetaminophen) are. Despite these exceptions, the handbook seems to be comprehensive enough for use in most laboratories. The entries are arranged alphabetically by the name by which they are best known. A Chemical Abstracts Service Registry Number index (not included in the parent handbook) and a synonym index provide additional access points to the entries.

The form of the entries differs from that in the parent handbook. They are much more readable for the layperson in this desk reference and do not contain the less relevant information, such as toxicity data and references to articles about the toxicity of the material. For each entry, the following information is given, if available: the material name, hazard rating (a rating of 1-3, with 3 the most hazardous), CAS Registry Number, NIOSH's Registry of Toxic Effects of Chemical Substances Accession Number, molecular formula (in Hill notation), molecular weight, properties of the material, synonyms (both English and non-English), OSHA Permissible Exposure, the American Conference of Governmental Industrial Hygienists Threshold Limit Values (the air concentration to which workers can be exposed for a normal 40-hour work week without ill effects), the German Research Society's (MAK) values (the workplace hazard potential), U.S. Department of Transportation Classifications (e.g., combustible liquids), and the Toxic and Hazard Review (a description of the toxicity and hazards of the material and directions for fighting a fire). Few entries contain all these data elements. However, most include at least the name, Registry Number, molecular formula and weight, properties, synonyms, and the Toxic and Hazard Review. Unique to this handbook are five chapters on chemical safety: "Safe Storage and Handling of Chemicals," "Respirators," "Selection of Chemical Protective Clothing," "Fire Protection," and "First Aid in the Workplace." These chapters give guidelines on maintaining a safety program. They are written by experts in the fields of occupational and chemical safety and include reference to articles for further study.

This reference is aimed at the worker in a laboratory setting who needs to know if a given material is toxic or hazardous, what conditions may make it toxic or hazardous, and what to do in case of an accident. However, note that first aid directions for a given material are not included. For information on materials not included, for more background information, or for toxicity data, the parent handbook, *Dangerous Properties of Industrial Materials*, should be consult-

ed. For specific handling procedures or first aid procedures, the Material Safety Data Sheet, available from the manufacturer of the material, should be consulted.   *DEBRA A. TIMMERS*

133. Managing Safety in the Chemical Laboratory. James P. Dux; Robert F. Stalzer. New York, NY: Van Nostrand Reinhold Co., 1988. 154p., bibliog., index. $28.95. 0-442-21869-9.

Managing Safety in the Chemical Laboratory is written for persons responsible for the management or safety of chemical laboratories. The authors, whose "experience in working in chemical laboratories spans four decades," are cofounders of a laboratory management and consulting company. Primarily beneficial for protection of the worker, safety management also offers benefits for the improvement of employee morale and reduction of insurance costs.

Some examples of chapters that relate to these issues are "The Safety Manual," "Protective Clothing," "Laboratory Design and Safety," and "Preparing for Emergencies." There are three appendices containing sections of government regulations: (1) "Hazard Communication Standard," (2) "Health and Safety Standards," and (3) "Superfund Amendments and Reauthorization." A bibliography and index are included at the end of the volume.

This small handbook would be useful as an on-the-spot reference source in the laboratory but also would be appropriate for college, university, and special libraries.   *SUE ANN M. JOHNSON*

134. Methods for Analytical Atomic Spectroscopy. 8th. ASTM Committee E-2 on Analytical Atomic Spectroscopy. Roberta A. Storer, ed. Philadelphia, PA: American Society for Testing and Materials, 1987. 1,125p., illus., index. $74.00. "PCN:03-502087-39. 0-8031-0994-6.

This publication is sponsored by the ASTM Committee E-2 on Analytical Atomic Spectroscopy, which was formerly known as Committee E-2 on Emission Spectroscopy. The first seven editions were published under the title *Methods for Emission Spectrochemical Analysis*. However, the title of this eighth edition has been changed to *Methods for Analytical Atomic Spectroscopy* to correspond with the name change of the Committee. A subcommittee of experts in the field of atomic spectroscopy has the responsibility of developing methods, practices, and terminology. This group also decides which methods and practices are outdated and should no longer be included in the compilation.

The purpose of this edition remains the same as that of earlier editions—to provide up-to-date methods and practices for the analytical laboratory. It covers both ICP and DCP plasma methods; spectrochemical analysis of both metallic and nonmetallic materials; and atomic absorption,

flame, X-ray, and optical emission spectometry. Included are the older, still-used methods as well as 25 new or revised standard methods and 17 new suggested methods.

The ASTM standard methods or practices are listed first, followed by the suggested methods. Both are arranged in alphanumeric sequence. In the case of the standard methods, the fixed designation, E-2, is given first, followed by the year the method was adopted and in parentheses the year of the last reapproval. The use of a superscript epsilson indicates an editorial change after the last revision or reapproval. The suggested methods are coded E-2 SM. This code is followed by numbers, e.g., 5-17. No explanation is given for the significance of these numbers. Throughout the book is the notation that the suggested methods have no official status but are included in the publication only for informational purposes. Each has been approved for inclusion by the chair of Committee E-2, the editorial committee, and a letter ballot of the sponsoring subcommittee. The name of the submitter and the person's place of employment is given for each suggested method.

Numerous aids are found througout the book to assist readers, including graphs, tables, and drawings. The standard methods portion includes a glossary of terms related to molecular spectroscopy, a second related to analytical atomic spectroscopy, and a third for statistical methods. It has an alphabetic subject index, which refers the reader to the various methods by alphanumeric numbers. In the front of the volume is a list of methods arranged by broad subject that refers to page numbers. This list is followed by the table of contents.

Several aids are also included to help the person who wants to submit a new suggested method or a comment about one of these methods. The foreword includes the official scope of Committee E-2; the names of the officers of the committee and its subcommittees are found at the end of the volume. The editorial guidelines for submission of suggested methods appear in the volume as Appendix I.

This will be a useful addition to library collections that serve chemists and engineers.

LUCILLE M. WERT

135. Thermodynamic Values at Low Temperature for Natural Inorganic Materials: An Uncritical Summary.
Terri L. Woods; Robert M. Garrels. Oxford University Press, New York: NY, 1987. 266p., bibliog. $18.95. 0-19-504888-1.

This reference presents literature values for thermodynamic data, including enthalpy of formation, Gibbs energy of formation, and entropy for over 1,000 inorganic species. Thermodynamic values for common elements, minerals, and related compounds (in various states: gas, liquid, solid, aqueous) are given for standard conditions (298.15 K degrees and 1 bar pressure). The list is uncritical in that various literature values are presented, giving the reader a range of values to use.

Each entry is referenced to the original article by listing a source number in the table. The full citation to the original article may be found by looking up that source number in the list of references at the end. An alphabetical bibliography is also given. This reference is useful to the researcher who needs thermodynamic values for inorganic compounds. It is especially useful in cases where values for minerals are needed, because it lists values for many minerals which are not found in critical tables. It must be remembered that these values are uncritical, and, in some cases, it may be better to consult critical tables such as Naumov's *Handbook of Thermodynamic Data* (National Technical Information Service, 1974); Rossini's *Selected Values of Chemical Thermodynamic Properties* (National Bureau of Standards, 1952); or Stull's *JANAF Thermochemical Tables* (National Bureau of Standards, 1971).

DEBRA A. TIMMERS

Histories

136. Essays on the History of Organic Chemistry in the United States, 1875–1955.
Dean Stanley Tarbell; Ann Tracy Tarbell. Nashville, TN: Folio Publishers, 1987 (c1986). 433p., illus., bibliog., index. $21.95. 0-939454-03-3.

This is a history of organic chemistry in the United States for the years 1875 to 1955. The "emphasis is placed on the growth of basic ideas of the science, on the interrelation of organic chemistry to other fields, on the contributions of key personalities, and to some extent on the institutional aspects of the development." Organic chemistry is the study of compounds that contain carbon and include such things as carbon monoxide, dacron, proteins, DNA, and viruses. The book is arranged as a collection of essays, each covering a period of time and a specific topic. Many of the time periods overlap depending on the topics, which include classical structural theory of organic chemistry, reaction rates and mechanism, valence, aromatic compounds, organic syntheses, organometallic compounds, stereochemistry, Walden inversion, instrumental revolution, polyisoprenes, and aromatic character.

In each of the essays, the topic or concept is described and illustrated, showing how it was discovered, why it is important, and, in some cases, what the future may hold. Although a chemical background would help, it is not totally necessary in order to understand what is being presented. The authors have done a commendable job in pulling together a great amount of information, synthesizing it, and writing the individual accounts. History of science students will find this a useful reference source of the development of organic chemistry in the United States. Academic libraries will want it for their history of science collections.

H. ROBERT MALINOWSKY

137. **A Guide to Archives and Manuscript Collections in the History of Chemistry and Chemical Technology.** George D. Tselos; Colleen Wickey, comps. Philadelphia, PA: Center for History of Chemistry, 1987. 198p., illus., index. price not reported. Center For History of Chemistry Publication, no. 7. 0-941901-05-Xpbk.

The present guide is an expanded and revised edition of the center's *Preliminary Guide to Archives and Manuscript Collections in the History of Chemistry*, compiled by George D. Tselos and Colleen Wickey, staff members of the recently established Center for the History of Chemistry. The center is jointly sponsored by the American Chemical Society, American Institute of Chemical Engineers, and the University of Pennsylvania. Its main purpose is "to discover and disseminate information about historical resources to encourage research, scholarship, and popular writing in the history of chemistry, chemical engineering, and the chemical process industries." The collections listed in the guide were compiled from several sources. The primary source was the *National Union Catalog of Manuscript Collections for 1959–1983*. This was supplemented by published guides for individual archival and manuscript collections.

The guide lists only collections of materials located in the United States. It includes collections of personal and professional papers of chemists, chemical engineers, biochemists, metallurgists, pharmacologists, and toxicologists. Other areas of coverage are the business records of chemical industries and records of academic departments or schools of chemistry and related areas. Not included in the guide are materials found in archives held by national, state, and local governments; archives held on the local premises of businesses and organizations; and collections of financial data. University archives are included only if they are combined with special collections or if the university has an archival or manuscript collection of a number of outstanding chemists.

The 1,241 consecutively numbered entries are arranged alphabetically by main entry: corporate name, surname of an individual, or name of a family. Each entry contains information on the size of the collection in terms of linear feet, number of items, or number of volumes. The type of record is indicated: business, papers, account books, personal papers, teacher papers, student notebooks, videocassette, blueprints, etc. Also given are the inclusive dates covered by the collection.

Entries for an individual contain a brief biography and birth and death dates. Corporate entries contain a brief description of the business, institution, or agency. Also provided is information about the accessibility of the collection: arrangement, indexing, finding aids, and whether it is open to use by outsiders. The guide is well indexed by three separate indexes: subject, name,

and geographical area. In the name and subject indexes, the references are to the entry number rather than to the page on which the entry is found.

This is the first time we have a tool that identifies the resource collections that can be used in the study of the history of chemistry; the Center for the History of Chemistry is to be applauded for its publication. It is hoped that the center will continue to expand and update it in the future. The guide will be of interest to historians of science, particularly to those interested in the areas of chemistry and chemical technology. It, also, will hold some interest for persons concerned with the history of their communities and regions if chemical industry has played a role in the development of the area. It is recommended for purchase by research libraries and others serving historians of science.   *LUCILLE M. WERT*

138. **William Cooper's A Catalogue of Chymicall Books, 1673–88: A Verified Edition.** Stanton J. Linden. New York, NY: Garland Publishing, Inc., 1987. 159p., illus., bibliog., index. $37.00. Garland Reference Library of the Humanities, v. 670. 0-8240-8557-4.

The catalogs were originally published in 1673, 1675, and 1688 by William Cooper, who was an author, editor, translator, publisher, bookseller, and auctioneer in seventeenth-century London. The preface states that they were the first catalogs of chemical, alchemical, medical, and hermetic books printed in English. Cooper's major interest as an author and publisher was in books dealing with chemistry, alchemy, medicine, and the occult. The catalogs were published to advertise the wares in his bookshop. In addition to his own catalogs, Cooper listed his books in his published works of other authors.

The purpose of the present authoritative edition is to provide specialists with an important source of authors and titles of scientific and occult materials and to provide a guide to the state of publication of such materials during that period. The 40-page introduction presents information that illuminates Cooper's career and the state of seventeenth-century book printing, publishing, and selling. The book is divided into three major parts: the introduction, *A Catalogue of Chymicall Books*, and two appendices. The introduction discusses Cooper's career as a publisher and bookseller, his role in the introduction of book auctioneering in England, his connections with the publication, *Philosophical Epitaph*, and the history of the publication of the catalogs. The material in this section is well documented with lengthy footnotes presenting the sources of information used in the introduction and noting sources which did not agree. In addition, at the end of the introduction is a list of abbreviations and symbols and a bibliography of sources.

The *Catalogue* portion of the book has 428 numbered entries arranged alphabetically by author surnames. This edition lists in a single alphabet the entries from all of the catalogs with the exception of "the third part of the 1675

edition which collected items on chemistry that were published in the *Philosophical Transactions of the Royal Society* during the first ten years." Linden verified all entries listed in the *Catalogue* by consulting the *Philosophical Epitaph*, and the catalogs of the Library of Congress, the Huntington Library, the Bodleian Library, and the British Museum. For incomplete entries listed in the original catalogs, Linden has added names of publishers, places of publication, volume sizes, and information about discrepancies between descriptions of entries in the various library catalogs. To confirm the existence of each item and to help in the identification and location of the items, Linden gives the call number of each from one major research library. One-third of the entries are concerned with alchemy and chemistry, 35 percent with medicine (general and chemical), 11 percent with magic, and 10 percent with baths and spas (mineral waters). The remaining entries deal with a wide range of subjects, from "America" to "Verse."

Appendix I contains supplementary entries from the lists included in other works published by William Cooper. Those listed in the *Catalogue* portion of the book are identified by an asterisk. Appendix II, "William Cooper and Works of Eirenaeus Philalethes," lists the works contributed to that author. The text that precedes the list discusses the various authors who might have used this pseudonym. References are given for the sources of information used in this appendix. The entries in the subject index do not refer to the textual content but only by entry number to the items listed in the *Catalogue of Chymicall Books*. This is a fascinating book that will be of interest to researchers in the areas of history of science (particularly chemistry and medicine), the occult, and the early English book trade. It is a must for the collection of research libraries serving these users, since it will lead them to sources of information needed in their research.

LUCILLE M. WERT

Manuals

139. Compendium of Analytical Nomenclature: Definitive Rules 1987. 2nd. Henry Freiser; George H. Nancollas. Boston, MA: Blackwell Scientific Publications, 1987. 279p., illus., index. $43.00. 0-632-01907-7.

The *Compendium of Analytical Nomenclature*, now in its second edition, is an effort to codify nomenclature and symbols used in analytical chemistry. It is based on a series of reports which were previously published in *Pure and Applied Chemistry*, the journal of the International Union of Pure and Applied Chemistry (IUPAC). As such it is part of a series of IUPAC manuals of nomenclature and symbols covering various areas of chemistry.

Although this edition incorporates seven additional reports, it is not complete, because there are still some sections which have yet to be written. The table of contents reflects this, indicating unwritten sections with an asterisk. Material, or-

ganized in sections (listed in the table of contents) each of which covers a specific subject area or analytical technique, is arranged in outline form, with numbered paragraphs. The volume includes an index of terms which refers the user to specific paragraphs rather than to page numbers.

This volume would be especially useful to individuals carrying out a research and writing program in analytical chemistry and would, therefore, be a significant addition to the library collection of any institution or organization supporting such programs.   *LOREN D. MENDELSOHN*

140. Quantities, Units and Symbols in Physical Chemistry. Ian Mills; Tomislav Cvitas; Klaus Homann; Nikola Kallay; Kozo Kuchitsu. Boston, MA: Blackwell Scientific Publications, 1988. 134p., bibliog., index. $32.50pbk. 0-632-01773-2pbk.

Quantities, Units and Symbols in Physical Chemistry is the successor to the *Manual of Symbols and Terminology for Physiochemical Quantities and Units*, which has been published in three editions. Produced by the International Union of Pure and Applied Chemistry (IUPAC), it codifies the quantities, symbols, and units used in physical chemistry. As such it is part of a series of IUPAC manuals of nomenclature and symbols covering various areas of chemistry. The purpose of these manuals is to "improve the international exchange of scientific information." The material is organized into several chapters as follows: (1) "Physical Quantities and Units"; (2) "Tables of Physical Quantities"; (3) "Definitions and Symbols for Units"; (4) "Recommended Mathematical Symbols"; (5) "Fundamental Physical Constants"; (6) "Properties of Particles, Elements and Nuclides"; (7) "Conversion of Units"; and (8) "References." The chapters and their sections are listed in the table of contents. Without a subject or term index, the book only provides an index of symbols, which is part of several pages of miscellaneous information at the close of the volume. These pages include the Greek alphabet and tables of conversion factors.

On the whole, this volume is well organized and easy to use; however, a subject index would be a helpful addition. This volume would be a significant addition to the library collection of any institution or organization supporting chemical research.   *LOREN D. MENDELSOHN*

Textbooks

141. Fundamentals of Clinical Chemistry. 3rd. Norbert W. Tietz, ed. Philadelphia, PA: W. B. Saunders Co., 1987. 2,020p., illus., index. $44.95. 0-7216-8862-4.

This textbook, in its third edition, is a very basic introduction to the many areas of clinical chemistry, intended for beginning undergraduate students in medical technology and clinical chemistry. Norbert Tietz is well known in the field and is the general editor of the textbook. He also has called upon experts in each area to author appro-

priate chapters. The 1987 *Fundamentals of Clinical Chemistry* is a companion to the 1986 *Textbook of Clinical Chemistry*, also edited by Tietz. Procedures which are briefly described in *Fundamentals* are often listed in detail in the *Textbook*. Consequently, it would be helpful to use both textbooks side by side. The new edition of *Fundamentals of Clinical Chemistry* includes greatly expanded chapters on newer analytical procedures and instrumentation. The index of the new edition has been improved with greater detail, and access points have been added to it. The flow of the chapters is more logical in the new edition. Those colleges and universities serving introductory programs in clinical chemistry and medical technology might consider adding *Fundamentals* and its companion, *Textbook*, as background references in those fields.

JODY KEMPF

142. **Problems in Chemistry.** 2nd rev. and expanded. Henry O. Daley, Jr.; Robert F. O'Malley. New York, NY: Marcel Dekker, Inc., 1988. 476p., index. $39.50. Undergraduate Chemistry: A Series of Textbooks, v. 10. 0-8247-7826-X.

This is a freshman textbook covering the fundamentals of chemistry in the form of brief discussions with worked examples and then a series of problems for the student to solve. The authors have chosen the problems with two concepts in mind: "1) to show students that much of what they learn in freshman chemistry is being used by chemists in their careers, and 2) to encourage students to read the chemical literature."

Answers to the problems are included in one of the appendices. Although not a reference book in the strictest sense, it can serve as a source of general chemical information for college and academic libraries.

H. ROBERT MALINOWSKY

Treatises

143. **Organic Reactions, Volume 35.** Andrew S. Kende, ed. New York, NY: John Wiley and Sons, Inc., 1988. 650p., illus., bibliog., index. $65.00. 0-471-83253-7.

Organic Reactions, now in its thirty-sixth volume, is an irregularly published series of volumes consisting of chapters which present comprehensive discussions of specific organic chemical reactions. The reactions under discussion must have wide applicability with emphasis placed on preparative aspects. Each chapter discusses a single reaction or phase of a reaction, and effort is made to present all known examples of the reaction. Each volume includes cumulative tables of contents and cumulative author and chapter/topic indexes covering all preceding volumes. Each chapter has a table of contents and a comprehensive list of references.

The past and present editors of this series have generally been among the most distinguished organic chemists from academia and industry. It would be difficult for a library to support a chemistry program in the absence of a current subscription to this source.

LOREN D. MENDELSOHN

COMPUTER SCIENCE

★

GENERAL SOURCES
Dictionaries

144. Dictionary of Computers, Information Processing, and Telecommunications. 2nd. Jerry M. Rosenberg. New York, NY: John Wiley and Sons, 1987. 733p., glossary. $39.95, $18.95pbk. 0-471-85558-8, 0-471-85559-6pbk.

The second edition of the *Dictionary of Computers, Information Processing and Telecommunications* by Jerry M. Rosenberg updates the first edition with the addition of terms and phrases relating to changes in the computing world, such as the spread of networking, teleconferencing, new transmission techniques, and the personal computer revolution. The dictionary includes over 12,000 entries including single terms, phrases, and acronyms, and the two-column format has been well chosen. Entries are printed in boldface type, and the concise definitions are printed in a slightly larger than normal typeface. The scope is broad, as is explained in the introduction. The editor states that he is aiming for two audiences—the experienced reader who is seeking a precise definition using the language of the discipline and the layperson who is seeking a more general definition. Most of the terms, however, are actually followed by only one definition. Definitions which have been taken from several named sources in the introduction are identified with a letter code.

On the whole, this dictionary is recommended as a useful addition to most public and college library reference sections, particularly because of the numerous inclusions of acronyms. It would be useful as well for scientific research collections, although in a more limited fashion, for quick reference.

One criticism of this dictionary addresses the inclusion of some terms which would appear unnecessary to either target groups. Terms, such as the prompts "HELLO" and "HELP" or such very general terms such as "owned," "owner," "junk," "accent," or the many phrases related to the telephone—for example, telephone company and telephone set do not require explanantions. The term "telephone" is defined as "a popular name for AT&T"!

A second criticism is that the reader might be led from one entry to another in a somewhat unnecessary fashion. This is in part because of the practice of defining a term with a form of the term itself. It is also a result of the overuse of the "see" reference when a definition seems to be called for at that particular entry point. One example would be the terms "BOOT" or "BOOTING" which both refer the reader to "bootstrap." "Bootstrap" is then followed by four definitions and a "see" reference. One of the definitions is to "use a bootstrap," and none of the definitions clearly defines that rather common computer word for the general reader.

The appendix, which provides the French and Spanish equivalents for those terms used in the dictionary, is a useful supplement. One suggestion for a third edition would be to include those languages closely tied to the sciences—German should definitely be included, and Russian and Japanese should be considered.

JOANNE M. GOODE

145. Dictionary of Computing and Information Technology. 3rd. A. J. Meadows; M. Gordon; A. Singleton; M. Feeney. London, Great Britain: Kogan Page, Dist. by Nichols Publishing Co., New York, NY, 1987. 281p., illus. $33.50. 0-89397-273-8.

As new electronic methods for the generation, transmission, and reception of information continue to be introduced, a need is created for sources of definitions of new terminology used in the field. The third edition of this well-known dictionary was designed to answer this need. Because of the continuing rapid changes of vocabulary in information technology and computing, three editions of this dictionary have been issued within a six-year period. The first was published in 1982 under the title *Dictionary of New Information Technology* and the second in 1984. The current edition introduces several hundred new terms and updates a number of the definitions. The dictionary now includes approximately 4,000 terms. All of the authors have background and experience in the fields of library and information studies or publishing. The terms and definitions in the dictionary were based on the periodical literature and advertising from the fields of computing and information technology published in the United Kingdom, continental Europe, and North America.

Because a term may have more than one use, the authors brought together and codified the "most important" specialized terms currently in use in the two fields. Emphasis is on information rather than on the techniques used for entertainment. Also, most of the terms that have restricted applications have been excluded. The dictionary is printed two columns to the page with the terms in boldface type and alphabetized letter by letter. Acronyms are printed in block capitals. The definitions vary in length from several lines to three columns. An italicized word within a definition serves as a cross-reference to that term. "See" references referring to the correct form of entry or to diagrams are printed in brackets after entries. Terms that are used in different formats, e.g., noun or verb, are usually listed only under one form. The dictionary is intended not only for nonspecialists but also for persons in other fields whose interests and work overlap with the field of computing and information technology. It will be useful in all types of libraries, as well as in office collections and businesses.   *LUCILLE M. WERT*

146. Dictionary of Information Technology. 2nd. Dennis Longley; Michael Shain. New York, NY: Oxford University Press, 1986. 382p., illus. $29.95. 0-19-520519-7.

The second edition of the *Dictionary of Information Technology* focuses on the terminology of the newest technologies: "computer networks and data communications, cryptography, expert systems, fifth generation computers, machine translation, microcomputers, on line information retrieval, Open System Interconnection, programming, and speech synthesis." The dictionary identifies in which field the term is used and includes how the words are used in different contexts. There are over 6,000 entries with more than 100 diagrams. Specialists have written extended entries in many of the fields covered. Acronyms are included with "see" references. Because of the 1986 publication date of the dictionary, there are terms created in the recent years that cannot be found in this work.

The British authors have extensive experience in the field of information technology and have adapted the second edition to their American audience by seeking assistance from an American professor of communications. The dictionary is an excellent reference book in the field of information technology up to 1986 and would be useful in public, college, university, and special libraries.   *JO BUTTERWORTH*

147. Historical Dictionary of Data Processing: Biographies. James W. Cortada. Westport, CT: Greenwood Press, 1987. 321p., bibliog., index. $49.95. 0-313-25651-9.

See full review under *Historical Dictionary of Data Processing: Technology.*

148. Historical Dictionary of Data Processing: Organizations. James W. Cortada. Westport, CT: Greenwood Press, 1987. 309p., illus., bibliog., index. $45.00. 0-313-23303-9.

See full review under *Historical Dictionary of Data Processing: Technology.*

149. Historical Dictionary of Data Processing: Technology. James W. Cortada. Westport, CT: Greenwood Press, 1987. 415p., illus., bibliog., index. $55.00. 0-313-25652-7.

This outstanding three-volume set, *Historical Dictionary of Data Processing: Biographies, —: Organizations,* and *—: Technology,* explores the history of data processing in organizations (companies, societies, and laboratories), individuals, and technology (hardware and software). Intended to provide basic factual material and historical interpretation, the three volumes include nearly 400 entries on all aspects of data processing from the earliest beginnings to the present decade. Entries range in length from short essays of a few hundred words to those of several pages, and each concludes with an excellent selection of bibliographical references for further reading. The author is senior marketing programs administrator for IBM and writes in a nontechnical, interesting, and informative style. He undertook an impressive amount of research for this work, which might be more appropriately termed an "encyclopedic dictionary."

The *Organizations* volume begins with a 44-page essay on the history of the data-processing industry, which serves to set the work as a whole in historical context and provide a unifying theme. In this volume, we learn about organizations ranging from IBM to the Homebrew Computer Club, an informal club in California consisting of many of the entrepreneurs such as Adam Osborne and Steve Jobs who later established microcomputer companies. Appendices list organizations by type and chronologically.

The *Technology* volume includes specific hardware (the Great Brass Brain was a tide predictor built before World War I, remaining in use for a generation) and software, as well as valuable overview articles such as those on operating systems, analog computers, and database management systems. The fascinating chronology in the appendix begins with the abacus, ca. 3000 BC.

The *Biographies* volume offers biographical information on more than 150 engineers, scientists, government officials, executives of manufacturing companies, and users. "Biographees range from ancient Europeans concerned about astronomy to software entrepreneurs of the 1980s." Appendices in this volume list individuals by date of birth and by profession. Each volume is separately indexed, but a clever cross-referencing system within the text refers the reader to other entries in the same and other volumes.

JULIE BICHTELER

Directories

150. Computers and Computing Information Resources Directory: A Descriptive Guide to Approximately 6,000 Live and Print Sources of Information on General and Specific Applications of Computers and Data Processing, Including Associations and User Groups, Consulting Organizations, Research Organizations and University Computer Facilities, Libraries and Information Centers, Trade Shows, Exhibits, and Association Conventions, Online Services and Teleprocessing Networks, Directories, Journals, and Newsletters. Martin Connors; John Krol; Janice A. DeMaggio, eds. Detroit, MI: Gale Research Co., 1987. 1,271p., index. $165.00. 0-8103-2141-6.

Those familiar with other Gale directories will immediately recognize the format and style of this one, which is itself largely compiled from a variety of other Gale publications. As its introduction states, modern computers have the power to "organize, massage, and manipulate huge quantities of data," and this work is a prime example of that ability.

In this case, the data being massaged are largely Gale's own, a way of cashing in on the information it has collected for other publications. This is not to say that the publication under review here is useless or bogus in any way. It is not. But potential buyers of the *CCIRD* should be fully aware that it is a digest of other Gale publications they may already own. However, the convenience of having in one location all the information on computers and related services will be worth the price to many users.

The work is divided into eight chapters and three indexes: "Trade and Professional Associations and User Groups"; "Consultants and Consulting Organizations"; "Research Organizations and University Computer Facilities"; "Journals and Newsletters"; and "Directories." Indexes include master name and keyword, geographic, and personal name listings. Information available for each entry is standard and varies widely, depending on the amount of data supplied by the source itself. The data included in this directory are of necessity, in such a rapidly evolving and changing field, somewhat uneven.

Its strength is probably in the realms related to business and commerce, rather than the more "bibliographic" information of interest to libraries. For this latter information, there are far better more comprehensive sources, such as *Ulrich's International Periodical Directory* and Bowker's *Computer Books and Serials in Print*. The *CCIRD* will not replace anyone's other computer directories, but it may nicely complement them. Libraries on a tight budget should review their needs carefully before acquiring this item, to see whether they require this particular solution.

FRED O'BRYANT

151. Directory of Statistical Microcomputer Software. Wayne A. Woodward; Alan C. Elliot; Henry L. Gray; Douglas C. Matlock. New York, NY: Marcel Dekker, Inc., 1988. 744p. $59.75pbk. 0-8247-7846-4pbk.

This is a comprehensive directory of over 200 microcomputer statistical programs. In 1985, the authors produced the first edition with 140 programs, many of which no longer exist. The field of statistical programs is fast growing, with constant changes. The authors, therefore, have tried to produce a directory that is up to date and containing information that will be of use to as many individuals as possible.

The programs are listed alphabetically by the name of the program. An address, phone number, and contact person are given, followed by a description of configurations, a listing of computers and operating systems, the amount of memory, storage medium, and price. A history of the product is given, including the date it was first released and the date of the most current version, along with the number of products currently in use. Also included in the listing are the documentation, type of data management and editing, access, graphics features, and product support. The statistical features that are supported are listed individually so that, at a glance, one can determine whether a particular package will be of any value.

There are several appendices that serve as an index to the directory. One covers vendors, listing programs by vendor name. Another is a program capabilities index that indicates which programs can perform certain statistical tasks. Finally, there is an appendix that matches programs with their computer or operating systems. Academic and special libraries serving those involved in using the microcomputer for statistical purposes will benefit from this comprehensive directory.

H. ROBERT MALINOWSKY

152. Encyclopedia of Information Systems and Services: An International Descriptive Guide to Approximately 4,100 Organizations, Systems, and Services Involved in the Production and Distribution of Information in Electronic Form. 8th. Amy Lucas; Annette Novallo; Nan Soper, eds. Detroit, MI: Gale Research Co., 1988. 2,338p. in 3v., index. $400.00. 0-8103-2532-2set. 0734-9068.

This eighth edition of the *Encyclopedia of Information Systems and Services* (*EISS*) continues Gale's tradition of comprehensive coverage of the electronic information and publishing industries. The *EISS* is intended to be an "international guide to organizations, systems, and services using computers and related new technologies to produce and/or provide access to information and data of all types in all subject fields." To meet this goal, the editors deliberately covered a broad spectrum of the electronic information industry. Consequently, the *EISS* provides information about database producers,

online host services, CD-ROM publishers, library and information networks, data collection and analysis centers, mailing list services, videotext/teletext information services, and a plethora of other information organizations and services.

All of the organizations described in the seventh edition were contacted and requested to update their listings. New organizations were also identified and contacted. As a result, this latest edition of the *EISS* contains thousands of changes in personal names, addresses, and other pieces of information regarding the over 4,100 entries. The *EISS* is divided into three volumes. Volume 1 contains descriptive listings for 2,800 information organizations located in the United States and its territories and possessions. Volume 2 provides descriptive listings of over 1,300 international and national organizations located in around 70 countries, excluding the United States. As could be expected, the majority of these entries are for organizations located in Canada and Europe. However, there is at least one entry for such countries as Bangladesh, Iceland, Qatar, and the Seychelles.

The listings in volumes 1 and 2 are arranged alphabetically by parent organization name. Each entry includes up to 17 categories of information such as organization name, address and telephone number; names of the director and contact; number of staff; and names of any affiliated organizations. Each listing also provides a description of the system or service; its scope; the number and types of input sources; any publications issued by the organization; the computer- based products and services; intended clientele; and availability of the system or service. Volume 3 provides consolidated indexes to the material in volumes 1 and 2. It contains eight categories of indexes: "Master" (i.e., a superset of the other seven indexes), "Data Bases," "Publications," "Sortware," "Personal Name," "Geographic," "Subject," and "Function/Service Classification." The last category is further subdivided into 24 separate classifications according to chief function or type of service provided. Entries are grouped under such headings as "Abstracting and Indexing," "Document Delivery," "Optical Publishing Applications," and "Photocomposition Services." This latest edition contains two new classifications: Mailing List Services and Transactional Services (e.g., making airline and hotel reservations).

CRAIG S. BOOHER

Encyclopedias

153. Encyclopedia of Lotus 1-2-3: A Complete Cross-Reference to All Macros, Commands, Functions, Applications and Troubleshooting. Robin Stark. Blue Ridge Summit, PA: TAB Books, Inc., 1987. 484p., illus. $29.95, $21.60pbk. 0-8306-7891-3, 0-8306-2891-6pbk.

This book presents complete information to Lotus users in an organized and logical manner. The descriptions are well written and liberally augmented with illustrations of the idea being presented. The book's primary function is as a reference guide for users, however, it is written to be useful as an introductory text for the beginner. The author does not assume prior knowledge of Lotus and defines and explains commands and concepts as they are introduced. Therefore, it is possible to use this book as a self-teaching text. Especially helpful is chapter 1, "Getting Started with Lotus 1-2-3" in which the author starts with basics of how to load the disk and get the program running. It allows a new user to begin working with Lotus in a very short period of time. It is, however, as a reference to the more experienced user that this book has its greatest value. At the conclusion of a section within a chapter, there is a cross-reference to a related command or function along with the page number on which it is found. These sections can serve as a reminder to the experienced user. To those users who are gaining proficiency with the program, these sections and this type of layout permits the user to easily increase his or her base of knowledge and utilization of the potential of the program. This is a valuable text to include in one's computer library. *ANNE PIERCE*

154. The MS-DOS Encyclopedia. Ray Duncan, ed. Redmond, WA: Microsoft Press, 1988. 1,570p., index. $134.95. 1-55615-049-0.

The *MS-DOS Encyclopedia*, edited by Ray Duncan, is a comprehensive resource primarily aimed at MS-DOS programmers. A secondary target group would be MS-DOS users seeking an in-depth explanation of commands and a broader understanding of the operating system and why it performs as it does. The first section presents an interesting history of the development of MS-DOS and its close relatives PC-DOS, Z-DOS, et al., as well as a description of the various versions of DOS. This section would be of interest to a general reader as well as to the more sophisticated. Section II, "Programming in the MS-DOS Environment," is the main thrust of the encyclopedia. The information presented is current to the changes connected with DOS 3.3 and the 80286, 80386 microprocessors. As throughout the book, comparisons to the OS/2 operating system are made. The choice of type size and layout design is very appropriate for this type of reference book. Boldface paragraph headers make it easy to skim for information, and screen displays and diagrams are included. Code and program examples are indented and printed in blue.

The third section is devoted to summaries of DOS commands. A key at the front of this section demonstrates how to optimize use of the summaries for both quick reference as well as more detailed information. Again, the design layout lends itself very well to the purpose of this section. Each command summary contains a purpose statement, syntax, description, examples of how the command can be entered, and explanations of error messages. The fourth section discusses programming utilities. Fifteen appendices follow as well as a subject index and a command and system call index. A statement in the preface indicates that updates will be forthcoming. A bul-

letin board service is available through Microsoft Press for readers who wish to make comments. Disks with the routines and code are available as companions to the book at $49.95 plus postage. The list of contributing authors and advisors represents a variety of backgrounds, and there are many familiar names from the programming world. The book is somewhat cumbersome at 1,569 pages; separation into two volumes might have been a better choice. It is sturdily bound with sufficient center margins which make it pleasant to use despite its size. The book is expensive; however, it would make a valuable and unique reference tool for most academic and corporate libraries as well as some large public libraries. *JOANNE M. GOODE*

Guides to the Literature

155. Computing Information Directory: A Comprehensive Guide to the Computing Literature. 5th. Darlene Myers Hildebrandt, comp. Federal Way, WA: Pedaro Inc., 1988. 893p., index. $139.95pbk. 0-933113-03-X. 0887-1175.

Computing Information Directory, formerly called *Computer Science Resources,* is by far one of the most comprehensive guides to the computer literature available. First published in 1981, it has become an annual since 1985. This comprehensive guide has developed into THE source of printed information for anything relating to computers, including software, hardware, and languages.

It is divided into 10 sections. The first section covers computing journals. Over 1,800 English-language journals are cited, including translations from other languages into English. It is arranged alphabetically by title of the journal and includes the following information: title, subtitle, note if ceased or suspended if applicable, date of first issue, date of last issue if ceased, place and publisher, frequency, cost, and ISSN. For a majority of the titles a brief note is included, and historical information is also provided. Section two is a special section covering university computing center newsletters. This section is in three parts: "Index to Computer systems at Academic Computer Centers," "University Computer Center Newsletters," and "Computer manufacturers." Under the first part, the various computer systems are listed followed by indications of which universities have that system. The newsletters are listed alphabetically by name of the institution. The remaining sections cover "Books, Biblios, Special Issues," "Dictionaries and Glossaries," "Indexing and Abstracting Services," "Software Resources," "Review Resources," "Hardware Resources," "Directories, Encyclopedias, and Handbooks," and "Computer Languages." Each of these sections lists sources alphabetically by title, giving full descriptive information and prices where applicable. Finally, there are six appendices: "Association for Computing Machinery (ACM) Special Interest Group (SIG) Proceedings, 1969-1986," "Computer-Related Curriculum Bib-

liography," "Bibliography of Career and Salary Trends in the United States, 1970-1987," "Expansion to the Library of Congress Classification Schedule, QA75, QA76, TK5101, TK5105: 1987 Draft and Subject Index," "Publishers' Addresses," and "Master Subject Index." Of particular interest for catalogers and reference librarians is the first appendix that lists ACM SIG proceedings. These proceedings can be very difficult to identify, but this listing should help a great deal with this problem. New to this edition is the Master Subject Index.

This outstanding work is highly recommended for any library that has an interest in computer science. It is well laid out, contains complete information, and is easy to use.

H. ROBERT MALINOWSKY

156. Selective Guide to Literature on Computer Engineering. Margaret Bean, ed. Margaret Bean; Sharon Balius; Sara Bowker; G. Lynn Tinsley; Carol Godt, comps. Washington, DC: Engineering Libraries Division, American Society for Engineering Education, 1985. 22p. $5.00pbk. Engineering Literature Guides, no. 1. 0-87823-100-5.

The first in a series of brief guides to the engineering literature, this annotated bibliography covers computer engineering and computer architecture for all sizes of computers. Publications limited to a specific type of microcomputer are excluded. About 100 titles, mostly published from 1980 to 1984 or serials which began earlier, are grouped by type: guides to the literature, indexing and abstracting services, databases, dictionaries, etc. Annotations are very brief and appear only for those works examined by the authors, who are engineering librarians. The lack of an index makes locating a specific title difficult; access is only through the table of contents, which lists broad categories.

Although highly selective and somewhat limited in coverage, this bibliography should be useful to engineering students and faculty, as well as to information specialists. A more substantial binding would have made it more useful. For the period of time covered, it supplements previous guides which include computer engineering literature such as Karen Quinn's *Guide to Literature on Computers* (American Society for Engineering Education, 1970) and Darlene Myers's *Computer Science Resources: A Guide to Professional Literature* (Knowledge Industry Publications, 1981). Inevitably, coverage overlaps somewhat with other guides in the same series, such as the *Selective Guide to the Literature on Computer Science.*

JULIE BICHTELER

157. Selective Guide to Literature on Computer Science. Rosemary Rousseau, comp. Washington, DC: Engineering Libraries Division, American Society for Engineering Education, 1985. 20p. $5.00pbk. Engineering Literature Guides, no. 3. 0-87823-103-X.

The third annotated bibliography in the *Engineering Literature Guides* series, in which titles are authored by engineering librarians, this work includes "material likely to be found in most libraries at academic institutions where Computer Science is taught." The bibliography excludes reference works solely on computer software and computer engineering, because these are the subjects of other guides in the same series. Approximately 90 titles, a few published as late as 1985, are grouped in broad categories—guides to the literature, bibliographies, indexing and abstracting services, databases, dictionaries, encyclopedias, standards and specifications, product directories, handbooks and manuals, etc. Brief annotations provide commentary on all titles except those in the section on yearbooks and surveys. The addition of author and title indexes would have improved access, which is limited to the table of contents.

University faculty and students, as well as information specialists, should find this highly selective bibliography helpful. The work supplements more extensive guides to the literature such as Darlen Myers's *Computer Science Resources: A Guide to Professional Literature* (Knowledge Industry Publications, 1981) and Karen Quinn's *Guide to Literature on Computers* (American Society for Engineering Education, 1970). Unfortunately, the flimsy binding will not hold up long under reference use.

JULIE BICHTELER

158. Selective Guide to Literature on Software Review Sources. Margaret H. Bean, ed. Margaret H. Bean; Theresa S. Lee; G. Lynn Tinsley; Carol Godt, comps. Washington, DC: Engineering Libraries Division, American Society for Engineering Education, 1986. 33p. $5.00pbk. Engineering Literature Guides, no. 8. 0-87823-108-0.

The eighth in this series of guides to the engineering literature concerns the rapidly expanding area of software evaluation for microcomputers. The guide, written and annotated by engineering librarians, provides information on review sources dealing with a variety of applications and systems. The 112 entries include directories (both evaluative and nonevaluative), journals, indexing and abstracting services, and online sources. The authors provide very useful commentary at the beginning of each section and substantial annotations, some of which are taken from publishers' announcements. Readers will especially appreciate the meticulous enumeration of criteria for inclusion and exclusion in the choice of titles. A title index would have been a welcome addition to access.

This guide will be valuable in all types of libraries, providing useful information on software review sources for librarians and patrons alike. Unfortunately, the binding detracts from the usefulness of the work. *JULIE BICHTELER*

Handbooks

159. AMIGA Programmer's Handbook, Volume II. 2nd. Eugene P. Mortimore. San Francisco, CA: SYBEX, Inc., 1987. 366p., illus., index. $24.95pbk. 0-895888-384-8pbk.

This handbook is volume II of a set designed as a detailed reference guide to Amiga computer programming. Volume I discusses the graphics-related Amiga library functions. Volume II, *AMIGA Programmer's Handbook*, discusses the 12 internal predefined devices of the computer: Audio, Translator/Narrator, Parallel, Serial, Input, Console, Keyboard, Gameport, Printer, Clipboard, Timer, and Track Disk. The information is applicable to the three currently available models of the computer, Amiga 1000, 500, and 2000, and provides information on all the commands for the 12 Amiga devices, including those that are new to the 1.2 software release. It is designed to provide information on the internal routines of the preprogrammed devices for use by programmers. It also will be of interest to program/owners, especially those who are familiar with the C assembly language or other high-level languages and anyone wanting information about AMIGA devices and their applications.

The handbook is organized for ease of use with lengthy introductions that discuss the Amiga computer, device system, programming environment, and summary. The introduction is followed by 14 chapters, the first two covering device input/output and programming facts, functions, and techniques that are general to all the devices. Each of the 12 remaining chapters is devoted to a particular device, divided into sections introducing the device's requirements and functions and providing information about its operations, structures, error codes, commands, and use. Several features are provided to aid the reader in using this handbook, including numerous diagrams of device operations. A note at the end of the introduction states that the "compiler and link" process are described more fully on the disk offered in the back of volume I. An appendix gives C language definitions of the Exec-Support library functions which are discussed in chapter 2.

Eugene P. Mortimore has authored a number of books on microcomputers and is president of Micro-Systems Analysis, Inc. Because this handbook is computer specific, it is not for every library collection. However, it should be in the collection of individual programmers, owners of Amiga computers, and libraries that have in-depth collections of computer materials.

LUCILLE M. WERT

160. A Comprehensive Guide to AI and Expert Systems Using Turbo Pascal. Robert I. Levine; Diane E. Drang; Barry Edelson. New York, NY: McGraw-Hill Book Co., 1988. 257p., illus., bibliog., index. $19.95pbk. 0-07-037476-7pbk.

This is an introduction to the basic concepts of artificial intelligence (AI) and expert systems written for advanced high school students, college students, and others who seek an understanding of AI. Robert I. Levine is senior research engineer for the Unisys Corp.; Diane E. Drang, director of educational services for the Jewish Board of Family and Children's Services; and Barry Edelson, scientific and technical editor and writer. The preface states that "All the book's information was extracted from popular texts in the field, expanded and then greatly simplified."

The chapters in the book are grouped into six sections: "Human and Machine Intelligence," "Inference Mechanisms—Tools for Machine Thinking," "Expert Systems—Knowledge Plus Inference," "Advanced Programming Techniques for Powerful Systems," "Advanced Knowledge Representation for Smart Systems," and "Languages Used in AI." The last section briefly describes two languages that are used to implement expert systems, PROLOG and LISP.

Each chapter is self-contained so that the need to refer to earlier chapters is eliminated when new concepts are introduced. Also, chapters that introduce new concepts contain hands-on programming examples. These sample programs are written in a subset of Turbo Pascal which permits them to be transported to any personal or home computer. Several features of the book are designed to aid the user in the understanding of the different concepts presented: diagrams throughout the text, an annotated bibliography and recommended reading list, and an in-depth subject index. Also, available for purchase are a tutorial disk, "The Intelligent Tutoring Series," and a disk containing all of the programs used in the book. The latter disk is for use with the IBM/PC. The book is recommended for high school and college libraries. *LUCILLE M. WERT*

161. Data Handling Utilities in Microsoft C. Robert A. Radcliffe; Thomas J. Raab. San Francisco, CA: Sybex, 1988. 520p., bibliog., index. $24.95. 0-89588-444-5.

This very useful book assumes prior knowledge and understanding of the C programming language and takes the reader further into the complexities and intricacies of programming in this language. The authors supply numerous source codes for data handling after first explaining their purpose and desirability. The goal here is to help programmers through many of their programming problems by showing them examples of how the C programming can deal with certain specific problems. Their approach is very useful and benefits those individuals who are programmers. The 12 appendices are useful for their quick availability of information, and the glossary is a quick reference to term definition. A listing of vendors of C products with addresses and phone numbers and costs could be very valuable. Indexes include source codes and other handy information, such as value range computations. This book is a very valuable addition to a pro-

grammer needing to use the C language. Written clearly, well-indexed, with numerous examples, it is worth adding to academic and special libraries. *JOHN NISHA*

162. dBASE III PLUS: Programmer's Reference Guide. Alan Simpson. San Francisco, CA: SYBEX, 1987. 1,029p., illus., index. $26.95. 0-89588-382-1.

Although the author expressly states that this reference book is written for people who "already know the basics of using dBASE III PLUS at the Assistant level" or for people who "have some familiarity with basic programming concepts with some language other than dBASE III PLUS," the first chapter discusses the basics of configuring and setting up the program. dBASE III PLUS is an extremely powerful yet complicated database manager, facts which highlight the value of this comprehensively written manual. Because this database program is very popular in the business world, Simpson has directed his discussions and applications accordingly. The book is divided into several parts; each part is then divided into chapters. This format assists the reader in identifying primary groupings of related topics, techniques, applications, and procedures. For example, part 3 includes two chapters that cover screen displays and printing formatted reports both of which involve similar questions. In addition, each chapter is arranged with a general discussion followed by a detailed outline of the specific commands with examples. The author anticipates other possible information needs of the reader when he appropriately directs the user to related material directly from the text with either the designation of "Tips" or "See also." The placement of these quick references saves the reader from immediately having to pore over the index to find further assistance. At the end of each chapter, there is a summary and additional reference statements pointing to other relevant sections of the book.

Other chapters cover such topics as the development and running of program applications, debugging, the commands and techniques for managing custom databases, programming enhancements, compilers and assembly languages, and solutions to programming problems. Alan Simpson tackles each set of applications or problems with his reader in mind. He not only outlines possible database structures but offers ideas of what type of information might be useful to store in these databases. Included appendices contain a list of all dBASE error messages, detailed steps on how to install dBASE III PLUS on both a single-user system and a networking system, and a chart of ASCII codes and symbols. Recommended for personal libraries and libraries with computer science collections. *CHARLES R. LORD*

163. DisplayWrite 3: Quick Reference Handbook. Gary C. Bond; Karen Cuneo. New York, NY: John Wiley and Sons, Inc., 1987. 185p., index. $22.95pbk. 0-471-85374-7pbk.

DisplayWrite 3: Quick Reference Handbook, written by Gary C. Bond and Karen Cuneo, was developed for the Center for Professional Computer Education. DisplayWrite 3 is a very powerful word-processing tool that can run on many different computer systems. This is a reference handbook for those people who have already received some training in the use of DisplayWrite 3.

It is arranged for quick and easy access through three sections: (1) "Menus/Screens," (2) "Reference Guide," and (3) "Index." Menus/Screens contains diagrams of the various menus or screens that appear during text editing with a page number for referral to more detailed information in the reference guide. The reference guide is the main part of the book, containing detailed instructions on the operation of each aspect of DisplayWrite 3. It also contains a listing of the Keystroke Commands with a brief explanation of that command. The index contains a fully cross-referenced alphabetical index of DisplayWrite 3 features and functions. A text codes chart is included following the index.

This is a very useful handbook to this word-processing program. It is easy to use and should be very useful to anyone who has access to the DisplayWrite 3 program. College, university, and large libraries will want a copy for reference purposes. *H. ROBERT MALINOWSKY*

164. The Dow Jones-Irwin Technical Reference Guide to Microcomputer Database Management Systems. George F. Goley, IV. Homewood, IL: Dow Jones-Irwin, 1987. 538p., index. $49.95.
0-87094-894-6.

DBMS—Database Management Systems—are the keys to the management of any microcomputer system. There are numerous DBMSs available, and each publisher of such a system has its own documentation. This particular guide is a detailed documentation of dBASE, R:BASE, and KnowledgeMan. It is a "reference tool that gives easy access to programming languages' verbs, syntax, and functions" and is intended for applications developers. In addition to being a guide to the above-mentioned DBMSs, it is a cross-reference resource to other DBMSs. Also, those who use BASIC, COBOL, Pascal, C, and other languages will be able to teach themselves a newer language. The book, through a concise and easy-to-read text, condenses the sometimes three- or four-volume vendor guides into one compact book covering "The Right Tool for the Job," "System Requirements," "Database File Creation," "Memory Variable Handling," "Operators," "System Environment," "Control Structures," "Debugging," "Menus," "Data Entry," "Reports," "The Future," "Command Syntax," "Pictures," "Using the Verb Reference," "dBASE III PLUS Command Verbs," "KnowledgeMan 2 Command Verbs," "R:BASE System V Command Verbs," "dBASE III PLUS Functions," "KnowledgeMan 2 Functions," "R:BASE System V Functions," "KnowledgeMan 2

Functions," "R:BASE System V Functions," "dBASE III PLUS Sample Application," "KnowledgeMan 2 Sample Application," and "R:BASE Sample Application."

This is a microcomputer reference tool with the information so arranged that it does not have to be read from cover to cover. It can also be used as a training guide for entry-level professionals. For public, college, academic, and special libraries, it is a recommended reference book for dBASE, R:BASE, and KnowledgeMan. *H. ROBERT MALINOWSKY*

165. The Electronic Business Information Sourcebook. John F. Wasik. New York, NY: John Wiley and Sons, Inc., 1987. 208p., bibliog., index. $24.95.
0-471-62464-0.

The purpose of this handbook is to bring together information about databases that are useful to businesses. Few companies or business professionals are aware of the hundreds of useful databases that are available. The author, therefore, considers this handbook to be "a primer on online database contents and applications." Preceding the introduction is a list of trademarks and service marks that are registered including BIOSIS, COMPENDEX, Dollars and Sense, Electronic Yellow Page, and Touch 'N' Shop. This is followed by a database subject index that indentifies what databases are available for a specific subject and in which chapters that database is discussed. There is a general index at the end of the book.

The rest of the book discusses the various databases, beginning with the very general ones and ending with the advanced applications. For each database, the vendor, costs, and features are presented in brief but complete discussions. Additional useful information is given when necessary. Four appendices cover "Annotated Bibliography"; "Information Brokers"; "Vendors, Services, and Trade Groups"; and "Selected Specialized Databases." These are all very brief in their content.

This is not a comprehensive handbook of databases for business but instead is a good introduction to what is available with information on where to go for further information. It is a book that would be useful in business libraries, public libraries, and to some extent in academic libraries. *H. ROBERT MALINOWSKY*

166. The Financial Manager's Guide to Microsoftware. Harvey L. Shuster; Ray D. Dillon. New York, NY: John Wiley and Sons, 1988. 303p., illus., index. $42.50. The Wiley/Ronald-National Association of Accountants Professional Book Series.
0-471-81176-9.

It is unfortunate when the strong points of any text are overwhelmed by poor design. Shuster and Dillon have authored a useful handbook for the businessperson, manager, or accountant who needs a guide through the maze of microcomputers and software. Their obvious familiarity and depth of knowledge is evident. Their advice is

tailored to the no-nonsense world of profit and loss. The emphasis is on software from spreadsheets to desktop publishing. There are chapters on user aids and documentation that are particularly good.

However, this is a book that only someone with a spreadsheet frame of mind could like. It is full of lists, checklists, charts, spreadsheets, etc., which are singularly unattractive. The format is cramped and confusing. The result is that the content is buried beneath a graphic chaos, and that is unfortunate because the material deserves better treatment. Furthermore can there be an accountant out there who does not know and use Lotus? For the rest of the readership, this is a good introductory text, but its usefulness is marred by the poor design. Although it does have these drawbacks, it would still be a useful addition to a public, academic, or special library reference collection. *GEORGE M. A. CUMMING, JR.*

167. IBM PC-DOS Handbook. 3rd.
Richard Allen King. San Francisco, CA: SYBEX, 1988. 359p., index. $19.95. 0-89588-512-3.

Another text dealing with PC-DOS, this handbook does an adequate job in presenting the necessary information to both programmers and users. The first nine chapters are of primary interest to those who wish to program. It is a straightforward presentation and assumes that the reader has little knowledge of the system. Therefore, it is very useful for the beginning programmer. The last eight chapters are of interest to the user and are independent of the beginning of the book.

Again, the author assumes little familiarity and takes the reader through keyboard functions, DOS commands, the line editor, and file naming and structuring among the topics. These are dealt with clearly and will help the reader to gain confidence in using the system quickly. There are eight appendices, varying in usefulness depending on what the reader needs from the system. The notes on different version numbers will be less useful than command summary to a user, although the function call summaries are handy for the programmer. Although this handbook does not seem to offer anything out of the ordinary, it is a useful resource to add to one's computer library if PC-DOS is the operating system one uses. *JOHN NISHA*

168. Master Handbook of Microcomputer Languages. 2nd. Charles F. Taylor, Jr. Blue Ridge Summit, PA: TAB Books, Inc., 1988. 499p., index. $26.95, $17.95pbk. 0-8306-0293-3, 0-8306-2893-2pbk.

The *Master Handbook of Microcomputer Languages*, formerly called *The Master Handbook of High-Level Microcomputer Languages*, describes the following "high-level" computer languages which can be used on microcomputers: Ada, BASIC, C, COBOL, Forth, Fortran, LISP, Logo, Modula-2, Pascal, PILOT, and Prolog. Taylor states that a high-level language is one where the "programmer doesn't have to be aware of and

keep track of many minute details about the computer being used." It may be helpful to think of these languages as being primarily machine-independent.

After a discussion of high-level languages, the author provides a general guide to the features of computer languages. This framework establishes a basis from which the reader can examine and compare the languages. Each chapter covers the program structure, data representation, arithmetic expressions, logical expressions, input and output, control structures, data structures, file handling, and graphic features of the individual languages. Information on how to obtain the programs is also included.

Although most readers of this book will probably be familiar with programming in one or more of these languages, it is still an excellent resource for a person who is just beginning to work in this area. In the summary chapter, Taylor compares each language in a concise chart by rating each of the features listed above. A word of wisdom is shared: "Which language is most suitable for a particular application depends on the nature of that application. There is no one universal language that is best for all applications."

It is in this spirit that the reader is invited to explore the strengths and weaknesses of each of these microcomputer languages. This is a highly recommended book for those who are not familiar with the various languages and should be an excellent reference source in college, university, and large public libraries. *CHARLES R. LORD*

169. The Professional User's Guide to Acquiring Software. John L. Connell; Linda Shafer. New York, NY: Van Nostrand Reinhold Co., 1987. 310p., illus., bibliog., index. $32.95. Van Nostrand Reinhold Data Processing Series. 0-442-21043-4.

This guide to the selection of computer software would be an asset to any computer user in a data management environment. It is part of a series of 11 publications on data processing by the same publisher. Chapters cover identifying the need for new software, cost estimation, evaluation techniques, quality control, personnel evaluation, and planning. Emphasis is on mainframe applications, however, one chapter is devoted to personal computers. An extensive bibliography, flow charts, tables, and an index make this a recommended book for large public libraries, as well as college and university libraries.
ANNE PIERCE

170. Programming the 80386. John H. Crawford; Patrick P. Gelsinger. San Francisco, CA: SYBEX, 1987. 774p., index. $24.95. 0-89588-381-3.

Here is an "everything you always wanted to know" book about the Intel 80386 microprocessor. Focusing on the chip's 32-bit features, the text treats the history of Intel microprocessors, data formats, data types, machine states and memory addressing, the complete instruction set, instruction set examples, memory management

and protection, interrupts and exceptions, operating system examples, and debugging support. The text is well organized and clearly presented with numerous examples illustrating each point. Several appendices compare the 80386, 80286, and 8086 chips, as well as the 80387, 80287, and 8087 numerics coprocessor chips. An index rounds out the work. The authors are also the developers of the 80386 processor, so they are intimately familiar with its features and well able to describe the fine points of its functioning. The book is recommended to anyone with a desire or need to learn the assembly level programming of the 80386 family of microprocessors. Academic and special libraries with research interests in the 80386 will want this as a reference source.

FRED O'BRYANT

171. Quattro: The Complete Reference.
Yvonne McCoy. Berkeley, CA: Osborne McGraw-Hill, 1988. 666p., illus., index. $24.95pbk. "Borland-Osborne/McGraw-Hill Business Series." 0-07-881337-9pbk.

Borland-Osborne has produced in one year three manuals by three different authors, all on the same software program, all covering, to a great extent, the same material. *Quattro: The Professional Spreadsheet Made Easy*; *Using Quattro, The Professional Spreadsheet*; and *Quattro: The Complete Reference* are interchangeable for most users of the integrated spreadsheet package. This last title, however, is possibly the most distinctive and certainly the most exhaustive of the three. *Quattro: The Complete Reference* abandons the tutorial approach found in the other books. McCoy accomodates beginning users, but her main objective is to present an encyclopedic guide to Quattro. Consequently, in this sense, McCoy's book is much more useful as a reference tool and much less useful as a how-to guide for beginners.

Notably, of the three books, McCoy's book comes closest to replacing the manual that comes with the software. For example, she provides an appendix with the printer codes necessary for telling the program what sort of printer you own. Also, she has thought to give us an appendix for common error codes encountered in the program. This is important information for users at every level and should be more common in manuals of this sort. One bit of information given in the package manual that is not present here is a table of performance statistics: program speed and hardware requirements. On the other hand, a table of program default settings is included, a useful and well-presented table.

The index for McCoy's book is surprisingly small. To some extent this defect is compensated for by the fine single-alphabet glossary of commands, macros and functions which appears in appendix A and by the detailed table of contents. The strength of the book, however, lies between the table of contents and appendices, in the exhaustive depth with which all of Quattro's capabilities are described. Section 1, "An Introduction to Quattro," is devoted to spreadsheet manipulation and consists of nine tersely written chapters puctuated by screen dumps and tables which summarize sections of text. McCoy seems to have left out nothing, including the lengthy treatment of the nearly always neglected subject of file maintenance.

Section 2, "Advanced Features," treats graphs and data management, and it is here that the distinction between McCoy's book and other Quattro books is most evident. It is a paradox that, although the database capabilities of integrated spreadsheet packages like Quattro are severely limited, they are nonetheless given short shrift in most spreadsheet manuals. McCoy does Quattro's limited capabilities their deserved justice. Advanced users will also appreciate the chapter on creating command-language macros, and the chapter on using add-in programs and customizing Quattro's menu screens.

This handbook is recommended for personal use as well as a reference source in public, collge, and university libraries. *STANLEY WILDER*

172. Turbo C: The Complete Reference.
Herbert Schildt. Berkeley, CA: Osborne McGraw-Hill, 1988. 908p., index. $24.95pbk. "Borland-Osborne/McGraw-Hill Programming Series." 0-07-881346-8.

The programming language C is not yet a household name as are Basic or Fortran, yet one has only to glance through the classified ads under "computer programmers" to see that C is well on its way to achieving an unprecedented degree of prominence as the language of choice for developers of new software applications. There are two principle reasons for C's burgeoning popularity: its incredible speed and its portability.

Portability refers to the ease with which a language allows applications to be adapted to run on more than one operating system. In other words, an exceptionally portable language such as C allows software developers to create programs which they can market to both Mackintosh and IBM users, for example. Borland's Turbo C software brings C programming out of its mainframe UNIX-environment home and makes it available to IBM microcomputer users. Herbert Schildt's *Turbo C: The Complete Reference* is likely to become an essential reference for Turbo C users.

At first glance, Schildt's volume seems short on access tools: its 900 pages of text are covered in only 13 pages of index and a subdivided table of contents. There are very few tables, no detachable ready-reference cards, and only one appendix, which does not lend itself to solving ready-reference questions. All important subjects and commands are referenced by the index or table of contents. Also, Schildt has not skimped on supplying sample algorithms and sections of code throughout the book. There is no question that sample code is more important than tables in books such as this.

Part 1 of the book is a guide to the language itself and includes chapters on C's handling of variables, functions, arrays, pointers, and the program control system, among other topics. These are the essential questions that any language addresses, and the information here is applicable to C in either mainframe or micro environments. Parts 2 and 3 are a guide to manipulating Turbo

C in particular. Part 2 covers the program's menus, screens, text editor, and graphics functions; part three is a useful reference to Turbo C's libraries and function conventions. Parts 4 and 5 cover the most advanced applications of C: artificial intelligence, language interpreters, software engineering, and debugging.

Each section is designed as a problem-solving guide, not as a tutorial, and this aspect of the book will enhance its usefulness to users over the long run. *Turbo C: The Complete Reference* is likely to become the standard work on C in the IBM microcomputer environment for some time to come. It is recommended for personal use as well as a reference source in public, college, and university libraries. *STANLEY WILDER*

173. Turbo C Programmer's Resource Book. Frederick Holtz. Blue Ridge Summit, PA: TAB Books, Inc., 1987. 247p., illus., index. $17.60pbk. 0-8306-3030-9pbk.

Both the novice and the experienced programmer will find *Turbo C Programmer's Resource Book* of value. It "takes you by the hand" for those knowing BASIC, and advanced concepts are included for the more experienced. Frederick Holtz is a professional programmer and C language instructor. Another Holtz manual is on the AT&T 6300. According to Holtz, "C has become the most popular programming language for general applications..." and he is advocating use of C by others by authoring this useful guide. Step-by-step instructions follow throughout the manual. Four functions are covered: (1) text handling; (2) high level math; (3) DOS/BIOS; and (4) graphics. Chapter one is devoted to "selling" Turbo C, with background information on its conception, in easy-to-read style. Each chapter ends with a summary. This book contains "basic introduction and practical applications" to Turbo C. Systems requirements include the IBM PC and close compatibles running DOS 2.0 and higher: IBM PC, XT, AT and almost every machine that is MS-DOS compatible. Included also is instruction on CBREEZE and HALO graphics. It contains many programming examples. *BERTA KEIZUR*

174. The UNIX Command Reference Guide. Kaare Christian. New York, NY: John Wiley and Sons, 1988. 361p., index. $23.95pbk. "A Wiley-Interscience Publication." 0-471-85580-4pbk.

Most computer tyros keep close to the terminal a well-thumbed rescue manual, that first printed source to grab when the unexpected leaps out at you from the other side of the screen. Usually it is not the tech manual that comes with the equipment or software but the plain written advice of the seasoned pro. In fact, there is a small industry that supplies printed help to the panic-stricken beginner. *The UNIX Command Reference Guide* is just such a manual.

Fifty of the most common UNIX commands are covered, arranged in seven general categories. The commands are arranged alphabetically within each group. There is a separate index with the commands arranged in one alphabetic sequence.

Each command is explained in sufficient detail. Most important, each entry is in clear and understandable English. That is the acid test for this sort of reference. If the user cannot clear up that on-screen mess with one quick read, then the guide is no better than the generally abysmal technical manuals that the manufacturers foist on the nonhacker world.

Christian has written several books on UNIX and knows the operating system. This book works well. It is also a good introduction to UNIX for the uninitiated. As a personal preference, the commands ought to be in one alphabetic sequence rather than split into seven different groups, but that is quibbling. This is an excellent user's aid and recommended for individuals as well as public, college, university, and special libraries. *GEORGE M. A. CUMMING, JR.*

175. VM and Departmental Computing. Gary R. McClain. New York, NY: McGraw-Hill Book Co., 1988. 253p., illus., bibliog., glossary, index. $39.95, $24.95pbk. "Intertext Publications/Multiscience Press, Inc."; Computing That Works. 0-07-044938-4, 0-07-44939-2pbk.

VM and Departmental Computing is another volume in the series, Computing That Works, which offers very practically oriented handbooks and manuals. This one focuses on the VM operating system and its optimal use in the departmental computing environment. The author, a manager with VM Software Inc., states a dual purpose in writing the book. The first is to provide an alternative source of information on VM to the technical documentation and, second, to then provide an understanding of VM in the context of a departmental computing situation. In his introduction, McClain points out the growing number of end users and the increase in interactive computing. He also suggests that a diminishing of the role of MIS systems as end users, rather than the technical staff, demands more involvement in computing planning and issues. The intended audience of the book, however, is both the technical staff and the end users.

Organized into 15 chapters, the book begins with an overview and history of VM and a discussion of the concept of departmental computing and then becomes more specific by addressing such issues as installation, security, managing workloads, applications, disaster recovery, and other practical topics. Each chapter finishes with a summary page. A table of contents and a subject index are included as well as a glossary and a bibliography. The glossary defines acronyms as well as phrases and terms and is particularly helpful. The bibliography is brief, but very current and draws from both the management and computing literature. The book would best fit a business-oriented collection rather than a computer science collection because of its emphasis on application. It is, therefore, recommended for a business library. *JOANNE M. GOODE*

How-to-do-it Books

176. 1001 Things to Do with Your Apple IIGS. Mark R. Sawusch; Dave Prochnow. Blue Ridge Summit, PA: TAB Books, Inc., 1987. 194p., illus., glossary, index. $18.95, $12.95pbk. 0-8306-0186-4, 0-8306-2886-Xpbk.

The authors state in the introduction that this book has three purposes: to help potential buyers to know possible applications that will run on the Apple IIGS; to help users to utilize the step-by-step programs within the book; and to help advanced users to create their own programs. It is as a listing that this book succeeds best; it contains a wonderful selection of possible things a user can do with this computer. But rather than being an information book, it is a marketing tool for TAB Books and Apple IIGS hardware. There are 21 programs written out that can be typed into the computer and run or can be ordered on a disk on the conveniently printed order form at the back of the book. A major problem of this book is that it entices rather than informs. Throughout the text are listings of various things the computer can do personally, professionally, and recreationally. There is even a listing from users who have written to the authors of 50 things they have done with their computers. Also included are listings of programs and books for the Apple IIGS. Most of the narrative could be condensed and Apple dealers should consider offering this book free rather than having the consumer purchase it. *ANNE PIERCE*

Manuals

177. CICS/VS Command Level with ANS Cobol Examples. 2nd. Pacifico A. Lim. New York, NY: Van Nostrand Reinhold Co., 1986. 582p., illus., index. $24.95pbk. Van Nostrand Reinhold Data Processing Series. 0-442-25814-3pbk.

Customer Information Control System/Virtual Storage (CICS/VS) is a teleprocessing monitor program developed by IBM. Teleprocessing monitors, as described by the author, "were the last link in the chain of developments that brought on-line capabilities to many commercial installations, freeing the application programmer to concentrate on actual applications." Lim first discusses the inadequacies of batch processing and on-line processing to appropriately respond to the growing need to have multiple terminals concurrently accessing the same files. Compilers were not able to deal with the complexity of this on-line environment, thus, the development of teleprocessing monitors. IBM's CICS has a command-level feature which "allows the programmer to code requests for on-line services in statements similar in format to those in the high-level language he is used to." The author uses ANS Cobol examples to illustrate the flexibility and features of this system.

The chapters include the following areas: the screen layout, creating maps, program components, linkage, processing, terminal input/output commands, file control commands, and program control commands. In addition, there are detailed discussions on the file inquiry program, file add program, file update program, file delete program, and file browse program. The final section covers general debugging techniques. This reference manual is "written primarily for ANS Cobol application programmers who are interested in or will be involved in writing CICS/VS application programs." But, the author does note that non-Cobol applications programmers may also be interested in the general discussion of the system. Libraries supporting computer science will want this reference manual as will individual users. *CHARLES R. LORD*

178. dBASE Compilers: A Programmer's Resource Book. Ken Knecht. Blue Ridge Summit, PA: TAB Books, Inc., 1988. 273p., index. $18.95pbk. 0-8306-2943-2pbk.

Clipper, Foxbase+, Quicksilver, and DBC III are examined in *dBASE Compilers.* The standard applications given with each compiler provide the reader with a basis of comparison for the compilation and linking process of each program. In addition to the four compilers, assistance with dBMAN and db/LIB are included.

Detailed comparisons made between dBASE and dBMAN are very helpful. The author describes dBMAN as "an interpreter similar to dBASEIII PLUS." Compilers can be used by programmers to increase performance speed, to establish stand-alone accessibility, to develop program security, and to enhance the primary program with additional commands and functions; however, a person should also be aware of the disadvantages to using compilers.

Knecht raises the following considerations: compiling and linking the application takes time; some applications require the use of overlays which require time and expertise; the compiler program may not have all of the dBASE program commands and functions; and debugging compiled files is potentially more difficult.

Although this book does not substitute for the program manuals, it does provide a snapshot comparison of these compilers and is recommended for university and special libraries. *CHARLES R. LORD*

179. dBASE Mac Programmer's Reference Guide. Edward C. Jones. Indianapolis, IN: Hayden Books, Howard W. Sams and Co., 1988. 300p., illus., index. $19.95. 0-672-48416-1.

dBASE Mac Programmer's Reference Guide is intended for the intermediate and advanced Macintosh user who wants to learn how to use dBASE Mac to design custom applications for others. Significant differences between this Macintosh database manager and its MS-DOS rivals are outlined, after which the author extensively outlines the many unique features of

dBASE Mac. The usual topics of database design, menu options, program control, and the like are thoroughly covered with the help of numerous figures representing Macintosh-style screens. How to convert dBASE III and similar programs to dBASE Mac is also covered

For the reader interested in how to employ dBASE Mac effectively, this book provides an inexpensive and useful tutorial. It is recommended for individuals and academic libraries supporting computer science. *FRED O'BRYANT*

180. **Hard Disk Management in the PC and MS DOS Environment.** Thomas Sheldon. New York, NY: McGraw-Hill Book Co., 1988. 238p., index. $24.95pbk. 0-07-056556-2pbk.

An excellent manual, *Hard Disk Management in the PC and MS DOS Environment* presents in a very readable, friendly format an overview of just about everything a hard-disk owner would need to know to maximize the potential of a hard disk. Optimally, one would want to read this manual before using a hard disk, because the organizational strategies and tips could save the owner a great deal of frustration down the road. Hard disk purchase is becoming much more commonplace even for first-time computer owners, and much of the mystery is dispelled with the use of a third-party manual such as this one. The author has included material and plenty of advanced techniques for the more experienced user as well.

In the first section, the author reviews the basics of DOS, hard-disk installation and set-up procedures and, using the analogy of an office system (which is continued throughout the book), explains the importance of creating directories and paths designed to fit the reader's specific needs, present and future. Chapters 5, 6, and 7 discuss ways to enhance the use of the disk by creating menus and help screens, utilizing batch files, and programming function keys. Throughout the book, the author supplies names of packaged software to accomplish the same techniques and mentions enhancements available to owners of DOS 3.3 and OS/2. The book includes an ample number of simple diagrams and uses shaded areas to explain commands and screen displays. The remainder of the book covers special topics such as OS/2, protecting data, backup tips and techniques, and other disk technologies. Five appendices follow which list very brief but useful information on specifications and supplementary disk and print materials. Rather than being classified as a standard reference book, this is a hands-on manual that one would want to keep by the terminal. The book is in paperback format and is one in a series titled Computing That Works. *JOANNE M. GOODE*

181. **PC Care Manual: Diagnosing and Maintaining Your MS-DOS, CP/M or Macintosh System.** Chris Morrison; Teresa S. Stover. Blue Ridge Summit, PA: TAB Books, Inc., 1987. 204p., illus., index.

$24.95, $16.95pbk. 0-8306-0991-1, 0-8306-2991-2pbk.

The *PC Care Manual* provides information on the care of every component in a typical PC system, including keyboard, monitor, printer, disk drive, and serial communication interface, with a chapter on each. It describes procedures for daily maintenance of a PC to keep a system in good working order. When trouble develops, it provides descriptive procedures in five BASIC program modules that instruct the reader in the use of a series of tests to check each component. Clear instructions are given to make the needed repairs or adjustments when the problem has been identified. If the problem is too complex, the reader is advised when to seek expert help. Instructions are systematically presented, and the book includes a detailed index for pinpointing sections for help. The manual will be of value in any personal or library collection where there are questions about PC system maintenance. *RAYMOND BOHLING*

182. **Power User's Guide to Hard Disk Management.** Jonathan Kamin. San Francisco, CA: SYBEX, 1987. 315p., index. $19.95. 0-89588-401-1.

One of the greatest mysteries any owner of more than the simplest personal computer system faces is how to install and use a hard disk system. Although the hard disk offers possibilities for sophisticated processing and information storage and management, it also presents challenges in its optimum use. Here in one easy-to-read-and-use book, Jonathan Kamin has assembled nearly everything a novice hard disk user needs to know to make the most of his or her system. The book is designed with the IBM family of computers (and compatibles) in mind, but its principles should apply to most hard disk systems.

Kamin takes a step-by-step approach, building on previous knowledge and examples as he goes along, while also pointing out common pitfalls and exceptions. The reader is led gently from initial installation to formatting, through the use of tree-structured directories, how to perform back-ups, and how to access and use RAM disks and extended memory. Additionally, Kamin discusses how to organize software tools, how to run RAM-resident software, and the how-to of system menus and security.

The book presents three fully developed hard-disk-system scenarios, which illustrate the concepts outlined in earlier chapters. Several appendices include software sources, an introduction to batch file programming, the ANSI.SYS device driver, differences between MS-DOS and PC-DOS, tables of ASCII characters, and a program for logging computer use. There is also an index of programs and a subject index.

Taken altogether, this is a welcome book to keep near one's machine for occasions when questions arise about how to get the best use from the system's hard disk. For the money, this is a small investment relative to the wealth of information it contains. Academic and special libraries may want several copies—some to circulate and a few to keep near their own office computers. *FRED O'BRYANT*

183. Quattro: The Professional Spreadsheet Made Easy. Lisa Biow. Berkeley, CA: Osborne McGraw-Hill, 1988. 600p., illus., index. $19.95pbk. "Borland-Osborne/McGraw-Hill Business Series." 0-07-881347-6pbk.

Judging from the proliferation of independently produced manuals on the use of popular software applications, it seems as if the market has found a solution to the, by now legendary, inability of software manufacturers to write clear, effective documentation to accompany their products. This is unfortunate for consumers who expect the quality of the software they purchase to be matched by the quality of its directions for use.

Lisa Biow's *Quattro: The Professional Spreadsheet Made Easy* is a fine example of the sort of documentation users should come to expect to be included with the software they buy. Quattro is an integrated spreadsheet package which, like Lotus 1-2-3, combines spreadsheet, graphics, and data management software. Intended for beginning users, Biow's book takes a tutorial approach throughout, yet users will also appreciate its usefulness as a ready reference for problems encountered in general use of the software.

There are many noteworthy aspects to the book's design. The index is ample and contains references to the primary commands used in the software, which cannot be taken for granted in software manuals. Four appendices follow the text: an alphabetical list of commands, directions on adapting the program to Lotus 1-2-3, a functions list, and directions for the most elementary aspects of the DOS operating system and disk management. Placing this information in an appendix is especially useful for the more advanced user who does not need this information in the main text. The function appendix is especially useful, because it contains explanations, often with examples.

The book also contains a detachable command card, which serves as a handy crib sheet to be kept close to the computer. Additionally Biow provides a scale drawing of the command template, which users place around the PF keys on the left hand side of the IBM keyboard. In truth, such drawings are useful primarily to users with pirated copies of Quattro, which is not a copy-protected program. Still, it would be helpful to have these templates produced on detachable cards, in color. It is surprising how many spreadsheet manuals take for granted that users understand the most important aspects of spreadsheet design. Although the column-row format of a spreadsheet looks simple, improper planning can have serious consequences when the time comes to generate graphs, data sets, or to adapt a spreadsheet for multiple uses. Biow provides a much-needed chapter on this topic. Unfortunately she waits until chapter 20 to do so, well after most users will have begun creating their own spreadsheets.

Biow's tutorial approach, combined with the excellent problem-solving tools, makes *Quattro: The Professional Spreadsheet Made Easy* a useful guide to use as well as a good source of ready reference to the Quattro software for academic and special libraries. *STANLEY WILDER*

184. Systems Programming in Turbo C. Michael J. Young. San Francisco, CA: Sybex, 1988. 403p., index. $24.95pbk. 0-89588-467-4pbk.

Written for C programmers working from every level of experience, this manual covers a variety of Turbo C (through release 1.5) advanced programming techniques. Young states that his book was written with the following people in mind: "First, for the applications programmer, it contains an extensive collection of software tools that can be immediately incorporated into a C program, significantly extending the Turbo C function library. Second, for the systems programmer, it provides complete source code listings and detailed explanations to exemplify many of the advanced features of Turbo C, and to show how to make optimal use of this language for systems-level programming."

The first chapter offers helpful procedures on how to install Turbo C and outlines the features of this programming system which are unique and in some cases not discussed in other basic C programming manuals. Other chapters include advanced programming, keyboard and console functions, printer functions, video functions, device-independent graphics, and memory-resident programs. The last chapter includes a library of string, time and date, and numeric miscellaneous fuctions for Turbo C. Appendices, bibliography and index are included.

The author uses a format that separates the application of a function or utility from the explanation of how it works. This style of organizing the material allows the user to easily identify and apply the function without having to understand the details of how it works. The functions are written to operate on IBM-compatible PCs, ATs, and the Personal System/2 series. Information on how to order a program disk is included. This is a good reference work for C programming for academic and special libraries.

CHARLES R. LORD

185. Troubleshooting and Repairing Personal Computers. Art Margolis. Blue Ridge Summit, PA: TAB Books, Inc., 1987 (c1983). 311p., illus., index. $16.95pbk. 0-8306-1539-3pbk.

This guide will direct the reader from the analog circuit into the digital circuit world with step-by-step instructions—from disassembly to reassembly. *Troubleshooting and Repairing Personal Computers* presents everything that is needed to know, from a hardware point of view, of the digital logic system for troubleshooting or repairing your PC. Art Margolis has picked up where the vendor manuals leave off—by not assuming anyone is "digital smart." Margolis, well known in the electronic service and electrical wiring field,

has written a manual to help users get over the frustration encountered when approaching digital schematics, perhaps with only a knowledge of analog circuitry. Chapters 1 through 8 deal with "chip changes"; chapters 9 and 10 launch into digital circuitry; and chapter 11 begins the more technical repairs. Remaining chapters deal with peripheral devices. Few pages lack explicit illustrations or diagrams. This guide is ideal for the experienced technician or the inexperienced computer owner. It would also be useful for large public libraries and academic and special libraries.

BERTA KEIZUR

186. Turbo Basic Programs for Scientists and Engineers. Alan R. Miller. San Francisco, CA: SYBEX, 1987. 276p. illus., index. $19.95. 0-89588-429-1.

Turbo Basic is one of the few programming languages that has the capability to assist a researcher in solving advanced problems on a microcomputer. The author outlines the purpose of the manual as both "to help the reader develop a proficiency in the use of Turbo Basic," and "to provide a library of programs useful for solving problems frequently encountered in scientific and engineering applications." The discussions and problem-solving exercises are appropriate for both the practitioner and upper-class engineering students who are familiar with such computer languages as Pascal, Fortran, or BASIC. The chapters in *Turbo BASIC Programs for Scientists and Engineers* include the following topics: evaluating and working within the limitations of the Turbo Basic Compiler, incorporating statistical tools, using vector and matrix operations, programming with the simultaneous solution of linear equations, sorting, and using curve-fitting equations. There are over 60 programs listed in the program index. A limited subject index is also included. Although there are certainly helpful discussions on the different applications, the strength of this volume when used as a textbook is in the number of detailed problem-solving programs. Students and teachers alike will find these programs very valuable. Academic and special libraries will find this a useful reference source on Turbo Basic programs.

CHARLES R. LORD

187. Turbo Pascal Toolbox. Frank Dutton. San Francisco, CA: SYBEX, 1988. 405p., index. $18.95pbk. 0-89588-472-0pbk.

As users become more knowledgeable about computers and programming, they quickly discover that there are many computing tasks which would make their work easier but are not included in their software packages. The decision has to be made either to find appropriate utility software or, if one has the programming skills, to write the necessary programs. Another solution is to use a resource such as *Turbo Pascal Toolbox,* which provides dozens of microcomputer programs and routines which have been road-tested and are ready to be incorporated into your work.

According to the author, this handbook is written for beginning programmers who have some knowledge of Turbo Pascal, intermediate-level programmers who can make excellent use of the documentation with selective applications, and professional programmers who may be first-time users of Turbo Pascal. It is also important to mention that "all the procedures and functions in this book are designed for use with Turbo Pascal versions 3.xx or 4.0 on IBM PC, XT, AT, and compatible computers using DOS 2.xx or later versions."

A brief outline of the contents includes the following programs, procedures, and functions: accessing DOS functions, performing both simple and complex directory searches, incorporating easy data-entry techniques, handling and storing dates and times, programming menus, manipulating and designing screens, and creating special characters and screen effects. In addition, there is a chapter of miscellaneous tools which include programs for changing the prompt, working with printers, utilizing file protection, and limiting program life.

Appendices include considerations for incorporating the procedures and functions into your programs and an alphabetical reference list. An index is also included. The tools which are identified as procedures and functions and the test programs are set aside from the text by using a window-format. The author also provides quick reference to those programs that are not compatible with all versions of DOS with a specially designated icon. This toolbox can save the reader countless hours of programming frustration and at the same time provide countless hours of tinkering enjoyment. Academic and special libraries with extensive computer science collections will need this excellent "toolbox."

CHARLES R. LORD

188. UNIX Utilities. Ramkrishna S. Tare. New York, NY: McGraw-Hill Book Co., 1987. 387p., bibliog., index. $34.95. 0-07-062879-3.

This is not a book for the beginning programmer or a person desiring to learn to program in the UNIX operating system environment. However, it is a good reference to the uses and capabilities of the major UNIX utility routines available in most versions of UNIX.

The primary purpose of this book is "to simplify utilities that may look cryptic and advanced in the reference manuals." The utilities are grouped based on their usage type. There are nine chapters dealing with file processing, debugging, language development tools, system development tools, database management systems, text formatting tools, data communications, writing tools, and miscellaneous utilities. Each utility is described in a simple manner with examples, and more advanced features are introduced by building on those examples.

The text is clear and easy to read (the entire book, by the way, was formatted using UNIX utilities). There are a few study problems for each chapter. The index is adequate, but the bibliography is rather sparse. Recommended as a good introduction to the use of UNIX utilities for academic libraries. *FRED O'BRYANT*

189. VAX Architecture Reference Manual. Timothy E. Leonard, ed. Bedford, MA: Digital Equipment Corp, 1987. 417p., illus., index. $38.00. "Order Number EY-3459E-DP"; "DECbooks." 0-932376-86-X.

VAX Architecture Reference Manual should be of value to at least three reader groups: systems architects, systems programmers, and faculty and students in computer science and electrical engineering. "This book introduces the design goals and terminology of the VAX family and then develops a precise description of the VAX instruction set, including those for memory management, exception and interrupt handling, and process control. The description of each instruction gives formats, operations, conditions codes, instruction-specific exceptions, op codes and mnemonics." This is a well-written, technical, and very comprehensive manual. An excellent index supplies a generous range of terms, code phrases, and acronyms for locating relevant sections in this manual that will be useful in large academic libraries and special libraries.

RAYMOND BOHLING

Textbooks

190. Introduction to Programming Using MODULA-2. Richard J. Sutcliffe. Columbus, OH: Merrill Publishing Co., A Bell and Howell Information Co., 1987. 551p., bibliog., index. $34.95pbk. 0-675-20754-1pbk.

This is a standard textbook introducing students and interested programmers to the MODULA-2 programming language. Beginners and experienced programmers alike should be able to use the book to become familiar with the concepts and syntax of MODULA-2. The chapter arrangement introduces topics in a logical and straightforward manner. There are short exercises after each of the 10 chapters (no answers given), as well as definitions and other sidebar material interspersed throughout the text, to provide interesting information and historical perspectives on computer programming. The text concludes with a series of 24 short appendices detailing specific peculiarities of MODULA-2, a brief bibliography, and an index. Overall, this appears to be an adequate text for the introduction and study of MODULA-2, aimed primarily at college classes and secondarily at the interested individual programmer. *FRED O'BRYANT*

191. Intuitive Digital Computer Basics: An Introduction to the Digital World. Thomas M. Frederiksen. New York, NY: McGraw-Hill Book Co., 1988. 253p., illus., index. $39.95, $24.95pbk. The McGraw-Hill Series in Intuitive IC Electronics. 0-07-021964-8, 0-07-021965-6pbk.

Intuitive Digital Computer Basics is designed to provide those with no prior training in electronics an introduction to the workings of electronic components and circuits, especially because these are used to construct modern digital computers. Although this is not a "textbook" in the classroom sense, the author presents many examples of simple circuits and expects the reader to set them up and examine how they work on a laboratory kit. Thus, to get the full benefit of the text, the reader must also purchase a lab kit such as the 200-in-One Electronic Project Kit from Radio Shack, which the author uses for examples throughout the text. The use of a digital multimeter is also necessary.

The text is clearly and simply presented with little or no mathematics required. The author discusses the workings of a wide variety of circuits and components, including TTL and CMOS technology, resistors, capacitors, diodes, transistors, serial and parallel digital communications, memory devices, microcodes, internal buses, internal clocks, and circuit logic. The illustrations are easy to read and understand, and an index facilitates the use of the text.

For persons with no previous electronics experience, this text would be a readable and thorough place to begin to understand the inner workings of modern computers and digital devices. For those with previous experience the book provides a good refresher and reference. Recommended for individuals and academic libraries that have an interest in this topic.

FRED O'BRYANT

192. Knowledge Acquisition for Expert Systems. Anna Hart. New York, NY: McGraw-Hill Book Co., 1986. 180p., illus., bibliog., index. $28.95. 0-07-026909-2.

"An expert system could be defined (Wellbank 1983): An Expert system is a program which has a wide base of knowledge in a restricted domain, and uses complex inferential reasoning to perform tasks which a human expert could do." This textbook covers in detail what is involved in an expert system. The ten chapters cover "The Nature of Expertise," "Programs as Experts," "System Analysis—A Comparison," "The Knowledge Engineer," "Fact-Finding by Interviews," "Reasoning and Probability Theory," "Fuzziness in Reasoning," "Machine Induction," "The Repertory Grid," and "Try It Anyway—Two Case Studies." A seven-page glossary helps with understanding this well-written textbook for academic libraries and those who want to know more about artificial intelligence and expert systems.

H. ROBERT MALINOWSKY

193. Mastering Turbo Pascal 4.0. 2nd. Tom Swan. Indianapolis, IN: Hayden Books, Howard W. Sams and Co., 1988. 774p., bibliog., index. $22.95. 0-672-48421-8.

This is a completely rewritten and expanded version of Swan's original text on Turbo Pascal. Over half the material is completely new, and all the rest has been extensively revised and improved to illustrate the many new features of Turbo Pascal version 4.0 and later. Swan's text and examples are written to perform on any IBM PC or compatible able to run Turbo Pascal. No prior experience is required.

Included are chapters describing the full range of Turbo Pascal abilities, as well as advanced techniques, pointers, separate compilation with units, color graphics, how to use a mathecoprocessor, and how to mix assembly language with Pascal. Complete "railroad diagrams" (syntax charts) are included along with an encyclopedia of all Turbo Pascal procedures and functions and sample programs for each. The text includes hundreds of tested examples, which may either be typed in from the keyboard or ordered from Swan in diskette form. Each chapter concludes with exercises, some of whose answers appear in the back of the book.

Swan, an accomplished and prolific author, has provided interested users and teachers with a Turbo Pascal "bible" which may be used and studied to good advantage. The price is right, and this text is recommended to all interested users as well as academic libraries supporting computer science. *FRED O'BRYANT*

194. Office Automation and Information Systems. Arnold Rosen. Columbus, OH: Merrill Publishing Co., Bell and Howell Information Co., 1987. 363p., illus., glossary, index. $27.95pbk. 0-675-20557-3pbk.

The purpose of this textbook is to replace those "that focus on a single entity and provide a balance between office automation and management information systems." It covers all aspects of office automation and information systems using an easy-to-understand text and good photographs. Each chapter ends with recommended projects to help better understand the concepts presented in the chapter. A typical chapter discusses the concepts in detail and ends with a summary, questions, and projects. It is divided into seven parts: "Introduction," "Management Mandates: Information and People," "The Information Processing Cycle: Systems and Hardware," "Software and Storage: Components and Technology," "Safety, Security, and Environment in the Office," "Personnel and Training," and "The Future." A nine-page glossary of the more commonly encountered automation terms is appended. The author is an authority on word processing, giving many seminars and authoring many books on the topic. In addition to being an excellent textbook on office automation, it also has reference value for large public and academic libraries in explaining concepts that may be hard to find. *H. ROBERT MALINOWSKY*

195. Programming the Z-80. 3rd rev. Rodnay Zaks. San Francisco, CA: SYBEX, 1987 (c1982). 624p., illus., index. $21.95. 0-89588-069-5.

This is a good book—even an excellent book—however, it is also six years old. In the fast paced world of microprocessor design, six years is equivalent to eons. This book is designed to teach its reader how to program the Zilog Z-80 microprocessor in its native machine language. The Z-80 is an eight-bit processor—now surviving in a world increasingly filled with 16-bit and even 32-bit machines. By today's standards the Z-80 is quaintly antique. Nevertheless, this processor remains prevalent in older personal computers and digital control devices.

Anyone who has access to a machine with a Z-80 processor in it and who wants to learn the intricacies of programming it at the assembly level would find this book clear and helpful. The book is built on the effective and time-honored principle of proceeding from the simple to the complex with each new concept augmenting and expanding what has already been learned.

The text covers topics such as Z-80 hardware organization, input/output techniques, the Z-80 instruction set and addressing modes, the theory and design of data structures, and numerous applications examples. Although the book claims to have over 200 "illustrations," these consist not of photographs but of clear and understandable diagrams illustrating how the Z-80 performs. The book is studded with exercises (answers not provided) and question-and-answer sections. Seven appendices list various conversion tables, instruction sets, and equivalence tables between the Z-80 and Intel 8080 processor, upon which it is closely modelled.

As the author states, learning to program in assembly language takes time—no matter what processor is studied—but the rewards for anyone so motivated can be great. This book is a handy reference to the Z-80 microprocessor and a well-organized textbook for either classroom or individual use. Recommended with the caveat that the Z-80 processor is not the "latest" thing on the market. *FRED O'BRYANT*

196. Programming Using the C Language. Robert C. Hutchison; Steven B. Just. New York, NY: McGraw-Hill Book Co., 1988. 529p., illus., index. $28.95pbk. McGraw-Hill Computer Science Series. 0-07-031541-8pbk.

Programming Using the C Language has been written so it can be used by programmers as an introduction to the C language and, also, as a college or university text for those already familiar with C. The authors are consultants at AT&T Bell Laboratories. Hutchison teaches UNIX system and C language classes to the technical employees, and Just consults on instructional software. The book comprises 18 chapters divided into the following seven parts: "Introduction"; "Data Types"; "Control Flow"; "Derived Variable Types"; "More on I/O, Pointers, Functions, and Variables"; "Complex Data Types"; and "Case

Studies." Each chapter begins with an overview and ends with a summary and a set of exercises. There are four appendices: "ASCII Chart," "ANSI C Proposals," "Using the C Language with the UNIX Operating System," and "Using the C Language with the MS-DOS Operating System." The text includes an index.

JO BUTTERWORTH

197. Turbo Pascal Programs for Scientists and Engineers. Alan R. Miller. San Francisco, CA: Sybex, 1987. 322p., index. $19.95. 0-89588-518-2.

The author's intention is that this book be used for two purposes, to assist the reader in becoming familiar with the Turbo Pascal language and to provide a library of programs useful for solving various scientific and engineering problems. This book succeeds in presenting useful programs in the following areas: statistics, vector and matrix operations, simultaneous solution of linear equations, curve-fitting, sorting, least-squares curve-fitting, equation solution using Newton's method, numerical integration, nonlinear curve-fitting, and other advanced mathematical topics. In all, 65 programs or procedures are given. Each is discussed in some detail prior to its solution being presented. Sample output is provided.

As a text presenting useful programming solutions to the specific problems included, this book succeeds very well. However, as a guide to learning to program (or learning to program better) in Pascal, it is somewhat less successful. Even its author points out that prior familiarity with a high-level programming language is necessary to make best use of the material included. So, rather than assisting the reader in becoming familiar with Turbo Pascal, this book seems to require familiarity with Pascal to get benefit.

Given this caveat, the programs presented are clear and useful and will be of value to persons who are doing problem solving in a Turbo Pascal environment. The book also includes brief ex-

ercises (with answers), a summary of Turbo Pascal, and an index. Where the interest exists, this would be a reasonable purchase for academic libraries.

FRED O'BRYANT

198. The 68000 Microprocessor: Architecture, Programming, and Applications. Michael A. Miller. Columbus, OH: Merrill Publishing Co., A Bell and Howell Information Co., 1988. 513p., illus., bibliog., index. $37.95. 0-675-20522-0.

This is an upper-level textbook covering the concepts and applications for the 68000 microprocessor family. It assumes that the student has some knowledge and background in the areas of digital electronics including number systems and binary codes; logic gates and Boolean algebra; registers, counters, and sequential logic; memories and memory systems; analog to/from digital conversion; and basic microprocessor computer concepts. The student should also have knowledge of bipolar and field-effect transistors as well as of basic circuit analysis.

The book is written in textbook fashion with the concepts discussed in detail. A glossary is included with each chapter, and numerous problem exercises are at the end of each of the 12 chapters: "Fundamentals of Microprocessor-Based Computers," "Introduction to the 68000," "The 68000 Instruction Set," "68000 Program Applications and Bus Cycle Timing," "Parallel Interfacing the 68000," "68000 Serial Data Interfacing," "Tutor 1.3 and EXORmacs Assemblers," "Exception Processing," "68000 Application Programs," "The 68000 Family of Microprocessors," "MC68020 32-bit Processor," and "68881 Floating Point Coprocessor and 68020 Module Support."

This is an excellent text and would be useful as a reference source for academic libraries because of its in-depth coverage and the glossaries within each chapter.

H. ROBERT MALINOWSKY

EARTH SCIENCE

★
GENERAL SOURCES
Abstracts

199. Geographical Abstracts: B—Climatology and Hydrology. Norwich, Great Britain: Geo Abstracts, Ltd., 1966-. v. 1- , index. Bimonthly. £43.50 per year. 0268-7887.

This was formerly called *Geo Abstracts: B—Climatology and Hydrology* and *Geographical Abstracts: B—Biogeography, Climatology, and Cartography.* See *Geographical Abstracts: G—Remote Sensing, Photogrammetry, and Cartography* for review.

200. Geographical Abstracts: C—Economic Geography. Norwich, Great Britain: Geo Abstracts, Ltd., 1966-. v. 1- , index. Bimonthly. £43.50 per year. 0268-7895.

This was formerly called *Geo Abstracts: C—Economic Geography.* See *Geographical Abstracts: G—Remote Sensing, Photogrammetry, and Cartography* for review.

201. Geographical Abstracts: D—Social and Historical Geography. Norwich, Great Britain: Geo Abstracts, Ltd., 1966-. v. 1- , index. Bimonthly. £43.50 per year. 0268-7909.

This was formerly called *Geo Abstracts: D—Social and Historical Geography.* See *Geographical Abstracts: G—Remote Sensing, Photogrammetry, and Cartography* for review.

202. Geographical Abstracts: E—Sedimentology. Norwich, Great Britain: Geo Abstracts, Ltd., 1972-. v. 1- , index. Bimonthly. £43.50. 0268-7917.

This was formerly called *Geo Abstracts: E—Sedimentology.* See *Geographical Abstracts: G—Remote Sensing, Photogrammetry, and Cartography* for review.

203. Geographical Abstracts: F—Regional and Community Planning. Norwich, Great Britain: Geo Abstracts, Ltd., 1972-. v. 1- , index. Bimonthly. £43.50 per year. 0268-7925.

This was formerly called *Geo Abstracts: F—Regional and Community Planning.* See *Geographical Abstracts: G—Remote Sensing, Photogrammetry, and Cartography* for review.

204. Geographical Abstracts: G—Remote Sensing, Photogrammetry, and Cartography. Norwich, Great Britain: Geo Abstracts, Ltd., 1974-. v. 1- , index. Bimonthly. £43.50 per year. 0268-7933.

Formerly called *Geo Abstracts: G—Remote Sensing, Photogrammetry, and Cartography,* this is part of the *Geographical Abstracts* series which provides abstracts to journal articles, proceedings, and articles in books. The seven sections contain abstracts to journal articles without author abstracts and abstracts to "relevant articles" submitted by authors. There is no complete list of journals abstracted; thus, a user has no guarantee of finding an abstract to an article from a specific journal. Timeliness of indexing varies, with many articles abstracted in the 1987 issues being published during the year, but abstracts to articles from 1985 and 1986 can also be found in the 1987 issues. Although abstracts are useful to researchers who have little or no time to look for relevant articles in the increasing numbers of published journals, an index which covers all of the major journals in a field on a timely basis is also important. Yearly cumulative keyword indexes for *Geographical Abstracts* are published, but until the cumulation for each year is issued, the user must depend on the general subject chapters in the various parts for subject access.

Many sources are available which provide regular, timely indexing of articles related to economic, historical, and social geography. *Public Affairs Information Service Bulletin; America: History and Life; Historical Abstracts; Index of Economic Articles;* and the *Journal of Economic Literature* are alternatives that provide regular, predictable coverage of journals in these fields. For coverage of landforms, quaternary, sedimentology, and other geological information, one should consult *Bibliography and Index of Geology* (however, it does not include abstracts). *International Aerospace Abstracts* indexes and abstracts a variety of materials on remote sensing and should be considered an important reference source in this field. *Meteorological and Geoastrophysical Abstracts* provides regular indexing and abstracting of publications on meteorology and serves as the chief index for the field.

Geographical Abstracts provides valuable information by abstracting individual papers from proceedings and providing keyword access to areas of physical geography and the earth sciences. However, because it does not abstract articles from a set group of publications on a timely basis, it is best used as a secondary reference source. Recommended for academic and special libraries. *LINDA R. ZELLMER*

205. Geographical Abstracts: A— Landforms and the Quaternary.

Norwich, Great Britain: Geo Abstracts, Ltd., 1960-. v. 1- , index. Bimonthly. £43.50 per year. 0268-7879.

This was formerly called *Geo Abstracts: A— Landforms and the Quaternary.* See *Geographical Abstracts: G—Remote Sensing, Photogrammetry, and Cartography* for review.

206. Geological Abstracts: Economic Geology, Geophysics and Tectonics, Palaeontology and Stratigraphy, Sedimentary Geology. Norwich, Great Britain: GeoAbstracts, Ltd., 1986-. v. 1- , index. Bimonthly. $428.00 per yr. all 4 sections; $114.00 per year per section. 0268-800X, 0268-7941, 0268-8018, 0268-8026.

Issued in four parts, each appearing bimonthly, this publication abstracts over 2,000 international journals, monographs, conference proceedings, and reports. Issues are arranged by subject, with each entry carrying full bibliographical information and either the author's or the contributor's abstract. Issues do not include indexes, but annual author and subject indexes are provided. Useful as a supplement to, but not a substitute for, *Bibliography and Index of Geology*, which provides more comprehensive coverage of the literature but carries no abstracts. This publication is also available online as DIALOG's Geobase file. There appears to be little overlapping between the citations in individual issues of the two publications. Because of its separation into parts by subject areas, each of which can be separately subscribed to, *Geological Abstracts* is ideal for the individual scientist or highly specialized research collection. *JUDITH B. BARNETT*

Atlases—Geographical

207. Concise Earth Book: World Atlas

Esselte Map Service AB Stockholm. Boulder, CO: Graphic Learning International Publishing Corp, 1987. 215p., illus. (col.), index. $11.95. 0-87746-101-5.

As its title implies, the *Concise Earth Book* is a brief version of its namesake. It attempts to fulfill much the same purpose as the parent edition and to do so in a compact size and format. Some of the features of the *Earth Book* were dropped, and some new material has been added. Maps comprise the initial 84 pages of the *Concise* edition. As in *Earth Book*, environmental maps,

employing the same 19 categories and other features are arranged in a geographical hierarchy by continent. No definition or explanation of the environmental classes is included. Each continental map is accompanied by a chart illustrating the flags of the nations on that continent. The continental maps and the regional maps are generally presented at scales which are 50 percent of those used in the parent edition. This small scale makes it difficult to read some of the place names in mountainous areas, and in many instances, the place names obscure the environmental coloring. A slim selection of world thematic maps is located at the end of this section. Although the types of projection are well chosen, they are rarely noted.

The second portion of this edition, 43 pages, is unique to the *Concise* version and is devoted to statistical data and relative location maps. The information recorded here includes both the popular and the official name of each country, its area, demographics, literacy rate, important cities, languages, religions, and monetary units. The currency of this data, which can rapidly become dated, is not given. A 79-page "Index of Names" concludes the atlas. Although the number of entries listed is not stated, it can be estimated to be over 19,000. A random selection of 15 features was checked in the index. All were listed and indexed correctly. Although slightly varied, the alphabetization and duplicate entry problems noted in the review of *Earth Book*, were present in the *Concise* edition as well. For example, the Gulf of Bothnia is indexed only under "Gulf," but the gulfs of Finland, Guinea, and Mexico are indexed under "Gulf" and "F," "G," and "M."

The *Concise Earth Book* cannot be recommended for top priority purchase. The extreme reduction in scale, as compared to *Earth Book*, decreases the value of the environmental maps. The new material, which is not unique, seems to play the role of "filler." This space could have been more productively used by including the continental thematic maps and a wider range of the world thematic maps which appeared in the parent edition. *MARSHA L. SELMER*

208. Earth Book: World Atlas. Esselte Map Service AB Stockholm. Boulder, CO: Graphic Learning International Publishing Corp, 1987. 327p., illus. (col.), glossary, index. $65.00. 0-87746-100-7.

This addition to the world atlas genre was produced to promote an "understanding [of] our planet and the interrelationships that exist between the human and physical environment," and to strengthen "our geographic literacy at home, office, and in our classroom or boardrooms." Graphic Learning International, a publisher of educational materials, believes that "the waste of resources, depletion of natural life forms, and erosion of human conditions are issues and phenomena that affect us all," and that "the very survival of humanity now depends upon the mutual respect for the Earth and its people."

Earth Book devotes its initial 95 pages to an "Encyclopedia of the Earth" which is organized under the philosophical elements, air, water, earth, and fire, to integrate the knowledge derived from a variety of disciplines. Each of the four subsections combines text (current through the early 1980s), illustrations, photographs, and maps to explain various aspects of the specific element's role in life on earth. Both the existence of this section and the style in which it is written point to the publisher's intent to provide an atlas for the educational and popular market.

The second section, 121 pages, is devoted to "The World in Maps." It utilizes the distinctive cartography of the Esselte Mapping Service, Sweden's largest commercial map publisher, a company which specializes in the production of maps and atlases for educational use. It is here that *Earth Book*'s departure from most other world atlases becomes evident. Based on the interpretations made possible by satellite imagery, it uses color and shading to depict the natural environment. The "Color Key," which appears inside the back cover, shows a total of 19 categories worldwide, with 15 nearly identical classes used for the United States. These categories are drawn from climate types (tundra), vegetation types (tropical rain forest), potential land use (arable land), and physical features (lava beds). This use of color requires a cognitive adjustment on the part of atlas users who are generally accustomed to the use of graduated hues to portray variations in height. In this atlas, elevation is depicted by shaded relief, spot heights, and spot depths. As indicated by the "Symbols" printed on the end paper, other major features included on the environmental maps are populated places, roads, railroads, boundaries, and hydrographic features

The maps are organized in a geographical hierarchy by continent. The environmental maps of the six developed continents, most at a scale of 1:25 million, are accompanied by a series of smaller-scale thematic maps. The topics covered are: political divisions, population (as of 1982), relief, annual rainfall, ocean currents, temperature, winds, climate, soils, organic production, and inorganic production. These are followed by larger-scale environmental maps of regions within that continent. That the *Earth Book* is intended for the American market is evidenced by its use of a 1:3 million scale for most of the regional maps of the United States. Most other regions are covered at a scale of 1:10 million. The concluding portion of this section is comprised of world thematic maps which both summarize the continental thematic maps and depict additional topics. Although there is no explanation of the map projections used, they are identified and have been appropriately chosen.

The final section of the atlas is a "Glossary" of the foreign, generic, geographical terms used in place names and an "Index" to about 57,000 place names. It is the index which proves to be problematic. Twenty-five place names, representing populated places, human-made features, hydrography, and physical features, were selected at random from the environmental maps and were searched in the index. All were listed in the index, and all but one were correctly indexed; however, this examination revealed some indexing problems. The method of alphabetizing hydrographic features is inconsistent. The Gulf of Bothnia and the Gulf of Guinea are indexed under "Gulf," but the Gulf of Finland and the Gulf of Mexico are indexed under "F" and "M." This inconsistency has resulted in duplicate entries for some hydrographic features. Lake-of-the-Ozarks is indexed under both "L" and "O," both with the same location reference; Lake-of-the-Woods is indexed under both "L" and "W," each with a different location reference.

Perhaps these problems result from the Swedish origins of the atlas. The direct translation of a Swedish index into an English index without additional editing may not have taken into account American conventions in place name alphabetization. A sample of eight additional place names, in the region adjacent to the one incorrectly indexed, leads one to conclude that there is a systematic longitudinal error in the location references for the features on pages 120 and 121. Wichita, whose location is given as H-5, is really in cell G-5; Bagnell Dam, whose location is given as N-4 is really in cell M-4. A thorough check of the indexing for the remaining U.S. plates revealed similar latitudinal errors on pages 124–125 and 130–131 Such a systematic error leads one to believe that the problem originated at some stage of a computerized indexing process.

The laudable goals set forth by the publishers of this atlas have not been fully met. The concept of the maps is excellent, but some of the environmental categories used may not be understood by users who have no training in either physical geography or the related sciences. The explanations and definitions of these categories must be sought throughout the "Encyclopedia" section. The educational mission of this atlas could be strengthened by the inclusion of page references to the relevant text as a part of the "Color Key." Map legibility is generally good—only in the Himalayas does the intensity of the shaded relief make the place names difficult to read. Because it provides environmental information at a level of detail not generally found in a world atlas, *Earth Book* can be recommended for purchase, but with the indexing caveats noted earlier. Because it is a complement to those that already exist, it should not be the only world atlas in one's collection.

MARSHA L. SELMER

Atlases—Science

209. **The Underground Atlas: A Gazetteer of the World's Cave Regions.** John Middleton; Tony Waltham. St Martin's Press, New York: NY, 1986. 239p., illus., bibliog., index. $16.95. 0-312-01101-6.

The Underground Atlas is a gazetteer to cave regions of the world. A brief preface, describing the contents and arrangement of the book, facilitates its use. Following the preface is a section on "The Continents," describing the overall distribution of karst and caves in the world. This is followed by the major portion of the book "The

Countries," which describes the occurrence of karst and caves in countries or geographic areas. Interspersed through the text are maps of karst regions in various countries and descriptions and maps of notable caves. Appendices contain lists of the longest and deepest caves, largest cave chambers, and various caving organizations. Appendix 3, "Further Reading," lists several sources (both books and periodicals) for additional information. Unfortunately, J. N. Jennings's *Karst Geomorphology* (Blackwell, 1985), a widely reviewed book on the topic, is not included in this list. The authors, John Middleton, a member of the British Cave Research Association, and Dr. Tony Waltham, a senior lecturer in geology at Trent Polytechnic, have done a commendable job on this book. With the exception of missing a key source in "Further Reading," this book is very informative. Recommended for college and university libraries supporting geology departments, especially those in areas with caves.

LINDA R. ZELLMER

Bibliographies

210. Death of the Dinosaurs and Other Mass Extinctions. David A. Tyckoson, ed. Gary Fouty, comp. Phoenix, AZ: Oryx Press, 1987. 96p., index. $16.00pbk. Oryx Science Bibliographies, v. 10. 0-89774-432-2pbk.

This unique bibliography covers the much-written-about theories of mass extinctions, with the death of the dinosaurs a major focus. The compiler has assumed that extinction is a fact of life prior to the entrance of man. It adequately covers a large number of theories, including asteroids as one method of mass extinction and man as another cause of extinctions. None of these theories have been proven, with the exception of extinctions brought about by humans.

Each article that is included has been reviewed by the compiler. The book is intended to be current and includes materials that are in the English language and readily available in libraries. Brief annotations are included for each entry. The entries are arranged in chronological order within eight topical areas: "Overview of the Mass Extinction Theory," "The Impact Theory," "Alternatives to the Impact Theory," "Periodicity in Extinctions," "The Impact Theory: People and Politics," "Man as a Cause of Extinction—Prehistoric Times," "Man as a Cause of Extinction—Present Times," and "Man as a Potential Victim of Extinction." An author index is included.

This bibliography gives good coverage on the theories of extinction and should be useful as a student resource in school and academic libraries.

H. ROBERT MALINOWSKY

211. Geologists and the History of Geology: An International Bibliography from the Origins to 1978: Supplement 1979–1984 and Additions. William A. S. Sarjeant. Malabar, FL: Robert E. Krieger

Publishing Co., 1987. 2v., bibliog., index. $162.50. 0-89874-939-5.

Geologists and the History of Geology is a bibliography of publications on the history of geology which were written in the Latin alphabet. Also included in the definition of "history of geology" is biographical information on geologists, most of whom are no longer living.

The book is divided into seven parts: an introduction; general works on the history of science, geology, subfields of geology and related subjects; historical information on societies, museums and other geology-related institutions; history of the petroleum industry; history of significant events (including newsworthy events, geological events, and expeditions) in the history of geology; biographical sections on geologists; and "Prospectors, Diviners and Mining Engineers." This is a two-volume supplement to the five-volume set of the same title which was published in 1980. The original set covered materials on the history of geology which were published prior to 1978; this volume presents materials omitted from the original five volumes and material published between 1979 and 1984.

Individual sections of the book are divided into separate parts; most include a section on general works followed by a section arranged alphabetically by subject, such as in the case of the section on the history of geology subdisciplines, or by geographic area. Arrangement of some sections (such as the United States) could be improved by further subdividing (alphabetically by state).

Although the book contains a massive amount of information, the most useful part is probably the biographical section. Brief biographies of geologists are presented with citations to biographical articles. The events section needs to be greatly expanded to include cross-references to individual geologists (as in the case of survey expeditions, such as Powell, Wheeler, or Hayden) or to government agencies. Some of the historical material might be obtained by using the subheading "history" in the *Bibliography and Index of Geology*. Information presented on geologic events such as earthquakes and eruptions could be obtained from the bibliography of a book on the topic. Because this book has such a small potential audience and is a supplement to the previous set, it is recommended primarily for libraries supporting geology programs which already own the five-volume set.

LINDA R. ZELLMER

Directories

212. U. S. Ocean Scientists and Engineers, 1987 Directory. 4th. American Geophysical Union, Washington: DC, 1986. 179p. $24.00pbk. 0-87590-238-3pbk.

The *1987 Directory of U.S. Ocean Scientists and Engineers* represents the fourth edition of a work first published in 1969 by the National Academy of Sciences. The present edition, prepared by the American Geophysical Union with

the aid of a grant from the National Science Foundation and with the cooperation of American Meteorological Society, American Society of Limnology and Oceanography, the Coastal Society, and Marine Technology Society, is nearly five times the size of the original edition, providing information on the scientific interests and employment of over 6,600 ocean scientists and engineers now teaching and working in the fields and subfields of marine science. Data were obtained in 1986 via a questionnaire to those included in the *Directory* and has been computerized for future updating. The *Directory* has three sections: (1) an alphabetical listing by surname; (2) a geographic listing by employer; and (3) a listing by areas of specialization. Entries in the name section are keyed to specialty areas by means of four-digit codes. The *Directory* is clearly printed on journal-quality paper and is generally easy to read, despite its small typeface. Typographical errors appear to be relatively few. The key to the specialty section follows rather than precedes the actual listing, making its use somewhat awkward. The *Directory* concludes with a copy of the "Directory Questionnaire" so that users may send in new or corrected information for future editions. The *Directory* appears in paperback and will require binding where heavy use is anticipated. Overall, this is a useful and competently produced directory to individuals in the marine sciences. It is recommended for individuals and collections where access to this information is needed.   *FRED O'BRYANT*

Encyclopedias

213. Album of Rocks and Minerals. Tom McGowen. New York, NY: Checkerboard Press, 1987 (c1981). 61p., illus. (part col.), index. $4.95pbk. 0-02-688504-2pbk.

Album of Rocks and Minerals is a well-written encyclopedic book that explains "how specific rocks and minerals were formed, why they look the way they do, and their many uses, both past and present."

This is not a comprehensive encyclopedia of rocks and minerals. It covers only 22 selections (quartz, copper, cassiterite, gold, silver, sulfur, hematite, calcite, halite, jade, galena, graphite, beryl, cinnabar, diamond, opal, ptchblende, corundum, turquoise, bauxite, sphalerite, and nickel) plus small thumbnail accounts of another 36. Each account is interestingly written with facts and folklore intertwined. The uses of each of the rocks and minerals is stressed, and the colorful illustrations by Rod Ruth add to the book.

School children will find this a good book to use when studying rocks and minerals, especially with the pronunciation guide at the back of the book. School and public libraries will find it a useful reference source.

H. ROBERT MALINOWSKY

214. Color Encyclopedia of Gemstones. 2nd. Joel E. Arem. New York, NY: Van Nostrand Reinhold Co., 1987. 308p., illus. (col.), bibliog., index. $49.95. 0-442-20833-2.

One of the most comprehensive sources of its kind, the *Color Encyclopedia of Gemstones* provides detailed information about all known naturally occurring gemstones. Included for each precious stone or family of stones are the chemical formula, crystallography, colors, luster, hardness, cleavage, density, inclusions, optics, birefringence, dispersion, pleochroism, spectral, luminescence, occurrence, size, and origin of name. There are several cross-references from unique gems to their families. One section of the book describes manmade gems and explains in detail the processes by which they are made. The most spectacular feature of the book is the collection of more than 300 color photographs of the stones described. In addition, the author provides detailed explanations of the meanings of the physical properties. Ease of use of this source is further enhanced by an index of all named gemstones. Public and academic libraries will find this a useful source of information on gemstones.   *EARL MOUNTS*

215. The Encyclopedia of Climatology. John E. Oliver; Rhodes W. Fairbridge, eds. New York, NY: Van Nostrand Reinhold Co., 1987. 986p., illus., bibliog., index. $89.95. Encyclopedia of Earth Sciences, v. 11. 0-87933-009-0.

Volume 11 of the long-running Encyclopedia of the Earth Sciences series is *The Encyclopedia of Climatology*, edited by John E. Oliver and Rhodes W. Fairbridge. The book discusses all aspects of climate studies through a collection of essays on various aspects of climatology. Arranged alphabetically, the essays vary in length, with topics ranging from discussions of "Acid Rain" to "Zones, climatic." The book also contains tables and diagrams (some of which were not quite legible in the review copy) to illustrate points. Bibliographies of varying lengths are available after most entries, a list of main entries precedes the body of the volume, and citation and subject indexes are available at the end of the book.

The preface of this volume states that it includes material on topics previously covered in the *Encyclopedia of Atmospheric Sciences and Astrogeology*, volume 2 of the series. Close examination reveals that 118 of the 212 entries were discussed in the earlier volume. Some of the "new" entries are merely edited versions of the originals, whereas others contain significant new information. Another significant work in the field of climatology is *World Survey of Climatology* (Elsevier 1969–1986), a now complete 16-volume set. Ten separate volumes of this set deal with climate in various areas of the world, and the remaining six contain material on heat balance climatology, statistical analysis, mechanisms, properties and distribution of climate, bioclimatology, agricultural climatology, city climatology, and atmospheric climatology. The user is faced with the choice between the extremely

comprehensive, albeit somewhat dated, *World Survey* and the less comprehensive, but more current, encyclopedia. The two sources are complementary; each is useful for certain questions on various aspects of climatology. *World Survey* can and should be used for comprehensive information, especially on world climates, but *The Encyclopedia of Climatology* is recommended because it is more current and will serve the needs of most students adequately.

LINDA R. ZELLMER

216. The Encyclopedia of Field and General Geology. Charles W. Finkl, Jr., ed. New York, NY: Van Nostrand Reinhold Co., 1988. 911p., illus., bibliog., index. $89.95. Encyclopedia of Earth Sciences, v. 14. 0-442-22499-0.

Van Nostrand Reinhold has published an encyclopedia of earth sciences in 15 volumes, of which this is volume 14. It is a companion to volume 13, *Applied Geology*, according to its preface; the subjects covered in the two volumes were to be covered in one, but size considerations deemed a division to be necessary. The subject matter of the two volumes overlaps, but *Field and General Geology* is concerned primarily with actual observation and measurement of geological phenomena, rather than with the application of geological knowledge to human endeavors.

The volume begins with several tables of conversion factors and a list of what measurement units are used in the different disciplines of geology. There are then approximately 135 articles, each signed and with a bibliography and cross-references, whenever appropriate, to other entries in this volume and others in the series. The thoroughness of discussion differs from article to article; all that were examined, however, provided enough detail for someone to understand the basic nature of the subject. The list of contributors includes names of geologists who are very well known in their fields (such as Daniel Merriam) and others who have little fame but who are known to be well versed in their specialty. The volume is well illustrated with both drawings and photographs, and tables are abundant.

A university or college with an active geology program would find this work a good addition to the collection; in fact, the whole series should be in the library. It is also a good addition to a civil engineering collection, along with the volumes on applied geology and soil science. Otherwise, a one-volume encyclopedia such as the *Cambridge Encyclopedia of Earth Sciences* (Crown, 1981) would be sufficient. *KENNETH QUINN*

217. The Encyclopedia of Structural Geology and Plate Tectonics. Carl K. Seyfert, ed. New York, NY: Van Nostrand Reinhold Co., 1987. 876p., illus., bibliog., index. $89.95. Encyclopedia of Earth Sciences, v. 10. 0-442-28125-0.

Comprising over 100 articles by 94 internationally recognized scientists, this authoritative work covers both classic concepts in structural geology and the relatively new field of plate tectonics. Faults, folds, strength of rocks, structural petrology, eustasy, continental drift, sea floor spreading, island arcs, and evidence against plate tectonics are among the topics included. The editor, a professor of geology at SUNY, Buffalo, and author of *Earth History and Plate Tectonics* (Harper and Row, 1979), has allowed personal viewpoints to be expressed by each author, resulting in some disagreements in interpretation and exposing the reader to the differing views found in the literature. Among the contributors are J. R. Heirtzler, A. E. M. Nairn, Peter Rona, and R. W. Van Bemmelen (inadvertently omitted from the list of contributors at the front of the volume). The book is intended for undergraduate and graduate students, researchers, and scientists in the academic world and industry. The volume is similar in format to the preceding volumes in the Encyclopedia of Earth Sciences Series. The signed articles, ranging in length from five to 70 pages, are alphabetically arranged and include figures and photographs as well as extensive bibliographic references to both the classic papers in the field as well as those published through 1984. Cross-references appear in boldface between entries, and "see also" references are given at the end of some articles. A list of main entry articles is found at the beginning of the volume. There is a subject and an author-citation index. It is recommended for undergraduate through specialized research collections in academic libraries. *JUDITH B. BARNETT*

218. McGraw-Hill Encyclopedia of the Geological Sciences. 2nd. Sybil P. Parker, ed. New York, NY: McGraw-Hill Book Co., 1988. 722p., illus., bibliog., index. $85.00. 0-07-045500-7.

The *McGraw-Hill Encyclopedia of Geological Sciences* has some serious flaws. In the preface, it implies that it covers geology, geochemistry, mineralogy, petrology, and geophysics. It does do a fair job of covering these fields, for the most part, and the list of contributors is very impressive (Stanley Runcorn and Cornelius Hurlbut are among them). The problem is not what is in the book, but what is not in the book.

In particular, paleontology is missing; there is not even an entry for dinosaurs. Some other omissions are glaring; for instance, the Alvarez theory of extinction by meteor impact does not appear in the book, despite it being a prominent subject over the last five years or so. The discovery that hot springs along midoceanic ridges can produce deposits of metallic sulfides has caused many deposits to be reclassified as having originated in that manner; however, there is not a mention of the phenomenon in this book.

The purpose of a second edition is to update information and to correct omissions and other mistakes. This has not been done in this edition. Although the information that is included is in good, readable form and for the most part accurate, the omissions do not make this a high

priority purchase unless it is to replace a worn copy of the first edition or to have a single-volume, relatively inexpensive encyclopedia of geology. *KENNETH QUINN*

Handbooks

219. Handbook of Holocene Palaeoecology and Palaeohydrology. B. E. Berglund; M. Ralska-Jasiewiczowa, eds. New York, NY: John Wiley and Sons, 1986. 869p., illus., bibliog., index. $114.00. "A Wiley-Interscience Publication." 0-471-90691-3.

Berglund and his assistant editor, M. Ralska-Jasiewiczowa of the International Geological Correlation Programme (IGCP), have compiled a work with broader application than the title would imply. Although the book is on paleoecological methods for analysis of organic deposits in lakes and mires, the techniques presented can be applied to older than Holocene deposits beyond the temperate zone represented here. The IGCP Project 58 devoted to "Palaeohydrology of the temperate zone" may have interest to researchers studying current ecological and climatological changes. Each chapter within each section has been prepared by a subject expert, and most show liberal use of charts, graphs, and tables. There are a few illustrations where appropriate. Each chapter has its own list of references at the end of the chapter, arranged alphabetically. The work is divided into eight sections: "Background to Palaeoenvironmental Changes during Holocene"; "Research Strategy," which focuses on past research and preliminary research planning; "Sampling and Mapping Techniques," which devotes extensive detail to mapping the past and present through a variety of techniques; and two sections on "Stratigraphical Methods" and "Dating Methods," which would assist researchers in selecting, evaluating, and applying techniques ranging from core sampling to radiomagnetic and paleomagnetic dating, tephrochronology for airborne volcanic materials which are spread over wide areas, and specific applications of dendrochronology in mires and in mountains to cross-date various species to verify the dating process.

The next two sections deal in varying detail with "Physical and Chemical Methods," with a focus on analyses of oxygen and carbon isotopes and phosphorous, nitrogen, pigments, and metals; and "Biological Methods," which gives analyses of pollen, charred particles, microfossils, algae, diatoms, fossil fruits and seeds, plant macrofossils, bryophyta, rhizopoda, cladocera, ostracoda, coleoptera, chironomida, and mollusca. Both field and laboratory techniques are given with presentation and interpretation of results. The final section, "Numerical Treatment of Biostratigraphical Data," gives reference instruction for data calibrations and statistical analysis and includes examples of computer programs. The subject index is adequate, though brief. One improvement would have been a name index, and there are a few cross-references. Format of chapters has some variation because of the use of 45 contributors, including most major names in the field; but the editors have maintained good use of headings, spacing, and varied typefaces to promote clarity of presentation. Format of references is consistent throughout. Certainly this work provides much detail for researchers in paleohydrology, paleoecology, and stratigraphic geology. Academic and special libraries serving these researchers will want this handbook. *JEAN E. CRAMPON*

220. Minerals of Mexico. William D. Panczner. New York, NY: Van Nostrand Reinhold Co., 1987. 459p., illus. (part col.), bibliog. $44.95. 0-442-27285-5.

Minerals of Mexico is a comprehensive source which describes in detail the occurrence of minerals throughout Mexico. The first part of the book presents in a clearly written manner the history and geology of the various mining districts. Included are black-and-white photographs of many of the mining camps and interiors of the mines. The second section of the book is an extensive bibliography (about 500 citations) of mineralogy sources published in English and Spanish as early as 1825 and as recently as 1985. The main section of the book is the catalog of minerals. For each mineral, the chemical formula and description are given followed by an alphabetical list of states and cities and the occurrences of the mineral within each. For most of the minerals, a color photograph is included. An appendix lists states, counties, and county seats. Public, academic, and special libraries will find this a useful reference source on minerals found in Mexico. *EARL MOUNTS*

221. Selected ASTM General Use Standards on Water. ASTM Committee D-19 on Water. Philadelphia, PA: American Society for Testing and Materials, 1987. 99p., illus. $24.00. "PCN:03-419088-16." 0-8031-1172-X.

The material in this handbook has been reprinted from the *Annual Book of ASTM Standards*, the accepted source for engineering standards set by the American Society for Testing and Materials. It contains the following Designations from the parent work: D 596, D 1066, D 1129, D 1192, D 1193, D 2777, D 3370, D 3856, D 4210, D 4375, and E 200. These are general water use standards covering definitions, reagent solutions, intralaboratory quality control, sampling, laboratory evaluation, and reporting of results. These standards are quite detailed. Using Designation D 1192, "Standard Specification for Equipment for Sampling Water and Steam," as an example, the standards include such information as scope, applicable documents, definitions, significance and use, sampling lines, valves and fittings, sample cooler or condenser, degassers, pumps, sample containers, sample labels, sample shipping containers, and shipping labels. This particular standard also includes "calculation of cooling surface requirements." The standards are very specific and detailed with many charts,

graphs, and calculations so that the user has few questions about how they are applied. Some standards include references. Libraries needing those standards pertaining to water will find this a valuable addition to their reference collections.

H. ROBERT MALINOWSKY

222. Tides, Surges and Mean Sea-Level. David T. Pugh. New York, NY: John Wiley and Sons, 1987. 472p., illus., bibliog., glossary, index. $101.00. 0-471-91505-X.

The cover of this technical handbook states that it is a "Handbook for Engineers and Scientists." Actually it is useful for many specialists including hydrographers, marine and coastal engineers, geologists who specialize in beach or marine sedimentation process, and biologists concerned with the ways in which living organisms adapt to the rhythms of the sea. "The aim of this book is to present modern tidal ideas to those who are not tidal specialists, but for whom some tidal knowledge is involved in their own professional or scientific field." The discussion of tidal phenomena can be found in many books as a chapter, but there are few books that cover the subject as comprehensively as *Tides, Surges and Mean Sea-Level* does.

In his introduction, David T. Pugh gives some historical background on the study of tides plus some definitions of common terms and some basic statistics of tides as time series. The main part of the book is divided into 10 fairly technical chapters covering "Observations and Data Reduction," "Forces," "Analysis and Prediction," "Tidal Dynamics," "Storm Surges," "Shallow-Water Dynamics," "Tidal Engineering," "Mean Sea-Level," "Geological Processes," and "Biology: Some Tidal Influences." These chapters are extremely well written, comprehensive, and accurate. Pugh has left out nothing.

A certain amount of mathematics is assumed, and mathematical equations are used throughout to explain the theories. There are a limited number of fairly good photographs and numerous charts, maps, and tables as well as an interesting quotation from literature at the beginning of each chapter. There are five appendices: "Filters for Tidal Time Series," "Response Analysis Inputs and Theory," "Analysis of Currents," "Theoretical Tidal Dynamics," and "Legal Definitions in the Coastal Zone." A 14-page list of references, glossary, and index complete this excellent book. Any research library concerned with coastal effects of the oceans and seas should have this book in its collection.

H. ROBERT MALINOWSKY

223. The Weather Almanac: A Reference Guide to Weather, Climate, and Air Quality in the United States and Its Key Cities, Comprising Statistics, Principles, and Terminology: Provides Weather/Health Information and Safety Rules for Environmental Hazards Associated with Storms, Weather Extremes, Earthquakes, and Volcanoes: Also Includes World Climatological Highlights and Special Features on Weather, Climate, and Society and on Ozone in the Upper Atmosphere. 5th. James A. Ruffner; Frank E. Bair, eds. Detroit, MI: Gale Research Co., 1987. 811p., illus. $110.00. 0-8103-1497-5. 0731-5627.

The *Weather Almanac* is "a reference guide to weather, climate and air quality in the United States." It includes statistical weather summaries as well as basic weather information, safety rules for weather-related and non-weather-related phenomena (floods, tornadoes, hurricanes, earthquakes, volcanoes, and even tsunamis), a review of weather and health, and information on air pollution and climate. Record-setting weather phenomena and weather and climate summaries for the world and selected cities in the United States are also presented. A brief discussion of the atmospheric ozone problem is also included.

In its fifth edition, *The Weather Almanac* is slightly larger than the fourth edition. This larger size has an advantage, because numbers on some of the summary maps and tables are more legible. Tables have been reformatted so that data are presented in six tables, rather than seven. Written city descriptions have been edited slightly. Tables of yearly precipitation, temperature, snowfall, and heating and cooling degree days have also been edited; 30 years' of data (1956-1985) are presented, instead of 40 years' of data (1944-1983). Because of this, libraries with students who may require long-term data are cautioned to retain the previous edition. This "new edition," which appears barely three years after the fourth, should be purchased with caution. Libraries may want to wait for the next edition if funds are tight. It is hoped that a more thorough revision will be done at that time. Clearer print is needed on both maps and tables. Some of the summary maps, presented on two pages separated by an inner two-inch margin, should be printed on a single page. Finally, it is hoped that new editions will be less frequent, published on a five-year schedule, eliminating the problem of frequently published editions with little (or less) new information.

LINDA R. ZELLMER

224. Weather of U. S. Cities: A Compilation of Weather Records for 281 Key Cities and Weather Observation Points in the United States and Its Island Territories to Provide Insight into Their Diverse Climates and Normal Weather Tendencies: Supplies Narrative Statements about the Various Cities' Climates, and Complements the Descriptions with Statistical Cumulations to Quantify "Normals, Means, and Extremes" for Each. James A. Ruffner; Frank E. Bair, eds. Detroit, MI: Gale Research Co., 1987. 1131p. $175.00. 0-8103-2102-5.

Weather of U. S. Cities is a collection of summary statistics for 281 cities in the United States and its territories drawn from NOAA weather records. Each city summary contains a narrative description of the weather in a given city and six tables of weather statistics. The size of the third edition has been increased, so tables are clearer and easier to read. The summary table entitled "Normals, Means and Extremes" has been reformatted and printed upright, rather than sideways, on one page, so the user no longer has to turn the book to read these statistics. Although the size has been increased, this new edition contains fewer data than the previous one; 30 instead of 40 years' of data are presented. Only two additional years of data are given in the third edition (1984 and 1985), and data from 1944–1955 have been eliminated. Thus, libraries purchasing the new set are advised to retain the previous edition. This set is useful because it provides summary statistics for major weather stations in the United States which would require several hours to collect from other sources. It is hoped that Gale would consider publishing this reference work less often, perhaps once every five years. *LINDA R. ZELLMER*

Histories

225. From Mineralogy to Geology: The Foundations of a Science, 1650–1830.
Rachel Laudan. Chicago, IL: The University of Chicago Press, 1987. 278p., illus., bibliog. $27.50. 0-226-46950-6.

A number of books on the history of geology have been published in the last five years. Rachel Laudan has added *From Mineralogy to Geology: The Foundations of a Science, 1650–1830* to the list. It is a survey of the early history of geology in Europe, especially on the continent. Laudan distinguishes between the causal aspects of geology, represented by the early mining schools in Germany, and historical geology, the reconstruction of events which resulted in the present land surface. Beginning with the early mineralogists, Laudan describes the early mineral classification schemes and how the allied science of chemistry led to Becker's theories on the occurrence of mineral deposits. Becker's ideas, which were expanded by eighteenth-century mineralogists, are compared with those of Linnaeus, who attempted to apply the taxonomic principles he developed for plants to minerals as well.

Laudan continues, tracing the influence of Abraham Werner on the development of geology as a separate field of study. Although he based much of his work on the principles of chemistry, Werner's development of the formation concept marks a turning point in the history of geology; from this time on, geology became a separate science, seeking to identify the history of the earth and the causes of present-day deposits. Although Laudan traces only the roots of geology as a science, she does a service by bringing to light the events preceding the British studies of the nineteenth century. Most basic geology classes refer to these studies, but few describe the events preceding them on the European continent. This book should be recommended reading for all advanced geology students and is recommended for all college and university libraries serving geology programs. *LINDA R. ZELLMER*

Manuals

226. The Marshall Cavendish Science Project Book of the Earth.
Steve Parker. Freeport, NY: Marshall Cavendish, 1988 (c1986). 43p., illus. (col.). $12.49. The Marshall Cavendish Library of Science Projects, v. 5. 0-86307-629-7.

"Science is all about discovering more about your world, finding out why certain things happen and how we can use them to help us in our everyday lives." This encyclopedic project book for young people covers our planet, earth. Through a series of chapters that explain the origin of the earth, the makeup of this planet, how the various continents were formed, volcanoes, magnetic poles, the makeup of rocks, minerals, physical makeup of the land, caves, fossils, animal regions of the earth, and mineral exploration, the student learns the basics of earth science. Experiments and projects are used to exemplify these concepts. The well-written text and good illustrations make this a useful reference book for school libraries and an excellent textbook for the classroom. *H. ROBERT MALINOWSKY*

227. The Marshall Cavendish Science Project Book of Water.
Steve Parker. Freeport, NY: Marshall Cavendish, 1988 (c1986). 43p., illus. (col.). $12.49. The Marshall Cavendish Library of Science Projects, v. 1. 0-86307-627-0.

This encyclopedic project book for young people covers water. The chapters systematically explain what water is made of, why we need water, how water works for man, ice, climatology, the endless cycle of water, physics of water, and why water is wet, the student learns why water is needed on earth for humans to survive. Through the use of experiments and projects the student is able to learn those concepts. This is a useful reference book for school libraries and, with its easily readable text and good illustrations, an excellent textbook for the classroom. *H. ROBERT MALINOWSKY*

228. Solutions Manual to Plate Tectonics—How It Works.
Allan Cox; Robert Brian Hart. Palo Alto, CA: Blackwell Scientific Publications, 1987. 64p., illus. $9.95pbk. 0-86542-329-6pbk.

Solutions Manual to Plate Tectonics—How It Works is a book that forces librarians to make an ethical decision: should the solution manual for a graduate-level textbook be added to a library's collection? This brief answer manual is not the

typical solutions book published for basic math, physics, and chemistry classes. Solutions to each problem in the book *Plate Tectonics—How It Works* are presented in the *Solutions Manual* in the form of simple answers, detailed diagrams or diagrams with explanatory notes, making it much better than most answer books. It is also unusual because it begins with an apology for errors in two tables in the first printing of the book. Libraries should check to see that their copy is a corrected copy. Blackwell has volunteered to replace defective first-run copies, or to print errata sheets, should the need arise. Closer examination of the *Solutions Manual* also reveals an error in the corrections to Table 7-3.

The book *Plate Tectonics—How It Works* is a very good addition to libraries supporting geology programs, because it teaches the mechanics of plate tectonics, rather than the theory. The answer manual is useful because it allows users to see whether they are understanding the material presented in the book. However, in a college or university which uses the book as a required text, the solutions manual is not recommended unless requested by the professor. *LINDA R. ZELLMER*

Tables

229. Tables for Microscopic Identification of Ore Minerals. 2nd rev. W. Uytenbogaardt; E. A. J. Burke. New York, NY: Dover Publications, Inc., 1987 (c1971). 430p., bibliog., index. $10.95pbk. 0-486-64839-7pbk.

Since its original publication in 1951, various editions of this work have served as a principal reference to those working in determinative mineralogy. The second revised edition, published in 1971 by Elsevier, doubled the number of ore minerals tabulated (500) in the first edition. Now, 15 years and many advances in ore microscopy later, Dover has reissued the 1971 edition with slight corrections, yet no additions. Mineralogists and research collections will regard the availability of the work as good news but should not look to it as an authority on data collected and verified since the 1960s.

The tables use reflectance (in air) and hardness (VHN or microindentation and polishing hardness) as the primary properties for determination of ore minerals. Reflectance data here were compiled from sources using a variety of controls and standards. Although much of the reflectance data are given in the four standard wavelengths, many are given in rounded off percentages or no data are given at all (80 ore minerals).

Here is where the work shows its age; quantitative methods in ore microscopy have become more advanced and standardized. The International Mineralogical Association's Commission of Ore Microscopy (IMA-COM) has, since the great majority of these data were collected, issued reflectance measurement standards. In addition, since the 1960s, significant progress has been made in reflected-light microscopy. These factors combine to improve the reliability and diagnostic value of reflectance measurements made since the compilations of these tables.

The VHN (Vickers Hardness Number) data given for a majority of the ore minerals prove to be more reliable because they were obtained by combining values measured by a small number of authorities independently working, such as Burke, Bowie and Taylor, and Young and Millman. The tables report no VHN data on 165 of the 500 ore minerals represented. Nonquantitative properties of color, bireflectance, anisotropy, internal reflections, and characteristic features such as cleavage, twinning, paragenesis, and crystal shape are also provided for nearly all entries.

Data were compiled from over 1,700 sources from the literature, and reference numbers are linked to the data in the tables. Carbon compounds are not included. As in any reference used for identification, arrangement of the data is essential to the value of having the data. Here the data are usefully presented in two major parts: first, by determinative tables which rank the ore minerals by quantitative properties and by classification, and second, by mineral descriptions, compiling all quantitative and qualitative property information on each ore mineral. Subarrangement of the latter section is by classification (e.g., tellurides, selenides, etc.), making it useful in paragenesis studies.

Mineral descriptions and data are clearly and systematically presented. The volume is sufficiently indexed. Noting its shortcomings, this corrected edition should still be purchased by any collection supporting mineralogy. Its role will be that of supplementing other more authoritative sources, such as Criddle and Stanley's *Quantitative Data File for Ore Minerals of the Commission on Ore Microscopy of the International Mineralogical Association* (1986).

JOHN T. BUTLER

Textbooks

230. Applied Hydrogeology. 2nd. C. W. Fetter. Merrill Publishing Co., Columbus: OH, A Bell and Howell Information Co., 1988. 592p., illus., bibliog., index. $43.95. 0-675-20887-4.

In this textbook for the new interdisciplinary field of hydrogeology, the author defines hydrology as the "study of water." The definition includes the occurrence, distribution, movement, and chemistry of all waters of the earth. Hydrogeology is defined as "the interrelationships of geological materials and processes with water." A professional working in the field often will have training in geology, hydrology, chemistry, mathematics, and/or physics. The book is intended to be an introductory textbook for an advanced undergraduate course or an undergraduate/graduate course and, also, a reference book for professionals working in the field.

The first edition of this title was published in 1980. This second edition has several new and expanded chapters. The emphasis of this edition is on site characterization, groundwater monitoring, contaminant hydrogeology, and computer modeling. The author is a faculty member at the University of Wisconsin-Oshkosh and began his career as a staff hydrogeologist for a consulting firm. More recently he has acted as a consultant to the U.S. Environmental Protection Agency. To collect information for the book, he drew heavily on his experience as a consultant, on recently published literature on the topic, attendance at professional meetings, and short courses. All new information was tested in the field before its inclusion in the second edition. In addition, the draft manuscript was reviewed by several specialists.

Past texts have tended to present an unbalanced view because of the authors' biases toward their own specialties. Fetter's goal in writing this textbook was to provide the practitioner with the tools of the field and to present a detailed perception of how each facet of this interdisciplinary field integrates with the others. The individual chapters can be divided into two categories. One discusses water, its chemistry, evaporation and precipitation, and runoff and stream flow. The second group discusses groundwater, the geology of occurrence, relationship to soil moisture, flow, regional flow, contamination, and development and management. The first chapter defines the various aspects of hydrogeology which are covered in the book and discusses fresh water in the United States, and lists sources of hydrogeological information. A chapter at the end of the book discusses field methods, techniques that can be used in the exploration for groundwater supplies, and their environment.

Throughout the chapters are case studies of actual situations to aid in the understanding of the occurrence and movement of groundwater in a variety of geological settings. A list of the case studies is found at the end of the table of contents. Extensive lists of references are found at the end of each chapter, and problem sets accompany some chapters. The book has a glossary of approximately 300 hydrogeological terms, and the appendix contains tables of unit conversions and various functions to solve problems in water chemistry, well hydraulics, and contaminant transport. This well-indexed book will be most useful as a textbook for students in the area of hydrogeology. It will find some use in specialized collections in the areas of geology, water resources, and natural resources.

LUCILLE M. WERT

231. Environmental Geology. 5th.
Edward A. Keller. Columbus, OH: Merrill Publishing Co., A Bell and Howell Information Co., 1988. 540p., illus. (part col.), bibliog., index. $34.95. 0-675-20889-0.

The fifth edition of *Environmental Geology* by Edward A. Keller appears barely three years after the fourth edition. A survey of the environmental aspects of geology which is meant as an introductory-level college textbook, it contains two additional chapters on soils and natural hazards but little other new information. The book contains seventeen chapters and two appendices divided into five sections. The flow of the book is logical, although material on basic geology should have been placed in the first, rather than third chapter. Because this book is similar to other books on the subject, particularly Donald R. Coates's *Environmental Geology*, content needs to be closely examined. Diagrams and photographs are, for the most part, clear, and used well to illustrate pertinent points. References are provided to direct the reader to further information. Few current, post-1985 (fourth edition) references are given. A number of the references are from popular sources (newspapers and popular magazines) rather than from scientific sources. In one particular instance, inaccurate diagrams and information from popular sources are presented as fact. This is disturbing because little information on the topic has been published and because information printed in a textbook should be accurate. Because of these inaccuracies, the less-than-logical flow of the book, and the lack of substantially new material, the new edition is not recommended.

LINDA R. ZELLMER

232. Regional Stratigraphy of North America. William J. Frazier; David R. Schwimmer. New York, NY: Plenum Press, 1987. 719p., illus., bibliog., index. $110.00. 0-306-42324-3.

Regional Stratigraphy of North America is a comprehensive survey of the geology, stratigraphy, and fossil record through geologic time. Unlike some historical geology books, which include several chapters of survey material before discussing historical geology, this book presents barely seven pages of introductory material prior to the second chapter, which is entitled "The Archean." The book is divided into chapters based on cratonic sequences (a group of stratigraphic units which developed over part of geologic time as a result of a specific set of geologic events), rather than on geologic time periods. Thus, the bulk of this 719-page book (which includes a bibliography and three indexes) discusses the history of large sections of the North American continent, rather than surveys of widely separated events that occurred during a specific time period.

Frazier and Schwimmer have written a book that will serve as both an upper-level textbook and a reference book for geologists. To do so, they have sifted through a tremendous amount of literature. In most cases, they have presented only one interpretation of a geologic event, even though that interpretation may be disputed. In some cases, such as the Quaternary, they have presented only a cursory survey and totally avoided discussion of possible controversies.

Thus, the book needs to be used with caution, keeping in mind that alternative interpretations of geologic sequences exist. Despite the fact that controversies are avoided, this book is highly recommended for colleges and universities with geology programs. Another problem that will have to be confronted is whether to place it in reference

(so it is available to the greatest number of patrons), or in the main collection (where it may be monopolized by a single patron). The ideal solution, albeit the more costly, is for the library to purchase two copies: one for reference and one for circulation. Libraries with geology programs may have to consider this solution, because this book will be in demand.   *LINDA R. ZELLMER*

Thesauri

233. Multilingual Thesaurus of Geosciences. G. N. Rassam; J. Gravesteijn; R. Potenza, eds. New York, NY: Pergamon Press, 1988. 516p. $95.00. 0-08-036431-4.

Multilingual thesauri are seen in many reference collections, but one wonders how often they are consulted. They would be useful if one were translating a paper and primarily needed a list of terms relevant to a particular discipline, or if one were consulting a map whose legend was in another language. Despite these reservations about multilingual thesauri, this one is very well constructed. Its creation was sponsored by the International Council for Scientific and Technical Information (ICSTI) and the International Union of Geological Sciences (IUGS). Approximately 5,000 terms are included and are listed alphabetically in their English version. Each term is assigned to one of 36 fields, 20 of which are geoscience fields, such as paleontology or structural geology; the other fields refer to classification domains, such as paleontology-systematics, or to cross-disciplinary concepts, such as methods or applications. The equivalents of the term in French, German, Italian, Russian, and Spanish then follow. The introduction and the index are presented in each of the six languages, making this tool equally useful in each of the languages. The Russian entries are in the roman alphabet, although the Russian introduction does use the cyrillic alphabet.

The *Multilingual Thesaurus of Geosciences* was a product of COGEODOC (Commission on Geological Documentation). It was edited by three members of the commission; one of the editors, G. N. Rassam, is the former chief editor of *Bibliography and Index of Geology*, an index of very high quality. This is an excellent selection for any library whose users access geological materials from several languages.

KENNETH QUINN

Treatises

234. Dense Chlorinated Solvents in Porous and Fractured Media: Model Experiments. English Language. Friedrich Schwille. Chelsea, MI: Lewis Publishers, 1988. 146p., illus. (part col.), bibliog., index. $55.00. 0-87371-121-1.

Dense Chlorinated Solvents in Porous and Fractured Media: Model Experiments has been translated from the German by James F. Pankow. (The original title was *Leichtfluchtige Chlorkohlenwasserstoffe in porosen und kluftigen Medien.*) Chlorinated hydrocarbons, such as carbon tetrachloride and trichloroethylene, are industrial chemicals of huge importance, most commonly as solvents. They are, accordingly, a very common constituent of groundwater contamination plumes. Computer models of the movement of these plumes are effective only if the properties of the constituents are known: the individual components move at different rates and behave differently in the various type of rocks through which groundwater flows. Physical models such as those discussed in Schwille's book are of immense help in constructing those computer models and in understanding groundwater contamination.

Friedrich Schwille is a retired hydrogeologist, and the translation of his book was recommended by Dr. John Cherry, a very well-known and respected hydrogeologist who was coauthor of a standard text in the field. The book is well illustrated with both drawings and color photographs, and the text is easy to understand if the reader has any background in the field of hydrogeology. In terms of both content and presentation, the book is of high quality, and it discusses a topic of considerable importance. It is highly recommended for the library of any university with groundwater studies in its curriculum.

KENNETH QUINN

235. Foraminiferal Genera and Their Classification. Alfred R. Loeblich, Jr.; Helen Tappan. New York, NY: Van Nostrand Reinhold Co., 1988. 2v., illus., bibliog., glossary, index. $199.95. 0-442-25937-9.

Loeblich and Tappan have produced an invaluable work in the systematic classification of Foraminifera. This is not merely a new edition of their *Treatise on Invertebrate Paleontology Part C Protista* published in two volumes in 1964 by the Geological Society of America and accepted as the reference source for both bibliographic compilation and systematic classification, but rather a new comprehensive treatise. The *Treatise* presented 1,192 genera with 1,267 names regarded as synonyms. *Forminiferal Genera and Their Classification* recognizes and illustrates 2,455 genera and 960 synonyms. An additional 208 genera are considered not recognizable based on present information, with 16 appearing too late to be included except for name and references.

In producing their new work, the authors have rechecked original references and investigated taxonomy. Some changes are a result of improvements in testing procedures for these organisms. The familiar taxonomic user notes from the previous set have been omitted to enable the work to focus solely on systematic classification. Morphologic descriptions are brief, and entries focus on test morphology, geologic series, and

geographic occurrence. The latter information has more detail than previously available in one place. Extensive work has been done to include new data.

In volume 1 following the main systematic sections, there are additional brief lists of "Family Group Taxa Based on Genera of Uncertain Status"; "Foraminiferal Genera of Uncertain Status"; "Unavailable Family-Group Names Used for Foraminifera"; "CladeGroups, Category Not Recognized by ICZN"; "Unavailable Generic Names Used for Foraminifera"; "Generic and Family Group Taxa Appearing Too Late for Inclusion"; "Apparently Described Genera for Which Reference Not Seen"; and "Generic Taxa Erroneously Regarded as Foraminifers." Even should a researcher disagree with decisions made by Loeblich and Tappan, the data presented here would be necessary to research an alternative opinion.

There is a 10-page glossary, a reference list including 3,456 citations in alphabetic order, and an extensive systematic index to both text and plates. This includes genera, supragenus, and species both valid and invalid. Text type is clear and easy to read; format and arrangement are appropriate, but some researchers may find the format cumbersome compared to the *Treatise*. The plates are in a volume separate from the text, thus requiring both volumes to be used at once. Plates have merely a plate number and item number attached, and cross-references to plate numbers are available from the text volume. The front of the plate volume lists plates in numerical order with titles, citation numbers, amount of magnification, source, and cross-reference back to the full text. Unfortunately, many plates are dark and difficult to examine in detail, perhaps a result of printing, or preprinting, of original material which has deteriorated in reproducible quality. Loeblich and Tappan's work, *Foraminiferal Genera and Their Classification*, is a representation of more than 60 years of professional experience and a culmination of two great careers in the field of Foraminiferal systematics.

JEAN E. CRAMPON

GENERAL SCIENCE

★
GENERAL SOURCES
Abstracts

236. International Development Abstracts. Norwich, Great Britain: Geo Abstracts, Inc., 1982-. v. 1- , index. Bimonthly. £56.00 per year. 0262-0855.

Published six times a year, this abstract journal is arranged by subjects that represent the broad range of international development activities—agriculture and rural planning; demography, population, and migration; economic conditions; economic policy and planning; education, training, and manpower planning; energy; the environment and natural resources; health, food, and nutrition; industry, institutional framework, and administration; international relations and cooperation; labor and management; politics; social policy; transport and communications; urban planning; and women and development. Each issue contains about 300 citations (with abstracts) to books, monographs, conference proceedings, and journal articles. Over 90 percent of the sources cited are written in English with abstracts written by the authors themselves. The editors do not clearly define the scope of journal coverage for this broad topic but have concentrated on professional journals such as the *American Journal of Agricultural Economics, Public Administration and Development, Journal of Rural Development* (Hyderabad), *Pakistan Development Journal, International Journal of Educational Development,* and the *International Journal of Health Services.*

A considerable time lag occurs between publication of the original article and its appearance in the abstract journal. For example, the fifth issue for 1987 contained few, if any, citings from that year, the majority dating from 1986 and a smaller percentage dating from 1985. This journal is produced by Geo Abstracts under the editorship of the staff of the Centre for Development Studies, University College of Swansea, Wales, which offers programs in health planning, regional development, social planning, and food and trade policy. An annual subscription includes the *International Development Index,* an author, subject, and regional index that includes relevant abstracts from the *Geographical Abstracts* journals. Online searchers can find these abstracts in GEOBASE, available through DIALOG. GEOBASE contains abstracts related to the worldwide literature on geography, geology, and ecology. Academic and special libraries will find this an excellent source of information on a wide range of international science topics.

LINDA JACOBS

Bibliographies

237. Scientific and Technical Books and Serials in Print 1988: An Index to Literature in Science and Technology. 15th. New York, NY: R. R. Bowker Co., 1987. 3v., index. $175.00. 0-8352-2362-0 set. 0000-054X.

First published in 1974, this important bibliographic index aims "to provide a comprehensive subject selection of Bowker's bibliographic database that would provide our users with current bibliographic and ordering information for in-print titles published or exclusively distributed in the United States." Up until 1978, it included only books, but since 1978 it includes serials. It covers all aspects of physical and biological sciences and their applications as well as all areas of engineering technology. Material for the books section of this index was taken from *Books in Print, Books in Print Supplement, Forthcoming Books,* and *Subject Guide to Books in Print.*

Each entry gives complete bibliographic information including price if available. Volume 1 is arranged by subject with entries within each subject arranged alphabetically by author. Volume 2 is the author index, and volume 3 includes three indexes: titles, serials, and publishers.

Although the information on books can be found in *Books in Print,* this set is useful in that it pulls all of the science and technology materials together and adds the serials. Special libraries with an emphasis on science or technology will want to have a copy of this for quick reference to find whether or not a publication is available.

H. ROBERT MALINOWSKY

Biographical Sources

238. Biographical Memoirs, Volume 57. National Academy of Sciences of the United States of America. Washington, DC: National Academy Press, 1987. 544p., illus., bibliog., index. $19.50. 0-309-03729-8.

"The *Biographical Memoirs* is a series of volumes, beginning in 1877, containing the biographies of deceased members of the National Academy of Sciences and bibliographies of their published scientific contributions." Each of the biographical essays is written by an individual who is familiar with the subject person, his or her discipline, and scientific career. A portrait of each entry is included, and a chronological listing of all of the individual's published scientific writings is at the end of the essay.

The biographical essays are interestingly written and include some childhood information along with all of the scientific achievements, education, awards, and honors. This volume covers the biographies of Arthur Francis Buddington, J. George Harrar, Paul Herget, John Dove Issacs, III, Bessel Kok, Otto Krayer, Rebecca Craighill Lancefield, Harold Dwight Lasswell, Jay Laurence Lush, John Howard Mueller, Robert Franklin Pitts, John Robert Raper, Karl Sax, Gerhard Schmidt, Leslie Spier, Hans-Lukas Teuber, and Warren Weaver. A cumulative index of the names of all subjects in volumes 1 through 57 is included. This is an outstanding series that has proven to be authoritative through the years, and a "must" for all academic libraries.

H. ROBERT MALINOWSKY

Directories

239. **Directory of Grants in the Physical Sciences 1987.** Phoenix, AZ: Oryx Press, 1986. 304p., index. $74.50pbk. 0-89774-334-2. 0890-541X.

The competition for funding in the area of research is becoming more and more complicated as funding sources diminish and the amounts of money decrease. This directory includes 1,561 programs "that provide funds for laboratory research, undergraduate education, scholarships and fellowships for graduate and postgraduate students, internships, and conferences in fields ranging from acoustics to zoology."

It is arranged in alphabetical order by the name of the program. Each entry is given an accession number used in the indexes. For each entry the following information is given: grant titles, accession number, grant description, restrictions, requirements, subject index terms, funding amount, *Catalog of Federal Domestic Assistance* number, and sponsor information, with key contact person and address.

Three indexes access the main part of the directory through a subject index, sponsoring organizations index, and sponsoring organizations, by type, index. All of this information is included on a database at The Oryx Press so that the latest information is always available.

This is an excellent directory that should be of use to researchers in all areas of physical science. Academic libraries will especially want a copy for referral purposes. *H. ROBERT MALINOWSKY*

240. **Directory of Special Libraries and Information Centers: A Guide to More than 18,500 Special Libraries, Research Libraries, Information Centers, Archives, and Data Centers Maintained by Government Agencies, Business, Industry, Newspapers, Educational Institutions, Nonprofit Organizations, and Societies in the Fields of Science and Engineering, Medicine, Law, Art, Religion, the Social Sciences, and Humanistic Studies.** 11th. Brigitte T. Darnay; Holly M. Leighton; Carol Southward, eds. Detroit, MI: Gale Research Co., 1988. 3v., index. $350.00. 0-8103-2696-5. 0731-633X.

It is hard to believe that any major library would not have a copy of this directory. In the past 25 years it has become the standard source of information on special libraries and information centers. Anthony T. Kruzas is the first editor and founder of this important directory. He had close to 10,000 entries in his first edition, as opposed to the some 18,000 in this edition.

The *Directory of Special Libraries and Information Centers (DSL)* includes special and research libraries, archives, information centers, other computer-based information retrieval systems, and many specialized collections of films, artifacts, computer tapes, and special objects. Its focus is on special collections, limited by subject matter or form, falling into five categories: (1) subject divisions, departmental collections, and professional libraries in colleges and universities, (2) branches, divisions, departments, and special collections in large public library systems, (3) company libraries, (4) governmental libraries, and (5) libraries supported by nonprofit organizations, associations, and institutions. It is arranged alphabetically by the name of the facility. At the end of the second part of volume 1 is a subject index that identifies major fields of interest for each library. In addition to volume 1, there are two other volumes to this set: volume 2—"Geographic and Personnel Indexes" and volume 3—"New Special Libraries," which is a periodic supplement to volume 1.

Each entry may have up to 24 separate bits of information about that library or information center: name of organization, name of library or information center, principal subject keyword, mailing address, telephone number, head of library or information center, founding date, number of staff, subjects, special collections, holdings, subscriptions, services, automated operations, computerized information services, networks/consortia, publications, special catalogs, special indexes, remarks, formerly, also known as, formed by the merger of, and staff names. Each is given a consecutive number to facilitate indexing. Canada is included as well as some key international libraries.

There are seven appendices in volume 1 covering "Networks and Consortia," "Regional and Subregional Libraries for the Blind and Physically Handicapped," "Patent Depository Librar-

ies," "Federal Information Centers," "Federal Job Information Centers," "United Nations Depository Libraries," and "Regional Government Depository Libraries." This is a highly recommended directory for public, academic, and special libraries.   *H. ROBERT MALINOWSKY*

241. International Research Centers Directory, 1988-89: A World Guide to Government, University, Independent Nonprofit, and Commercial Research and Development Centers, Institutes, Laboratories, Bureaus, Test Facilities, Experiment Stations, and Data Collection and Analysis Centers, as Well as Foundations, Councils, and Other Organizations Which Support Research. 4th. Darren L. Smith; Summer A. O'Hara, eds. Detroit, MI: Gale Research Co., 1988. 1,567p., index. $360.00. 0-8103-4362-2. 0278-2731.

The *International Research Centers Directory* (*IRCD*), now in its fourth edition, has expanded to two volumes and over 6,000 listings, an increase of about 40 percent in entry coverage over the previous edition. Its subtitle adequately describes its scope and coverage: a current, continuing worldwide guide for government, university, independent nonprofit, and commercial research and development activities in a wide range of subject areas. An excellent companion to the publisher's *Research Centers Directory*, which focuses on the United States and Canada alone, *IRCD* covers all other countries and thereby complements the other Gale Research publications in this area.

Criteria for inclusion are that each entry must be a research and development installation; institute; center; laboratory; bureau; testing station; data collection and analysis center; statistical center; or a foundation, council, or government ministry that conducts, sponsors, administers, or coordinates major research programs and/or facilities within a given country. Descriptive data were collected from questionnaires and official publications of the centers. The editors point out, however, that the information included is in no way an evaluation of the importance of the center. Conversely, omission is not an indication of unimportance.

Organization is in two parts. An "International" section is followed by "Countries" in alphabetical order. The international section describes "multinational organizations that span and coordinate research projects and related activities involving participation by members in several countries." Volume 1 contains the international section and the entries for "Afghanistan" through "Japan"; volume 2 has "Jordan" through "Zimbabwe" and the indexes. Except for the 72-page international section, the volumes follow an alphabetical arrangement by country and, within a country, by organization name. There are cross-references for centers located within a country but listed in the international section.

Each entry for a research center follows the same pattern, beginning with a sequential entry number. Seventeen possible items of information are available for each entry. A sample entry guide lists and explains them. In addition to the expected organization names, addresses, phone numbers, and directors, there are several informative sections, including parent organizations, a research description with specific fields of interest, the existence of special facilities, a list of publications and information services, recurring programs (such as seminars, symposia), laboratory facilities, and affiliations or subsidiaries.

New features of the fourth edition include the use of diacritics where possible in foreign names, publications, and addresses and the inclusion of telex numbers in addition to telephones as part of the basic header for each entry. Three indexes provide access by entry number. The name and keyword index allows alphabetical access by name, subject, and keyword in names. It also includes prior names' tracings and centers whose functions have been absorbed by others, with reference to the current unit by entry number. A country index is arranged alphabetically by country, then alphabetically by research unit. The subject index includes terms derived from Gale Research's *Research Centers Directory* and Library of Congress subject headings, as well as from terms used by the respondents. "See" and "see also" references are included in this index. References within a subject are alphabetical by country.

This edition provides expanded coverage and, used in complement with other publications or by itself, is a welcome addition to the reference resources available to the researcher.

R. GUY GATTIS

242. Life Sciences Organizations and Agencies Directory: A Guide to Approximately 8,000 Organizations and Agencies Providing Information in the Agricultural and Biological Sciences Worldwide. Brigitte T. Darnay; Margaret Labash Young, eds. Detroit, MI: Gale Research Co., 1988. 864p., index. $155.00. 0-8103-1826-1.

Life Sciences Organizations and Agencies Directory (*LSOAD*) is the most recent addition to Gale's Organizations and Agencies Directories series. Its mission, as stated in the preface, "is to provide a convenient, one-step source to organizations and agencies in the life sciences in the United States and abroad." *LSOAD* attempts to cover all aspects of the biological sciences (excepting those concerned more directly with medicine or the health sciences), agricultural sciences, natural resources and environmental sciences, nutrition and food science, and biotechnology. Much of the material is derived from other Gale publications.

Entries are arranged in 18 chapters which can be grouped as follows: associations, botanical gardens, computer informaton services, consulting firms, educational institutions, libraries and information centers, research centers, state acad-

emies of science, state and federal government agencies, federal government grants and programs, and U.S. foreign agricultural officers. Content of the entries varies from chapter to chapter, but all provide names of contact persons, addresses, and telephone numbers. Additional pertinent information is provided for each organization. As a rule, these entries provide all the essential information for each organization discussed.

Access to the 7,662 entries is through the table of contents and a keyword and organizational name index. The "Chapter Scope Notes" section provides very useful annotations to the table of contents, with helpful information on the arrangement of the entries in each chapter. An "Index Notes" section at the beginning of the volume provides helpful information on using the index, although it might be more useful if this page were located at the beginning of the index.

Of all the chapters, the most comprehensive are those listing associations and libraries or information centers. The weakest chapter is that listing educational institutions. The contents of this chapter are extremely limited, listing only those institutions which have forestry or veterinary medicine programs or which are participants in the National Sea Grant College Program. The absence of more general programs in the biological or agricultural sciences severely limits the usefulness of this chapter. Of additional concern is the absence of some major research laboratories from the research centers chapter (Jackson Laboratory and Woods Hole Oceanographic Institute, for example). Although no directory can be entirely comprehensive, one at least expects that the most important research centers will be listed.

One other problem concerns authority control in the index. The "Index Notes" page explicitly states that "the user should check under all variant name forms in the master Index." As a result, the user frequently must look in several places in the index to locate the organization of interest. In addition, the keywords (identifiable by the presence of parentheses) frequently do not adequately index the organizations. It is often necessary to search under a permuted form of the name of an organization instead of under the keyword where one would expect to find it. Finally, the entries in the chapters listing computer information services could be improved by listing the vendors through which the information services are available. Although the problems listed above are significant, it is likely that they will be dealt with in later editions of this work. As it stands, *LSOAD* would be a useful addition to any library serving the life sciences research community.

LOREN D. MENDELSOHN

243. **Pacific Research Centres: A Directory of Organizations in Science, Technology, Agriculture, and Medicine.** 2nd. London, Great Britain: Longman Group UK Ltd., Dist. by Gale Research Co., Detroit, MI, 1988. 517p., index. $300.00. Reference on Research. 0-582-01608-8. 0952-4568.

Need to know the name of a New Zealand research group working on cystic fibrosis or the telephone number and business affiliations of a Japanese firm conducting research on insulated cables? Who is the research director and what are the research activities of the Bicol River Basin Development Programme located in the Phillipines? The editors of this volume intend to facilitate "the flow of technological information between western nations and the Orient." For this purpose, they have compiled a directory of both public and private research centers located in the western Pacific region in such countries as Austalia, New Zealand, Brunei, the People's Republic of China, Indonesia, Japan, Thailand, Vietnam, Taiwan, and Korea.

This volume should interest technical directors, scientists, business executives, journalists, librarians, engineers, and conference organizers. The broad range of research interests represented spans the industrial and technical fields, the life sciences, and agriculture. Research center activities may focus on biogas, bionics, cheese, concrete, dyes, ergonomics, graphite, pest control, steel pipes, or space probes. The University of Sydney in Australia, the University of Science in Malaysia, and the National Chung Hsing University in Taiwan are representative entries from the academic sector. The Fujisawa Pharmaceutical Company in Osaka, Japan; the Liquid Fuels Management Group in Wellington, New Zealand; and the Energy Research Institute in Beijing represent private industry, consultancies, and official research centers, respectively.

First published in 1985, this volume updates 40 percent of the entries from the first edition, based on returned questionnaires distributed in 1987. Additionally, the Longman editorial team has updated entries from other sources. The editors plan to publish future editions at "regular intervals." Each entry begins with the name of the research center as known within that country, followed by an English translation. The entry includes address, telephone number, telex number, facsimile number, product range, affiliation with other research centers, current director, departments, number of research staff, annual research expenditures, a summary of current and proposed research activities, publications, and organizations for which the center undertakes substantial research.

The editors have arranged the entries in alphabetical order—first by country and then by the original name of the research center. An index provides access by title of the research center, both in the original language and in its English translation, and also includes commonly used acronyms. A subject index, using the controlled language of the British Standards Institution's *Root Thesaurus*, identifies institutions by specific research activities. Large public libraries, academic libraries, and special libraries will find this a valuable source of information for research facilities in the Pacific Ocean region.

LINDA JACOBS

244. Science and Technology in the USSR. Michael J. Berry, ed. Harlow, Great Britain: Longman Group UK, Ltd., Dist. by Gale Research Co., Detroit, MI, 1988. 405p., illus., bibliog., index. $95.00. Longman Guide to World Science and Technology. 0-582-90053-0.

This work represents the sixth volume in the ongoing series from Longman designed to survey the scientific activities and organizations of the various major countries and regions of the world. Despite its rather hefty price, the book contains much of value to recommend itself.

There are two major sections, the first providing a general overview of the organization of Soviet science and technology, which covers the history, scientific administration, and planning processes, the Academy of Sciences, the ministerial network, and higher educational system. Part 2 treats each of the major research and technical areas, listing current research efforts and the major organizations working in each field. Most chapters end with brief bibliographies pointing the reader to further sources on each topic.

Information about the workings of Soviet science is never easy to come by. This book places a wealth of data in one convenient source. The reader is led through the maze of developments following the 1917 revolution, which affected science as much as any area of Soviet life. This information helps create a better understanding of the reasons that Soviet science functions as it does. A concluding postscript brings this information up to date with a discussion of the July 1987 decree of the Central Committee and the Council of Ministers for Science and Technology "On Increasing the Role of the State Committee for Science and Technology in the Management of Scientific and Technical Progress" and subsequent developments.

Major portions of the book are devoted to each area of specialization—aeronautics, computers, agriculture, metallurgy, etc.—giving brief summaries of current research, along with lists of organization names and addresses, directors names, and publications. This directory information alone may make the purchase of the book worthwhile to anyone who needs rapid and convenient access to the names and addresses of Soviet scientific organizations. Librarians will likewise welcome the information on how to locate and verify periodical titles. All in all, this is a comprehensive and useful book, which should certainly find its way into all major scientific libraries. It is recommended. *FRED O'BRYANT*

Encyclopedias

245. Exploring Our World: Deserts. Terry Jennings. Freeport, NY: Marshall Cavendish, 1987. 48p., illus. (col.), glossary, index. $14.99. 0-86307-820-6.

With the need to have a set of guides and handbooks covering various aspects of the nature of this planet, this book, as a part of a successful six-volume series, shows the ecological relationships of the desert life-forms and geographical features of the world's deserts through the use of 70 color photographs, figures, illustrations, maps, a clear introduction, and informative diagrams. It is a welcome new book, especially for young geographers and laypersons. It was originally published in the United Kingdom in 1986 as *The Young Geographer Investigates: Deserts* by Oxford University Press. Although there are books discussing the arid areas on the world's surface, the majority of them are for adults and professionals. This work presents a general picture of the characteristics of the world's deserts. It covers in clear narration locations; the different types of deserts; animals and inhabitants; reasons deserts are formed; history; water and oasis; minerals; and planning of deserts. Special features include a file of facts, methods for further research, practical experiments and projects for the reader to undertake, sources for further information, and a glossary of terms. For a broad overview, this is a very welcome book for young people. School and public libraries should find it useful as a reference source. *ZE-HUI NIU*

246. Exploring Our World: Mountains. Terry Jennings. Freeport, NY: Marshall Cavendish, 1987. 48p., illus. (col.), glossary, index. $14.99. 0-86307-819-0.

This book was originally published in the United Kingdom by Oxford University Press in 1986 under the title *The Young Geographer Investigates: Mountains*. It is an encyclopedic volume covering the historical and geological development of mountains and their environment. It is well written for the school child, with colored illustrations, pleasant layout, and good use of boldface titles. There is a brief index as well as a table of contents. Special features include a file of facts, methods for further research, practical experiments and projects for the reader to undertake, sources for further information, and a glossary of terms. This volume and the other five in the six-volume set, *Exploring Our World*, will be very useful in school and public libraries. *H. ROBERT MALINOWSKY*

247. Exploring Our World: Polar Regions. Terry Jennings. Freeport, NY: Marshall Cavendish, 1987. 48p., illus. (col.), glossary, index. $14.99. 0-86307-823-0.

This book was originally published in the United Kingdom by Oxford University Press in 1986 under the title *The Young Geographer Investigates: Polar Regions*. This 48-page encyclopedic volume is for elementary school children and covers the historical, geographical, meteorological, anthropological, and biological aspects of the Arctic and Antarctic circles in a unified manner. The subdivisions are well chosen, addressing the main issues of interest to the inquiring mind. Sections include "Why Are the Arctic and Antarctic So Cold?" "Glaciers," "Modern Polar Explorers,"

"The Alaskan Pipeline and Permafrost," and other topics. It has two-column text with quality color illustrations and photographs on good quality paper. The binding is durable, the layout is pleasant with good use of bold titles, and there is a brief index as well as a table of contents. Special features include a file of facts, methods for further research, practical experiments and projects for the reader to undertake, sources for further information, and a glossary of terms. This book should intrigue and fascinate the user.

YVES KHAWAM

248. **Exploring Our World: Rivers.** Terry Jennings. Freeport, NY: Marshall Cavendish, 1987. 48p., illus. (col.), glossary, index. $14.99. 0-86307-818-4.

This book was orginally published by Oxford University Press in 1986 under the title *The Young Geographer Investigates: Rivers*. It is a book about rivers presented in simple language and uncomplicated sentence structures. The numerous color photographs, cartoon drawings, and color drawings are well detailed and clearly labelled to illustrate explanations in the text. The book effectively allows the student's feet "to get wet" with explanations of the uses of water, the importance of rivers in history, and a description of the water cycle. The text covers the sources, the life, and the forms of rivers. It also touches on water pollution, drinking water from river water, water power, and the various plants, animals, and fish that live in the rivers. A number of major rivers around the world are described, including the Rhine, Amazon, and Nile. A fact file summarizes some of the major points covered in the various sections. Also included are tasks and questions that can be completed in a library, field research projects, and further sources of information. This is an attractive book that is well written and inviting to read; essential for school and public libraries. *BARBARA KRUSER*

249. **Exploring Our World: Temperate Forests.** Terry Jennings. Freeport, NY: Marshall Cavendish, 1987. 48p., illus. (col.), glossary, index. $14.99. 0-86307-822-2.

This book was originally published in the United Kingdom by Oxford University Press under the title *The Young Geographer Investigates: Temperate Forests*. It is one of six books in the series *Exploring Our World* presenting clear, practical introduction to physical and human geography of temperate forests. It discusses the kinds of trees found in a temperate forest, their life cycles, and how they are used by humans. Jennings features attractive full-color maps, vivid photographs, drawings, and informative diagrams with large boldface words. A list of books, magazines, agencies, and other places for the inquisitive reader to seek out more information is included as well as a glossary. High school libraries and public libraries will find this title, as well as the others in the series, useful for reference.

DANIEL Y. H. WONG

250. **Exploring Our World: Tropical Forests.** Terry Jennings. Freeport, NY: Marshall Cavendish, 1987. 48p. illus. (col.), glossary, index. $14.99. 0-86307-821-4.

This book was originally published in the United Kingdom by Oxford University Press under the title *The Young Geographer Investigates: Tropical Forests*. It is one of six books in the series *Exploring Our World*, presenting a clear, practical introduction to the physical and human geography of tropical forests. Jennings features highly attractive full-color maps, vivid photographs, drawings, and informative diagrams with large boldface words. A list of books, magazines, agenicies, and other places for the inquisitive reader to seek out more information is included as is a brief glossary. High school and public libraries will find this a useful reference source.

DANIEL Y. H. WONG

251. **Milestones in Science and Technology: The Ready Reference Guide to Discoveries, Inventions, and Facts.** Ellis Mount; Barbara A. List. Phoenix, AZ: Oryx Press, 1987. 150p., bibliog., index. $29.50. 0-89774-260-5.

This encyclopedic work covers basic discoveries and inventions in all areas of science and technology. It is arranged alphabetically by the name of the invention or the discovery followed by a brief description of the topic, including dates the event occurred, names of the people involved, and the nationalities of the persons named. A broad subject indicator is assigned to each of the entries. Each entry also includes reference to "additional reading" so that the user can pursue a more detailed discussion of the topic. A complete list of these references is included at the end of the volume. To locate the information, a personal name index, chronological index, geographical index, and field of study index are included.

Students, laypersons, reference librarians, and practitioners working outside their own fields of expertise will find this a useful encyclopedia for brief information about 1,000 significant events in science and technology. A recommended book for school, public, and academic libraries.

H. ROBERT MALINOWSKY

252. **The New Illustrated Science and Invention Encyclopedia.** Reference. Westport, CT: H. S. Stuttman, Inc., Dist. by Marshall Cavendish, Ltd., Westport, CT, 1988. 27v., illus. (col.), index. $399.95. 0-86307-491-Xset.

The *New Illustrated Science and Invention Encyclopedia* has several very positive features. The "Thematic Index," which includes scientific invention milestones (timeline form), and a glossary (as well as a regular alphabetical index) are quite valuable and user-friendly. Also, the volumes on inventors and inventions are a great "quick reference." Another bonus is the "Careers in Science" volume, which gives brief but fact-packed sections of information on many occupations. Besides these specialty features, this set

contains concise, easy-to-use information on most inventions in which today's middle school student would be interested. For many, this book would be all that is necessary for the answer to a quick question or short report; for others, it would whet the appetite and be just the beginning.

There is very little on the negative side. Some of the changes in this edition are initially confusing if one is used to using earlier versions. Also, the only reference to the specialty volumes, other than the indexes, is in volume 1 of the set. This might lead students to miss some useful information.   *ANNE BEST*

Guides to the Literature

253. **Core List of Books and Journals in Science and Technology.** Russell H. Powell; James R. Powell, Jr., eds. Phoenix, AZ: Oryx Press, 1987. 134p., index. $35.00. 0-89774-275-3.

"The purpose of this book is to bring together prime choices of English-language books and journals in the areas of contemporary science and technology." It covers agriculture, astronomy, biology, chemistry, computer science, engineering, geology, mathematics, and physics. The compilers also included only those books that have been published in the 1980s, which means that some earlier published classics will not appear.

It is strong on textbooks and excludes books that were proceedings of meetings, seminars, symposia, or congresses. All journals selected were chosen from those covered by a major indexing or abstracting service. The entries are arranged alphabetically by title within broad subject disciplines. Books and journals are listed separately. All books have complete bibliographical data, including price and a brief annotation. Three indexes—author/editor, title, and subject—aid in locating a specific title. This is an excellent general guide to the literature that should benefit school, public, and academic libraries in assessing their collections in science and technology.

H. ROBERT MALINOWSKY

254. **Scientific and Technical Information Sources.** 2nd. Ching-Chih Chen. Cambridge, MA: The MIT Press, 1987. 824p., index. $55.00. 0-262-03120-5.

This publication represents a unique source and the most complete listing of sources in science and technology. Each of the 23 chapters is devoted to a type of reference source, and within each chapter, sources are listed according to the various branches of science and technology. Most of the entries include annotations of varying lengths as well as citations to reviews of the items in journals and other books. Unfortunately there are problems with both the arrangement of the entries and the entries themselves which detract significantly from the value of this resource. Within each broad subject area, entries are arranged alphabetically by title, and, even though the arrangement is definitely preferable to listing

by author or main entry, as the editor indicates, the name of the author or the main entry is printed in boldface, which tends to belie this notion. The broad subject categories are not listed alphabetically, which would be preferable for most users, but instead according to a classification scheme. These two basic features of the resource are especially troublesome to the scientists or engineers who represent one of the target user groups. A better arrangement would be alphabetically by broad subject area with the name of the author or main entry in a less conspicuous type font.

There are several kinds of inaccuracies and inconsistencies in the citations themselves. The same title is sometimes cited more than once with variations in the form of the title, the author, or the publication date and with different annotations. In some cases, titles which are not self-explanatory have no annotations. The annotations are usually accurate and descriptive. In other instances, titles are not verifiable. An example of this is the title for the book *The Literature Matrix of Chemistry* which becomes here *The Literature of Matrix Chemistry*. Authors' names are also sometimes invented. "Tukey," a well-known statistics editor, is found here as "Turkey." The index is thorough, but there are omissions, and occasionally wrong page numbers are given. Citations are often incomplete regarding the number of volumes, and mention is sometimes made of only some of the volumes published. Some of the items appear in the wrong subject areas; e.g., a history of the microscope is included in the section on computer engineering. And there are important omissions, such as the *IES Lighting Handbook*, to name only one. The attempt to produce a resource of such monumental scope is admirable; however, a better approach would have been to produce a series of resources devoted to specific sciences and technologies. This would require a less formidable editing effort.   *EARL MOUNTS*

Handbooks

255. **Science and Technology in France and Belgium.** E. Walter Kellermann. Harlow 07 Great Britain: Longman Group UK, Ltd., Dist. by Gale Research Co., Detroit, MI, 1988. 131p., tables, maps. $100.00. Longman Guide to World Science and Technology. 0-582-00084-X.

If keeping abreast of the newest international advances in science and technology is important, the series Longman Guide to World Science and Technology, provides a tool for accomplishing this on a country-by-country basis. With other volumes still on the drawing board, Longman has already published guides to the Middle East, Latin America, China, Japan, the U.S., and the U.S.S.R..

This volume on France and Belgium provides an overview of government policies designed to promote science research and development. The author discusses, in detail, the funding and activi-

ties of various government agencies and their relationships to various private laboratories and educational institutions devoted to research and development in the sciences. He lists current major research areas and explains the underlying reasons for the current policies. Basing his comparisons on the trade balance in high technology, the author feels that France has a chance to overtake the United Kingdom in the "world league of science and technology." He points to "French successes in software, in rocketry, in space physics and in exports of defense materials—the latter achieved at a much smaller expense of [research and development] than in the United Kingdom." Successes in fusion research and international collaborations have led to "commercially successful applications."

The government of France believes that "in order to stay competitive in world markets," it has "to expand its scientific-industrial base," "achieve a more equitable demographic distribution by regional development," and "give priority to the support of innovative ideas." The main areas of French research efforts include biotechnology, electronics, technology for developing countries, nuclear energy, space, oceanography, agriculture, and town planning. France supports "key programmes" in science and technology, regional development, technology transfer, and "the strengthening of small and medium-sized industries." The government has adopted a policy of "valorization," which validates "the results of [research and development] in terms of market requirements and practicable technology transfer." Cooperative projects between industrial and government laboratories constitute regional "poles" of "organically linked technological and educational growth."

A scant 18 pages of the book devotes itself to science policy in Belgium. Belgium's Science Policy Office "notes Belgium's relative strength in the medical sciences, in fundamental biology, in the chemical and pharmaceutical industry and in telecommunications, but finds weaknesses in the industry, namely in the mechanical sector, in electronics and advanced software." Recently the government initiated programs in energy, terrestrial and marine resources, waste recovery, aerospace, computers, biotechnology, and international cooperation. The Belgian government believes "that long term solutions for industrial survival require an expansion of selected science and technological fields in Belgium's education system."

"Businessmen, technical salesmen, science policy makers, embassy officials," and European observers will find this book full of useful data. Tables supply information on educational institutions, the finance of research and development, and the distribution of research personnel both regionally and within various industries. The author also includes addresses of various general information services and government agencies in both France and Belgium. A useful handbook for academic and special libraries.   *LINDA JACOBS*

Histories

256. The Inventors: Nobel Prizes in Chemistry, Physics, and Medicine. Nathan Aaseng. Minneapolis, MN: Lerner Publications Co., 1988. 79p., illus., glossary, index. $9.95. 0-8225-0651-3.

This well-written and interesting little book for young people briefly covers eight inventions that were recognized with Nobel Prizes—the X-ray, radio, EKG, phase contrast microscope, transistor, radiocarbon dating, laser, and CT scan. Each of the inventions or discoveries is covered in a separate chapter, with a brief discussion on the apparatus or discovery and interesting comments on the developers. There is a portrait of each scientist, with well-selected photographs adding to the text. This fascinating book is highly recommended for school and public libraries.

H. ROBERT MALINOWSKY

Yearbooks/Annuals/Almanacs

257. McGraw-Hill Yearbook of Science and Technology, 1988: Comprehensive Coverage of Recent Events and Research as Compiled by the Staff of the McGraw-Hill Encyclopedia of Science and Technology. New York, NY: McGraw-Hill Book Co., 1987. 518p., illus., bibliog., index. $60.00. 0-07-046183-X. 0076-2016.

The *McGraw-Hill Encyclopedia of Science and Technology* is such a vital reference source that it is hard to conceive of a public, college, or university library not having a copy. There are reasons for it being so valuable; prominent among these are the completeness of coverage and a writing style that makes the articles interesting despite the necessary use of a specialized vocabulary. Science and technology are very fast-moving fields, however, and certain articles can become outdated in only a year or two. Consider, for instance, what was learned about comets in 1986 as Halley's comet was probed. The results were too late to be included in the 1987 edition of the encyclopedia except in a very incomplete and preliminary fashion. That is why yearbooks can be a valuable resource. The cutoff date for the 1988 *McGraw-Hill Yearbook of Science and Technology* was fairly early in 1987, judging by the absence of any mention of one of the most prominent scientific events of that year: the discovery of the supernova 1987 Shelton in February. That consideration aside, the yearbook does a good job of updating information on a wide range of topics. Most articles end with a bibliography, with the citations mainly from the 1984-1986 period, and are signed.

This yearbook is the same high quality of the encyclopedia that it supplements and should be purchased by those who own the *McGraw-Hill Encyclopedia of Science and Technology*, namely public, college, and university libraries. As with any yearbook, it should be realized that it is not comprehensive and that there can be a considerable lag between the deadline for inclusion of material and the date of publication.

KENNETH QUINN

MATHEMATICS

★
GENERAL SOURCES
Exam Reviews

258. Barron's How to Prepare for Advanced Placement Examinations: Mathematics. 3rd. Shirley O. Hockett. New York, NY: Barron's Educational Series, 1987. 525p., index. $9.95pbk. 0-8120-3876-2pbk.

"This book is intended for students who are preparing to take one of the two Advanced Placement Examinations in Mathematics offered by the College Entrance Examination Board." It covers Calculus AB, which is a full-year course in elementary functions and introductory calculus, and Calculus BC, which is an intensive full-year course covering the calculus of functions of a single variable, including the calculus topics of AB plus infinite series and differential equations. The main part of the book is a "topical review" of topics contained in Calculus AB and BC, including carefully worked out illustrations. The next section contains sample questions found in an AP Calculus Examination, and this is followed with practice examinations for Calculus AB and BC. Also included are actual examinations that have already been administered. The answers to all of these examinations are given at the end of the book. School, public, and college libraries will find this a well-used book on their reference shelves. *H. ROBERT MALINOWSKY*

Handbooks

259. A Handbook of Fourier Theorems. D. C. Champeney. New York, NY: Cambridge University Press, 1987. 185p., illus., bibliog., index. $39.50. 0-521-26503-7.

Professor Champeney, in his *A Handbook of Fourier Theorems*, has given applied mathematicians and engineers a work that they can refer to for rigorous statements of the most important theorems in Fourier theory. The author's intended audience is not the "mathematics specialist," and, therefore, he does not include proofs with the theorems. The reader is anticipated to have had undergraduate classes in mathematics for engineers and physical scientists. The handbook is divided into sixteen chapters with chapters 1 through 5 giving introductory and background

information and terminology in mathematical analysis and chapters 6 through 16 giving explanations and examples of the theorems. The author has included an excellent bibliography, and the index provides references to abbreviations and mathematical symbols. A professor at the School of Physics and Mathematics at the University of East Anglia, the author has recognized the communication gap that often exists between mathematicians and engineers. In this handbook, he has helped to lessen the problem and created a very good reference tool for engineers and engineering librarians.
JO BUTTERWORTH

Histories

260. A Concise History of Mathematics. 4th rev. Dirk J. Struik. New York, NY: Dover Publications, Inc., 1987. 228p., illus., bibliog. $7.95pbk. 0-486-60255-9pbk.

Struik's book may be turned to as a model of concise histories. Authoritative, succint, and highly readable, it has undergone four American editions and many overseas editions and has been translated into German, Russian, Spanish, Italian, Chinese, and other languages. The new fourth edition has been updated by a summary of the twentieth century up to World War II and the computer era and by extensive enrichment of the bibliographic material. Also, the mathematical histories of several countries have been revised as a result of translators' addenda or the author's work in preparing translations to other languages.

A historian, mathematician, and professor emeritus at Massachusetts Institute of Technology, Dirk J. Struik displays a strong grasp of the development of mathematics from earliest man to the present, with emphasis on the ways in which mathematics has been influenced by historical events and by other disciplines. The chronological organization of the book naturally allows an introductory focus on several early cultures—the Ancient Orient, Egypt, Greece, and India—before broadening to a century-by-century view, with great thinkers in the field generating important mathematical developments. In addition to its value as a solid mathematical history source, the work has great reference value as a biographical source. It is also a starting point for further reading on mathematical topics or mathematicians, through the exhaustive bibliographic ma-

terial in the introduction and at the ends of chapters. Most libraries with an interest in mathematics will find it an indispensable resource.

MARILYN VON SEGGERN

Textbooks

261. College Board Achievement Test: Mathematics Level I. 5th. Morris Bramson. New York, NY: Prentice Hall Press, 1987. 248p., illus. $7.95pbk. "An Arco Book." 0-668-06595-8pbk.

College Board Achievement Test: Mathematics Level I is a reference textbook for the mathematical information required for the level one achievement test in mathematics. It is primarily designed for students in grades 11 and 12. It would also be useful to mathematics teachers at the high school level as a curriculum guide for courses that prepare students for the level one examination or for guidance personnel who counsel students in the choice of an appropriate examination.

The author, Morris Bramson, has taught mathematics at the high school and university levels. He has worked with students who must take the examination, and he is familiar with the content that is necessary at the college level. Divided into five sections, the first includes a description of the level one examination in mathematics, directions on how to use the book as a tool for review, and directions for taking the examination. The content description accurately reflects the guidelines for the test, which is published by the College Board. The explanation of the exam includes question formats, advice on guessing answers, and the rules for administering the exam. Because many students are not accustomed to multiple-choice testing in mathematics, more ideas about test-taking strategies would have been useful in this section, e.g., working backward from the selection of possible answers.

Section 2 of the book is a very complete, detailed outline of content areas. All areas included in the examination are carefully covered. The author also included several topics that are not topics for the level one exam. For example, the level one examination includes basic trigo-nometry from the right triangle. The content review includes trigonometric functions of the general angle, identities, graphing, and oblique triangles. These are topics for the level two examination.

Section 3 includes detailed explanations of the topics listed in part 2 and illustrative problems with solutions. The content areas are fully explained. The sample questions cover the entire range of topics at a level of difficulty which is comparable to that of the exam. Once again, several extra topics are included: logarithms, vectors, and sequences. The problems are not given in multiple-choice formats, as they are found in the exam, but the selection of problems is good. Part 4 presents more practice problems which are arranged by topic. The solutions to the problems are included at the end of the section.

The final section of the book contains six sample achievement exams in mathematics. The tests are written in the same format as are the actual College Board Examinations, i.e., 50 multiple choice questions of various types. Sample questions do include problems where the correct answer is "none of these" or "cannot be determined from the information given." Teachers should explain to students that some figures are not drawn to scale and that the figure for question two in sample test one is incorrectly labelled. Answer sheets are provided which involve shading the circle of the correct answer but which are not duplicates of the official answer sheets. Part 5 also includes well-constructed post-test activities, and the official scoring system and percentile rankings are described. The students are asked to complete diagnostic checklists to determine which topics they should review in more depth before taking subsequent sample examinations.

Students and teachers will find this textbook to be a useful reference in preparing for the level one mathematics achievement test. A careful explanation of the review process, complete content guides, carefully chosen sample exercises, well-prepared sample tests, and helpful posttest diagnostic tools make this textbook useful for independent review by the student or for in-class review by the teacher. A student who successfully completes the review will be more than prepared for the level one achievement test in mathematics.

SANDRA K. DAWSON

MEDICINE

★

GENERAL SOURCES
Bibliographies

262. Encyclopedia of Health Information Sources: A Bibliographic Guide to Approximately 13,000 Citations for Publications, Organizations, and Other Sources of Information on More than 450 Health-Related Subjects. Paul Wasserman; Suzanne Grefsheim, eds. Detroit, MI: Gale Research Co., 1987. 483p. $135.00. 0-8103-2135-1.

The title page of this new Gale publication indicates what the complete coverage of the *Encyclopedia of Health Information Sources* really is: "Includes: abstract services and indexes; annual, reviews, and yearbooks; associations and professional societies; bibliographies; directories and biographical sources; encyclopedias and dictionaries; handbooks and manuals; online data bases; periodicals; popular works and patient education; research centers, textbooks and general works; and other sources of information on each topic." The bibliography is intended to be a quick source of information on a particular subject. There are some 450 subjects under which the various items are listed. All printed materials are post-1980.

To use this excellent reference source, one must first consult the detailed table of contents to locate the topic that is being researched. Numerous cross-references are included to guide the user to the right term. Once the term has been identified, the user can now proceed to the indicated page to see what is available. This is a highly selective list and is not meant to be comprehensive. Each entry gives only the vital information for identification: title, address, frequency, and year. There are no annotations. Under Acquired Immunodeficiency Syndrome one finds a very brief collection of entries: abstract services and indexes (5); associations and professional societies (6); bibliographies (2); handbooks and manuals (1); online databases (6); periodicals (2); popular works/patient education (1); research centers, institutes, and clearinghouses (7); and textbooks and general works (5). Additionally, there is a "see also" reference to immunologic disorders. Obviously, these are only a small percentage of sources available, but they are some of the better-known sources and would be a good start for someone who wanted background information on AIDS.

All in all, this is a welcome new general guide to materials in the field of medicine. It should be of great use to the general public as well as to the student and researcher who want only limited information. The publisher indicates that interim updates between editions are being considered.

H. ROBERT MALINOWSKY

263. A Research Guide to the Health Sciences: Medical, Nutritional, and Environmental. Kathleen J. Haselbauer. Westport, CT: Greenwood Press, Inc., 1987. 655p., glossary, index. $49.95. Reference Sources for the Social Sciences and Humanities, no. 4. 0-313-25530-9.

A Research Guide to the Health Sciences is the fourth title in the Greenwood Press series, Reference Sources for the Social Sciences and Humanities. It is intended for researchers at all levels, though it focuses on the needs of students and on those professionals working outside their particular specialty.

The over 2,000 cited sources are arranged in four major parts: Part A—"General Works Arranged by Type;" Part B—"Basic Sciences Supporting Clinical Medicine;" Part C—"Social Aspects of the Health Sciences;" and Part D—Medical Specialties." Each section is introduced with a narrative overview, followed by listings of sources which are in turn followed by a narrative discussion of strengths and weaknesses. The cutoff date for works discussed in this research guide is the end of 1986.

Titles of the works appear in boldface type and are underlined, which aids in locating them on the pages of difficult-to-read computer-generated type. An extensive title/keyword index also assists the user, especially by designating major discussions of a source with boldface numerals. There is a short glossary which defines format types.

The author was comprehensive in her selection of sources and provides excellent narratives discussing them. Unfortunately, the arrangement of narrative/source listing/narrative/etc., does not lend itself to quick referral, especially in a format where the type causes eye strain. In spite of the format, it is still recommended for medical school libraries.

LINDA VINCENT

Biographical Sources

264. Medical Biographies: The Ailments of Thirty-Three Famous Persons. Philip Marshall Dale. Norman, OK: University of Oklahoma Press, 1987 (c1952). 267p., bibliog., index. $10.95pbk. 0-8061-2046-0pbk.

This biographical work, which analyzes "in the light of modern medical understanding" the physical ailments and causes of death of an ages-wide gamut of famous persons, considered chronologically, is written in a lively, narrative style. The author, a physician and former chief surgeon of a large international corporation, intended the work for the general reader, adding that "The eager appetite...for biographical fare, coupled with recent popular interest in medical science, has seemed to make it timely."

The use of anecdotes, quotations from autobiographical materials and other primary documents, and the author's philosophical comments make this a very readable and enlightening volume. Adding to the enjoyable quality of the book are interjections, such as "His marital record reads like a war casualty list," [Charlemagne's] and "Cellini's panicky reaction to illness suggests that he was basically something of a coward." Among the medical biographies included are that of Buddha, Henry VIII, Samuel Pepys, George Washington, Lord Byron, and James Garfield. The author treats the subject with sensitivity but not morbidity. Dale concludes with the statement, "He who has conquered the fear of death has, in a humanistic sense, conquered death itself." The volume includes an extensive "References and Sources" listing in alphabetical order by name of those whose biographies are included, and it has a subject index. Public and academic libraries will find this a useful circulating reference work.

JUANITA A. CUTLER

Dictionaries

265. Dictionary of Medical Eponyms. B. G. Firkin; J. A. Whitworth. Park Ridge, NJ: The Parthenon Publishing Group, Inc., 1987. 592p., illus. $48.00. 0-940813-15-7.

To quote the authors of this welcome reference tool, "Eponyms are unfortunately like the Joneses—no one can keep up with them." Nevertheless, Firkin and Whitworth significantly contribute to a better understanding and appreciation of the origin and context of medical eponyms with the publication of the *Dictionary of Medical Eponyms*. Medical eponyms are terms that most often, but not always, reflect the name or names of those persons responsible for the discovery of diseases or syndromes or for the development of tests or techniques. Some also refer to locations of disease occurrence, whereas others simply refer to anatomical features, both normal and abnormal.

In this book, Australians Firkin and Whitworth alphabetically list eponyms commonly used in the practice of internal medicine. Included with nearly every entry is the term or phrase, a definition in technical medical parlance, and a brief historical vignette explicating attribution, origin, and significance. Lacking are references to the literature where the term or phrase was originally coined. Included are black-and-white photographs for many entries depicting the persons for whom terms are named.

Given the general dearth of available medical eponym reference books, this thorough dictionary is destined to be appreciated, certainly by medical librarians, but also by health care practitioners and laypeople interested in the language of medicine.

GERALD J. PERRY

266. Logan's Medical and Scientific Abbreviations. Carolynn M. Logan; M. Katherine Rice. Philadelphia, PA: J. B. Lippincott Co., 1987. 673p., bibliog. $22.50. 0-397-54589-4.

This is a comprehensive compilation of over 20,000 medical and scientific abbreviations, initializations, acronyms, and symbols that have been drawn from over 60 sources, including "medical dictionaries, approved hospital lists, medical journals and texts, submissions from various medical specialties and various other technical listings, and from many hundreds of dictated consultations, history and physical reports, surgeries, and discharge summaries." One of the primary reasons for compiling this dictionary is that the authors encountered much inconsistency among the many sources, with jargon and abbreviations unique to one facility. Each abbreviation may have from one to many definitions. Only the identifying word or phrases are given, with no actual definition beyond this identification.

The authors are making a plea for consistency and have outlined what they feel should be the rules for using abbreviations. This is an admirable job and one that will take time for acceptance. Abbreviations used or devised by practicing physicians and other health care workers have often depended on personal preference; and trying to make these individuals adhere to a set of rules when writing reports, charts, and the like, is almost hopeless. Nevertheless, this attempt has produced an exceptionally useful dictionary to be used in conjunction with a dictionary that gives definitions. In addition to the straight alphabetical abbreviations, there is a section on symbols, covering the many meanings for the same symbol. There are also separate sections of abbreviations used for chemotherapy regimens, Latin terms used in charting and prescriptions, cancer staging abbreviations, and the elements. A short bibliography is included. The authors are interested in adding to this list and have included a form for submission of new abbreviations with their identifications or old abbreviations with new identifications. A very useful book for persons having to interpret reports or charts, it is recommended for public, college, academic, and special libraries.

H. ROBERT MALINOWSKY

267. The Medical Word Finder: A Reverse Medical Dictionary. Betty Hamilton; Barbara Guidos. New York, NY: Neal-Schuman Publishers, Inc., 1987. 177p. $45.00. 1-55570-011-X.

The vocabulary of medicine may be considered a marvel of eloquence and precision or terribly intimidating and confusing, relative to one's familiarity with the field. Authors Betty Hamilton and Barbara Guidos have listed in *The Medical Word Finder: A Reverse Medical Dictionary* 10,000 terms and phrases used in common parlance, along with the technical medical terminology to which they refer.

For both the layperson and the health professional needing to jog his or her memory, this word finder is an extremely handy supplement when using a medical dictionary, especially when embedded in medical definitions are terms and concepts equally as indefinable as the term originally searched. Hamilton and Guidos have included terms and phrases both broad and specific in scope. Words or phrases used for diseases, body parts, drugs, sensations, physical and emotional states, pathogens, symptoms, and treatments, as well as other concepts, are included.

For each entry, synonyms, related terminology, and, when appropriate, suffixes and prefixes used to qualify the meaning are given. Terms and phrases are alphabetized letter by letter. Particularly useful are lists of medications referred to in common usage by generic name. When looking up antihistamines, for example, one finds technical names for 36 different drugs.

This reverse medical dictionary is highly recommended for both public and medical library reference collections.   *GERALD J. PERRY*

Directories

268. ABMS 1987 Directory of Certified Internists. 2nd. Evanston, IL: American Board of Medical Specialties, 1987. 1,900p., index. $59.95. 0884-6448.

This is the second edition of the official directory of physicians certified as specialists (also called diplomates) in internal medicine. It is published by the American Board of Medical Specialties in collaboration with the American Board of Internal Medicine. The biographical data in the directory are provided, in most cases, by the physicians themselves, and the certification is verified by the American Board of Internal Medicine. In the cases where the physician did not provide any data, only the name and address appears; in a few cases, the physician chose not to be included.

The directory contains a biographic data section and a companion geographic cross-reference index. Physicians are listed alphabetically in the biographic data section with their complete record. In the geographic index, the physicians are listed, by name only, under their state and city. The physicians who are certified in one of the subspecialties of internal medicine will also be listed, by name only, in the separate subspecialty

geographic indexes which follow the main geographic index. In the general information section, there is a list of the 23 certification boards with addresses, a list of abbreviations, and a description of the American Board of Internal Medicine, including the members of the board, the areas of certification offered, the prerequisites for certification, and the procedures for applications.

The easy-to-use layout of the volume is pleasing, despite the terribly small print. This directory uses the same arrangement as the other specialty board directories published by ABMS. The *ABMS Compendium of Certified Medical Specialists* also uses this arrangement and actually incorporates the information from each of the individual specialty directories into one comprehensive set. Thus any library owning the *ABMS Compendium* would not want to purchase this or any of the other specialty directories.

This directory (or its related directories or the *Compendium*) is recommended for not only health sciences libraries, but also for public libraries, who have patrons wanting to check the credentials of their physicians or to locate a specialist in a particular city.   *JOAN LATTA KONECKY*

269. Barron's Guide to Medical and Dental Schools. 3rd. Saul Wischnitzer. New York, NY: Barron's Educational Series, Inc., 1987. 324p., illus., tables, maps, index. $9.95. 0-8120-3842-8.

Preparation for a career in medicine or dentistry often begins in high school, with students assiduously making career decisions early to enhance their chances of success in the rapidly changing, highly competitive medical and dental school marketplaces. Author Dr. Saul Sischnitzer, director of Health Careers Consultants of Queens, NY, has published the third edition of *Barron's Guide to Medical and Dental Schools* as a career decision-assistance tool for such students.

Included are data, mostly in easy-to-read tabular form, regarding current (as of publication) admissions to American Medical Association (AMA)-accredited medical schools, American Dental Association (ADA)-accredited dental schools, and American Osteopathic Association (AOA)-accredited osteopathic schools.

Profiles of each school include admissions requirements, faculties, and education expenses. Unique to the third edition is expanded attention to enrollment data regarding minorities and women. This information may have applications beyond those immediately apparent in this successful and inexpensive guide. Necessarily, most of the data covered will become dated. Additionally, each school's in-depth profile averages only about half a page of text, aside from the tables, charts, and maps. Still, this guide is highly recommended for undergraduate and high school library reference collections, career counselling services including high school guidance, and, of course, prospective students of medicine or dentistry. Educators in and students of the health professions may also benefit from the statistical data cumulated in the guide. Subject and school name

indexes, as well as appendices including sample admissions application forms and a listing of relevant professional societies, further enhance its usefulness. *GERALD J. PERRY*

270. Canadian Handbook of Medical and Surgical Specialists: Featuring Specialists of the Royal College and Corporation Professionnelle des Medécins du Québec.

Don Mills, Ontario: Southam Communications Ltd., 1987. 353p., index. $137.00pbk. 0835-250X.

This is the "First Annual Edition 1987-88" of a directory of specialists in the medical and surgical fields located in Canada. Each specialist is listed under one or more of 59 specialties. Included in the entry are full name, address, place and date of graduation, and telephone number. A geographic index lists just the names under each specialty within each province. Finally, the last part of the directory lists newly certified specialists as of 1986.

This is strictly a name directory and is intended to help the user locate a physician in a particular specialty in Canada either locally or nationally. The publishers warn, however, that individuals can be prosecuted if any attempt is made to sell, loan, or lease this list. The directory will find its greatest use in large academic and research libraries where referrals are made, but large public libraries may also find a use for the book. *H. ROBERT MALINOWSKY*

271. Canadian Medical Directory. 33rd.

Don Mills, Ontario: Southam Communications, Ltd., 1987. 658, 179, 95p. $125.00. 0068-9203.

This directory contains a warning not to use the contents of this directory to develop a mailing list of any kind for any purpose. Contents are "for the sole use of the purchaser." Sections include "How to Use the Directory with Abbreviations," "Certification," "Contents," and "Physicians of Canada (white pages)." Three sections are separated into white pages with alphabetical physician biographies; blue pages listing physicians by province and town (omitting "interns, hospital residents, retired doctors, and medical officers of the armed forces"); and buff pages with miscellaneous information, including "those doctors who graduated from Canadian medical schools in 1986" (they are not listed in the white or blue pages), "medical officers of the armed forces," and information "about health departments, universities, etc." Not every specialty collection will need this regional volume. *JAMES L. CRAIG*

272. Directory of Biomedical and Health Care Grants 1988. Phoenix, AZ: Oryx Press, 1988. 408p., index. $74.50pbk. 0-89774-383-0. 0883-5330.

This is another directory from the GRANTS database maintained by Oryx Press containing 2,316 health-related funding programs ranging from laboratory investigations to programs that are designed to study the needs of society in health care delivery. "Special efforts have been made to increase coverage of such areas as clinical and programmatic studies in gerontology and mental health; clinical studies of the cause, detection, and elimination of cancer; health care delivery and maintenance; programs researching all areas relating to Acquired Immunodeficiency Syndrome (AIDS)."

The entries are arranged in alphabetical order by name of grant program. Each entry includes grant title, accession number (used in indexing), grant description, restrictions, requirements, application/renewal data, funding amount, *Catalog of Federal Domestic Assistance* page number, and sponsor information, including contact persons and address and telephone number. Three indexes cover subject, sponsoring organizations, and sponsoring organizations by type.

This is an excellent guide for individuals and institutions to locate sources of funding. It is highly recommended for all academic and research libraries. *H. ROBERT MALINOWSKY*

273. Federal Health Information Resources. Melvin S. Day, ed. Herner and Co., comp. Arlington, VA: Information Resources Press, 1987. 246p., index. $29.50. 0-87815-055-2.

The federal government sponsors hundreds of agencies, some well known, others obscure, which may provide information and reference inquiry services in the health care field. The information from these agencies may go untapped because of a lack of awareness. This volume identifies and describes the "major sources of biomedical and health information produced or maintained by agencies or contractors of the Federal Government." The information has been drawn from *Health Information Resources in the Federal Government*, DIRLINE (Directory of Information Resources Online), records of the National Referral Center of the Library of Congress Science and Technology Division, and directories maintained by federal departments and agencies.

The well-written source descriptions are arranged alphabetically by the official resource name within 72 broad subject areas. Each entry lists the parent or generic name for the agency; address; telephone number or numbers;, and description of the subject scope, services, and, when applicable, databases, publications, or holdings. These descriptions are the key to the usefulness of the book, alerting the user to the availability of professional materials, consumer items, publication lists, databases which may be available commercially, and those sources which provide literature searching or accept reference inquiries. An appendix alphabetically lists the databases described in the main entries, giving complete descriptions and availability. The subject/title index is cross-referenced and provides multiple access points to the information sources and services. The agency/organization index includes the pri-

mary resource names as well as those that were mentioned in the main entries. The acronym list is limited to those used in the book. This should prove to be a useful book for those who are not intimately familiar with a major portion of the potential information sources, databases, and publications available from the federal government. *JOAN LATTA KONECKY*

Encyclopedias

274. Encyclopedia of Medical Organizations and Agencies: A Subject Guide to More than 11,000 Medical Societies, Professional and Voluntary Associations, Foundations, Research Institutes, Federal and State Agencies, Medical and Allied Health Schools, Information Centers, Data Base Services, and Related Health Care Organizations. 2nd. Anthony T. Kruzas; Kay Gill; Robert Wilson, eds. Detroit, MI: Gale Research Co., 1987. 975p., index. $185.00. 0-8103-0324-8.

The *Encyclopedia of Medical Organizations and Agencies* (*EMOA*) "is intended to provide a single, comprehensive source of subject-classified information on health-related organizations and agenicies." These organizations and agencies include national and international associations, state and regional associations, foundations and funding organizations, educational and training programs, state government agencies, federal government agencies, information and database services, and research centers and institutes. There are over 10,500 entries in this second edition of *EMOA*.

The arrangement is by 78 subject areas that include the following categories: disease conditions, medical specialties, social health programs, and special aspects of medicine in general. All 78 subjects are listed in the table of contents and should be consulted as the first step in using this encyclopedia. There is also a subject cross-index that lists topics and indicates in which of the 78 subject categories one can find information on that topic. Each entry has full name, address, telephone number, and name of key administrator. Additional information may include year founded, staffing, members, regional or local groups, publications, description of the agency, research description, and computer-based products and services. There is a master name and keyword index. Although many of the entries in this encyclopedia can be found in the *Encyclopedia of Associations*, it is a recommended reference source because of its comprehensiveness. Researchers and librarians will find it very useful in locating information about a facility in the health-related fields. *H. ROBERT MALINOWSKY*

Exam Reviews

275. Barron's How to Prepare for the Medical College Admission Test MCAT. 5th. Hugo R. Seibel; Kenneth E. Guyer. New York, NY: Barron's Educational Series, Inc., 1987. 396p., illus., tables. $8.95. 0-8120-2989-5.

The Medical College Admission Test (MCAT) is frequently used to gauge a student's competitive ability to succeed in medical school. The MCAT attempts to objectively measure knowledge of the basic sciences, as well as ability to critically analyze and read. Some educators believe students with high scores have a good chance of surviving the rigors of medical school. Authors and medical school professors Dr. Hugo Seibel and Dr. Kenneth E. Guyer have published the 5th edition of the popular *Barron's How to Prepare for the MCAT (Medical College Admission Test)* to assist students preparing for this challenging test.

Included in the 5th edition are suggested test-preparation and test-taking strategies, reviews in science (biology, chemistry, and physics) and mathematics, and most important, four model examinations. The authors suggest that students take the sample tests as though they were the actual MCAT and use their scores to determine areas of weakness for additional preparation and study. According to Seibel and Guyer, "By proper and careful preparation utilizing all possible modes, the individual is simply presenting his or her true potential for the study of medicine." The guide is well organized, to the point, and written in an easy-to-read, crisp style. Undergraduate libraries, medical school educators, and potential students of medicine are the recommended audiences for this inexpensive, useful Barron's guide. *GERALD J. PERRY*

276. Rypins' Questions and Answers for Boards Review: Basic Sciences. Edward D. Frohlich, ed. Philadelphia, PA: J. B. Lippincott Co., 1987. 179p. $16.95pbk. 0-397-50823-9pbk.

This is a question-and-answer review supplement to the fourteenth edition of the textbook, *Rypins' Medical Licensure Examinations.* The sole purpose of this review text is to "provide to the person preparing for licensure examination an opportunity for pretesting and a means to identify areas of possible weakness so that he or she may return to the review chapters and other source material as necessary."

After a brief introductory chapter that explains the purpose of the examination and the forms of the test questions, 11 contributors have put forth a series of questions covering anatomy, physiology, biochemistry, general microbiology, immunology, pathology, pharmacology, public health, community medicine, and behavioral sciences. The answers are given at the end of each section of questions. This is a recommended book

for any medical, university, or special library where individuals will be involved in taking the medical licensure examinations.

H. ROBERT MALINOWSKY

Handbooks

277. The Essential Book of Traditional Chinese Medicine. Liu Yanchi. Kathleen Vian; Peter Eckman, eds. New York, NY: Columbia University Press, 1988. 2v., illus., bibliog., index. $40.00 v. 1, $80.00 v. 2. 0-231-06520-5 set, 0-231-06196-X v. 1, 0-231-06518-3 v. 2.

This excellent encyclopedic handbook of Chinese Medicine was translated by Fan Tingyu and Chen Laidi and took some six years to produce as a cooperative project between the People's Medical Publishing House and the United States-China Educational Institute. The information in *The Essential Book of Traditional Chinese Medicine* is a result of documented information developed over the past 4,000 years by scholars and medical practitioners in China using herbs, acupuncture, and intelligent lifestyle practices.

Liu Yanchi is an associate professor and head of the teaching section on the basic theory of traditional Chinese medicine at the Beijing College of Traditional Chinese Medicine and a contributor to the *China Medical Encyclopedia* and the *Digest of Traditional Chinese Medicine*.

Volume 1 covers theory with eight chapters on "Traditional Chinese Medicine: An Orientation," "*Yin-Yang* and the Five Elements," "The *Zang-Fu* System," "The Channel System," "Etiology: The Cause of Disease," "Pathogenesis: The Course of Disease," "Methods of Examination," and "Diagnosis: The Differentiation of Symptom-Complexes." This is a fascinating book, with every chapter interestingly filled with Chinese medical knowledge.

The chapter on causes of diseases is divided into exogenous and endogenous. Exogenous causes are environmental and related to wind, cold, heat, dampness, dryness, and fire. On the other hand, endogenous causes are emotional and related to the emotions of joy, anger, sadness, pensiveness, grief, fear, and fright. Each of these environmental excesses and emotions "affects a specific *zang* organ. Accordingly, if the cause of the illness is known, the location of the disease can be determined. Conversely, if the location is known, the cause can usually be deduced."

Each chapter has the same format—general information, specific information, and summary. Excellent line drawings are used to help explain the procedures that are used.

Volume 2 covers clinical practice and has Barbara Gastel as an editorial consultant. Here the author discusses general treatment and treatment of specific diseases, most of which are common, such as pneumonia, common cold, nephritis, impotence, mumps, dysmenorrhea, and appendicitis. The seven chapters cover "Treatment: Principles and Basic Methods," "Chinese Medicinal Herbs: Basic Concepts and Common Examples," "An Introduction to Traditional Chinese Prescriptions," "Treatment of Some Common Conditions," "Treatment of Common Communicable Disease," "Treatment of Common Gynecologic Disorders," and "Treatment of Other Selected Conditions."

The chapter on herbs includes a detailed chart listing the various herbs that are used, giving property, flavor, channel acted on, action, indication, amount, and remarks. This chart is to be used in conjunction with the one in the chapter on prescriptions, which gives the prescription followed by composition, amount, action, indication, and remarks. The chapters on specific diseases are fascinating, describing each disease by giving etiology, essentials of diagnosis and treatment, and then the specific treatments through the use of herbs and acupuncture.

There are two appendices in the second volume covering names of herbs and names of prescriptions. These are both lists with no reference to any part of the text.

The Essential Book of Traditional Chinese Medicine "provides a never-before-attempted integration of Western scientific and Eastern traditional concepts." Although the use of Chinese medicine is not totally accepted by medical professions in other parts of the world, there is great merit in studying this field. Researchers are more and more using the knowledge of herbs that have been used for centuries in China and determining why they work the way they do.

In view of all of this, this book is "an enlightening and very readable introduction—for the physician, scholar, or lay reader—to the system of medical care believed in and used by one quarter of the world's population." This is a must book for all medical libraries and one that would be useful for its readable content in public and academic libraries.

H. ROBERT MALINOWSKY

Histories

278. The Age of Miracles: Medicine and Surgery in the Nineteenth Century. Guy Williams. Chicago, IL: Academy Chicago Publishers, 1987 (c1981). 234p., illus., index. $16.95, $8.95pbk. 0-89733-286-5, 0-89733-285-7pbk.

Guy Williams chronicles the amazing evolution of medicine from the time of internal "humours," treatment by blistering, cauterising, purging, and bleeding via leeches to the dawning of scientific medical investigations and modern medical practices. Written in a lively and informative manner, the book covers a wide range of topics, including anaesthetics, antiseptics, orthopedics, birth, hospitals, the mentally ill, xrays, and blood transfusions. The text will give the reader a vivid understanding of the medical and social environment of the time and a healthy appreciation of modern medicine. Simply put, this book will fascinate the average reader and the health professional alike and is enthusiastically recommended for public, college, and university libraries.

JOAN LATTA KONECKY

Manuals

279. Accreditation Requirements Manual of the American Association of Blood Banks. 2nd. Randi V. Rosvoll; Leonard Boral; Kenneth J. Fawcett; Grace M. Neitzer; Jerry Kolins; Robert L. Thurer; Margaret E. Wallace; Richard H. Walker, eds. Arlington, VA: American Association of Blood Banks, 1987. 198p., index. $18.00pbk. 0-915355-33-7.

"Requirements and recommendations for AABB accreditations of blood banks and transfusion services are established by the National Committee on Inspection and Accreditation in cooperation with the Committee on Standards and the Committee on Histocompatibility Testing and approved by the Board of Directors." There were a sufficient number of changes in the twelfth edition of the *Standards for Blood Banks and Transfusion Services* that required a new edition of the Inspection Report Form (IRF) and a second edition of the *Accreditation Requirements Manual of the American Association of Blood Banks*. This detailed manual is to help in filling out the Inspection Report Form. It resembles the form with the same corresponding numerals and letters. Each question in the form is quoted verbatim in the "Items" and then followed by "References" which refer to the specific sections of the *Standards*. Following this are "Explanations" which detail what the facility must do to meet accreditation requirements or could do to meet the recommendations. Throughout the manual are instructions "must," "shall," and "are to be." These indicate requirements that have to be met. In the case of recommendations, the word "should" is used.

This is a unique and essential book for any blood bank that is requesting accreditation. Its use is in those facilities, but libraries serving medical schools and large academic libraries should have a copy of this manual for reference purposes.   *H. ROBERT MALINOWSKY*

280. Guide to Planning and Managing Multiple Clinical Studies. Bert Spilker. New York, NY: Raven Press, 1987. 410p., illus., index. $36.50. 0-88167-264-5.

The third volume in a group of guides to clinical studies and written for a broad range of investigators, including researchers from private practice, academic institutions, private industry and government institutions, this book is designed to be a practical review, from a nonstatistical viewpoint, of the factors involved in planning, conducting, and managing multiple clinical studies.

Dr. Spilker begins by discussing the steps and procedures for choosing a project or drug for evaluation; establishing the goals, strategies, and approaches; and designing the total project based on those elements. He then describes techniques for recording, coding, processing, and analyzing the data retrieved from multiple studies and for integrating efficacy and safety data. The initial section concludes with a discussion of the ethical issues involved in single and multiple clinical studies. The second section is devoted to the issues related to the study of surgical procedures and medical devices, the development of orphan drugs, and studies involving special patient populations or worldwide projects. In addition, considerations are presented for planning pharmacokinetic or combination drug studies and for comparing different dosage forms or treatment modalities.

Management of multiple studies is the focus of the third section, including coordination and allocation of resources, coordination of multiple projects, and monitoring of multiple clinical studies. The author also makes recommendations for managing the costs of projects, clinical studies, and drug development. The guide, logically planned and very well written, clearly reflects the experience and knowledge of the author, who is the head of the Department of Project Coordination at the Burroughs Wellcome Company. Very practical information, with an emphasis on pharmaceutical clinical studies, is accompanied by an abundance of helpful tables and illustrations. The appendix consists of the Declaration of Helsinki, several tables, and several application forms, and the tables include a summary of the contents of a new drug application, a registration dossier for the European Economic Community, and an IND application. There is a lengthy references section and a subject index.

Because of the wealth of practical information, the clarity of the writing, and the excellent organization of the information, *Guide to Planning and Managing Multiple Clinical Studies* will be extremely useful for the investigators, sponsors, and managers of multiple clinical studies. This guide is recommended for collections that support clinical research, especially pharmaceutical clinical research.   *JOAN LATTA KONECKY*

281. The Marshall Cavendish Science Project Book of the Human Body. Steve Parker. Freeport, NY: Marshall Cavendish, 1988 (c1986). 43p., illus. (col.). $12.49. The Marshall Cavendish Library of Science Projects, v. 6. 0-86307-630-0.

This encyclopedic-project book for young people covers the human body. Through a series of chapters that explain the body system, touch, hearing, the brain, hormones, sex, body maintenance, lungs, heart, blood, digestion, urinary tract, bones, and muscles, the student learns how the human body is put together and functions. Experiments and projects are used to exemplify how some of the body parts work. The well-written text and good illustrations make this a useful reference book for school libraries and an excellent textbook for the classroom.

H. ROBERT MALINOWSKY

Textbooks

282. **Invitation to Medicine.** Douglas Black. New York, NY: Basil Blackwell, Inc., 1987. 223p., illus., glossary, index. $19.95pbk. Invitation Series. 0-631-14765-9pbk.

This book was intended to display the "flavour of medical practice" and was written for students considering medicine as a career, scientists working on medical research, and laypersons with an interest in medicine. Divided into three parts, the book begins with a discussion of the broad scientific principles which underlay the practice of medicine, including genetics, environment, and pathology. Part 2 addresses "the application of medical knowledge to individual patients," or in other words, the actual practice of medicine. This set of chapters reviews diagnosis, history taking, physical examination, management, drug treatment, surgery, and patient communication. Public health and the relationship between medicine and society in general are presented in the third part. The book also contains a section of suggested further readings, a glossary, and an index.

Invitation to Medicine contains a quantity of good information, but the word choice and the style of writing are too ponderous for much of its intended audience. In addition, the British orientation may be considered a drawback by many American readers. Good intentions notwithstanding, the volume does not manage to convey a taste of the excitement and the frustrations found in the practice of medicine. It is a good overview, but not recommended for precollegiates or the average layperson, or the libraries which serve these patrons.   *JOAN LATTA KONECKY*

Thesauri

283. **MASA: Medical Acronyms Symbols and Abbreviations.** 2nd. Betty Hamilton; Barbara Guidos. New York, NY: Neal-Schuman Publishers, Inc., 1988. 277p. $45.00. 1-55570-012-8.

MASA is a collection of 32,000 acronyms, symbols, and abbreviations used in all major medical specialties and related fields. This edition includes 12,000 more entries than were in the first edition. The list was compiled by extensive exploration of medical journal literature as well as medical books. Items of recent use are included along with long-standing terms and phrases. The arrangement is alphabetical, and numbers are ignored in any entry for the sake of filing by the letters. Differences affecting the filing system occur in letters with periods, acronyms with numbers, and lower-case letters; for example the order would be "DP," "D.P.," "2,4-DP," "dp."

Additionally, all known uses of a set of letters are included (EC has 37 uses), "see" references direct the user from less preferred forms to preferred forms, and "see also" references link related forms. At the end of the acronyms and abbreviations is a short list of symbols.

There would be a number of uses for this book. The reference shelf would be one, but the book should also be in offices, medical departments, and record rooms. It is a very useful tool.

LILLIAN R. MESNER

★
INDIVIDUAL DISEASES
Atlases—Medical

284. **Adnexal Tumors of the Skin: An Atlas.** Kinya Ishikawa. New York, NY: Springer-Verlag, 1987. 133p., illus., bibliog., index. $75.00. 0-387-70019-6.

Kinya Ishikawa's *Adnexal Tumors of the Skin: An Atlas* is a special visual survey of the microstructure of skin appendages and associated tumors. Included are 116 photomicrographs taken by the author depicting in great detail both normal and pathologic or tumorous cellular structures.

The first section of this high-quality atlas contains normal structure illustrations of the hair follicle, apocrine, and eccrine glands. The second section illustrates follicular, sebaceous, aprocine, eccrine, and basal cell tumors, drawn from the author's clinical experience. Included with each micrograph are references to the medical literature for the past five years. Each illustration may be easily found using either the detailed table of contents or subject index.

The high-contrast, micrographic illustrations, complemented by their convenient organization, make this book a recommended reference tool for both dermatologic oncologists and medical library collections supporting research in skin cancer.

GERALD J. PERRY

285. **Kaposi's Sarcoma: A Text and Atlas.** Geoffrey J. Gottliev; A. Bernard Ackerman, eds. Philadelphia, PA: Lea and Febiger, 1988. 330p., illus. (part col.), bibliog., index. $98.50. 0-8121-1041-2.

Until the advent of AIDS, Kaposi's sarcoma was an infrequently encountered disease. Now, however, it is a frequent AIDS-related disease, and the number of new books on the subject is few. Gottliev and Ackerman have compiled an excellent text with accompanying atlas that should be useful in hospitals, clinics, physicians' offices, and medical libraries. The text is technical and intended for researchers and physicians.

The atlas's photographs are clear and well described, with many in color. Twenty-six contributors have helped to put together this authoritative reference source. The titles of the 18 chapters give an excellent insight to what is covered in this book: "A Riddle within a Puzzle," "Kaposi: The Man and the Sarcoma," "On the Name Kaposi," "Atlas of the Gross and Microscopic Features," "Atlas of the Gross and Microscopic Features of Simulators," "Systemic Manifestations," "Electron Microscopy," "Conventional Kaposi's Sarcoma in Africa," "Atypical Aggres-

sive Kaposi's Sarcoma in Africa," "Kaposi's Sarcoma in AIDS," "Infectious Diseases in AIDS," "Immunopathologic Aspects," "Malignant Lymphoma in AIDS," "Ultrastructural Changes in AIDS and Their Etiologic Significance," "Immunosuppressive Therapy," "Epidemiology," "Autopsy Findings in Patients with AIDS," and "AIDS and Ethics."

The last chapter on ethics is excellent and sums up the problems that the medical profession has in dealing with AIDS or any other infectious disease: "As in many ethical issues, a precise solution may never emerge, but as in all ethical discourse, respect for the dignity and freedom of individuals and concern for a fair and just distribution of burdens and benefits must rule the debate." This is a highly recommended book for university and medical libraries.

H. ROBERT MALINOWSKY

Bibliographies

286. **A.I.D.S.: A Bibliography.** Robert D. Reed, comp. Saratoga, CA: R and E Publishers, 1987. 64p. $10.00pbk. 0-88247-757-9pbk.

This is a small selective bibliography of material on AIDS. Most of the information is available in public libraries and some in school libraries with any academic library having all of the information. It is arranged alphabetically, by author within five chapters: "Magazine Articles," "Newspaper Articles," "Books," "Selected Publications by Date," and "A Scientific and Clinical Bibliography." An excellent historical bibliography intended to supplement all other bibliographies that have been published.

H. ROBERT MALINOWSKY

287. **AIDS 1987 (Acquired Immune Deficiency Syndrome).** David A. Tyckoson. Phoenix, AZ: Oryx Press, 1988. 159p., index. $29.50pbk. Oryx Science Bibliographies, v. 11. 0-89774-434-9pbk.

The Oryx Science Bibliographies are intended to bring to the reader the most recent references on current issues. This is the third bibliography published by Oryx Press covering AIDS. Each entry has been evaluated and fully annotated. References are those that are most readily available at public and college libraries.

AIDS 1987 contains 637 citations to articles about AIDS. The entries are within 31 separate topical sections including medical and health care, transmission, AIDS in women and children, IV drugs, AIDS in minorities, AIDS blood test, AIDS in schools and colleges, AIDS hysteria, costs, AIDS in correctional facilities, funding, and AIDS in animals. The entries have been well selected and present an excellent cross-section of what has been published during the period covered. These bibliographies are highly recommended for public and academic libraries as well as school libraries. *H. ROBERT MALINOWSKY*

Directories

288. **A.I.D.S.: How and Where to Find Facts and Do Research.** Robert D. Reed. Saratoga, CA: R and E Publishers, 1986. 40p., bibliog. $4.00pbk. 0-88247-758-7pbk.

This is a small, very brief, and very general list of sources for locating information on AIDS. It gives, prior to the sources, brief information on what AIDS is, how it affects the body, symptoms, statistics, and other bits of information. There are only 39 general sources, but they are sources that most laypeople and school children will be able to locate and use. This book should be supplemented with current information, but it is still a useful directory for school and public libraries. *H. ROBERT MALINOWSKY*

Encyclopedias

289. **Communicable Diseases.** Thomas H. Metos. New York, NY: Franklin Watts, 1987. 96p., illus., bibliog., glossary, index. $9.90. A First Book. 0-531-10380-3.

This is an excellent book for schools and public libraries covering the various communicable diseases from common cold to the deadly AIDS. It is extremely well written in clear language for secondary school children and adults with no scientific background. The facts are presented briefly but completely and illustrated with excellent photographs.

The 12 chapters cover "Causes of Communicable Diseases," "Classification of Diseases," "Defense Mechanisms," "Diagnosis and Treatment," "Diseases You Can Catch Again and Again," "Childhood Diseases," "Medical Progress can be Hazardous to Your Health," "The Great Plagues and Epidemics," "Deadly Communicable Diseases," "Sexually Transmitted Diseases," "Is It Catching?" and "The Future." The four pages devoted to AIDS are particularly well written and the facts are given with no reference to any one high-risk group. Instead it indicates what AIDS is, how it is contacted, and what is being done to combat it including mention of the drug AZT.

A very good glossary of terms associated with communicable diseases is included. This is a highly recommended book for school and public libraries. *H. ROBERT MALINOWSKY*

290. **Questions and Answers on AIDS.** Lyn Robert Frumkin; John Martin Leonard. Oradell, NJ: Medical Economics Books, 1987. 190p., bibliog., glossary, index. $21.95pbk. 0-87489-461-1pbk.

There are numerous books published on AIDS, each in its own format and each covering the topic from a different perspective. Frumkin and Leonard have put together a unique "encyclopedia" of information about AIDS, gleaned from all of the past research, in question-and-

answer format. It was intended for health care workers, but its format and the amount of information that it contains makes it a welcome book for any library.

The questions and answers are grouped within ten chapters: "What Is AIDS?" Definitions and the Origin of the Syndrome," "How AIDS Manifests Itself: The Five Principal AIDS-Related Conditions," "Who Gets AIDS? Groups at Risk," "How AIDS Is Acquired: Modes of Transmissibility" "Exposure to the AIDS Virus: The Meaning of Antibody Positivity," "Protecting the Individual and Health Care Worker," "The Epidemiology of AIDS," "Research and Funding for AIDS," "Resource Centers for AIDS Information and Support," and "Epilogue."

At the end is a list of references, each keyed to a question that was asked, thus documenting the answers to the questions. A glossary and index round out this book. It is hoped that the authors will publish a new edition as new information surfaces on this disease. It is important that everyone be kept abreast of the latest developments in the research of AIDS. This book is intended for those working in the health professions, but it has great value as a reference book in all types of libraries.

H. ROBERT MALINOWSKY

Field Books/Guides

291. **Stroke: A Guide for Patient and Family.** Janice Frye-Pierson; James F. Toole. New York, NY: Raven Press, 1987. 211p., illus., bibliog., glossary. $25.00, $18.50pbk. 0-88167-279-3, 0-89004-637-9pbk.

"Stroke remains one of the least manageable illnesses in the United States....More than 175,000 individuals die of stroke every year, making it the third leading cause of adult death in the United States." Several good books on dealing with stroke have been produced for the general, interested public, but a wide audience continues to need this type of book. For them, this book is a good starting point. It introduces the relevant areas with controlled amounts of detail, references, addresses of contacts, and complex terminology, so that further information seeking is helped. The authors are a neurological nurse clinician and a neurologist, James F. Toole, who is author of an important text on cerebrovascular disorders. The contributors' skills reflect the scope of the book; they include a rehabilitation nurse, a physical therapist, an architect, a speech language pathologist, a psychiatrist, and an ophthalmologist. *RUTH LEWIS*

Handbooks

292. **AIDS: Etiology, Diagnosis, Treatment, and Prevention.** Vincent T. DeVita, Jr.; Samuel Hellman; Steven A. Rosenberg, eds. Philadelphia, PA: J. B.

Lippincott Co., 1985. 352p., illus., bibliog., index. $42.50. 0-397-50697-X.

Although books on AIDS become dated very quickly, this one is still useful in that it gives a good amount of historical information that is needed to fully understand the complexities of this disease. The editors "have attempted to provide a comprehensive source of information on all aspects of AIDS written by the clinicians and scientists who have made the central contributions in this field." Each of the chapters is authored by a specialist who has worked with AIDS: (1) "The Epidemiology of AIDS and Related Conditions"; (2) "The Etiology of AIDS"; (3) "Prospects for Diagnostic Tests, Intervention, and Vaccine Development in AIDS"; (4) "Immunologic Features of AIDS"; (5) "Pathologic Features of AIDS"; (6) "Infectious Complications of AIDS"; (7) "Kaposi's Sarcoma in AIDS"; (8) "Malignant Neoplasms in AIDS"; (9) "The AIDS-Related Complex"; (10) "Treatment of immunologic Disorders in AIDS"; (11) "Safety Precautions for Dealing with AIDS"; (12) "Psychosocial Issues in AIDS"; and (13) "AIDS: A General Overview."

This is a technical book for clinicians and not one that the layperson will be able to understand. It was accurate at the time of publication; but it should be emphasized that, with any book that deals with AIDS treatment, one should always seek out supplemental texts and handbooks to make sure of having the latest information. A bibliography of 199 entries and an index round out this useful handbook, recommended for retrospective coverage of AIDS information.

H. ROBERT MALINOWSKY

293. **AIDS: The Ultimate Challenge.** Elisabeth Kubler-Ross. New York, NY: Macmillan Publishing Co., 1987. 329p. $17.95. 0-02-567170-7.

Elisabeth Kubler-Ross is an exceptional person who, for the past 20 years, has been involved in caring for terminally ill patients, both adults and children. She states, "My goal has been, and still is, to educate health-care professionals as well as clergy to become more familiar with the needs, concerns, fears, and anxieties of individuals (and their families) who face the end of their lives." In 1969 she published *On Death and Dying*. It is not surprising that she embarked on writing a book about AIDS. Her insights into dying are remarkable, and the wealth of information that she presents in this book should be of great value to counselors of every type, whether medical or clergical or personal volunteer.

In *AIDS: The Ultimate Challenge* she brings her experience in working with people, many of who lived only a short time after she met them. She also brings an insight into the problems of people who work with AIDS patients in finding housing, health care, compassion, and companionship. The the chapters cover "Working with AIDS Patients," "Parents of Children with AIDS," "Children and AIDS," "Babies with

AIDS: We the People," "Letters to the Editor," "Finding a Way," "Women and AIDS," "Letting Go," "The Birth of Support Systems," and "AIDS in Prison."

Her most pressing concern is making sure that no one is unwanted just because they have AIDS. To die alone and unloved by those who you have lived with is one of the saddest things in a lifetime. Especially sad, she says, is when babies are unwanted by their mothers, shunned by foster families and adoption agencies, and left unmothered and unloved to die alone in hospitals.

This is a truly moving book that should be on the shelves of every public library. Its ramifications go beyond AIDS patients to anyone who is terminally ill.   *H. ROBERT MALINOWSKY*

294. AIDS: What Does It Mean to You?
Rev. Margaret O. Hyde; Elizabeth H. Forsyth. New York, NY: Walker and Co., 1987. 116p., illus., glossary, index. $12.95. 0-8027-6699-4.

AIDS: What Does It Mean to You? is a popularly written, general information handbook. Its primary audience is young adults who may have many misconceptions about AIDS. It is written in straightforward language covering history, possible causes, research needs, and the epidemic of fear. There are chapters on "Living with AIDS," "Avoiding AIDS," "Plagues in Other Times," "AIDS International," "An Epidemic of Fear," "Medical Progress," and "AIDS: A Challenge for the Future."

Margaret O. Hyde and Elizabeth H. Forsyth are excellent writers for young people and have collaborated on other books, including *Know Your Feelings, Suicide: The Hidden Epidemic,* and *Terrorism: A Special Kind of Violence.* All school and public libraries should have several copies of this very well-written book.
          H. ROBERT MALINOWSKY

295. A.I.D.S.: Your Child and the School. Danek S. Kaus; Robert O. Reed. Saratoga, CA: R and E Publishers, 1986. 24p., illus., bibliog. $3.00pbk. 0-88247-756-0pbk.

This small handbook is meant to be a guide for teachers and parents. It presents the facts in a straightforward text that is accurate and easy to understand. Information included covers the definition of AIDS, symptoms, statistics, AIDS by ethnic group and gender, who is at risk, guidelines for admitting children with AIDS to schools, and some common sense and caring information.

This is a very useful small book that should be made available to all parents who have a concern about AIDS in schools. School and public libraries will want to have several circulating copies as well as one on reserve.
          H. ROBERT MALINOWSKY

296. AIDS, Acquired Immune Deficiency Syndrome, and Other Manifestations of HIV Infection: Epidemiology, Etiology, Immunology,

Clinical Manifestations, Pathology, Control, Treatment and Prevention.
Gary P. Wormser; Rosalyn E. Stahl; Edward J. Bottone, eds. Park Ridge, NJ: Noyes Publications, 1987. 1,103p., illus., bibliog., index. $98.00. 0-8155-1108-6.

The commendable purpose of this book is embodied in its dedication "to the health care workers throughout the world engaged in the management of victims of HIV infection—that they may render, with compassion, the highest level of care possible." The editors intend the book to be a "reference source for the essential information needed by most practitioners and specialists," unexpectedly confronted with the need to treat AIDS patients. They also explain that this comprehensive overview was needed because of the large volume of literature on the subject and the special nature of many of the clinical problems experienced by AIDS patients, combined with a lack of familiarity with such conditions on the part of some health care professionals. *AIDS* is an enormous volume that may provide more information than is needed by most practicing physicians. Dr. James W. Curran, director of the AIDS Program, Centers for Disease Control, concluded that the reference "provides comprehensive and up-to-date information on the subject. This text should greatly assist health care workers in fulfilling their dual roles as providers of care and sources of accurate information to patients." Contributors include physicians practicing in regions where AIDS has been prevalent, physicians and scientists in public health agencies extensively involved with AIDS control, and basic scientists in laboratories working on either HIV or animal retroviruses with related properties.

A detailed Table of Contents provides a convenient guide to location of information. Five major subject areas are identified, and the titles of the 54 sequentially numbered chapters as well as all subheadings within each chapter are listed. The largest section, part 4, about 40 percent of the book, is devoted to "Clinical Manifestations." These 21 chapters address the nature of the HIV infection (general considerations, in the central nervous system, and in children); specific opportunistic infections; malignancies; and involvement of pulmonary hepatic and neurological systems. Other chapters include "Treatment and Prevention of HIV Infection," "Epidemiology" and discussion of retroviruses and HIV origins.

Some features of the citation of references are somewhat inconvenient for the investigator interested in the primary literature. Occasional statements made in various chapters are not clearly documented. This may be because of lengthy discussions of tables in which the source is identified, but in other cases no source indicated in the table. Unfortunately, references are indicated numerically, in order of appearance—requiring the reader to consult the list to identify authors and making reference retrieval by author name tedious. No more than three authors are given in any citation, thus further complicating contributor identification and retrieval. The book has been described as an "interim report," and the editors add a "Notice" to the reader to be aware

of subsequent reports which may contradict findings reported in this document or alter the interpretation of current information. They further caution that each individual must bear the responsibility for recognition of risks and take prudent action to avoid exposure to, or transmission of, the virus. This is a recommended treatise for academic and special libraries. *PAT JAMESON*

297. AIDS and the Law. William H. L. Dornette, ed. New York, NY: John Wiley and Sons, 1987. 375p., index. $95.00. "Wiley Law Publications"; Medico-Legal Library. 0-471-85740-8.

"AIDS is not spread by common everyday contact but by sexual contact. There is great misunderstanding resulting in unfounded fear that AIDS can be spread by casual, non-sexual contact. This misunderstanding and associated fear in turn have caused much controversy and litigation." The main purpose of this book is to educate lawyers, physicians, and laypersons in the laws as they affect those exposed to or infected with the AIDS virus. U.S. Surgeon General C. Everett Koop's sensible yet sensitive foreword urges that "We are fighting a disease, not people," a fact often forgotten in the furor over the disease.

William H. L. Dornette has done an excellent job of presenting the facts behind AIDS. He also does a remarkable job in explaining the medical background of infections with the AIDS virus and the associated legal issues to members of the legal and medical professions and also to the nonprofessionals. The 15 detailed and well-written chapters cover subjects from "The Medical Background,""Introduction to the Law," "AIDS in the Workplace," "Educating the Infected Child," "Housing the AIDS Victim," "AIDS and the Family," "Discrimination against the Handicapped," to "Negligence and Intentional Torts."

This is by far one of the most readable, straightforward accounts of AIDS and its legal ramifications. The author has brought together in one volume a wealth of information that is educational, accurate, and timely. Anyone who has any doubts about AIDS should be guided to this excellent book to learn the facts. All libraries should have a copy on their reference shelves.

H. ROBERT MALINOWSKY

298. AIDS and the Law: A Guide for the Public. Harlon L. Dalton; Scott Burris; Yale AIDS Law Project, eds. New Haven, CT: Yale University Press, 1987. 382p., illus., index. $22.50, $7.95pbk. 0-300-04077-6, 0-300-04078-4pbk.

"This book is about law. It is not, however, aimed exclusively (or even primarily) at those who are steeped in the law. Rather, it is meant for whoever has a professional need to come to grips with the legal issues spawned by the AIDS epidemic—for educators, counselors, legislators, policy makers, law enforcement and corrections officials, public health officials, health care providers, social service providers, research scientists,

employers, employee representatives, insurers, providers of goods and services, social workers, social scientists, social activists, representatives of interest groups, the staffs of drug treatment programs, members of AIDS support groups, and, of course, lawyers for any and all of the above."

To put together a book that would appeal to each and every one of these groups was a monumental task for the Yale AIDS Law Project, composed of members of the Yale Law School community. They have written the book with as little legal jargon as possible. They have also tried to cover as many of the topics as possible but admit that some topics have only been glossed over, and others have not had their own chapters, such as AIDS and the Latino communities and licensing of new drugs.

The 20 chapters are contained in six parts: "The Medical Background," "Government Responses to AIDS," "Private Sector Responses to AIDS," "AIDS and Health Care," "AIDS in Institutions," and "Confronting AIDS: The Problems of Special Groups." The first three chapters covering the medical background of AIDS are excellent. Other chapters cover topics that are at the forefront of many people's minds, such as AIDS in the schools, prostitution, AIDS in the workplace, housing, lawsuits, insurance, military, prisons, the Black community, and the lesbian and gay community. A glossary, bibliography, and brief index are included.

This book is recommended for all public and academic libraries and also for personal purchase by those who want to be educated with facts and not fallacies about AIDS.

H. ROBERT MALINOWSKY

299. AIDS in the Workplace: Legal Questions and Practical Answers. William F. Banta. Lexington, MA: Lexington Books, D. C. Heath and Co., 1988. 257p., index. $27.95. 0-669-15334-6.

William F. Banta is a director and shareholder with the law firm of Kullman, Inman, Bee and Downing, a professional corporation based in New Orleans, Louisiana. Doing a good job of bringing together a vast amount of information related to AIDS and the workplace, Banta reviews "all the laws potentially applicable to AIDS in the workplace—federal, state, and local—and then analyzes specific AIDS issues in a practical fashion, identifying the pros and cons of alternative approaches and recommending courses of action that are both consistent with legal cases and protective of legitimate business interests."

In the first chapter, he covers the general considerations of medical facts, social, ethical and legal considerations, business concerns, and health care. The second chapter discusses the various federal laws that have an impact on AIDS victims, and the third chapter explains the state and local laws. Union, arbitration, and compensation are covered in the fourth chapter; and the controversial pros and cons of testing for AIDS are covered in the fifth chapter. The sixth chapter presents case studies on the practical side of AIDS in the workplace with examples, situations, and written documents. The last two

chapters cover the sticky legal aspects of preventing AIDS-related claims and the employer's defenses against AIDS-related claims. All chapters are well written and understandable for the layperson yet written for the professional.

There are five appendices covering "U.S. Government Guidelines on AIDS"; "Health Care Materials: Policies, Forms, Guidelines, and Checklist"; "Employer Policies, Procedures, and Checklist"; "Legal Documents Pertaining to AIDS"; and "Sources of Further Information Concerning AIDS." This is an excellent book giving specific, timely suggestions on what to do about AIDS in the workplace. Personnel managers in any firm would be wise to have a copy of this book for reference as would libraries at all levels, from public to special.

H. ROBERT MALINOWSKY

300. **Clinical Infectious Diseases.** Russell J. Stumacher. Philadelphia, PA: W. B. Saunders Co., 1987. 723p., illus., bibliog., index. $19.95. 0-7216-2137-6.

Dr. Russell J. Stumacher, director of the Infectious Disease Unit, the Graduate Hospital and clinical associate professor of medicine, University of Pennsylvania School of Medicine, both in Philadelphia, has published in *Clinical Infectious Diseases* a guide for the clinical diagnosis and treatment of infectious, or communicable, diseases. The book, developed by the author from teaching materials for an elective course, is a practical handbook designed to assist everyday decision making in an infectious-disease clinic.

Introductory chapters outline general aspects of infectious disease, including overviews of microbiology, antimicrobial agents, antibiotics, and antibiotic resistance. Subsequent chapters address specific pathologic conditions including gram-negative bacteria and septic shock, pneumonias, urinary tract infections, infectious arthritis, sexually transmitted diseases, the acquired immunodeficiency syndrome (AIDS), and infectious complications of other diseases and conditions, including renal disease, cancer, and alcoholism. This book is very easy to use. Its organization in outline form and thorough indexing provide quick access to brief information. Users, primarily medical students, hospital residents, and clinicians, will certainly need to consult additional sources for more detailed data on the variety of infectious diseases patients may have. However, as a starting point and introduction to a rather complex amalgam of conditions, this book is highly recommended.   *GERALD J. PERRY*

301. **Handbook of Hematologic and Oncologic Emergencies.** Janice P. Dutcher; Peter H. Wiernik, eds. New York, NY: Plenum Medical Book Co., 1987. 354p., illus., bibliog., index. $49.50. 0-306-42646-3.

"This handbook of hematologic and oncologic emergencies provides a compact, concise, yet comprehensive guide to the management of a variety of difficult clinical situations." Twenty-four authorities have written 29 chapters covering a wide variety of topics concerned with emer-

gencies in cancer and blood emergencies. A typical chapter will begin with an introduction that presents some history on the emergency. This is followed by clinical presentations, danger signs, and other diagnostic evaluations. Finally, treatment and outcome are given.

The chapters cover "Syndrome of Inappropriate Antidiuretic Hormone Secretion and Hyponatremia," "Acute Tumor Lysis Syndrome," "Hypercalcemia in Malignancy," "Tumor-Associated Hypoglycemia," "Renal Failure Related to Drugs and Disease," Cerebral Herniation Syndromes," "Epidural Spinal Cord Compression," "Pain Treatment," "Psychiatric Emergencies in Oncology," "Meningeal Leukemia and Carcinomatosis," "Hyperleukocytosis in Leukemia," "Thrombocytosis," "Bleeding and Coagulopathy," "Transfusion Reactions," "Fever and Infection," "Typhlitis and Related Acute Gastrointestinal Problems," "Obstipation," "Intestinal Obstruction," "Biliary Tract Obstruction," "Spontaneous Splenic Rupture," "Urologic Emergencies in Oncology," "Priapism," "Malignant Pericardial Diseases," Pulmonary Emergencies in Oncology," "Dermatologic Emergencies," "Oncologic Emergencies in Ophthalmology," "Pediatric Oncologic Emergencies," "Acute Drug Reactions and Anaphylaxis," and "Management of Chemotherapy Tissue Extravasation."

This is a required handbook in any hospital where hematologic and oncologic emergencies might arise. Students will find this a very informative text for the study of emergencies associated with cancer and blood. Finally, any medical library will want to have this excellent handbook as part of its reference collection.

H. ROBERT MALINOWSKY

302. **Handbook of Skin Clues of Systemic Diseases.** Paul H. Jacobs; Todd S. Anhalt. Philadelphia, PA: Lea and Febiger, 1987. 123p., glossary, index. $9.95. 0-8121-1095-1.

This handbook is intended for the practicing physician. It is designed as an aid for diagnosing diseases. "Clues," mentioned in the title, are the skin abnormalities that are often present for specific diseases. For instance, ulcerations of the skin may be a sign of several diseases including diabetes.

The book is divided into 20 chapters arranged alphabetically by skin clue. Each chapter presents a skin lesion type and lists the systemic diseases that are associated with it. Only lists are presented; no details or explanations are given. For example, chapter 20 covers "Vitiligo," which medical dictionaries describe as a loss of pigmentation resulting in smooth light patches of skin. Under "Vitiligo," which is not defined in the book, is a list of seven "suspect" diseases. Addison's disease, pernicious anemia, and thyroid disease are three of the seven listed. On the page across from each suspected disease is information on what symptoms to look for and what tests to order. Across from pernicious anemia, "serum B12, antiparietal

cell antibody, anti-intrinsic factor antibody, Schilling test, occasionally bone marrow, neurologic examination for posterior column abnormalities" are listed.

The index is good. It would have been helpful, however, if the glossary were more extensive. It should be noted that no illustrations, color or otherwise, are included. This book is certainly no substitute for a treatise on the subject, e.g. *Dermatology in General Practice*. For more details, the physican would turn to the treatise.

Handbook of Skin Clue of Systemic Diseases succeeds in its intended role as a starting point, and the price is very reasonable. Academic and medical libraries will want a copy for their reference shelves, but the practicing physician will need a copy in his office.

FRANK R. KELLERMAN

303. **Healing AIDS Naturally.** Laurence Badgley. San Bruno, CA: Human Energy Press, 1987. 411p., illus., bibliog., index. $14.95pbk. 0-9141523-00-4pbk.

Although natural healing is frowned upon by the medical profession when it is used to treat life-threatening diseases, it is important that such books be brought to the attention of the general public. AIDS is a fatal disease. As a result, individuals will grasp on to anything that will help them live a little longer in hope that a cure can be developed before they die.

Keeping in mind that any natural therapy treatment should be undertaken under the guidance of a licensed health care practitioner, this manual can be of some comforting value to terminally ill individuals. Eighteen chapters cover background information on the immune system and the various natural therapies, including vitamins, minerals, homeopathy, herbs, wheat grass, algae, mushrooms, and acupuncture. The discussions are detailed, accurate, and to the point. "Natural therapy practitioners have a place in the spectrum of healing arts, and indications are that there will be an increasing demand for their services in the future."

This is a good reference source for natural therapies and one that would be useful in public libraries as well as research libraries that are interested in the applications of natural therapy treatment. In no way should an individual avoid contacting a reputable physician for treatment. However, after diagnosis, individuals with a life-threatening disease such as AIDS may wish to acquaint themselves with natural therapies.

H. ROBERT MALINOWSKY

304. **Respiratory Disease.** Anne E. Tattersfield; Martin W. McNicol. New York, NY: Springer-Verlag, 1987. 287p., illus., bibliog., index. $57.50. Treatment in Clinical Medicine. 0-387-16209-7.

Books in the series Treatment in Clinical Medicine aim to fill a gap between standard textbooks and current research reports, covering pathophysiology and drug therapy in the same volume. Drs. Tattersfield and McNicol, both practitioners of chest medicine in England and involved in research on drug action and optimizing therapy, are well qualified to produce such a volume for respiratory diseases.

The first half of the volume deals with etiology, physiology, diagnosis, prevention, and practical management of the main respiratory conditions. The second half discusses the administration, side effects, mode of action, selection, and dosage of drugs used for respiratory diseases, including corticosteroids, oxygen therapy, respiratory stimulants, and cough suppressants.

Books of this type are increasingly important for practitioners on many levels who do not have the time to digest all the relevant literature yet need a practical synthesis. The references at the end of each chapter are current through about 1986.

RUTH LEWIS

Histories

305. **AIDS.** Jay E. Menitove; Jerry Kolins, eds. Arlington, VA: American Association of Blood Banks, 1986. 106p., bibliog., index. $14.00. 0-915355-26-4.

AIDS consists of papers presented at the AIDS Technical Workshop held in San Francisco in November of 1986 as part of the 39th annual meeting of the American Association of Blood Banks. Although the date of this publication makes it out of date for current research information, it is an excellent brief history of the AIDS crisis up to 1986. The five papers that are included are "Etiology, Epidemiology and Natural History of AIDS and HTLV-III/LAV Infection"; "Immunologic Aspects of AIDS"; "Clinical Features of the Acquired Immune Deficiency Syndrome"; "Testing for HTLV-III/LAV"; and "HTLV-III Antibody Positive Blood Donors: Test Significance and Donor Notification." This book should be in academic and medical libraries as a supplemental history on AIDS.

H. ROBERT MALINOWSKY

306. **AIDS: The Facts.** John Langone. Boston, MA: Little, Brown and Co., 1988. 247p., notes. $8.95pbk. 0-316-51412-8pbk.

"AIDS is a topic in motion, a subject difficult to keep up with and propelled along as much by an unrelenting media and public-agency blitz as by the momentum of the virus itself, which fast outruns the deluge of new research reports and updates that appear weekly in the scientific journals and are presented at the same feverish pace at symposia here and abroad."

John Langone, a scientist for more than 25 years, has pulled together a great amount of information on AIDS and presented it in a readable manner for the layperson and student as well as for the researcher. His style of writing is easy to follow, and once one begins reading a chapter, there is a tendency to complete it without stopping.

The book is actually a thumbnail history of the virus as reflected in the chapter headings: "What Is AIDS?," "What Are the Symptoms?," "What Is the AIDS Virus?," "Where Did the Virus Originate?," "How Does the Virus Cause Infection?," "How Did the Virus Go from Monkeys to Humans?," "How Contagious is AIDS?," "How Easily Is AIDS Transmitted between Men and Women?," "AIDS in Africa and Haiti," "The Role of Cofactors," "Can AIDS Be Conquered?," "Preventing AIDS," "When Someone Has AIDS," and "AIDS Testing." The chapter on preventing AIDS is very frank and discusses safe sex practices, including the use of condoms which is controversial in some religious circles.

Suggestions for helping someone who has AIDS are especially well written with such hints as do not avoid him, touch him, cry with him, laugh with him, be prepared for him to be angry, and send him a card that simply says "I care." Libraries should have several circulating copies of this excellent book as should counselors, school teachers, and the clergy. Highly recommended.

H. ROBERT MALINOWSKY

307. The AIDS Reader. Loren K. Clarke; Malcolm Potts, eds. Boston, MA: Branden Publishing Co., 1988. 330p., bibliog. $14.95pbk. 0-8283-1918-9pbk.

"This book is one of the first to take a comprehensive view of the new virus and will substantially contribute to public understanding of the virus, the disease and the epidemic. It provides, for the first time, a collection of material which both traces the evolution of the disease, and places it in the present perspective of what we know and what we do not."

This is a historical reader. Its editors have selected those published articles that best trace the evolution of the disease. The articles are arranged within seven sections: "What Is AIDS?," "The Virus and the Community," "Where Is AIDS Going?," "Stopping AIDS," "The Epidemic Comes Home," "We Get Letters," and "The Agony of Africa."

This is an excellent book and recommended for all libraries. There is no index, but there is a detailed table of contents. There are six pages of suggested readings that include books and articles. *H. ROBERT MALINOWSKY*

308. Respiratory and Infectious Disease. Wrynn Smith. New York, NY: Facts on File Publications, 1987. 226, illus., bibliog., glossary, index. $35.00. A Profile of Health and Disease in America. 0-8160-1458-2.

This is a historical overview of respiratory and infectious diseases in America. It also contains "current statistics on the incidence, prevalence, and mortality of major diseases within one of the major specialties," in this case, respiratory and infectious. It includes "data for different geographical areas within the United States and around the world, the length of a hospital stay and treatment costs, discussions of major controversies, and information on the use of various medicines and surgical procedures." Five separate

areas are covered: acute respiratory and infectious disease, chronic lung disease, enteric infections, sexually transmitted diseases, and nosocomial and other infections.

Acute respiratory and infectious disease covers acute and chronic lung diseases which are among the leading causes of death and disability in the United States annually. Specific diseases covered include pneumonia, influenza, tuberculosis, whooping cough, diphtheria, legionnaires' disease, and common cold. Through the use of numerous graphs, charts, and maps, epidemiologic information is given as are trends and patterns based on age, sex, and cultural groupings. This type of information is presented for each of the other sections.

Chronic lung disease covers chronic bronchitis, emphysema, asthma, hay fever, occupational lung disease, and black lung. Enteric infections include salmonellosis, botulism, shigellosis, brucellosis, tularemia, trichinosis, and toxoplasmosis.

With AIDS in the news, the section on sexually transmitted diseases should be of interest. In addition to AIDS, chlamydia, gonorrhea, herpes, and syphilis are covered. Finally, nosocomial and other infections (measles, rubella, polio, mumps, encephalitis, Reye's syndrome, chicken pox, meningococcal infections, scarlet fever, infectious mononucleosis, rabies, tetanus, plague, leprosy, typhus fever, and malaria) are discussed.

This is not intended to be a comprehensive source of information. It is, however, an excellent overview for each disease and a good place to start for statistical material.

H. ROBERT MALINOWSKY

Manuals

309. AIDS: A Manual for Pastoral Care. Ronald H. Sunderland; Earl E. Shelp. Philadelphia, PA: The Westminster Press, 1987. 76p., bibliog. $6.95pbk. 0-664-24088-7pbk.

Counseling persons with AIDS is one of the most difficult tasks that any pastor has to encounter. Many religions have taken this counseling very seriously. Sunderland and Shelp have put together an excellent small manual that should help all pastors with this difficult task.

After a brief introduction that gives some historical facts, the authors cover "Confronting the Reality of AIDS," "Grief Recgonition and Response," "Pastoral Care of People with AIDS," and "Ethical Issues." Throughout the book the three caring responses are stressed—reality, empathy, and support. The ethical issues are especially inspiring in that the authors try to stress that pointing the finger is rampant, even in the pastoral community, and that "This tendency ought to be resisted, both by the people of God and by society." The book ends with three case studies that serve as an example of how to handle certain situations.

This is a recommended book for all pastors and one that public libraries will find useful on the reference shelf and as a circulating book.

H. ROBERT MALINOWSKY

310. **AIDS: A Self-Care Manual.** AIDS Project Los Angeles. Betty Clare Moffatt; Judith Spiegel; Steve Parrish; Michael Helquist, eds. Santa Monica, CA: IBS Press, 1987. 306p., biblio., glossary, index. $24.95pbk. 0-9616605-1-1.

This outstanding book "is comprehensive in its coverage of many of the concerns and needs of people exposed to the AIDS virus, diagnosed with AIDS symptoms, worried about AIDS, and grieving for those who have been diagnosed with the disease." It is intended to help concerned individuals learn all they can about the AIDS epidemic.

Section 1, "AIDS: An Overview," covers the facts and fiction associated with AIDS. Section 2, "A Socio-psychological Perspective," treats such topics as stress, the family's response, youth at risk, when a friend has AIDS, and terminal illness. "A Medical Perspective" is presented in section 3 and includes the antibody test and coping with ARC. The treatment and prevention of AIDS and Kaposi's sarcoma and alternative therapies are covered in the fourth section, "Treatment: A Therapeutic Perspective." Section 5 is the prevention section covering safe-sex guidelines, alcohol, drugs, and AIDS, and the risks and concerns of women. "A Self-Care Perspective" is covered in section 6, giving information on nutrition and exercise, dental care, general hygiene, home care, symptom management, and the side effects of Kaposi's sarcoma. The practical aspects of AIDS care are covered in section 7 with subjects such as power of attorney for health care, benefits, insurance, and social services. Section 8, "A Spiritual Perspective," and section 9, "A Healing Perspective," cover the spiritual and mind healing processes. Finally, section 10 contains numerous lists of self-care resources and includes a glossary, lists of AIDS-related organizations and hotlines, bibliography of books, tapes, and resource material, and self-care forms and charts.

This is a highly recommended book for anyone who has AIDS or ARC and for libraries, counselors, concerned others, clergy, and the general public. AIDS education is essential and this manual will do much to help educate individuals.

H. ROBERT MALINOWSKY

311. **AIDS and the Church** Earl E. Shelp; Ronald H. Sunderland. Philadelphia, PA: The Westminster Press, 1987. 151p., bibliog. $8.95pbk. 0-664-24091-7pbk.

The church's response to the AIDS crisis has been varied. On the whole, however, it has been positive in dealing with the disease and less positive in dealing with the sexuality of many of the individuals who have contacted the disease. This conflict has troubled a great many of the clergy and in some cases has shut out those who seek help from the church. Shelp and Sunderland have written a "manual" for the clergy in hopes that it "will help the people of God to respond to this expanding public health crisis in an informed and loving manner."

The book is filled with documented information about AIDS and the people who have died of AIDS. All sides of Christian morality are brought out including the highly volatile statements of Jerry Falwell. The authors acknowledge some religious leaders' beliefs that the epidemic is a way of God punishing the homosexual community. They, however, make a great effort to educate the people at all levels by stressing that "AIDS is not the fault of any particular group of persons." They conclude with the statement: "The situation of people with AIDS is one of unprecedented and chronic grief, demanding an intense pastoral response from the religious community. AIDS is an unparalleled crisis, not only because it remains incurable but because of the stigma that quickly attaches to the disease, and those it afflicts, and its potential to destroy lives and communities."

This is a "must" book for every pastor and a very useful reference book in public libraries.

H. ROBERT MALINOWSKY

312. **The Essential Asthma Book: A Manual for Asthmatics of All Ages.** Francois Haas; Sheila Sperber Haas. New York, NY: Charles Scribner's Sons, 1987. 298p., illus., bibliog., glossary, index. $16.95. 0-684-18592-X.

Written primarily for a lay audience, this book "is an excellent attempt to bridge the gap between what doctors know about asthma and what patients need to know to cope with their disease intelligently. It is well organized, cogently phrased, comprehensive, and, most important, accessible to the layman." The authors are particularly well qualified, and each brings his or her own individual strengths to the book. Francois Haas, a mild asthmatic, earned his Ph.D. in pulmonary physiology from New York University School of Medicine and is presently a member of the faculty and director of the Pulmonary Function Laboratory. He has studied respiratory physiology and lung mechanics for a number of years, and the results of his research have been widely published in the scientific literature. Sheila Sperber Haas, his wife and coauthor, earned a Ph.D. in psychology from the City University of New York, where she studied the relationship between personality types and breathing styles. She has written on a variety of scientific topics for professional and general audiences.

Using easy-to-understand nontechnical language, the functions of the respiratory system are explained, and the reader is introduced to asthma and its several causes. The many tests used to diagnose asthma are carefully outlined, and detailed information regarding drug therapy and the various physical therapies that can be used to control asthma are discussed. Of special usefulness are chapters on how to deal with potential asthma emergencies, stress and anxiety management, the pregnant asthmatic, and coping with the special problems of juvenile asthmatics. A glossary of medical terminology is included as

well as three appendices which provide names and addresses of organizations that can provide additional sources of information about asthma, a nationwide list of summer camps for asthmatic children, and a bibliography of sources for additional reading.   *RICHARD EIMAS*

313. Manual of Gastroenterologic Procedures. 2nd. Douglas A. Drossman, ed. New York, NY: Raven Press, 1987. 266p., illus., bibliog., index. $21.00. 0-88167-303-X.

The second edition of the *Manual of Gastroenterologic Procedures*, edited by Dr. Douglas A. Drossman, associate professor, Division of Digestive Diseases, University of North Carolina School of Medicine, Chapel Hill, is a convenient reference guide for those training in the diagnosis and care of patients with gastroenterological disorders. According to Drossman, the *Manual* provides a standardized approach for learning the "indications, contraindications, preparations, techniques, and complications of most all the gastroenterological procedures." Essentially a how-to book, this manual is organized in outline form, taking the reader through the indications, contraindications, equipment and patient preparation requirements, techniques, and interpretation of data for a wide variety of common procedures.

The second edition, following the first by five years, has been updated and expanded in scope to include many recent technological and procedural developments in diagnosing and treating digestive diseases. Nine new chapters have been added, including discussions of the bentiromide test of pancreatic function, balloon dilation of strictures, percutaneous endoscopic gastrostomy, the coagulation of bleeding gastrointestinal lesions, and the placement of nasobiliary catheters and endoprostheses. Additionally, a section dealing with the special considerations of pediatric patients has been added. Intended as a reference guide and starting point for physicians, students, and nurses, this handy manual features updated references at the end of each procedure chapter for additional information. A complete index providing access to both procedures and disorders further enhances this book. Given its overall brevity, this book is primarily recommended to practicing clinicians, students, and nurses. However, hospital libraries and departments may find this manual a convenient addition to the collection.   *GERALD J. PERRY*

314. Mortal Fear: Meditations on Death and AIDS. John Snow. Cambridge, MA: Cowley Publications, 1987. 92p. $6.95pbk. 0-936384-49-2pbk.

This is not really a reference book but a book that a reference librarian would suggest for an individual who is having some difficulty in coping with the realities of AIDS and its consequence, death. It is a collection of five meditations on AIDS and three articles discussing death. Although the casual reader may find this book very frightening, those seeking some solace will find it a book of inspiration. The publishers are of the Society of St. John the Evangelist, a religious community for men in the Episcopal Church. Their aim is to provide books that will enrich the religious experience and challenge the minds of their readers. This is a recommended book for public libraries.   *H. ROBERT MALINOWSKY*

315. TNM: Classification of Malignant Tumours. 4th rev. P. Hermanek; L. H. Sobin, eds. New York, NY: Springer-Verlag, 1987. 197p., illus., index. $15.00. 0-387-17366-8.

The TNM is an international scheme for the classification of cancer based on the extent or stage of the disease process. The system describes in standardized coding format the extent of tumor growth and whether the cancer has spread to other parts of the body. The rationale for having an international language to classify cancer cases is to provide universally understood standards to eliminate ambiguity when communicating clinical experience. Additionally, the classification system helps clinicians plan for treatment, determine prognosis, and evaluate results.

The fourth, fully revised edition of the *TNM Classification of Malignant Tumours*, published by Springer-Verlag for the International Union against Cancer, is the key reference tool for interpreting the TNM coding system and represents the concerted efforts of the national TNM committees from the United States, Canada, the United Kingdom, France, Italy, Germany, and Japan. Included in the new edition are classifications for previously unclassed tumors, enhancements of existing codes and the elimination of variations in the scheme. Use of this tool assumes knowledge of the system and certainly knowledge of cancer development, though introductory statements explain briefly the notation scheme. Because of its almost universal acceptance, the TNM Classification is a necessary addition to biomedical reference collections, particularly those supporting cancer research, treatment, and education.   *GERALD J. PERRY*

316. Training Guide to Cerebral Palsy Sports: The Recognized Training Guide of the United States Cerebral Palsy Athletic Association. 3rd. Jeffery A. Jones, ed. Champaign, IL: Human Kinetics Books, Human Kinetics Publishers, Inc., 1988. 240p., illus., bibliog., index. $20.00pbk. 0-87322-125-7pbk.

This volume is a comprehensive and well-documented treatment of the subject of working in sports with people who suffer from cerebral palsy. Divided into four major parts, part I gives the introduction to the field. There is a brief history of the movement, a section on *les autres* (the others), disabled individuals included in the programs of the U.S. Cerebral Palsy Athletic Association, and a careful classification of the athletes according to their levels of physical competence. Part II treats the training and preparation of the athletes: the role of the therapist, general training

principles, general fitness and exercise, interval training and record keeping, flexibility exercises, optimal nutrition, and the basics of wheelchairs for maximal functioning.

Part III goes into a number of specific activities, among which are archery, horse riding, handball, soccer, field events, and bowling. Consideration is given to principles of coaching, types of training, techniques and strategies, and various types of equipment for each activity. Part IV considers the future of the field and the development of new programs and success strategies. At the end of the book, there is a good glossary of medical terminology to help the beginner. There are also three good appendices with information on other handicapped sports organizations, the staff of the U.S. Cerebral Palsy Athletic Association, and wheelchair manufacturers.

This book was put together by people who are active and experienced in this field. Even though it is devoted primarily to sports for cerebral palsy athletes, its principles apply to sports for other disabled persons. In its present brief form, it constitutes a good manual or handbook. In an expanded form, it would be an excellent textbook on this subject. *LILLIAN R. MESNER*

Textbooks

317. **Know about AIDS.** Margaret O. Hyde; Elizabeth H. Forsyth. New York, NY: Walker and Co., 1987. 68p., illus., bibliog., index. $10.95. 0-8027-6738-9.

Children have a natural curiosity about disease, but with AIDS there is more than a curiosity. There is a concern of just how it might affect them. This excellent textbook covers all areas of AIDS in an easy-to-read style that is full of facts. It does not go into great detail, but it gives enough information that the student should not go away confused about AIDS.

The ten chapters cover "AIDS: A New Problem," "Who Gets AIDS?," "Viruses and AIDS," "Where Did AIDS Come From?," "The Geography of AIDS," "AIDS at Home and in School," "A Death in the Family," "The Search for a Cure," "Who Needs a Test for AIDS?," and "AIDS in Your Future." This is a recommended book for school and public libraries.
H. ROBERT MALINOWSKY

318. **Teaching about A.I.D.S.: A Teacher's Guide.** Danek S. Kaus; Robert D. Reed. Saratoga, CA: R and E Publishers, 1987. 75p., bibliog. $6.50pbk. 0-88247-766-8pbk.

R and E Publishers have brought together information from several of their other publications to form a teacher's guide for teaching a section on AIDS in the schools. It is well written, factual, and up to date through 1987.

The first part covers "What Is A.I.D.S?" and gives an explanation of AIDS, symptoms, how it affects the body, statistics, AIDS in children, and other facts about AIDS. The second part is a series of questions and answers about AIDS. The third part pertains to teaching about AIDS and covers "Reducing Your Chances of Contracting AIDS."

This is a good book and one of which teachers should be aware. Also included is a list of general sources and resources that can be of supplemental help for a section on AIDS that is to be taught in the school classroom. School libraries will want a copy for their reference shelves.
H. ROBERT MALINOWSKY

319. **Textbook of Pediatric Infectious Diseases.** 2nd. Ralph D. Feigin; James D. Cherry. Philadelphia, PA: W. B. Saunders, Co., 1987. 2,412p. in 2v., illus., bibliog., index. $150.00 set, $75.00 v. 1, $75.00 v. 2. 0-7216-1372-1 set, 0-7216-1370 v. 1, 0-7216-1371-3 v. 2.

As in the 1981 edition of this textbook, the authors are attempting to provide a reference which comprehensively discusses the infectious diseases of children. New information in this area is accumulating at such a rate that every section in this second edition required revision, and entire chapters were rewritten. In addition, 27 sections were added to cover diseases which, in many cases, were unknown in 1981 and, in some cases, were identified even as this volume was being created. The textbook was created using the combined expertise of 164 contributors from 60 institutions and eight countries. This hefty textbook is divided into seven parts; the two predominant parts are "Infections of Specific Organ Systems" and "Infections with Specific Microorganisms." Sections cover a variety of elements including pathology, etiology, microbiology, epidemiology, and prognosis, but special emphasis is placed on the clinical manifestations, differential diagnosis, and therapy.

Additional parts in the textbook cover "Host-Parasite Relationships and the Pathogenesis of Infectious Diseases"; "Infection Control"; "Chemotherapy, Including Antimicrobial Agents, Antiviral Agents, and Immunization"; "Prevention of Infectious Diseases, Including Prevention in Day Care and Human or Animal Bites"; and "Aids to the Diagnosis of Infections." Special sections discuss interactions of infection and nutrition, immunologic responses to infection, microbial virulence factors, perinatal bacterial infections, intrauterine infections, opportunistic infections, acquired immunodeficiency syndrome, toxic shock syndrome, and other topics. The rapid changes in the field of pediatric infectious diseases create a high demand for a comprehensive textbook containing the newest information. This textbook meets this demand with a massive amount of information enhanced by a readable layout and a generous supply of high quality black-and-white photographs, illustrations, tables, and figures.

The sections are well written, logically organized, and generally have lengthy bibliographies. The two-volume set has an index in the back of each volume, and the page numbers are marked clearly on the spine of each volume, fine details

which increase the usefulness of any multivolume set. The second edition of *Textbook of Pediatric Infectious Diseases* will be a significant resource for students and clinicians interested in pediatrics or infectious diseases and, therefore, a quality addition to medical library collections.

JOAN LATTA KONECKY

320. Understanding AIDS. Ethan A. Lerner. Minneapolis, MI: Lerner Publications Co., 1987. 64p., illus., glossary, Index. $9.95. 0-8225-0024-8.

This well-written book for children is one of few books available that gives straightforward facts about AIDS education. It could be used as a textbook in schools for a classroom discussion of AIDS and its impact on everyone. Dr. Lerner, an immunologist, is at the Harvard Medical School and has had an interest in educating young people on the facts of AIDS.

He presents the information in five "facts" chapters: "Swollen Glands: Facts about Infections," "Different Folks: Facts about Sexuality," "Andy's Appendix: Facts about Transfusions," "Adding Insult to Injury: Facts about Hemophilia," and "Getting High: Facts about Drug Abuse." Each chapter is presented in storybook fashion that is interesting to read and yet very factual, beginning with a story that leads up to the information being presented in that chapter. Two additional chapters are life history accounts of individuals who have AIDS and how it affected the people around them.

This is a very positive, "upbeat" book that should be easy for young people to understand. The terminology has been defined in such a way that it is not too simple and not too technical. For example, "The virus which causes AIDS knocks out your immune system—like kryptonite knocks out Superman. It kills most of your T cells, which makes your immune system deficient, or less effective."

This is a must book for school and public libraries. Adults will be well advised to read this book for a quick, factual education about AIDS, gays, and drugs. *H. ROBERT MALINOWSKY*

Treatises

321. Plain Words About AIDS. 2nd. William Hovey Smith, ed. Sandersville, GA: Whitehall Press-Budget Publications, 1986. 199p., illus., glossary, index. $17.50pbk. 0-916565-09-2pbk.

This is not a reference book in the truest sense, but it does give some historical facts about AIDS that should be of value in any reference collection. Of particular reference value is the 41-page glossary. The book, itself, contains presentations of the 1986 International Conference on AIDS, June 23-25, Paris, France; the 1986 Seventh National Lesbian-Gay Health Conference, March 13-15, Washington, DC; the 1985 International Conference on AIDS, April 14-17, Atlanta,

Georgia; releases from the Centers for Disease Control, June 5, 1981–July 4, 1986; and articles from the national wire services through October 20, 1986.

This is a technical book, and its format may not be appropriate for the layperson. Topics covered in the eight chapters are "The Body's Defense Systems," "AIDS, What It Is and Is Not," "The AIDS Epidemic," "The AIDS Virus," "Transmission of AIDS," "Prejudice and Passion," "Precautions for Health Workers," "and "Political and Legal Aspects." With a specialized glossary of AIDS-related terms, this is a good resource book for academic and special libraries.

H. ROBERT MALINOWSKY

★
INTERNAL MEDICINE
Atlases—Medical

322. Anatomical Dissections for Use in Neurosurgery: Volume 1. Wolfgang Seeger; H. R. Eggert. New York, NY: Springer-Verlag, 1987. 313p., illus., bibliog. $128.00. 0-387-81998-3.

The author is a microneurosurgeon with more than 10 years of such experience, with seven similar atlases previously published. This work will be published in two volumes. It is based on a one-year dissection course; drawings originate from the dissections. As anyone knows who has done dissections, all systems are viewed almost simultaneously during dissection. Because blood vessels may have more branches than shown in atlases, it is inevitable that some will be destroyed along with other tissues as the dissection is done. Consequently, two brain specimens will be needed when using this volume. With the first, the blood vessels and leptomeninges are removed so "the young neurosurgeon" can concentrate study on the cerebrum and the cerebellum. In a second brain, with the vessels and leptomeninges intact, the relationship between surface structures, the ventricular system, and deep-seated parts can be shown. The drawings are intended to show only the essential structures and are based on actual specimens. This work is highly specialized and really intended for use in the laboratory. Libraries holding the earlier titles may wish to have this one, also, or to purchase based on user demand. *JAMES L. CRAIG*

323. Anatomy: A Regional Atlas of the Human Body. 3rd. Carmine D. Clemente. Baltimore, MD: Urban and Schwarzenberg, 1987. 439p., illus. (col.), index. $42.50. 0-8067-0323-7.

Clemente's *Anatomy: A Regional Atlas of the Human Body* is a collection of very elegant anatomical illustrations. Most of the figures in the book came from the Johannas Sobotta and Edward Pernkoph collections. In addition to these are added figures from Benninghoff-Goerttler

textbooks and radiographs from Dr. Lothar Wicke of Vienna. They make up a comprehensive and extensive collection of high quality, meticulous demonstrations of the human body. The table of contents displays the organization of the book and is an excellent way to find one's way into the atlas. Seven parts separate the body into regions, and each part has eight to 16 subdivisions. By simply looking over the two pages, a reader can get to the set of plates for the region desired. In addition to this, however, the book contains an extensive index with both broad anatomical areas (e.g., arm, aorta) and very specific anatomical areas (e.g., renal calyx). Under each entry the indexer has listed all of the plate numbers that demonstrate the particular structure. Every possible means of access seems to be provided for the user.

Each section and subsection of the atlas contains anywhere from two to 25 plates with copious labelling of structures. Each plate is accompanied by notes pointing out less apparent features of the drawings. In this third edition, 96 new plates were added and six eliminated. These elaborated the sections on the thorax, abdominal, and pelvic areas as well as the sections on the limbs and neck and head. Formerly black-and-white illustrations were done in color, which enhances the beauty of this book. Dr. Clemente has intended this atlas to aid the students learning anatomy in any of the health-related disciplines rather than only medical students. The book uses anatomical terms, which should not hinder a beginner, because anatomy requires the use of a discrete terminology. This atlas would be an excellent tool for learning anatomical terminology and would be a worthy addition to any collection in a health science institution. It would, also, serve as a good tool in a public collection.

LILLIAN R. MESNER

324. Atlas of Axial, Sagittal, and Coronal Anatomy: With CT and MRI.

A. John Christoforidis. Philadelphia, PA: W. B. Saunders Co., 1988. 563p., illus. (part col.), index. $125.00. 0-7216-1278-4.

With the introduction of computed tomography and magnetic-resonance imaging techniques, study of the human body and routine diagnostic studies are no longer limited to invasive examination or indirect visualization with contrast media. The axial views of anatomy, and occasionally the coronal and sagittal planes, are very important in computed tomography although all three planes are regularly used in magnetic-resonance imaging. Thus, sectional anatomy as correlated with radiographs, computed tomograms, and magnetic-resonance images provides a tridimensional view of anatomical structures. The objective of this atlas is to present a "comprehensive and stereoscopic view of the human body by comparing photographs of anatomical sections in different planes with computed tomograms and magnetic-resonance imaging."

The atlas is divided into six sections—"Head and Neck," "Spine," "Thorax," "Abdomen," "Pelvis," and "Extremities"—each of which has a short introduction discussing the related technical

factors and a bibliography. The plates are very clearly labelled and 76 anatomic sections are in color. The quality of the images is excellent, although radiologists will recognize that these are early magnetic-resonance images and do not display the resolution of state-of-the-art magnetic-resonance imaging. Each set of images is accompanied by a line drawing depicting the plane from which the sections were taken. Finally, there is an index to the images in the back of the volume. The atlas is a beautifully labelled, well-organized reference source which should be valuable to radiologists, anatomists, and a wide variety of students and specialists for teaching, learning, clinical, and surgical purposes. Medical libraries should consider adding this tool to their collections.

JOAN LATTA KONECKY

325. Atlas of Blood Cells: Function and Pathology. 2nd. D. Zucker-Franklin; M. F. Greaves; C. E. Grossi; A. M. Marmont.

Philadelphia, PA: Lea and Febiger, 1988. 777p. in 2 v., illus. (part col.), bibliog., index. $225.00. 0-8121-1094-3.

The *Atlas of Blood Cells: Function and Pathology* is a comprehensive pictorial and textual resource in two volumes, for the research-oriented pathology laboratory and for the clinician who needs an insightful clarification of the complex properties of blood cells and ramifications of abnormalities of this system. It is "not intended as a disease-oriented" aid to diagnosis or treatment. The purpose of the atlas is "to incorporate the many diverse technical advances which have provided entirely new insight into the biology, immunology and enzymology of normal and malignant hematopoietic cells."

Since the time of the first edition, new technology has progressed even more rapidly and, in areas such as flow cytometry, has been extensively applied to routine analysis of cell populations. This second edition reflects many changes in some of the rapidly progressing areas of the field, such as biochemical and molecular genetics. Nineteen experts have contributed to the production of a very current, authoritative, and thoroughly documented treatise presenting a comparative, multidisciplinary characterization of cells based on techniques ranging from the traditional histochemical methods to recent, high-technology, molecular aspects of the analysis of blood cells in both normal and pathological conditions.

The subjects are presented in 12 chapters, of which eight are devoted to specific types of cells, and four either to topics of general relevance or of special importance. The two largest chapters contain separate parts by different authors, with independent bibliographies: "Lymphocytes" (334p., 594 references), and "Erythrocytes" (104p., 369 references). "Lymphocytes" contains four parts: (1) normal lymphocytes; (2) lymphoproliferative disorders; (3) lymph nodes, reactive and neoplastic conditions; and (4) acquired immune deficiency syndrome (AIDS). The "Erythrocytes" chapter has three parts: (1) general aspects of erythropoiesis, (2) abnormalities of red cell production, and (3) abnormalities of red

cell destruction. Introductory chapters include "In vitro and in vivo hematopoiesis" and "Cellular identification and markers." Concluding chapters are "Nonrandom chromosome changes in hematologic diseases" and "Aplastic anemia and hematopoiesis after marrow transplantation." Separate chapters devoted to other classes of cells include "Neutrophils," "Eosinophils," "Basophils," "The mononuclear phagocyte system," "Megakaryocytes," and a new one to this edition, "Dendritic cells." Various types of disease states considered for each type of cell include malignancies, hereditary conditions, infections, and chemical- or drug-induced disorders.

Other impressive revisions or additions include clarification of the system of identifying cell markers, use of antibody and nucleic acid probes, capabilities and applications of flow cytometry, effects of interleukins on various cells, and the impact of oncogenes. Many photographs and illustrations of the first edition were retained and are supplemented by even more new ones to depict cytomorphology and cytochemistry as observed by light and fluorescence microscopy as well as scanning and transmission electron microscopy. Most of the light and fluorescence micrographs are in color. The physical quality of the publication is outstanding.

This atlas is reminiscent of the exquisite *Encyclopedia of World Art*, also printed in Italy. Both are works of art themselves—beautiful and informative. This reference is highly recommended for any person, department, or library with an interest in hematology, oncology, pathology, or related subjects. *PAT JAMESON*

326. An Atlas of Surgical Exposures of the Extremities. 2nd. Sam W. Banks; Harold Laufman. Philadelphia, PA: W. B. Saunders Co., 1987. 414p., illus. $65.00. 0-7216-1531-7.

Nearly 30 years have elapsed since the publication of the first edition of *An Atlas of Surgical Exposures of the Extremities*. Fifteen printings of the first edition over that stretch of time attest to both the popularity and practicality of this atlas. However, advances including arthroscopic, vascular-reconstructive, and joint-replacement surgery and the concomitant development of new techniques for dissection (while retaining function to facilitate these operations) have prompted the authors to issue a second edition. The new edition includes many illustrations from the first, as well as 13 new depictions, specifically of bone, joint, and vascular exposures. As with the previous edition, narrative has been kept to a minimum. Authors Banks and Laufman let the drawings speak for themselves. Of particular note is the method utilized to produce the illustrations. Anatomical dissections of cadavers were performed, during which each step of the surgical process was photographed. The photographs were then labelled and reproduced by artists in half-tone drawings. The atlas is organized by anatomic region, including the shoulder area, the elbow, wrist and hand, hip joint, femur, knee, ankle, and foot. For each exposure drawing, the indication or rationale for the operation is given, followed by the drawings depicting the process. Intended as a guide and atlas of both common and unusual surgical exposures, the second edition is highly recommended for orthopedic and vascular surgeons, students of general surgery, and the reference collections of medical libraries. *GERALD J. PERRY*

327. Atlas of the Human Brain in Section. 2nd. Melville P. Roberts; Joseph Hanaway; D. Kent Morest. Philadelphia, PA: Lea and Febiger, 1987. 134p., illus., bibliog., index. $35.00. 0-8121-1030-7.

This atlas provides cross-sections of the entire human brain in three planes and, thus, fills something of a gap in the field of brain atlases. The entire human brain is covered, with some cross-sections of the spinal cord. There are approximately 60 plates, each consisting of a large, clear photograph of a section (one-and-one-half times life size), completely labelled for anatomical detail, along with a small drawing showing the orientation of the section. The introduction recounts the methodology used to prepare the sections. Coronal (transverse) sections are presented first, followed by horizontal, sagittal, and then microscopic sections. An index to the plates covering anatomical details is provided. The photographs of sections are clear, with very good contrast between gray and white matter, and are very well labelled. There are enough sections to provide complete coverage of the gross anatomy of the brain. No overall or surface views of the brain are provided, although these are readily available in other sources. Among other brain atlases, *Structure of the Human Brain, a Photographic Atlas* (New York, Oxford University Press, 1976) presents more histological detail but much less general anatomical coverage. The *Atlas of the Human Brain* (Amsterdam, Elsevier, 1978) illustrates the brain surface well but has fewer sections. The *Atlas of the Human Brain in Section* would be an excellent addition to the collections of academic and medical libraries. *JOHN LAURENCE KELLAND*

328. Computer Tomography of the Head and Neck. Thomas H. Newton; Anton N. Hasso; William P. Dillon, eds. New York, NY: Raven Press, 1988. 460p., illus., bibliog., index. $135.00. "A Clavadel Book"; Modern Neuroradiology, v. 3. 0-88167-392-7.

Computer tomography has revolutionized diagnostic radiology and become one of the primary procedures in use, especially in the particularly difficult areas of the head and neck. The editors attempted to create a referenced textbook with "state-of-the-art imaging and important clinical, epidemiological and pathological data." They have succeeded admirably, with the aid of an international panel of contributing authors.

In a progressive manner, the book reviews the normal anatomy and then presents the pathology for each section of head and neck imaging. Beginning with the skull base and vault, the text moves to the temporal bone and mastoid, then to the

paranasal sinuses, nasal cavity and facial bones, and finishes with the orbit, the oropharynx, the nasopharynx, the cervical soft tissues, and the larynx. Each of the 12 chapters is completed by an impressive reference section, and the volume has a detailed index.

Even though the text is very well written with a great deal of useful information, the most impressive elements of this volume are the extremely high quality images with their descriptive notes and clear labels. This text will be heavily consulted by students and practitioners alike, especially in the areas of anatomy and radiology. This text is highly recommended.

JOAN LATTA KONECKY

329. Living Anatomy: A Working Atlas Using Computer Tomography, Magnetic Resonance and Angiography Images. Robert A. Novelline; Lucy Frank Squire. Philadelphia, PA: Hanley and Belfus, Inc., Dist. by C. V. Mosby Co., St Louis, MO, 1987. 117p., illus. $29.00. 0-932883-03-6.

This is an atlas of photographs of computed tomography and magnetic resonance scans of the human body, covering the chest, abdomen, pelvis, head, face, spine, and neck, in an oversized, softcover, spiral-bound manual. The format is a step-by-step problem-solving program—with each photograph a problem is posed. Turning the page exposes the answer and discussion. Although designed primarily for teaching anatomy to laboratory students, this book has additional uses. *Living Anatomy* is an atlas of CT and MR scans that serve as normal references by clinicians. Also, with this atlas, student physicians can practice CT and MR scan image interpretation.

CAROLYN DODSON

330. Pocket Atlas of MRI Body Anatomy. Thomas H. Berquist; Richard L. Ehman; Gerald R. May. New York, NY: Raven Press, 1987. 97p., illus., bibliog. $12.50. 0-88167-282-3.

Magnetic-resonance imaging (MRI) has dramatically enhanced the physician's ability to diagnose and interpret anatomical findings. Thoroughly high-tech, MR images allow trained interpreters to view both surface and deep structures from a variety of planes or vantage points.

Berquist, Ehman, and May's *Pocket Atlas of MRI Body Anatomy* is a small, handy guide to extracranial anatomy, featuring 96 MR images of major body organs of the upper and lower extremities, chest, abdomen and pelvis. This *Pocket Atlas* particularly emphasizes the extremities, documenting in great detail anatomic features of the wrist, hand, knee, and ankle. Each image is labelled to highlight depicted anatomical features. Additionally, a line drawing accompanies each image, illustrating the plane and level of the image, essentially putting the image in perspective vis-à-vis the body part analyzed.

Given its brevity and compactness, this atlas may not be considered appropriate as a reference guide to extracranial MRI. It lacks any narrative discussion of included images, and there is no consideration of the MRI process or rationale for its use in given situations. However, it is clearly a convenient reference tool for physicians and students as an aid to interpreting MR images and is, therefore, also recommended as a medical library reference tool for those collections supporting clinical practices, residencies, and internships.

GERALD J. PERRY

Encyclopedias

331. Encyclopedia of Neuroscience. George Adelman, ed. Boston, MA: Birkhauser, 1987. 2v., illus., bibliog., index. $125.00. "A Pro Scientia Viva Title." 0-8176-3335-9.

Encyclopedia of Neuroscience is "intended to help enhance [the almost explosive growth of neuroscience in the past 25 years] and to encourage wider interest in the most intriguing of all scientific questions, How does the brain, the organ of mind, understand itself?" The first general encyclopedia of neuroscience contains more than 700 articles alphabetically arranged. It is intended for all interested in neuroscience from the generalist to the specialist, and from undergraduate to professor.

The term "neuroscience" is used in its widest possible sense, but the coverage is not exhaustive. Each entry may be read on its own and usually includes a section of "Further Reading." The index and a series of "see also" references at the end of each narrative entry should help the reader locate relevant material. Most entries have a minimum of two "see also" references. The entries are listed in the front of each volume. This is an outstanding purchase. *JAMES L. CRAIG*

Handbooks

332. Cardiac Emergency Care. 3rd. Edward K. Chung, ed. Philadelphia, PA: Lea and Febiger, 1985. 415p., illus., bibliog., index. $35.00. 0-8121-0978-3.

The preface of *Cardiac Emergency Care* states that the handbook covers common cardiac emergencies and that it is designed to be clinical, precise, and practical. This third edition, broadly revised since the last edition five years ago, reflects the significant changes in the therapeutic aspects of cardiac emergencies. Also, new chapters, such as "Coronary Artery Spasm," have been added. The book has 25 chapters in all, and the scope is wide. Besides covering specific conditions or emergencies, chapters cover diagnostics, therapeutics, coronary care units, and nursing aspects of cardiac emergency care.

All but one of the 26 contributors are medical doctors and many are faculty at universities. One contributor is a registered nurse. The highly technical writing is obviously intended for professionals. The chapters are well documented. The book's best uses would be in hospital and health school libraries and in cardiac care units. Also, nurses and other professionals who are involved with cardiac emergenicies will find this a very useful personal reference source.

LILLIAN R. MESNER

333. Current Clinical Practice. Franz H. Messerli, ed. Philadelphia, PA: W. B. Saunders Co., 1987. 831p., bibliog., index. $49.95. 0-721601460-4.

Current Clinical Practice is a collection of patient management strategies "used by the specialists and superspecialists of the major private clinics in the United States." Indeed, these are the contributors to this book. The format is brief, with short tables and selective reference lists, with a few references as current as 1986. The index and internal organizations make this a handy reference book for a practicing physician. Chapters on common complaints and diseases and several less common ones are arranged in 13 specialty sections and a section on general patient care. Several chapters include a discussion of socioeconomic aspects, an unusual feature for clinical handbooks.

RUTH LEWIS

334. The Respiratory System. 2nd. Alfred P. Fishman, ed. Bethesda, MD: American Physiological Society, 1985. 4v. in 6, illus., bibliog., index. $136.00. Handbook of Physiology: A Critical, Comprehensive Presentation of Physiological Knowledge and Concepts, Section 3. 0-683-03244-5 v. 1, 0-683-01522-2 v. 2, 0-683-05334-5 v. 3, 0-683-03039-6 v. 4.

Section editor Alfred P. Fishman has brought together an array of experts in the respiratory sciences for the second edition of the exhaustive American Physiological Society's *Handbook of Physiology: A Critical, Comprehensive Presentation of Physiological Knowledge and Concepts, Section 3: The Respiratory System*. This section was called *Respiration* in the first edition. According to Fishman, the objective of this undertaking was to provide a thorough contemporary understanding of respiratory physiology within a respiratory biology context. Fishman's goal of moving toward a structure-function understanding of the respiratory system, integrating disciplines, with a basis in cellular biology, is realized in this section.

The previous edition, published over 20 years ago, focused on the function of the lungs in light of traditional organ physiology. Coverage included lung volumes, ventilation, gas exchange, and mechanisms and control of breathing. The second edition aims to represent a field of knowledge that has grown dramatically and draws attention to the lung as an organ within a physiological system. This edition also includes the lungs' additional systems.

Reflecting the orientation of the editors, chapters new to the second edition include coverage of the development and growth of the lungs, pulmonary circulation, pulmonary metabolism, and pulmonary immunologic defense mechanisms. Expanded coverage in scope has resulted in a physical expansion of the section into four volumes, two with two parts each: volume 1, "Circulation and Nonrespiratory Functions"; volume 2, parts 1 and 2, "Control of Breathing"; volume 3, parts 1 and 2, "Mechanics of Breathing"; and volume 4, "Gas Exchange."

The *Handbook* series is rightfully recognized as standard library reference material. Chapters within the sections are thoroughly referenced and as reviews of the literature, provide important links to additional information. The *Handbook*'s usage can be difficult, however, because of the lack of a comprehensive index for each section's volumes. Indeed, subsequent to the expansion of coverage in the new edition, problems of access have been exacerbated. Additionally, as a result of its exhaustive treatment, this handbook functions more as a treatise than as what one traditionally thinks of as a handbook.

Libraries, departmental collections, and interested health professionals who have collected the previous edition will certainly want to update their holdings with this new version. Highly recommended, the section on the respiratory system remains a standard reference tool, both for librarians and library patrons. *GERALD J. PERRY*

335. Science of Stretching. Michael J. Alter. Champaign, IL: Human Kinetics Books, Human Kinetics Publishers, Inc., 1988. 243p., illus., bibliog., index. $26.00. 0-87322-090-0.

This is a welcome new book for the exercise enthusiast and for coaches who teach physical education and training. Prior to the appearance of Michael Alter's *Science of Stretching*, information about stretching was difficult to locate, fragmented, and, in most cases, too technical.

"The purpose of this text is to provide you with a general survey of current knowledge on flexibility in terms of its limitations and optimal development." Part I, "Factors Related to Felxibility and Stretching," is the biggest part of the book, analyzing the Factors related to flexibility and stretching. The first chapter is an overview, whereas chapter 2 covers "Contractile Components of Muscle: Limiting Factors of Flexibility." The remaining 10 chapters in this first part are "Connective Tissue as a Limiting Factor of Flexibility," "Mechanical and Dynamic Properties of Soft Tissues," "The Neurophysiology of Flexibility: Neural Anatomy and Neural Transmission," "Osteology and Arthrology," "Social Facilitation and Psychology in Relation to Stretching," "Potpourri," "Relaxation," "Muscular Soreness: Its Etiology and Consequences," "Types and Varieties of Stretching," and "Stretching Concepts."

Part II, "Functional Anatomy," contains three chapters that cover the anatomy of the human body—lower extremity and pelvic girdle, vertebral column, and upper extremity. All 15 chapters are very well written and include review questions and answers as well as recommended readings. Actually, this handbook could be used as a textbook. Finally, Part III contains 242 stretching exercises. These include basic, as well as more advanced, exercises. Each is explained in a step-by-step manner and includes a line drawing to illustrate how to accomplish the exercise. The book ends with an 11-page list of references, an author index, and a subject index. This is a highly recommended book for school, public, academic, and special libraries. Individuals who are into physical fitness should definitely have this book. *H. ROBERT MALINOWSKY*

336. **Standards for Blood Banks and Transfusion Services.** 12th. American Association of Blood Banks, Committee on Standards. Arlington, VA: American Association of Blood Banks, 1987. 52p., index. $10.00pbk. 0-915355-35-3pbk.

Standards for Blood Banks and Transfusion Services is the most authoritative handbook for blood banking, with the twelfth edition including major changes relating to donor testing, labelling, and record keeping as well as including new material on prevention of transfusion-associated AIDS and interim measures to attempt to reduce the risk of transfusion-associated non-A, non-B hepatitis.

This is a very concise handbook covering various procedures in collecting, testing, and storing blood. The first part, "Donors and Donor Blood," covers donor selection; collection of blood from the donor; therapeutic phlebotomy; preparation of blood components; testing donor blood; labelling; and conditions for storage, transportation, and expiration of blood and components. The rest of the handbook covers specific areas of blood banking including plasmapheresis, cytapheresis, compatibility testing, transfusion of blood and components, transfusion complications, autologous blood, records, and histocompatibility testing.

This well-written handbook is a must for all laboratories concerned with the collecting, testing, and storing of blood. *H. ROBERT MALINOWSKY*

Manuals

337. **Manual of Immunoperoxidase Techniques.** 2nd. Robert J. Wordinger; Ginger W. Miller; Donna S. Nicodemus. Chicago, IL: American Society of Clinical Pathologists, 1987. 98p., illus. $12.00. 0-89189-232-X.

Research in immunology is advancing quickly, as has been demonstrated in recent years by the rapid proliferation of numerous diagnostic and laboratory techniques available to researchers, pathologists, and medical technologists. The second edition of the *Manual of Immunoperoxidase*

Techniques, by Wordinger, Miller, and Nicodemus, a handbook of immunoenzyme techniques, attempts to update current knowledge in this highly specialized field in an easy-to-follow, clear and concise manner.

As stated in the preface to the first edition, the *Manual* grew out of a series of workshops, sponsored by the American Society of Clinical Pathologists, concerned with basic immunology and immunoperoxidase techniques. The second edition retains the first edition's objective but includes expanded data regarding recently developed techniques including the applications of recombinant protein-A, the inhibition of endogenous peroxidase activity, and the uses of proteolytic enzymes. These procedures and concepts will be understood by a well-defined coterie of technicians. However, the *Manual* does include an introduction to some basic tenets of immunology for the relatively uninitiated, including an explanation of antibodies, antigens, and enzymes and their immunologic roles.

According to the authors, this handbook is intended for a technical audience. Given that caveat, it may be recommended as a reference tool for those libraries and departmental collections supporting advanced research in immunology and the diagnostic applications of immunoenzyme techniques for the delivery of patient care. *GERALD J. PERRY*

338. **Practical Microscopic Hematology: A Manual for the Clinical Laboratory and Clinical Practice.** 3rd English. Fritz Heckner; H. Peter Lehmann; Yuan S. Kao. Baltimore, MD: Urban and Schwarzenberg, 1988. 120p., illus. (part col.), bibliog., index. $23.00pbk. 0-8067-0813-1pbk.

Practical Microscopic Hematology, subtitled "A Manual for the Clinical Laboratory and Clinical Practice," is an updated English translation of the 1986, sixth edition, of *Pracktikum der Mikroskopischen Hamatologie*. Previous English-language editions of this work were completed in 1980 and 1982. This book is a practical laboratory manual written to assist those clinicians, technicians, and students who diagnose blood disorders by visually inspecting blood or bone marrow smears under a microscope. Although it permits the user to compare and constrast healthy and diseased cells, it is not a comprehensive text in hematology, and the authors have intentionally omitted any discussion of the diseases illustrated. The myriad schematics, figures, tables and, photomicrographs (both color and black-and-white) form the substantive base of this 120-page manual. Around this core, "the text gives only such information considered necessary for an understanding of the illustrations." The book begins with a description of good smearing technique and concludes with a brief three-page index and one page of references. In between are three sections: "Normal Morphology of Blood and Bone Marrow Cells," "Morphology of Pathologic Changes," and the appendix. Both sections on morphology mention basic rules and erythropoiesis, granulopoiesis, thrombopoiesis,

lymphopoiesis, monocycle macrophage, and the reticuloendothelial system. Normal morphology includes useful comparative charts on the differential diagnosis of the 12 most important cells (based on the criteria of cell type, shape, nucleus, nucleoli, and cytoplasm) and the cytochemistry for 18 normal cell types (based upon the criteria of peroxidase, nonspecific esterase, periodic acid Schiff, and acid phosphatase). The section on diseased states includes the French/American/British classification of leukemias, a table on myeloproliferative diseases, and extensive information on the anemias. The appendix contains detailed procedures on how to prepare special stains, illustrations of bone marrow, and other information.

The new edition includes some new materials and text, plus color photographs have replaced some of the black-and-white illustrations. This and previous editions were written under the primary authorship of F. Heckner, Director of Clinical Hematology at Germany's Einbeck Hospital. H. P. Lehmann and Y. S. Kao, of the Louisiana State University Medical Center in New Orleans, contributed toward the English translation of the text. This book is most appropriate for the collections of medical/health science libraries and those special/academic libraries that support strong research/laboratory programs in human blood disorders. *BILL COONS*

339. Regional Guide to Human Anatomy. Alan Twietmeyer; Thomas McCracken. Philadelphia, PA: Lea and Febiger, 1988. 194p., illus. $12.95pbk. 0-8121-1103-6pbk.

This is a gross anatomy laboratory manual for students in allied health fields. The material is elementary and the illustrations sketchy. There is a list of common Greek and Latin roots, but no index. The arrangement is in four units covering upper limbs; thorax, abdomen, and pelvis; head and neck; and lower limbs. Each includes exercises for skeleton, musculature, circulatory system, and other systems of that region.
CAROLYN DODSON

340. Techniques in Surgical Casting and Splinting. Kent K. Wu. Philadelphia, PA: Lea and Febiger, 1987. 269p., illus., bibliog., index. $24.95pbk. 0-8121-1076-5pbk.

Despite the great importance of casts, cast-braces, and splints in orthopedic surgery, there has been a dearth of sources which discuss the techniques of casting and splinting in depth. In response to this need, Dr. Wu has designed a comprehensive encyclopedic textbook on both plaster and fiberglass casts, casting materials, and casting techniques, written in "cookbook" style. The text uses everyday English, avoiding the technical terms, and has a multitude of step-by-step photographs and drawings which illustrate the process of making each type of cast.

The first chapters contain a review of the history of casts and cast-braces, the advantages and disadvantages of various casts, and the practical uses of casts. This is followed by the materi-

als and equipment, basic casting techniques, and the complications of casts. The subsequent chapters progress logically from the simplest casts to the most difficult and elaborate casts, such as the Minerva and Risser casts. For each cast, Dr. Wu has noted the indications, the required cast materials, the patient's position, and a description of the technique for applying the cast. The chapters also have extensive bibliographies, and the volume has an index. This slender book is well designed, with a wealth of information that is not replicated elsewhere, and should be suitable for many levels of experience, from the paramedic and the athletic trainer to the medical students, residents, and the orthopedic surgeons. Dr. Wu's textbook is recommended for the collections which support these professionals. This textbook will likely be the "worn bible" in many professional offices. *JOAN LATTA KONECKY*

Textbooks

341. Basic Neuroscience: Anatomy and Physiology. Arthur C. Guyton. Philadelphia, PA: W. B. Saunders Co., 1987. 393p., illus. (part col.), bibliog., index. $34.95. 0-7216-2061-2.

Combining the principles of neuroanatomy and neurophysiology into one textbook, the author creates an integrated approach to the structure and function of the nervous system for a broad audience of students from many disciplines. The volume begins with a section on the gross anatomy of the nervous system, highlighted by clear four-color illustrations of the important neuroanatomical structures. The subsequent chapters detail the functional anatomy and physiology of each element of the nervous system. The textbook incorporates a wide range of subjects, including membrane biophysics; neuromuscular transmission; sensory receptors; somatic sensations; motor functions; cerebral cortex function; theories about activation of the brain, behavior, and sleep; the autonomic nervous system; the senses of vision, hearing, taste, and smell; the neural control of respiration, body temperature, and other body functions.

The chapters are visually easy to read and easy to scan thanks to the distinctive red headings which identify specific segments of text. The book is also enriched with many detailed two-color pictures and a lengthy bibliography. In addition, there is an index at the end of the book.

As a basic textbook, this volume does a credible job of integrating neuroanatomy and neurophysiology into one cohesive discipline without overwhelming the student with unwarranted depth and detail. It is a very well-designed textbook and should be considered for collections supporting studies in the life sciences, especially biology, anatomy, physiology, psychology, and medicine. *JOAN LATTA KONECKY*

342. Elias-Pauly's Histology and Human Microanatomy. 5th. Peter S. Amenta, ed. New York, NY: John Wiley and Sons, 1987.

566p., illus. (part col.), index. $38.00. "Wiley Medical Publication." 0-471-83646-X.

This new edition of a classic histology text is a beautiful book. The 4th edition was published in 1978 and has been used extensively as a textbook. The editor of this revision, Peter S. Amenta, is well qualified for this work and has written a histology text of his own. He is faithful to the successful format of previous editions and to the nomenclature of the 1983 edition of *Nomina Anatomica*. There are carefully labelled photographs and diagrams on nearly every page. These clearly reproduced illustrations, many of them in color, are the most valuable feature of this volume and make it a useful reference book for medical libraries, as well as a textbook.

RUTH LEWIS

343. Essentials of Surgery. David C. Sabiston, Jr. Philadelphia, PA: W. B. Saunders Co., 1987. 1,249p., illus., bibliog., index. $39.95. 0-7216-1820-0.

Intended to be a more compact version of the medical school standard, *Textbook of Surgery: The Biological Basis for Modern Surgical Practice*, this volume was designed to provide thorough but concise coverage of the basic principles and concepts of surgery, with ample descriptive detail and an emphasis on the "information regarded as essential for all medical students regardless of career choice." Ninty-one highly respected contributors present chapters on the surgical specialties of cardiothoracic surgery, neurosurgery, plastic and maxillofacial surgery, orthopaedics, otolaryngology, gynecology, and urology.

A sampling of the initial chapters demonstrates the comprehensive nature of the textbook: "Homeostasis and Shock," "Fluid and Electrolyte Management," "Preoperative Preparation," "Hemostasis," "Nutrition and Metabolism," "Anesthesia," "Wound Healing and Management," "Burns," "Surgical Infections," "Surgical Complications," "Transplantation," and "Surgical Aspects of Viral Hepatitis and Acquired Immunodeficiency Syndrome." The chapters are clearly written, well organized, and have a concluding bibliography. The chapters also are enhanced by a special section, "Selected References," which contains an annotated list of the more important references on the chapter topic, to be used for self-study by the student.

The textbook also contains a superb number of high quality photographs, figures, and tables. *Essentials of Surgery* exhibits a commendable attention to detail and quality. It has accomplished its goal of providing a compact yet comprehensive resource for medical students and should be considered a good choice for a medical school library collection.

JOAN LATTA KONECKY

344. Leavell and Thorup's Fundamentals of Clinical Hematology. 5th. Oscar A. Thorup, Jr. Philadelphia, PA: W. B. Saunders Co., 1987. 1,013p., illus.

(part col.), bibliog., index. $75.00. 0-7216-5679-X.

This is the fifth edition of a textbook that strives to present a clinically oriented, comprehensive discussion of hematology "that will help students, residents, and practicing physicians keep up with increasing demands for self-study." The volume places a strong emphasis on diagnosis and treatment to maintain its clinical usefulness yet also presents the fundamentals of hematologic physiology as a foundation to the understanding of blood disorders.

The first seven chapters explain in detail the normal structure and function of the blood system components, the immune system, and the processes of hemostasis and coagulation. The succeeding chapters discuss the various hematologic disorders and transfusion therapy. As a result of the many changes made in the field of hematology since the last edition published in 1976, there were extensive revisions made to most chapters, and some chapters were completely rewritten. The chapters have lengthy, not predominately contemporary, bibliographies, and there is a detailed index in the back of the volume.

There are several colored illustrations and a fair number of black-and-white tables, photographs, and illustrations. A classic textbook which has been extensively updated, this edition, like its predecessors, will be useful for students and practitioners alike and may be considered a worthwhile addition to a hematology or medical textbook collection. *JOAN LATTA KONECKY*

345. Skin Surgery. 6th. Ervin Epstein; Ervin Epstein, Jr. Philadelphia, PA: W. B. Saunders Co., 1987. 676p., illus., bibliog., index. $95.00. 0-7216-1809-X.

Epstein and Epstein's *Skin Surgery* is a minor medical classic, a standard reference for both dermatologists and dermatologic surgeons. The sixth edition of this popular book, published nearly 30 years after the first, differs considerably from the previous fifth edition, yet still remains the useful resource it has always been.

Featuring contributions from over 50 topical specialists, *Skin Surgery* surveys a wide range of dermatologic surgical considerations. Introductory chapters address broad concerns including principles of office surgery, local anesthetics, basics of skin surgery, and outpatient emergencies. Subsequent chapters focus specifically on various procedures and techniques such as laser surgery, chemical peeling, liposuction, hair transplantation, and electrosurgery. Additionally, specific dermatologic diseases, including cancer, nevi, warts, and acne, receive special attention.

The fifth edition consisted of two volumes and, according to Epstein and Epstein, was too long, with some repetition of material. This new edition has been reduced to a single volume and has been considerably updated through rewriting and the contributions of many new authors. Of particular interest is the inclusion of data considered by the editors to be "beyond the skills of dermatologists or even dermatologic surgeons." Accordingly, this information is included for readers' edification and to demonstrate the advances

made in recent years in the understanding and treatment of skin diseases. The sixth edition of *Skin Surgery,* as a survey of technical, surgical approaches to skin diseases, is highly recommended for medical reference collections.

GERALD J. PERRY

Treatises

346. **Blood Group Systems: Rh.** Virginia Vengelen-Tyler; Steven R. Pierce, eds. Arlington, VA: American Association of Blood Banks, 1987. 103p., illus., bibliog. $16.00. 0-915355-41-8.

Blood Group System: Rh is a collection of papers which were presented at a workshop in 1987. The purpose of this workshop was to "present the state of...knowledge of the Rh system." The individual papers discuss the biochemistry, genetics, immunology, and serology of the Rh system. Previous contributions by the editors and the authors of the individual papers indicate that this book should be a significant contribution to the primary literature in its subject area and a valuable addition to medical libraries.

LOREN D. MENDELSOHN

347. **New Developments in Biotechnology: Ownership of Human Tissues and Cells.** Philadelphia, PA: Science Information Resource Center, J. B. Lippincott Co., 1988. 168p., illus., bibliog., glossary. $34.00. 0-397-53002-1.

In 1987, the United States Office of Technology Assessment (OTA) initiated a series called New Developments in Biotechnology. *Ownership of Human Tissues and Cells* is the first volume. Biomedical research has been utilizing cells from healthy and diseased patients for many years. In the more recent past, techniques for manipulating cells and tissues have advanced to the point that researchers and companies may be using them to develop commercial feasible products. The government (the courts and Congress) has indicated that altered forms of cells may be patentable. The question arises about the ownership of cells, tissues, and body parts. Is the donor of the cells that were used to create a cell line for research or the raw material for a gene-cloning procedure entitled to economic compensation? Who owns the cells? This report by the OTA Task Force outlines the issues, the possible solutions, and ramifications of the possible solutions.

Questions in the areas of economics, law, and ethics need to be addressed. The first chapter is "Summary, Policy Issues, and Options for Congressional Action." Chapter 3, "The Technologies," explains hybridoma technology and other innovations. "The Interested Parties," chapter 4, discusses the roles and concerns of donors, researchers, and the biotechnology industry. It includes a table of products being developed from human cells and tissues that may be available over the next few years. Legal, economic, and ethical issues are discussed in the succeeding

chapters. Two handy appendices are included: "Code of Federal Regulations, Part 46, Subpart A: Basic Policy for Protection of Human Research Subjects, Department of Health and Human Services" and a "Glossary."

Ownership of Human Tissues and Cells makes a valuable contribution to the debate on this subject. It is appropriate for many audiences and readable on the nontechnical level. This publication is also available from the Government Printing Office for $7.50pbk (s/n 052-003-01060-7; SuDocs No. Y3.T22/2:2 B 52/4/v.1)

FRANK R. KELLERMAN

348. **The Pituitary Gland.** C. R. Kannan. New York, NY: Plenum Medical Book Co., 1987. 594p., illus., bibliog., index. $79.50. Clinical Surveys in Endocrinology, v. 1. 0-306-42506-8.

The purpose of the first volume in this new series is "to integrate the current knowledge in this dynamic field with the existing body of information already available to the clinician. The chapters in this book attempt to portray current research information seen through the eyes of a clinician." This book has brought together a great amount of information on the pituitary gland and presents it in such a way that the researcher can find information easily. Included with the text are some 1,500 citations to references on the pituitary gland. After a brief history of the pituitary gland in chapter 1, the author discusses the gland itself, diseases, and functions in the remaining 20 chapters: growth hormone, acromegaly and gigantism, hypopituitary dwarfism, thyrotropin and thyrotropin-releasing hormone, inappropriate pituitary thyrotropin secretion, central hypothyroidism-trophoprivic hypothyroidism, adrenocorticotropic hormone, Cushing's disease, isolated ACTH deficiency, LH and FSH (the gonadotropins), hypogonadotropic hypogonadism, prolactin, prolactin-secreting tumors, hypopituitarism, tumors of the pituitary, pituitary apoplexy, the empty sella syndrome, arginine vasopressin, diabetes insipidus, and syndrome of inappropriate secretion of ADH.

Dr. Kannan is chairman of the Division of Endocrinology and Metabolism in the Department of Medicine at the Cook County Hospital in Chicago, Illinois, and a clinical associate professor of medicine at the University of Illinois at Chicago. His highly specialized book will be a welcome addition to the literature on the pituitary gland and will be a needed reference source in medical libraries. *H. ROBERT MALINOWSKY*

★
NURSING
Biographical Sources

349. **Who's Who in American Nursing, 1986–1987.** 2nd. Jeffrey Franz; Pamela A. Vandervort-Jones, eds. Owings Mills, MD: Society of Nursing Professionals, 1987.

615p., index. $69.95. 0-940863-03-0. 0740-7912.

This is a selective biographical directory laid out in the very familiar alphabetical format of other Who's Who directories. In the preface, the managing editor states that the goal is to cite the top 10,000 professionals in the country (U.S.). Although he does not say how many individuals are represented in this directory, he does explain that this edition has 3,000 more entries than were in the first edition. The elements of "reference interest" in selecting entries for inclusion are weighted. Weights are assigned to education achievement, advancement to positions of responsibility, contributions to nursing literature and research, honors, awards, fellowships and special appointments, demonstrated leadership in professional organizations, professional certification by national organizations, and achievement in the nursing field beyond that of the majority of contemporaries.

Entries include personal and family information; educational institutions and degrees; areas of work and practice; positions held; awards and honors given; and certifications, publications, research, and offices held. The basic document is an alphabetical-by-last-name arrangement, but it also contains an index that is classified by areas of practice. Within the classes, the arrangement is by state and then alphabetical. These are added points of access.

There are other types of nursing directories, but apparently not any based on selection criteria of achievement. This would be a very good investment for a reference collection that covers the nursing field. LILLIAN R. MESNER

Directories

350. **Doctoral Programs in Nursing, 1986–1987.** New York, NY: National League for Nursing, Council of Baccalaureate and Higher Degree Programs, 1987. 13p. $4.95pbk. Publication no. 15-1448. 0-88737-351-8.

Doctoral Programs in Nursing, 1986–1987 is the most recent annual issue of a small directory produced by the National League for Nursing. The directory is a very simple listing of programs in 26 states and the District of Columbia. Its four columns give only the institution's name and address, the name and title of the resident administrator, the areas of study covered, and the degree awarded. A brief paragraph at the head of the first column mentions that the league does not accredit these programs, and it directs the user to get the appropriate information about programs from the individual institutions. This pamphlet would be an appropriate addition to the reference collection in any nursing school.

LILLIAN R. MESNER

Handbooks

351. **Applied Psychiatric—Mental Health Nursing Standards in Clinical Practice.** Mary Jane Schirger Krebs; Kenneth H. Larson, eds. New York, NY: John Wiley and Sons, 1988. 388p., bibliog., index. $22.95. "A Wiley Medical Publication." 0-471-84530-2.

"This text is written for nurses in general hospital settings, ambulatory services, and home-care and psychiatric facilities as a guide for managing challenging clinical issues through the utilization of psychiatric nursing standards." The past ten years have placed many demands on the practicing nurse. "Not only must the complexities of the equipment and medications used be mastered, but also the nurse must effectively approach the patient's reactions to illness." To this end, the profession has developed standards that can be applied to every situation that the nurse may encounter. This book is intended to assist the nurse in the practical application of these standards in daily clinical practice so that its "use and experience will foster confident independent functioning."

The 14 contributors have produced a book that is detailed and full of information concerning clinical nursing procedures and practices. Part I is an introduction, "Professional Practice Standards and Nursing," with two chapters outlining "The Nursing Process" and "Theory: A Basis for Practice." These two chapters form the background for the two remaining parts. Part II, "Management Issues in Clinical Practice," covers 10 situations that a clinical nurse may encounter—coping, anxiety, noncompliance, disorientation and confusion, splitting, depression and suicide, manipulation, thought disorders, control, and eating disorders. Part III, "Future Directions," gives some insight on the factors that will influence future nursing practice. Each chapter is similar in format beginning with a brief introduction followed with the various standards that apply to the specific situation. The standards are fully described in a well-written text. Each chapter concludes with an outline of the guidelines that have been discussed by way of the standards.

This is an excellent handbook and should be available to all nurses who are involved with psychiatric care. It is well written, easy to use, and accurate. H. ROBERT MALINOWSKY

352. **Nurses' Handbook of Health Assessment.** Janet Weber, ed. Philadelphia, PA: J. B. Lippincott Co., 1988. 401p., illus., bibliog., index. $15.95. 0-397-54685-8.

The *Nurses' Handbook of Health Assessment* is designed by its author to remind students and practitioners of what must be covered in the assessment process. She warns that nothing in the book is covered in depth and that other sources have to be referenced to cover an area more thoroughly.

The book's coverage is very comprehensive. The first two chapters cover the overview of the nursing history, the history's components, and questions used to elicit subjective information from the patient. Chapter 3 gives the guidelines for performing the physical assessment, and the remaining 15 chapters give the techniques and procedures for assessments on all of the body systems. For each procedure, a normal finding is given along with deviations from normal findings. Pediatric variations on findings are also given. Many line drawings illustrating procedures are present where appropriate. At the end of the handbook, there are 11 appendices that give common laboratory values, developmental norms, growth charts, immunization tables, recommended dietary allowances, a sample history, height and weight charts, and an eye chart.

The design of the handbook enhances its usability. Except for the first two chapters, the rest of the book is in outline form with columns and tables. It is pocket-sized so that it can travel with the user and it has section headings in red ink. It is a good design for quick reference. By utilizing the table of contents, the user can go directly to the system under concern and flip into the specific area wanted. The second finding option is a very comprehensive index.

Meant for professionals familiar with its technical language, the handbook assumes basic knowledge of anatomy and general medical terminology. It most probably would be used by practitioners, but it would also be a good addition to a reference collection in a nursing library or nursing unit. *LILLIAN R. MESNER*

Manuals

353. **Manual of Home Health Care Nursing.** Joleen Walsh; Carol Batten Persons; Lynn Wieck. Philadelphia, PA: J. B. Lippincott Co., 1987. 592p., illus., glossary, index. $29.75pbk. 0-397-54616-5pbk.

Diagnosis-related groups (DRGs) have drastically changed the way hospitals are reimbursed for hospitalized Medicare patients. These patients are discharged earlier, resulting in a rapid increase in the need for home health care nursing services. The professional nurse must adapt acute-care techniques, formally found only in the hospital setting, for use in the home care situation. This book presents a large variety of acute-care techniques, with emphasis on teaching the client and the family members to perform as many self-care activities as possible. Clearly detailed are techniques in the areas of activity/rest, circulation, elimination, food/fluid, hygiene, monitoring/surveillance, safety, and ventilation, ranging from bed making and low back pain exercises to Holter monitors, enteral feeding tubes, and foley catheters.

The book methodically reviews nursing assessment, diagnosis, intervention, and health promotion counseling elements for each technique, using a clear and logical outline format. Age-specific modifications are provided for the tech-

niques when appropriate. Other helpful segments include documentation, with suggestions for record keeping, a health teaching checklist, a product availability segment, which lists products and vendors, and a short list of selected references. There are several useful appendices, including home health care resources, teaching/learning principles and strategies, home health care certification form, isolation precautions, self-help devices, and environment for the visually impaired. There are also a glossary, a list of medical abbreviations, and a subject index.

This book can be recommended for its useful outline format, the detailed, clearly written information and the abundant illustrations found throughout the text. The manual will be a very useful guide for practitioners in the home health care field and nurses who teach self-care techniques to patients. It should be highly considered for most, if not all, nursing and home health care collections. *JOAN LATTA KONECKY*

Tables

354. **Nursing Student Census with Policy Implications, 1986.** Peri Rosenfeld. New York, NY: National League for Nursing, Division of Public Policy and Research, 1987. 86p. $18.95. Publication no. 19-2175. 0-88737-356-9.

The *Nursing Student Census with Policy Implications, 1986* is a collection of statistics covering nursing students and nursing education. It is based on an annual census done by the National League for Nursing. The preface mentions that this title is a "module" or part of a larger document called *Nursing Data Review* that, like this, is issued on an annual basis.

It is a well-organized work with clear, standard, statistical tables. Each broad subject area has an executive summary preceeding it that gives interpretations of the statistics and, where appropriate, points out historical trends. There is also a general summary of the Nursing Student Census for the year of 1986. Finding the statistics desired is easily done by consulting the table of contents.

The preface points out that the league is committed to a 100 percent response rate on certain portions of the census and that approximately a 90 percent response rate is obtained on others. This is impressive. Review of a few back issues demonstrates that data from new areas of concern, such as male and minority students, are being added. Historical tables seem to indicate that the census is evolving; most tables go back to the 1960s. The census covers all levels of education for RNs. At a very modest price, this document would be an excellent addition to the reference collections in medical, academic, and special libraries as well as in large public libraries. These statistics would be of value to educators, administrators, policymakers, and anyone concerned with supply and demand of nurses. *LILLIAN R. MESNER*

Textbooks

355. Techniques in Clinical Nursing: A Nursing Process Approach. 2nd. Barbara Kozier; Glenora Erb. Menlo Park, CA: Addison-Wesley Publishing Co.., Health Sciences Division, 1987. 1,024p., illus., bibliog., glossary, index. $37.75. 0-201-11755-X.

Techniques in Clinical Nursing is a text covering the techniques used in daily patient care. The second edition is updated and revised as well as elaborated. It is now in three parts: the text (which is smaller), the instructor's manual, and the student's checklist. Many more techniques are added, as are illustrations of the techniques. Also, a new appendix contains variations of the techniques for care in the home situation. The text is organized so that the chapters can be used in any order. This gives flexibility to variant teaching situations. After the first four chapters, which are general orientation chapters, chapters 5 through 42 cover specific technique areas. Each chapter starts with an outline, objectives, and terms with their definitions. Then the chapter is divided into the specific techniques covered in that area (e.g., intravenous therapy had 15 specific techniques). The scope of the text does not take in specialty areas such as intensive care or orthopedic care. Meant for basic nursing techniques, it is very thorough and well illustrated, however, and gives numerous examples to illustrate each technique. The underlying philosphy and rationale for the techniques are also presented. This is a high quality text for nursing students on any level of nursing training or education. *LILLIAN R. MESNER*

★
PHARMACOLOGY/PHARMACY
Bibliographies

356. Drug Abuse Bibliography for 1984. Polly T. Goode, comp. Troy, NY: The Whitston Publishing Co., 1987. 729p., bibliog., index. $63.50. 0-87875-322-2.

Goode's *Drug Abuse Bibliography for 1984* is the 14th annual supplement of *Drugs of Addiction and Non-Addiction, Their Use and Abuse: A Comprehensive Bibliography, 1960–1969*, originally compiled by Joseph Menditto. Included are unannotated citations to international literature about drug abuse for 1984, drawn from over 50 major abstracting and indexing sources, including the *Alternative Press Index, America: History and Life, Catholic Periodical and Literature Index, Dissertations Abstracts International, Index Medicus, Popular Periodical Index, Sociological Abstracts*, and *Women's Studies Abstracts*. These titles represent only a brief sampling of the wide variety of sources drawn on to compile this exhaustive resource tool.

The bibliography is divided into four sections with addendum indexes and lists. The first section includes references to books, monographs, and pamphlets arranged alphabetically by author. The second and third parts include citations to periodical literature and consist of a title index where citations are arranged alphabetically by title and a subject index where references are arranged alphabetically by subject term. Included in preliminary pages are lists of the subject terms used and journals cited. Finally, an author index covering all included references completes this bibliography. The practice of using unannotated published bibliographies for topical research has waned in recent years, greatly because of the advent and widespread use of computer-assisted database searching to retrieve subject-specific, tailored lists of references to the literature. However, this bibliography series remains highly recommended as a resource for drug-abuse literature in great part because of the nearly Herculean effort of its editors in thoroughly gleaning relevant citations from so many important and varied sources. Public library clientele and medical and related health area professionals will all find this bibliography a practical, easy-to-use tool for accessing drug-abuse literature. *GERALD J. PERRY*

Dictionaries

357. Organic-Chemical Drugs and Their Synonyms (An International Survey). 6th rev and en. Martin Negwer. New York, NY: VCH Publishers, 1987. 3v., index. $125.00. 0-89573-550-4 set.

This international standard reference work on organic-chemical drugs contains close to 9,000 entries with about 80,000 synonyms. The main part of the dictionary includes drugs that had come to the author's attention by the end of 1982 with the supplement including those to the end of 1985. The drugs are arranged in two columns by the concept of incremental molecular formulas. For each drug, the following information is given: consecutive number for indexing purposes, molecular formula with arrangement according to Hill, Chemical Abstracts Service index number (CAS number), structural formulas, systematic names, references marked by "R," synonyms marked by "S," and the drug's concern or therapeutic use, marked by "U." References may be the "identity of the salts derived from the acids or bases listed" or "certain additives in some drugs and the necessary links with other consecutive numbers."

With such a volume of information, there have to be multiple access points. The author has provided these in the third volume through the use of several indexes. The first index is a group index that lists the drugs under one or more of 2,200 catchwords. Following each catchword is a list of the consecutive numbers that refer to that catchword. The CAS number index is simply a listing of the various CAS numbers with the cor-

responding consecutive number. Finally, the synonym index is the key to accessing the first two volumes when one has a drug name in hand. This includes tradenames, slang, and technical names.

This is a highly recommended set of books for academic, medical, and special libraries that serve researchers in drug research. It has become one of the standard sources for the identification of drugs that are mentioned in the literature.

H. ROBERT MALINOWSKY

358. USAN and the USP Dictionary of Drug Names: A Compilation of the United States Adopted Names (USAN) Selected and Released from June 15, 1961 Through June 15, 1987, Current USP and NF Names for Drugs, and Other Nonproprietary Drug Names.
Mary C. Griffiths; Carolyn A. Fleeger; Lloyd C. Miller, eds. Rockville, MD: United States Pharmacopeial Convention, Inc., 1987. 708p., illus. $69.50. 0-912595-23-3. 0090-6816.

The U.S. Food and Drug Administration has stated that interested persons may rely on *USAN and the USP Dictionary of Drug Names* for the established name for any drug in the USA. This authoritative dictionary lists the nonproprietary names, brand names of research-oriented firms, investigational drug code designations, and Chemical Abstracts Service registry numbers for drugs. The nonproprietary names include the USAN (U.S. Adopted Names), compendial names (National Formulary or U.S. Pharmacopeia), or, for drugs not currently recognized in the U.S., the international nonproprietary names (INN). The USAN program is the formally organized effort authorized to produce "simple and useful nonproprietary names for drugs" while the drug is still in the investigational stage.

This 25th annual edition has returned to the original format of a single main list that includes all names in one alphabetic sequence. The main list is followed by a grouping of the USAN, USP, and NF names into categories of pharmacologic activity. The four appendices contain the Guiding Principles for Coining U.S. Adopted Names for Drugs; Chemical Abstracts Service registry numbers and National Cancer Institute numbers; molecular formulas; and the names and addresses of domestic firms concerned with compounds for which USAN have been selected. This dictionary is the authoritative list of established names for drugs in the U.S. and should be included in the collection of any library which has a need for drug names, especially medical and pharmaceutical libraries. The return to the original format makes this volume easier to use than were the previous two editions.   *JOAN LATTA KONECKY*

Directories

359. The Sourcebook for Innovative Drug Delivery: Manufacturers of Devices and Pharmaceuticals, Suppliers of Products and Services, Sources of Information.
Santa Monica, CA: Canon Communications, Inc., 1987. 152p., index. $75.00. 0-9618649-0-7.

Drug delivery methods are rapidly changing, and it is becoming more difficult to keep up with these advances and their suppliers. *The Sourcebook for Innovative Drug Delivery*, the first centralized source for information on these methods, is the result of a joint effort between Canon Communications, Inc., the publisher of *Medical Device and Diagnostic Industry* (*MD&I*) and *Medical Product Manufacturing News* (*MPMN*), and Tyson Consulting Group, Inc., a health care consulting firm with over 40 years' experience in health product development. The first of four major sections lists manufacturers of drug delivery devices and pharmaceuticals. This is an alphabetical list of eight categories, from aerosols to transdermal products, which is then further subdivided into the subcategories: (1) drug delivery devices and (2) pharmaceuticals. Manufacturers' names are then alphabetically listed in their appropriate categories. According to the introductory material, most companies chose the categories and subcategories under which they are listed.

Section 2, "Suppliers of Products and Services to Device and Pharmaceutical Manufacturers," lists companies who provide equipment, supplies, etc. for the manufacturing of drug delivery systems. Companies are arranged under eight categories some of which include: chemical and biological materials; plastics; and research and development. The third major, and most useful, section is a directory of over 300 companies involved in innovative drug delivery. Entries include name, address, telephone number, parent company, and key personnel. Also included are the company's primary area of business, e.g., service supplier, device manufacturer, etc. The last section includes sources of information including governmental agencies, online databases, professional and trade organizations, publications, and universities. Addresses are provided for agencies, universities, and organizations. The online databases section includes a brief description of the databases including addresses for three major vendors.

The book also includes some introductory material explaining the purpose of the book, how to use the sourcebook, and an overview of the drug delivery techniques. The two most valuable sections are 3 and 4: the directory and the sources of information. Some of the category indexes of the first two sections are not very specific, and users will have to be familiar with the drug-delivery industry to successfully use these sections. The book is well organized and accurate and, according to the publisher, took a period of almost two years to compile, edit, and update. Because this is a relatively new industry, the number of com-

panies is likely to become much larger than those included in this work. The publishers mention later editions of this work but do not say how often to expect new editions. To remain effective, *The Sourcebook for Innovative Drug Delivery* will have to be updated on a regular basis. The drug-delivery industry as well as medical and academic libraries will find this a useful work especially if the publishers attempt to keep the information as current and accurate as possible.

KIMBERLEY M. GRANATH

Handbooks

360. **Drug Facts and Comparisons.** St Louis, MO: Facts and Comparisons, J. B. Lippincott Co.., Philadelphia, PA, 1988. 2,283p., index. $69.50. 0-932686-88-5. 0277-9714.

The 1988 edition of the classic text *Facts and Comparisons*, renamed in 1982 to *Drug Facts and Comparisons*, continues the tradition of objective drug information arranged for comparison of similar drug products. This edition has been revised substantially, with 24 new drugs, hundreds of new drug products, dosage forms and formula changes, and more than 35 new tables and diagrams. The information found in each of the monographs is based on the current FDA-approved package literature and supplemented from the biomedical literature. In addition to the FDA-approved indications and dosage recommendations, other established or potential uses are included under the heading "Unlabeled Uses."

Organized by therapeutic class, there are 12 chapters which are further divided in groups and subgroups to allow products of similar content or use to appear together. Each chapter begins with a complete outline of the information included. The drug monographs provide general information for groups of related drugs or specific information for a particular generic drug. Each of the monographs is divided into nine sections: "Actions," "Indications" (and "Unlabeled Uses"), "Contraindications," "Warnings and Precautions," "Drug Interactions," "Adverse Reactions," "Overdosage," "Patient Information," and "Administration and Dosage." Following each monograph are the product listings, grouped by dosage form or strength, with a cross-reference to the prescribing information in the drug monograph at the top of the page. Additional details are provided by indication codes at the left side of the page showing distribution status, prescription, or over-the-counter; sugar-free liquid preparations; and FDA-controlled substances. The "Cost Index" reflects the relative wholesale cost for equivalent amounts of the products. The alphabetical index lists references for all the drugs by their generic name, brand name, synonyms, common abbreviations, and therapeutic group names. This excellent work continues to provide a wealth of information for pharmacists, other health professionals, and interested laypersons. It is a valuable reference tool in medical, academic, and public libraries. *JOAN LATTA KONECKY*

361. **Handbook of Drugs for Tropical Parasitic Infections.** L. L. Gustafsson; B. Beermann; Y. A. Abdi. Philadelphia, PA: Taylor and Francis, 1987. 151p., illus., bibliog., index. $44.00, $20.00pbk. 0-85066-403-9, 0-8506-404-7pbk.

Tropical parasitic diseases affect millions of humans. Drug therapy for these diseases is an important part of the solution to this problem. In recent years, several texts on tropical medicine and atlases of human parasitology have been published. The *Handbook of Drugs for Tropical Parasitic Infections* provides information on over 40 drugs used to treat tropical parasitic infections. The information is based on handbooks of pharmacology, texts on tropical medicine, and quite current journal articles.

The work was supported by grants from the Swedish Agency for Research Cooperation with Developing Countries (SAREC). All three authors are medical doctors. Lars L. Gustafsson is at the Karolinska Institute, Huddinge University Hospital, Stockholm, Sweden. Bjorn Beerman is with the Department of Drugs, National Board of Health and Welfare, Uppsala, Sweden. Yakoub Aden Abdi is on the faculty of medicine in Mogadishu, Somalia.

This well-organized handbook contains a list of abbreviations; a brief section of drug recommendations, arranged by type of protozoal or helminthic infection; and a useful index. The body of this monograph consists of essays on selected drugs, arranged alphabetically by generic name. The clinical pharmacology for each drug has been emphasized. Information on chemical structure, pharmacology and mechanism of action, pharmacokinetics, clinical trials, adverse effects, and dose schedule is provided. Each essay includes a detailed list of references. The handbook is a convenient resource for a clinical doctor who is treating tropical parasitic infections and a useful reference source in medical libraries.

SHARON GIOVENALE

Manuals

362. **Current Good Manufacturing Practice for Finished Pharmaceuticals (CGMPRs).** 7th. Langeloth, PA: Keystone Press, 1987. 58p. Handi-Regs no. 8001. not available. 0-940701-10-3pbk.

This small manual is a reprint of part 211 of FDA regulations that cover "Current Good Manufacturing Practices for Finished Pharmaceuticals." Part 211 includes 10 subparts: A—"General Provisions," B—"Organization and Personnel," C—"Buildings and Facilities," D—"Equipment," E—"Control of Components and Drug Product Containers and Closures," F—"Production and Process Control," G—"Packaging and Labelling," H—"Holding and Distribution," I—"Laboratory Controls," J—"Records and Reports," and K—"Returned and Salvaged Drug Products." These are the word-for-word regulations with no additional information. Drug manu-

facturing firms will have the full regulation, but this small reprint of specific sections should be useful in special libraries.

H. ROBERT MALINOWSKY

363. **Manual of Antibiotics and Infectious Diseases.** 6th. John E. Conte, Jr.; Steven L. Barriere. Philadelphia, PA: Lea and Febiger, 1988. 392p., bibliog., index. $19.50. 0-8121-1107-9.

The *Manual of Antibiotics and Infectious Diseases* is a reference work intermediate, both in size and scope, between the "pocket manuals" familiar to most clinicians and a comprehensive treatise. Contemporary examples of these two extremes include the 50-page *Pocket Manual of Antimicrobial Agents* (10th ed., 1987), published by the Mayo Foundation; or the 133-page *Handbook of Antimicrobial Therapy* (1986), published by the Medical Letter; and the 1,678-page treatise, *The Use of Antibiotics: A Comprehensive Review with Clinical Emphasis* (4th ed., 1987, Lippincott). All are established publications, with similar subject fields, and they tend to be revised with a frequency inversely proportional to their size.

Frequent revision is desirable for books about this continually changing field. This edition reflects new information and emphasis appropriate for recent progress in treatment or prevention of some diseases, and changes in the incidence of, and types of infectious diseases—especially those associated with the acquired immune deficiency syndrome (AIDS) and other sexually transmitted diseases.

As stated in the preface, "this manual is designed for students, housestaff, practicing physicians, and other health professionals involved in the day-to-day care of patients with infectious diseases." It provides data about clinical applications of antimicrobial agents in a quick reference format but also has annotations about rationale, precautions, concerns, and principles of use to support judgments required under unusual circumstances or in case of complications. Furthermore, the authors consider therapeutic and prophylactic use of not only antibiotics (natural and synthetic agents for use against a variety of microbial forms), but also immunobiological reagents both for passive and active immunization against bacteria and virusus.

The contents are organized in seven sections: "Antibiotics," "Empiric Antibiotic Therapy," "Therapy of Established Infection," "Antibiotic Susceptibilities," "Prophylactic Antibiotics," "Availability and Clinical Use of Immunobiologic Agents and Antiparasitic Drugs," and "Sexually Transmitted Diseases—Treatment Guidelines." In the previous edition, a separate section was devoted to hepatitis, but the topic is now treated in section 4. under a segment dealing with materials "...distributed by the Centers for Disease Control." The book provides a large amount of pertinent data in a volume of manageable size. However, incomplete indexing makes access to some information difficult.

The reference lists tend to be short, general in nature, and mostly contain citations prior to 1985. The information appears to be current in most cases, even though specific documentation is rare or absent. There is no mention of the use of azidothymidine (AZT, Retrovir, or Zidoudine) for AIDS patients. There is only a two-page discussion of AIDS, confined to methods for prevention of infection, and a few comments about testing for antibody against human immunodeficiency virus. Treatment of opportunistic infections is, thus, relegated to whatever information is given about specific antibiotics.

Although it will not "fit in a pocket," physicians and nurses would find personal copies of this book useful for quick reference and frequent review, because of its concise presentation of data on a large range of subjects, accompanied by helpful brief explanations. This book is strongly recommended for libraries serving hospitals or schools of medicine or nursing. *PAT JAMESON*

★
PUBLIC AND SOCIAL HEALTH
Biographical Sources

364. **Sigmund Freud.** Paul Roazen, ed. New York, NY: Da Capo Press, Plenum Publishing Corp, 1985. 186p., bibliog. $9.95pbk. Da Capo Series in Science; "A Da Capo Paperback." 0-306-80292-2pbk.

This paperback includes a biography of Sigmund Freud and selected essays by 10 contributors discussing various aspects of Freud's theories and concepts. It is an interestingly written book with the emphasis on the essays rather than on the biography. Essays include "Critique of Neo-Freudian Revisionsism," by Herbert Marcuse; "The Superego and the Theory of Social Systems," by Talcott Parsons; and "Freud and Woodrow Wilson," by Paul Roazen.

H. ROBERT MALINOWSKY

Dictionaries

365. **AIDS to Zits: A "Sextionary" for Kids.** Carole S. Marsh. Bath, NC: Gallopade Publishing Group, 1987. 28p. $4.95pbk. 1-55609-210-5pbk.

This is by far one of the most interesting and well-written dictionaries of sex terms for children. The definitions are frank and may offend some people, but they are written in the terminology that a child can understand. For example: "Anus, anal sex: The anus is the opening you have a bowel movement from when you go to the bathroom. Anal sex is putting a man's penis into a man's or woman's anus...." "Horny (HOR NEE): A slang word that means you feel like you want to have sex." "Ejaculation (E JACK U LAY SHUN): When semen squirts from a boy's penis during orgasm." And then there are the some-

what humorous, but advising, definitions such as "Baby: What people who have sex without any birth control often end up with!" This definition goes on to describe sexual intercourse, fertilization, pregnant, and the responsibility of being a parent.

The dictionary is, therefore, more than just a dictionary. It gives advice on safe sex and what happens if one does not practice safe sex or birth control. This is a highly recommended book for school libraries and public libraries. However, schools may encounter some resistance from some parents and should be honest with them if asked about the nature of the book.

H. ROBERT MALINOWSKY

366. **Dictionary of Terms Used in the Safety Profession.** 3rd. Stanley A. Abercrombie, comp. Des Plaines, IL: American Society of Safety Engineers, 1988. 72p. $25.00. 0-939874-79-2.

The field of safety draws its practitioners from more than 30 disciplines and subject areas. As a guide to this multidisciplinary field, the American Society of Safety Engineers has produced its third edition of the *Dictionary of Terms Used in the Safety Profession*. The arrangement is straightforward. Following prefaces to the first, second, and third editions, boldface terms and their concise definitions are arranged alphabetically. The clear type and boldface headings are visually appealing.

Excellent appendices accompany the text. Appendix A offers well over 200 listings of professional associations, institutes, societies, and federal government agencies, all with their acronyms. Appendix B contains key relevant sources for additional terms and definitions, and Appendix C, appearing at the end of the book, contains quick reference tables for metric conversions.

This work is an excellent ready-reference tool for both practitioner and layperson. The definitions are succinct and easily understood. Good cross-references are provided, and acronyms are explained. Especially helpful are the ANSI standard numbers that are supplied within the definitions where applicable. Public, academic, and special libraries will find this a useful reference source on safety. *LINDA VINCENT*

Directories

367. **1987 Directory of Community Blood Centers.** Arlington, VA: American Association of Blood Banks, 1987. 141p., maps. $35.00pbk.

This is a directory of 130 community blood centers in the United States. Each entry contains a map showing the area of coverage of the center, a brief history, names of key personnel, address, number of employees, financial information, special services, area of service, number of blood draws, governing body, association membership, computer services, components/products produced, publications, and research projects. Three appendices list addresses and telephone numbers of Council of Community Blood Centers members, American Red Cross Blood Services, and American Association of Blood Banks members. This directory is projected to be revised on a biennial basis. Medical libraries will find it a useful directory for their reference shelves.

H. ROBERT MALINOWSKY

Field Books/Guides

368. **A Family Affair: Helping Families Cope with Mental Illness.** Committee on Psychiatry and the Community, Group for the Advancement of Psychiatry. New York, NY: Brunner/Mazel, 1986. 104p., bibliog. $17.50, $9.95pbk. Report no. 119. 0-87630-444-7, 0-87630-443-9pbk.

Trends in the delivery of mental health care, including the social and public policy of deinstitutionalization, advances in medication research and therapies, and national concern for health care cost containment, are having a profound impact on the families of the mentally ill. According to the authors of *A Family Affair: Helping Families Cope with Mental Illness*, these trends are causing a redirection of the responsibility for care to the families of the mentally ill. The ramifications of this redirection are discussed in this insightful, powerful and at times harrowing book. The Group for the Advancement of Psychiatry, Committee on Psychiatry and the Community, attempted to determine the nature and scope of problems family members had in their care-giver roles. To study these problems, the group sought help from newspaper columnist "Dear Abby," Abigail Van Buren, who in 1982 placed an ad in her column soliciting letters from families of the mentally ill describing their problems caring for stricken family members. The resulting letters, with interpretive analysis, comprise the substance of *A Family Affair*.

The committee documents the various stages through which families progress living with and caring for a mentally ill parent or child, relying extensively on first-hand accounts from the solicited letters. Also documented are the frustrating problems families face when they reach out for support from trained professionals and community service agencies. Finally, recommendations are made for improved communication and cooperation between families and professional mental health care-givers. This book cannot be recommended as a reference book; it is doubtful this was the intent of the authors. It is recommended, however, for patient, mental health, and public library collections as a poignant discourse on an important public health issue.

GERALD J. PERRY

369. **Health Enhancement for America's Work Force: Administrator's Guide.** YMCA of the USA. Champaign, IL: Human Kinetics Publishers, Inc., 1987. 69p. $10.00pbk. 0-87322-107-9pbk.

Through the many years of its existence, the YMCA has developed an acclaimed series of health and fitness programs reinforced by a training and certification program for its health enhancement staff and its nationwide network of facilities. The YMCA is in a powerful position to provide well-developed corporate health education and fitness programs. This manual is designed to assist the local YMCA staff in implementation of such corporate health enhancement programs in the YMCA facilities. The volume gives an overview of the standardized YMCA programs, which are described more thoroughly in *Health Enhancement for America's Work Force: Program Guide* (see entry 370). One chapter explains how to develop a plan of operation, including program content, quality control, evaluation procedures, legal aspects and liability, program siting, management and staffing, budget, and collaboration with other community organizations. Another chapter discusses how to sell the fitness programs to corporations, with brief sections on market analysis, preparing for the sales presentation, making the presentation, and promoting the program to the employees. In addition, there is a chapter that assists local YMCAs that would like to become broker YMCAs, which are partners in a nationwide network of YMCAs using standard programs, pricing strategies, and administrative procedures to serve multisite companies such as IBM, Honeywell, and Pfizer. There are five appendices consisting of sample forms: health enhancement evaluation, an agreement between the YMCA and a company, a letter to the CEO of the YMCA, a fact sheet, and a response form. Although well written and usefully organized, the guide is severely limited in scope. Only someone who is in charge of developing or promoting a fitness program, most likely a YMCA program, would find this book to be valuable.

JOAN LATTA KONECKY

370. Health Enhancement for America's Work Force: Program Guide. YMCA of the USA. Champaign, IL: Human Kinetics Publishers, Inc., 1987. 150p., bibliog. $10.00pbk. 0-87322-106-0pbk.

The fitness craze is spreading to the workplace, as management realizes that improved employee fitness reduces rising corporate health care costs, sick leave, and disability, while it increases employee morale and productivity. Thousands of corporations are offering fitness programs to their employees, spending over $2 billion annually for such programs. Statistics show that some corporations have realized a return of from $3 to more than $5 for every dollar invested. Many of these corporations prefer to provide their fitness programs through independent organizations such as the local YMCAs.

The intent of this book is to provide guidelines for the local YMCA and to standardize the YMCA programs which may be included in corporate health enhancement packages. The volume outlines the elements of a corporate fitness program: preprogram health screening and informed consent, fitness testing and exercise plans, and education programs. The first chapter includes copies of a health screening form, a cardiovascular risk factor estimate, a medical clearance form, and two informed-consent forms. The fitness testing section presents measurements for four areas of fitness: cardiorespiratory fitness, body composition, flexibility, and muscular strength and endurance and has a nice series of tables depicting the test norms by sex and age. The chapter on exercise plans discusses the appropriate exercises for the four basic fitness areas with five levels: starter, beginner, intermediate, advanced, and expert plans. The final chapter examines suggested education programs such as a fitness education minilecture series, weight management, stress management, and smoking cessation. There are two appendices in this program guide. The first appendix contains 41 exercise planning charts, and the second provides 20 detailed outlines of the five-minute fitness education minilectures. Although this program guide will be most useful to its intended audience of YMCA personnel, it should also prove informative for anyone involved in fitness programming, corporate or otherwise. The program guide, although very narrow in scope, successfully combines concise, authoritative text with useful forms, tables, and charts.

JOAN LATTA KONECKY

371. Traditional Foods Are Your Best Medicine: Health and Longevity with the Animal, Sea and Vegetable Foods of Our Ancestors. Ronald F. Schmid. Stratford, CT: Ocean View Publications, 1987. 270p., bibliog., index. $17.95. 0-941145-03-4.

Through the careful study of history, anthropology, research, and his own clinical experience, Dr. Ronald Schmidt, a licenced, practicing, naturopathic physician, demonstrates and exposes how modern agricultural practices, and processed refined foods have contributed to the many serious diseases that plague our society today. Traditional diets included more fiber, less vegetable fat and saturated animal fat, and more polyunsaturated animal fat than do modern diets. Schmid shows how reverting back to the more "primitive" diets of traditional cultures can aid in our body's own treatment and cure of gastroenterological disease, cancer, colds, allergies, yeast infections, arthritis, heart disease, and many other conditions. Written for the layperson, this book serves as a general guide to a healthier life through understanding and practicing better nutrition. Some sections seem too general, but for those interested in additional sources of general or technical information, a lengthy bibliography has been provided. Also, the book is enhanced by five appendices covering information on seafood, lab tests, food irradiation, explanation of California's Organic Food Act of 1982, and exercise and sports. This is an easy-to-read reference source for school, public, and college libraries as well as for home libraries.

JELENA RADICEVIC

Glossaries

372. American Psychiatric Glossary.
6th. Evelyn M. Stone, ed. Washington, DC:
American Psychiatric Press, Inc., 1988.
217p. $19.95, $10.00pbk. 0-88048-275-3,
0-88048-288-5pbk.

This glossary has become a standard reference work, valuable for many levels of users. The sixth edition includes more than 200 new or revised entries and incorporates the nomenclature of the revised third edition of the *Diagnostic and Statistical Manual of Mental Disorders* (1987), DSM-III-R. Definitions of words and phrases are arranged alphabetically and the words are printed in italics if they are defined elsewhere; this is a helpful feature, especially for laypersons. This volume also includes a list of abbreviations and tables on commonly abused drugs, drugs used in psychiatry, legal terms, neurologic deficits, psychologic tests, research terms, and schools of psychiatry.   *RUTH LEWIS*

Handbooks

**373. Basic Handbook of Training in
Child and Adolescent Psychiatry.** Richard L. Cohen; Mina K. Dulcan. Springfield,
IL: Charles C. Thomas Publisher, 1987.
513p., bibliog., index. $72.50. 0-398-05368-5.

This excellent handbook is intended to fill a gap in the literature of child psychiatric training and education. The editors did a thorough literature search and were unsuccessful in identifying one textbook in the English language that covers this topic. Using 23 contributors, this handbook has been written to provide a historical and developmental perspective on the field; description of theoretical concepts; accounts of personal experiences; a practical guide to the planning, updating, and administration of programs; assitance in licensing and certification; and descriptions of what the editors consider to be good training with a long-term positive impact on the quality of education in child psychiatry.

After a lengthy introduction, the book covers, in four sections, "Guidelines for Training in Specific Areas," "Approaches to Program Direction," "Administrative and Policy Issues," and "Other Education and Training Programs in Child and Adolescent Psychiatry." A final section offers a conclusion.

The text is very detailed, and the editors have maintained a consistency between the articles. At the end of each chapter is an extensive bibliography. At the end of the book is an author index to each of the citations in these bibliographies. There is also a detailed subject index to topics covered within the text.

This is an exceptional book written by an outstanding group of professors. The editors are well-known professors. Dr. Cohen is from the Western Psychiatric Institute and Clinic at the University of Pittsburgh, and Dr. Dulcan is the chief, Child and Adolescent Psychiatry Department at Emory University. Any academic library with educational programs in child psychiatry should have this handbook, and individual practicing psychiatrists who specialize in child psychiatry will find this a must for their personal reference shelves.   *H. ROBERT MALINOWSKY*

374. Handbook for Hospital Secretaries.
Ann E. Lobdell. Chicago, IL: American
Hospital Publishing, Inc., American
Hospital Association, 1987. 187p., illus.,
index. $19.95pbk. 1-55648-006-7pbk.

"The job of a hospital secretary is a challenging and demanding one," with 24-hour, seven-day-a-week demands. The hospital secretary may report to any individuals, including physicians, administrators, engineers, researchers, professors, or special scientists. There are certain protocols that are required in the hospital settings as well as special reports that have to be prepared. If the hospital is in a teaching setting, there is the requirement to prepare research papers and do a certain amount of library research or bibliographical verification. This handbook covers all areas very well and is intended to be used on a continuing basis. It is divided into two parts. Part 1 covers the basic skills or office procedures. This includes telephone manners, handling office visitors, mail, appointment calendars, filing systems, using the medical library, dictation, and typing letters and memos. Part 2 covers advanced skills and special documents. This section is the most detailed because it covers those types of functions that may not be used on a daily basis. It covers formal reports, schedules and agendas, conference notices, meeting minutes, call schedules, curricula vitae, case protocols, study protocols, meeting abstracts, journal manuscripts, and bibliographic citations. Even though this handbook is meant for hospital secretaries, there is much information that would be of use to secretaries in general.
            H. ROBERT MALINOWSKY

**375. Handbook of Behavioral Medicine
for Women.** Elaine A. Blechman; Kelly D.
Brownell. New York, NY: Pergamon Press,
1988. 503p., bibliog, index. $75.00. Pergamon
General Psychology Series, no. 149.
0-08-032383-9.

The *Handbook of Behavioral Medicine for Women* discusses subjects that you would expect in any handbook of behavioral medicine: life cycle issues, diseases, and behaviors contributing to good health or health problems. But, the discussions are from the perspective of the physical, social, and psychological circumstances of women. There are also chapters on research methodology and theoretical issues in the treatment of women. The editors and most of the contributors are clinical psychologists who have produced here a valuable synthesis of important facts, unanswered questions, and relevant references from many sources.   *RUTH LEWIS*

376. Handbook of Community Health.

4th. Murray Grant. Philadelphia, PA: Lea and Febiger, 1987. 277p., illus., bibliog., index. $19.50pbk. 0-8121-1083-8pbk.

The purpose of this book is to provide an introduction to public health and preventive medicine. Each chapter outlines the major topics for one area within community health. "Health Statistics," "Medical Sociology," "Maternal and Child Health," "Occupational Health," and "Health Manpower" are a few of the 21 chapters. Tables of statistics, often from the National Center for Health Statistics, are included to illustrate main points in most chapters. For example, the chapter "Accidents" has the table "Death Rates for Motor Vehicle Accidents."

The first edition appeared in 1967. This is the fourth edition, and the monograph has been kept up to date with the inclusion of issues of the 1980s. For instance, Individual Practice Associations (IPAs) have joined Health Maintenance Organizations (HMOs) in the chapter, "Health Care Delivery." However, just a couple of paragraphs are devoted to IPAs. Descriptions and explanations are brief, uncomplicated, and easy to understand. The book is not directed toward the specialist. Because the chapters are short, the table of contents would be an adequate guide to the statistical material. However, the index is good and does include entries for the information in the tables.

Over the years *The Handbook of Community Health* has proven its usefulness. It is included in the "Selected List of Books and Journals for the Small Medical Library" (Alfred N. Brandon and Dorothy R. Hill, *Bulletin of the Medical Library Association* 75(2):155, April, 1987) and is among the books on that list that are "suggested for initial purchase." *FRANK R. KELLERMAN*

377. Healing Yourself during Pregnancy. Joy Gardner. Freedom, CA: The Crossing Press, 1987. 210p., illus., index. $10.95pbk. 0-89594-251-8pbk.

Joy Gardner has authored several books on natural healing, in which therapy for health problems is based on herbal or natural remedies. In *Healing Yourself during Pregnancy*, Gardner has collated cures, predominantly based on food or herbal preparations, for nearly 30 common health complaints during pregnancy. Introductory chapters focus on preparation for conception and pregnancy. The third chapter shares the title of the book and is a cache of cures. Ailments discussed cover a wide range and include anemia, colds, headaches, morning sickness, stress, toxemia, and varicose veins. Final chapters cover labor and birth preparation.

Thoroughly indexed, this new age health care book includes several interesting appendices including sources for additional information, safety and toxicity data for herbs used in recommended preparations, and information about homeopathy. Clearly alternative in scope, presentation, and tone, this book is recommended for patient education, obstetric and gynecological clinic library, and public library reference collections. Those in-

terested in natural medicine and holistic or new age philosophy and practice may also take an interest in this author and her work.

GERALD J. PERRY

378. The Healthy Male: A Comprehensive Health Guide for Men (and the Women Who Care about Them). Maureen Mylander. Boston, MA: Little, Brown and Co., 1987. 293p., index. $12.95. 0-316-59368-0.

The primary purpose of this guide is to promote health and prevent disease in men. It is a "book about bodybuilding—not the Arnold Schwarzenegger type, but the type that fends off diseases that kill men in their prime, extends life expectancy, and makes men well and fit." Each of the three sections of the book covers a particular area in a man's life. The first part, "Lifestyles for Men," covers those areas that help to promote good health—nutrition, getting and staying fit, good and healthy sex, sleep and rest, and managing stress. This section outlines some ways to avoid the pitfalls of poor health. The well-written and concise text covers diet, dieting, eating habits, an excercise plan, common sports injuries, good sex, bad sex, sex misinformation, condoms, safe sex, STDS, sleep and rest, and ways to combat stress.

Part 2, "Diseases of Men," discusses heart disease, cancer, accidents, stroke, emphysema, chronic bronchitis, suicide, cirrhosis, homicide, and diabetes. Each disease or cause of death is discussed with suggestions on how to prevent it. One chapter, "Life's Problems, Small and Large," covers those sometimes embarrasing problems that happen to everyone—bad breath, gas, borborygmus (noisy gut), body odor, dandruff, diarrhea, ear wax, potbellies, baldness, beards, skin problems, suntan, athlete's foot, jock itch, prostate problems, and other problems. And, finally, this section ends with a chapter on drugs and addiction, including information on alcoholism, hangovers, cocaine, marijuana, smoking, and love.

The last part, "The Bottom Line," tells the results of staying healthy—"Growing Older and Better" and "Why Men Die Young, or How to Add Eight (or More) Years to Your Life." This chapter gives a better outlook on life and a good feeling that maybe the reader can learn something from the previous sections. All in all, this is an excellent reference book for any male and one that should be on library reference shelves. It contains much information that is scattered throughout many other books.

H. ROBERT MALINOWSKY

379. Listen to Your Body: A Head-to-Toe Guide to More than 400 Common Symptoms, Their Causes and Best Treatments. Ellen Michaud; Lila L. Anastas; Prevention Magazine. Emmaus, PA: Rodale Press, 1988. 525p., index. $27.95. 0-87857-728-9.

Written by a registered nurse and a writer on children's issues and then checked for accuracy by a referral board of specialists in fields ranging from obstetrics to urology and orthopedic surgery, *Listen to Your Body* is intended to provide a popular guide to symptoms of body systems and their malfunctions. The authors stress at several points in the text that this information is not intended to serve as a substitute for professional medical attention. Nineteen chapters cover all major body systems (such as bones, joints, and muscles and abdomen and digestive organs), with the final section focused on "whole body symptoms." These latter entries include variation in appetite, sensitivity to cold, fatigue, drowsiness, insomnia, and fever. The format of each entry covers common causes of the symptom involved, possible accompanying symptoms, and recommended medical treatment. A detailed index is provided, with medical jargon used sparingly. An exception is the heading for AIDS, where the reader is referred to "Acquired Immune Deficiency Syndrome." This guide is suitable for the reference collections in public and large undergraduate libraries.

ROBERT B. MARKS RIDINGER

380. Sports Medicine: A Practical Guide. Nathan J. Smith; Carl L. Stanitski. Philadelphia, PA: W. B. Saunders Co., 1987. 238p., illus., bibliog., index. $19.95. 0-7216-1167-2.

Smith and Stanitski's *Sports Medicine: A Practical Guide* is an easy-to-read introduction to a variety of topical medical issues in athletic training. Written for coaches and trainers, as well as for health care practitioners, this concise guide focuses on both the preparticipation evaluation of the athlete's health status and the quick diagnosis and management of sports-related injuries. Broad chapters include discussions of the counseling of both athletically active children and their parents, women in sports, and pediatric sports injuries. Most chapters in this brief introductory guide focus on a variety of common sports-related injuries, featuring the specific musculoskeletal joints or regions commonly injured; the knee, shoulder, head, spine, and neck. For each body part, the authors discuss the anatomy of the region, common causes for injury, the biomechanical processes involved, and diagnostic and treatment considerations.

Additional chapters discuss nontraumatic medical and behavioral conditions that may impact on athletic performance, including the abuse of drugs, nutritional considerations, chronic health conditions, and problems attributed to both heat and cold. Rounding out this useful book are select position statements from the American College of Sports Medicine representing the opinions of this professional organization on a variety of topical issues such as the use and abuse of alcohol and anabolic steroids by athletes, recommendations regarding the quantity and quality of exercise for attaining and retaining physical fitness, programming for reasonable weight-loss, and the participation of women in long-distance running. Certainly not a complete

reference text, this practical guide, none the less, is a useful introduction, particularly for coaches and trainers, to some very important medical considerations in the training of athletes. The inclusion of the American College of Sports Medicine's position papers additionally enhance the usefulness and authority of this guide. Libraries supporting physical education and recreation programs in particular may wish to add this handy book to their reference collection.

GERALD J. PERRY

381. Vaccination Certificate Requirements and Health Advice for International Travel: Situation as of 1 January 1987. Geneva, Switzerland: World Health Organization, 1987. 83p., illus., bibliog., maps, index. $7.80pbk. 92-4-158012-7pbk. 0254-296X.

This small volume, published annually, is a valuable reference for people planning international travel and those who advise them. It is updated by the World Health Organization's *Weekly Epidemiological Record*. About half of the book deals with the geographic incidence and appropriate precautions for yellow fever and malaria. There is easy access to geographical areas by internal arrangement and an indexing, including a subject approach. *RUTH LEWIS*

Manuals

382. Code Blue: Cardiac Arrest and Resuscitation. Mickey S. Eisenberg; Richard O. Cummins; Mary T. Ho, eds. Philadelphia, PA: W. B. Saunders, Co., 1987. 244p., illus., bibliog., index. $18.95. Blue Book Series. 0-7216-1822-7.

"This book supplements and complements the material in the most recent edition of the *Textbook of Advanced Cardiac Life Support*, as well as the *1986 Standards and Gudelines for Emergency Cardiac Care*," both issued by the American Heart Association. The editors warn that it is not to be used as a text on the subject. Although it resembles a text in format, the book dwells more on therapy and management for cardiac arrest rather than on physiology and pathology. Its chapters are authored by 11 individuals, and there is no standard organization to the chapters. The information tends to be very concise and brief, however, so it lends itself to being used as a manual. It seems designed more for study before or after the emergency rather than at the time of the emergency. The fastest route of access is through the index, which is fairly thorough. The "see" and "see also" referencing that is so liberal in the index, however, is what makes it seem less of an emergency manual than a subject background manual. The manual was meant for those involved with a cardiac arrest and is best suited for the collection of a professional program in the health field. *LILLIAN R. MESNER*

383. **How to Persuade Your Lover to Use a Condom...And Why You Should.** Patti Breitman; Kim Knutson; Paul Reed. Rocklin, CA: Prima Publishing and Communications, Dist. by St Martin's Press, New York, NY, 1987. 83p. $4.95pbk. 0-914629-43-3pbk.

This is a "straight from the hip" book on the use of condoms to prevent sexually transmitted diseases, including AIDS. It covers all aspects of sex in general terms and the need to protect oneself with a condom, regardless of the type of sex one is having. This is not a book for the prudish, but then again most everyone who is sexually active will find no surprises in what is being discussed.

The first part covers sex in today's society and why it is important to use condoms. The second part is the tricky part, "How to Persuade Your Lover to Use a Condom." It delves into the question of when to bring up the subject, responses, monogamy, and safe sex. The third part, "...And Why You Should," gives the facts about AIDS, gonorrhea, syphilis, herpes, nongonococcal urethritis, venereal worts, and HTLV-I. It then covers the facts about birth control. The last part talks about condoms, in general, in terms of questions and answers. Almost every possible question has been raised about condoms, with frank answers given.

This is a highly recommended book for any sexually active individual and one that would be useful on any reference shelf of school and public libraries. *H. ROBERT MALINOWSKY*

384. **The Paramedic Manual.** Michael K. Copass; Mickey S. Eisenberg; Steven C. Macdonald. W. B. Saunders Co., Philadelphia: PA, 1987. 288p., illus., index. $19.95. 0-7216-1862-6.

This manual was designed for prehospitalization emergency personnel. Its coverage is step-by-step advice for conditions including medical emergencies, traumas, obstetric and gynecological emergencies, psychiatric emergencies, and environmental emergencies. The changes in this second edition include new 1986 guidelines for cardiac-arrest management and revisions in all chapters to reflect state-of-the-art standards of care. In addition, materials on the role of the paramedic, approaches to the patient, and trauma scores are added. The arrangement of the contents is generally in a standard format that allows for quick accessing of the information in the chapters for emergencies: introduction, history, examination, assessment, management, and additional comments. The chapters on the roles of the paramedic, the approach to the patient, triage, and exposure of emergency personnel to communicable diseases are in simple, outline form. The chapters on differential diagnosis, drugs commonly prescribed to patients, and drugs administered by paramedics are in tabular form. The introductory note states that the manual's format "allows maximal retrieval of relevant information in a minimum of time." The full outlining of the contents and the completeness of the index allow for this. Indexing is done for anatomical terms, conditions or diseases, drugs, traumas, symptoms, medical terminology, and slang terminology. The manual's authors based this authoritative handbook on their work with the Medic-1 Program of Seattle. It should be part of any health science reference collection.

LILLIAN R. MESNER

385. **Practical Guide to the Care of the Medical Patient.** Fred F. Ferri. St. Louis, MO: The C. V. Mosby Co., 1987. 591p., illus., bibliog., index. $19.95. 0-8016-1661-1.

Dr. Fred Ferri's *Practical Guide to the Care of the Medical Patient* is a pocket-sized handbook of brief, concise, practical, and quick information for handling medical (as opposed to surgical or obstetric) patients. The book's coverage is very broad within the medical specialty. There are lists of medical record abbreviations, outlines of information on 150 commonly used drugs (preparations, dosages, actions, contraindications, and precautions), outlines of instructions on note keeping, outlines of procedures for examinations, and medical emergency procedures (e.g., poisonings). Sections devoted to disease systems, such as hematology, have subsections dealing with the etiology, diagnostics, and causes of specific diseases, such as microcytic anemia. Appendices are included at the end of the book that give additional tables of drug and conversion data and various other types of drug information. Access to the book may be gained through either the table of contents or the index depending on whether the person's approach is general or specific. The language of indexing is medical, however, so it would be difficult for beginning level students to use this source.

The contributing authors are all from St. Vincent's Medical Center in Bridgeport, Connecticut. Most of them are professors of medicine at Yale, New York Medical College, and the University of Connecticut. In the preface, the authors state that the manual was meant for medical students, residents, practicing physicians, and other allied professionals. Although its best use would be in the practitioner's pocket, it also would be useful on a reference shelf in a medical facility.

LILLIAN R. MESNER

386. **Travel and Tropical Medicine Manual.** Elaine C. Jong. Philadelphia, PA: W. B. Saunders, 1987. 323p., illus., bibliog., index. $19.95. 0-7216-1194-X.

This manual, kept handbook size by use of quite small print, is full of information for health care providers who see travelers or immigrants. "The contributing authors represent a wide diversity of clinical experience but are unified by their dedication to the practice of medicine relevant to geographic exposures of their patients." Most of the contributors are connected with the University of Washington, Seattle; the editor is director of the Travel and Tropical Medicine Clinic there. Section 1 deals with general problems of travelers, including malaria prevention, jet lag, and advice to pregnant travelers and those with chronic

medical conditions. Condensed information on epidemiology, etiology, clinical features, diagnosis, treatment, and prevention on several parasitic, infectious, and dermatologic problems follow. There is a chapter on fish and shellfish poisoning. The appendix lists travelers' clinics in the United States and Canada and contacts and textbooks for additional information. There is a subject index. Recommended for libraries that furnish travel and immunization reference services.

RUTH LEWIS

387. **Wilderness Medicine.** 3rd. William W. Forgey. Merrillville, IN: ICS Books, Inc., 1987. 151p., illus., bibliog., index. $7.95pbk. 0-934802-37-8pbk.

Dr. William W. Forgey's third edition of *Wilderness Medicine* is a medical emergency handbook written in plain English for the traveler who finds him- or herself in unfortunate circumstances in isolated locations where immediate access to professional health care is unavailable. Forgey, trustee of the Wilderness Education Association, emergency room physician, and director of the Emergency Medicine Department of the Ross Clinic, Merrillville, Indiana, has included in this paperback essential first-aid advice for travelers in the wild and pretrip medical kit preparation recommendations. Suggestions for improvements made by users of previous editions of the book enhance this third edition. Common and unusual medical problems are addressed by the author, ranging from the merely irritating to life-threatening, and include frostbite and hyperthermia, bites and stings, infections, pain, wounds, and fractures.

To use this popular first-aid book, readers must consult the book's "Instant Reference Clinical Index," where symptoms, body parts, therapies, diseases, and injuries in scientific and lay-language and drugs by brand and generic name are listed with convenient cross-references. Since the book lacks a table of contents and is arbitrary in organization, the use of the index is a must.

This fact-packed reference source is highly recommended pretrip reading for wilderness travelers. Public libraries may also wish to purchase it, though its usefulness is certainly best appreciated at the point of need. *GERALD J. PERRY*

Textbooks

388. **Essentials of Athletic Training.** Daniel D. Arnheim. St. Louis, MO: Times Mirror/Mosby College Publishing, 1987. 381p., illus., bibliog., index. $27.95pbk. 0-8016-0335-8bpk.

Athletic training, according to author Daniel Arnheim, fellow of the American College of Sports Medicine and professor of physical education, is a subspecialty within sports medicine providing the necessary link between medical practitioners and coaches and athletes for the prevention and care of sports-related injuries. Arnheim's *Essentials of Athletic Training* is an introductory

text, written for physical education and/or coaching students, focusing on elementary principles of biomechanics and the prevention and immediate care of injuries to the musculoskeletal system.

Working from the premise that most common sports injuries are preventable, Arnheim discusses in introductory chapters the development of athletic training as a subspecialization and the relationship between physical conditioning and the wearing of protective equipment with the prevention of injuries. Subsequent sections of this text cover the etiology of sports injuries, including how injuries occur and are diagnosed, the body's responses to injury, and general management observations. Finally, Arnheim discusses specific sports injuries, related anatomy, methods of prevention, and techniques for providing necessary quick care. This volume is nicely illustrated, thoroughly indexed, and easy to use. It is a good introduction to basic tenets of injury prevention and care. However, the book cannot be recommended as a reference tool, despite its clear and organized presentation, in great part because of its brevity. For a more substantive discussion of sports medicine and athletic training, readers are referred to the author's sixth edition of *Modern Principles of Athletic Training*.

GERALD J. PERRY

★
SPECIAL MEDICAL DISCIPLINES
Atlases—Medical

389. **EEG Atlas for Anesthesiologists.** Ina Pichlmayr. New York, NY: Springer-Verlag, 1987. 414p., illus. $129.00. 0-387-18092-3.

The electrical potentials of the human cortex change depending on levels of brain activity, and are influenced by the body's state of metabolism. According to lead author Ina Pichlmayr of *EEG Atlas for Anesthesiologists*, anesthetic agents as well as disease processes produce interpretable changes in the metabolic-functional activity of the brain that can be mapped using the electroencephalogram (EEG). The interpretation of EEG findings relative to sedation and recovery and the practical use of this data by anesthesiologists are the subjects of this atlas survey. Changes in a patient's EEG can be interpreted by the trained anesthesiologist to determine depth and stage of sedation. Additionally, instances of inadequate anesthesia and points in time when the patient begins to recover and physiological processes normalize, all can be monitored using the EEG. EEG monitoring, thus, can be used to ensure greater safety in the delivery of care to the surgical patient.

Included in this atlas are examples of EEG spectral analyses documenting a variety of clinical findings. The volume is divided into three segments, including an introduction, chapters on the effects of anesthetic medications on EEG and perioperative factors influencing brain activity, and, finally, chapters on EEG as a monitoring

methodology. Each chapter includes a set of figures illustrating EEG findings both before and after the administration of an anesthetic agent and interpretation of findings, with relevant recording conditions. Unfortunately, this otherwise useful atlas lacks an index, and the table of contents is too cursory to guide the reader easily to data regarding specific medications, processes or perioperative conditions.

Anesthesiologists responsible for EEG interpretation during surgical and gynecological procedures will find this atlas a useful, if not altogether convenient, tool. Libraries may opt to include this volume in their reference collections, particularly if supplemented by Pichlmayr's publication *The Electroencephalogram in Anesthesia* (Springer-Verlag, 1983). *GERALD J. PERRY*

Exam Reviews

390. Comprehensive Review of Respiratory Care. 2nd. William V. Wojciechowski; Paula E. Neff. New York, NY: John Wiley and Sons, 1987. 642p., illus., bibliog., index. $25.95pbk. "A Wiley Medical Publication." 0-471-83090-9pbk.

The first edition of this book, entitled *Comprehensive Review of Respiratory Therapy*, was written primarily for persons preparing for the National Board for Respiratory Care exam. "This edition is more comprehensive and encompasses the matrices of both national examinations," the Entry Level Certification and Advanced Practitioner Examination (NBRC), as well as the curriculum content areas listed in the *Essentials and Guidelines of an Accredited Educational Program for the Respiratory Care Practitioner,* implemented by the Joint Review Committee for Respiratory Therapy Education. Several content areas, a mathematics section, and clinically oriented questions have been added to this edition. The format is a section of questions on a broad subject, followed by a section of referenced answers and a list of references. *RUTH LEWIS*

391. Rudolph's Pediatrics: A Study Guide. Robert H. Pantell; David A. Bergman; Maureen T. Shannon; Benjamin W. Goodman, Jr., eds. Norwalk, CT: Appleton and Lange, 1987. 266p. $29.95. 0-8385-8487-X.

This volume is a study guide developed for use with the eighteenth edition of the textbook *Pediatrics* by Abraham M. Rudolph. There are 29 chapters of questions which are directly linked with the 29 chapters in Rudolph's *Pediatrics*. Each chapter is followed by an answer section, which frequently provides further explanations of the answer. There is a list of abbreviations in the back of the volume. This is a nicely designed study guide but, as such, has little use as a reference tool and cannot be recommended as such. *JOAN LATTA KONECKY*

Handbooks

392. AIDS: A Guide for Dental Practice. Charles E. Barr; Michael Z. Marder. Chicago, IL: Quintessence Publishing Co.., Inc., 1987. 127p., illus. (part col.), bibliog., index. $34.00pbk. 0-86715-133-1pbk.

This text is intended to assist dentists in providing routine dental care to AIDS patients and recognizing and treating the oral manifestations of AIDS. It is a handbook outlining the various preventive procedures that one must take to ensure the safety of the patient, the dentist, the assistants, and other patients. Throughout, the book stresses that AIDS cannot be contacted by casual encounters. The eight chapters cover "Epidemiology of AIDS," "Etiology of AIDS," "Immunology of AIDS," "Oral Manifestations of AIDS," "Clinical Manifestations of Marker Diseases Associated with AIDS," "AIDS: Control of the Dental Office Environment," "Treatment of the AIDS Patient," and "Ethical, Legal, and Social Responsibilities of the Dentist."

This is a highly technical book intended for the dentist or researcher. Each chapter has an extensive bibliography, charts, diagrams, and numerous photographs, most in color. Academic and medical libraries will want a copy of this specialized book for their reference collections.
H. ROBERT MALINOWSKY

393. Handbook of Clinical Psychoneuroendocrinology. Charles B. Nemeroff; Peter T. Loosen, eds. New York, NY: Guilford Press, Dist. by John Wiley and Sons, New York, NY, 1987. 502p., illus., bibliog., index. $60.00. 0-471-91768-0.

Psychoneuroendocrinology, simply put, is the science which explores the interactions of hormones and the central nervous system with respect to their involvement in psychiatric disorders. Nearly 20 years ago, it was concluded that the brain has a powerful influence on the functions of the endocrine systems. Recent research indicates that the neurotransmitter systems of the brain, which regulate the secretion of various hormones, also may be implicated in depressive illnesses, schizophrenia, and other major psychiatric disorders.

This handbook attempts to provide a comprehensive overview of the current knowledge in this field as perceived by a broad sampling of psychoneuroendocrinologists from a wide variety of institutions. The book examines the psychiatric aspects of endocrine disease, endocrine disturbances in psychiatric disorders, and the effect of psychiatric treament on neuroendocrine parameters. In addition, there are chapters on the effect of the hormones on behavior, premenstrual syndrome, stress, aging, and the fetus. Each chapter concludes with a lengthy list of references, and there is an index. The field of psychoneuroendocrinology is advancing so rapidly that neither this volume nor any other can truly claim to represent the current state of knowledge. However, this handbook is logically organized and

provides concise chapters on a wide variety of topics in the field. It should be a valuable addition to those special and medical libraries that have clientele in psychiatry, endocrinology, neurology, and related areas.

JOAN LATTA KONECKY

394. Handbook of Office and Ambulatory Gynecologic Surgery. Philip D. Darney, ed. Oradell, NJ: Medical Economics Books, 1987. 226p., illus., bibliog., index. $26.95. 0-87489-393-3.

This book was written for gynecologic surgeons and their co-workers, who may want to extend their skills to office, clinic, or surgicenter. The trend to ambulatory settings will continue for "operations with few complications, with predictable findings and outcomes, and of short durations," partly because it results in lower medical costs. There are introductory chapters on the general practice of ambulatory surgery, anesthesia and analgesia, and patient counseling. These are followed by chapters written by Darney or one of nine other contributors with relevant experience with specific procedures. They discuss patient selection, counseling and preparation, equipment needed, complications, and postoperative care, as well as operative procedures. References follow each chapter. This would be a useful reference source in any medical library as well as the private physician library. *RUTH LEWIS*

395. A Manual of Clinical Obstetrics. Nancy Whitley. Philadelphia, PA: J. B. Lippincott Co., 1985. 778p., illus., bibliog., index. $39.50. 0-397-54258-5.

This work is intended as a quick-reference manual to obstetric procedures, useful to nurses, nurse-midwives, family-planning nurse-practitioners, medical students, and others working in obstetrics. The author cautions in the preface, "Some of the procedures described in the *Manual* are appropriate only to individuals with advanced training and certification, such as nurse-midwives and family planning nurse-practitioners."

The volume is in three main sections: Antipartum; Intrapartum; and Postpartum and Family Planning. The chapters are in a modified outline format to give quick access to required information, with much of the information presented in clearly defined tabular form. Included throughout are notes and clinical alerts (in boldface type), definitions of medical terms, objectives, rationale and background information, and extensive references to both books and periodicals.

The work is extremely well illustrated with drawings, diagrams, and photographs. Following several chapters are supplemental sections containing full-size analysis forms, scoring sheets, evaluation sheets, personal and medical history questionnaires, etc. The information is technical and covers every aspect of clinical obstetrics including emergency procedures, information on drugs and medical devices, family planning and setting up and assisting at obstetrical surgical procedures. The well-defined index gives quick access and designates both figures and tabular information. A useful manual for medical libraries. *JUANITA A. CUTLER*

396. Medical Factors and Psychological Disorders: A Handbook for Psychologists. Randall L. Morrison; Alan S. Bellack, eds. New York, NY: Plenum Press, 1987. 374p., bibliog., index. $50.00. 0-306-42425-8.

The interplay between biological and psychological factors is the subject of an expanding body of literature impacting on our understanding of illness, therapeutics, and prevention. Nonmedically trained clinical psychologists are increasingly involved in the treatment of and research into a variety of medical and psychiatric disorders and require access to information regarding the relationships between medical disorders and psychological variables. Morrison and Bellack's *Medical Factors and Psychological Disorders: A Handbook for Psychologists* is a handbook for clinical psychologists reviewing clinical findings with relevant psychobiological and physiological considerations for a wide variety of both medical and psychiatric disorders.

Collated in this handy manual are state-of-the-art reviews contributed by a variety of field experts. Introductory chapters address the essentials of biopsychology, behavioral medicine, and pharmacotherapy. Part 2 covers a wide variety of common psychiatric disorders including anxiety, affective, psychosexual, and organic mental disorders. Part 3 addresses nonpsychiatric disorders including lower back pain, arterial hypertension, neurological disorders, cancer, headache, and eating disorders. For each disease state or disorder, contributing authors address psychological and biomedical factors contributing to the etiology or maintenance of the problem and intervention modalities, both pharmacologic and psychotherapeutic. Each chapter summarizes and reviews pertinent issues and includes references to relevant literature.

Enhanced with a very easy-to-use table of contents, this manual is a convenient reference source for psychologists. Chapters tend to be cursory, and given the extensive scope of this undertaking, readers will probably use this manual only at initial stages of investigation. Still, Morrison and Bellack's handbook is recommended for both departmental and library reference collections serving psychologists and clinicians in training. *GERALD J. PERRY*

397. A Woman's Body: The New Guide to Gynecology. Niels Lauersen; Steven Whitney. New York, NY: Putnam Publishing Group, 1987. 562p., illus., index. $14.95pbk. "A Perigee Book." 0-399-51337-Xpbk.

A Woman's Body: The New Guide to Gynecology covers every aspect of women's health issues in a lively, frank, nontechnical narrative which is perfectly balanced with relevant medical jargon to

enable a woman to understand the procedures and clinical terminology she will encounter in the course of a gynecological examination. It is also a guide to normal female body functions. The intent is not self-treatment but to help the woman to intelligently cooperate with the physician in attaining her optimal physical well-being.

The coverage is thorough, with discussions ranging from the psychological and physiological aspects of self-examination to the same aspects of the gynecological examination, all aspects of the reproductive process including contraception, abortion, birthing, sexual function and dysfunction, sexually transmitted diseases including AIDS, cancer, and other diseases. A section on intelligently choosing a gynecologist precedes a chapter which describes in straightforward terms exactly what to expect in the course of a gynecological examination, from the questions which should be asked to the actual procedures of the pelvic examination.

The frank, sensitive discussion is detailed to the point of including a description of the speculum, not only as a medical instrument but also in terms of the physiological and psychological aspects of its use. Dr. Lauersen, a gynecologist and clinical professor of obstetrics and gynecology at New York Medical College, collaborated with Steven Whitney, a writer and journalist, in the writing of the book. The volume is well illustrated with drawings and photographs and includes an index but no bibliography. In addition to personal libraries, public and academic libraries will find this to be a good guide for the reference shelves.

JUANITA A. CUTLER

Manuals

398. Common Skin Disorders: A Physician's Illustrated Manual with Patient Instruction Sheets. 3rd. Ernst Epstein. Oradell, NJ: Medical Economics Books, 1988. 200p, 110p, illus. (col.), index. $35.95pbk. 0-87489-440-9pbk.

This is the third edition of a concise and practical manual of skin diseases commonly found in North America. It was written by a practicing dermatologist for use by nonspecialist physicians. The four introductory chapters cover general dermatologic topics of particular interest, including topical corticosteroids, diagnosis of dermatitis, treatment of dermatitis, and dermatologic surgical techniques. Each disease unit presents a brief discussion of the disease, followed by a detailed review of the treatment that Dr. Epstein prefers, based on success rate, simplicity, and expense. Dr. Epstein frequently comments on his reasons for preferring a particular treatment or technique.

Following this section are two appendices, one which explains microscopic examination for fungus and another which lists the sources for common dermatologic products. The next portion of the volume contains nearly 100 color illustrations, depicting "mini case histories," with informative descriptions. The final segment contains a collection of over 50 patient information sheets, which may be freely reproduced, each including space for personalizing by the physician. The introduction to the manual explains why and how to use patient information sheets effectively and gives recommendations for writing more sheets. The patient information sheets are well done and, when combined with the illustrations and the treatment units, make this a unique and useful volume.

This manual is replete with practical pointers, definitive explanations, and the advice born of experience which is missing from the textbooks. This would not replace those textbooks but would be used on a day-to-day basis in a clinical practice. Highly recommended for its intended audience, practicing physicians, especially dermatologists and family practitioners, it also should be considered for collections used by medical residents and students in clinical rotations and might be useful in consumer health collections because of the clear explanations of treatment and the excellent patient information sheets.

JOAN LATTA KONECKY

399. Manual of Adult and Paediatric Medical Oncology. S. Monfardini; K. Brunner; D. Crowther; S. Eckhardt; D. Olive; S. Tanneberger; A. Veronesi; J. M. A. Whitehouse; R. Wittes, eds. New York, NY: Springer-Verlag, 1987. 401p, illus., bibliog., index. $38.00pbk. 0-387-15347-0.

Previously published by the International Union against Cancer as the *Manual of Cancer Chemotherapy*, this edition, as the new title testifies, has attempted to expand to become a manual of the essential data from all aspects of oncology. The authors intended that this manual "will prove of help in the development of the fight against cancer, particularly in developing countries, by providing an updated and relatively unexpensive source of reliable information." The new, broader scope required that there be an extensive reorganization of the original manual, to incorporate the additional information on "preclinical data, epidemiology, and non-medical management." The manual retained the unique "telegraphic" style of the first edition, in which the information is presented almost exclusively in table and graph form, to promote rapid access to the data.

The first part of the manual, "General Aspects," reviews the principles of hormone therapy and immunotherapy and presents data on the initial patient evaluation and treatment response, planning of chemotherapy, management of complications, and the chemotherapeutic drugs. The second part addresses the management of adult malignancies by site, with emphasis upon chemotherapy, and the third part deals with the therapy of pediatric malignancies, also by site.

The chapters in the fourth part cover the tumors of particular concern in African and Asian countries. Each chapter finishes with a short bibliography entitled "Further Reading." In addition, there are three short appendices: "Common Abbreviations for the Most Widely Used

Cytotoxic Agents," "Nomograms for Determination of Body Surface Area from Height and Weight," and "List of Recommended Publications for Further Reading." The manual also contains a number of clear, very detailed illustrations.

This manual is an excellent ready-reference source of the basic information on medical oncology. The "telegraphic" style of the manual is easily readable and allows this small handbook to contain a vast amount of data. This would be useful for any practitioner in medical oncology, both specialists and nonspecialists, and in the reference collections of health science libraries.

JOAN LATTA KONECKY

400. Neonatology: A Practical Guide.

3rd. Alistair G. S. Philip. Philadelphia, PA: W. B. Saunders Co., 1987. 559p., illus., bibliog., index. $19.95. 0-7216-2126-3.

Neonatology as a pediatrics subspecialty is a recent phenomenon. Over the past quarter-century, however, there has been a dramatic expansion of the informational knowledge base practitioners in this field need to draw on, in great part as a result of the advances of new technologies and our greater understanding of diseases of the newborn infant. Philip's third edition of *Neonatology: A Practical Guide* is an introduction to this knowledge base, providing clinicians with a practical guide to the diagnosis and treatment of both common and unusual neonatal disorders. This useful handbook is divided into five sections, beginning with an analysis of general concepts and common problems in the care of the newborn infant and continuing on to address the assessment and diagnosis of disorders, procedural techniques, including two quizzes to test the reader's retention of presented data, and finally an extensive section of appendices including how-to information on specific techniques and charts, tables, and graphs of rates and values.

The third edition includes a variety of new chapters, among them coverage of hepatitis B transmission, the role of car seats in preventing trauma, congenital tuberculosis, and rationale for discouraging expectant mothers from smoking. Revised chapters include sections on neonatal infections and intracranial hemorrhage. As with previous editions, this handy book is easy to use, compact, and clearly a reference point for brief, guide-like information. Finally, section references and a thorough index make this book a recommended handbook for clinicians caring for the newborn and for clinic, departmental, and medical libraries.

GERALD J. PERRY

401. Parasites: A Guide to Laboratory Procedures and Identification. Lawrence R. Ash; Thomas C. Orihel. Chicago, IL: American Society of Clinical Pathologists, 1987. 328p., illus. (part col.), bibliog., index. $37.00. "ASCP Press." 0-89189-231-1.

Medical laboratory personnel associated with parasitology will find this manual very useful. Ash and Orihel, professors of tropical diseases at the schools of public health of two prominent universities, have written "a resource that encompasses the basic technology required for detection of parasites and identification of their diagnostic stages in feces, blood, other body fluids, and tissues." Following lists of procedures, figures, tables, and plates, a short preface and an introduction comprise the first and longer of the two major parts of the manual, "Preparation and Examination." This section begins with chapters on collection, preservation, and examination of feces, aspirates, body fluids, and urine. Other chapters discuss the examination of tissues and blood, parasite cultures, procedures for AIDS-related parasitic infections, and serodiagnostic procedures. The concluding chapter in this section briefly discusses quality control and laboratory safety.

In the first major section, 98 of the 100 procedures described in the volume provide detailed instructions for the preparation of fixatives and culture media, staining and sample collection techniques, among others. Typically, each procedure begins with a short background discussion that is followed by a list of reagents, step-by-step instructions, comments on possible problems, and reference to the original description of the procedure. Not all parasitic laboratory procedures currently in use have been included because the authors have selected "the procedures which [they] consider to be the most reliable and which pose the fewest problems in their application."

The second major section of the manual, "Identification and Diagnosis," summarizes in five chapters much of the basic information needed to identify the common parasites found in humans. Beginning with a chapter on microscopy, the section covers protozoa and nematode, cestode, and trematode infections and finishes with a discussion of artifacts in specimens which can be confused with parasites. The second section includes 17 pages of plates (some in color) to assist in the identification of parasitic organisms.

The manual concludes with an extensive list of references cited in the volume and an excellent bibliography of recommended current publications for the library and laboratory. Libraries developing collections in medical parasitology will find the list of textbooks, manuals, atlases, and journals very helpful. The manual's index includes references to the plates, tables, and illustrations as well as to the text. Although this volume is recommended for library collections with interests in medical parasitology and tropical medicine, this spiral-bound manual will be most useful for personnel in the medical laboratory.

STEVEN L. SOWELL

Textbooks

402. Fundamentals of Orthopaedics. 4th. John J. Gartland. Philadelphia, PA: W. B. Saunders Co., 1987. 442p., illus., bibliog., index. $24.95. 0-7216-1413-2.

Gartland's fourth edition of *Fundamentals of Orthopaedics* continues in the tradition of previous editions introducing students to the scope and intrinsic concerns of the field of orthopedics. According to Gartland, one in seven Americans

suffers from some form of musculoskeletal impairment, costing society in excess of $65 million in lost earnings and medical expenses. In addition, according to Gartland, more medical time is utilized to treat musculoskeletal disorders than conditions of any other body system.

The focus of Gartland's *Fundamentals* is on disorders, deformities and trauma to the musculoskeletal system. The introductory chapter provides an overview of the clinical realm of the profession; bone and bone development; diagnostic procedures, including physical and radiologic examination as well as laboratory procedures; and treatment tools such as plaster, braces, dressings, and surgical procedures. Subsequent chapters address a wide range of congenital and acquired deformities of the musculoskeletal system, fractures, trauma, infections, tumors, and neuromuscular conditions. Unlike previous editions, for this book Gartland has recruited contributors, including John Dowling, Eric Hume, and Philip Marone.

Students and those requiring a concise yet complete introduction to the field of orthopedics will find this book a handy, nicely organized and easy-to-use introduction to the essentials. Library reference collections supporting treatment and training programs relative to the musculoskeletal system will certainly be interested in this new edition of Gartland's.   *GERALD J. PERRY*

PHYSICS

403. Dictionary of Effects and Phenomena in Physics: Descriptions Applications Tables. Joachim Schubert. New York, NY: VCH, 1987. 140p., bibliog. $24.95. 0-89573-487-7.

This English edition of the 1984 German book, *Physikalische Effekte*, has been supplemented with additional references to Anglo-American literature and definitions of effects. The author treats physical effects comprehensively by fleshing out the dictionary section with a short essay on the historical use of physical effects terminology and by adding numerous tables, a chronology, and a bibliography.

The definitions and explanations are aimed at a "broad readership" but do assume some familiarity with and knowledge of physics and the related terminology. Approximately 400 effects and phenomena are presented in 100-200 word definitions which describe the effect; its relationship to other effects; and give the full name, dates, and profession of the scientists who are responsible for naming or discovering the effect (for more recently discovered phenomena, the biographical information often is missing). Citations to literature of the original discussion, as well as references to additional reading, are listed.

The effect is also given a code which assigns it to one of 37 categories within the field of physics (e.g., atomic and quantum physics, electrokinetics, low temperatures, nuclear physics, optics). A section of tables, appearing as an appendix to the dictionary, lists effects in each of the 37 categories by cause-effect-name and suggests literature from the bibliography which would supply further reading. The definitions are lucid and concise. The background information added to

the definitions gives the book more substance and adds to its reference value. Recommended for all libraries used by students and researchers in the physical sciences. *MARILYN VON SEGGERN*

Manuals

404. The Marshall Cavendish Science Project Book of Light. Steve Parker. Freeport, NY: Marshall Cavendish, 1988 (c1986). 43p., illus. (col.). $12.49. The Marshall Cavendish Library of Science Projects, v. 4. 0-86307-628-9.

This encyclopedic project book for young people covers the field of light. Through a series of chapters that discuss what light is, seeing with the eye, the sun, electric lights, lasers, laws of reflection and refraction, lenses, telescope, microscopes, and rainbows, the student learns the basics of light. Experiments and projects are used to exemplify these basics using a well-written text and good illustrations. It is a useful reference book for school libraries and an excellent textbook for the classroom.
H. ROBERT MALINOWSKY

405. The Marshall Cavendish Science Project Book of Mechanics. Steve Parker. Freeport, NY: Marshall Cavendish, 1988 (c1986). 43p., illus. (col.). $12.49. The Marshall Cavendish Library of Science Projects, v. 3. 0-86307-625-4.

"Science is all about discovering more about your world, finding out why certain things happen and how we can use them to help us in our everyday lives." The field of mechanics is covered in this project book. The laws of motion, energy, friction, machines, power, mechanics in the house, engine power, and machinery are outlined in a systematic way, using experiments and projects as well as an excellent text and good illustrations. It is a good reference source for school libraries and an excellent textbook for the classroom. *H. ROBERT MALINOWSKY*

TECHNOLOGY

★
GENERAL SOURCES
Bibliographies

406. Bibliographic Guide to Technology 1987. Boston, MA: G. K. Hall and Co., 1988. 2,037p. $325.00. 0-8161-7062-2. 0360-2761.

G.K. Hall publishes annual *Bibliographic Guides* in 20 fields with a wide range of subject areas. The *Bibliographic Guide to Technology 1987* covers engineering in the following areas: industrial, structural, civil, transportation, hydraulic, sanitary, highway, mechanical, electrical, nuclear, and mining. In addition to these engineering fields, this *Guide* also includes literature references on patents and trademarks, environmental technology, aeronautics and astronautics, metallurgy, chemical technology, food processing and manufacture, and production management. *Guides* are annual subject bibliographies, compiled from publications cataloged by the Research Libraries of the New York Public Library (NYPL) and the Library of Congress. The two-volume *Technology 1987* guide includes "all languages and all forms—non-book materials as well as books and serials." Complete LC cataloging information is provided for each title, along with ISBN and the identification of NYPL holdings. Cataloging follows the AACR format.

Arranged in one alphabetical sequence, access is by main entry (personal author, corporate body, names of conferences), added entries (co-authors, editors, compilers), titles, series titles, and subject headings. The full bibliographic information appears in the main entry, whereas all secondary entries have only abbreviated or condensed entries. Subject headings are in boldface and capitalized. A "Sample Entry" in the Preface illustrates the format followed in the publication. Selections for the *Technology 1987* guide are based on the items cataloged during the past year (1987) by the Library of Congress and the Research Libraries of NYPL in the "technology" field. Selection of titles from LC MARC tapes is "based on the LC classification *T* (except *TT* and *TX*). Entries from NYPL are selected on the following basis: if an NYPL record contains two or more subject headings which correspond to two or more LC subject headings from MARC technology records, then that NYPL record is

included...."Subject headings for all technology records in the entire MARC database are scanned to provide comprehensive selection of NYPL titles."

This *Guide* provides a wealth of information for libraries, librarians, and scholars and researchers. It can serve as a valuable selection and acquisition tool and as an aid in cataloging, and special subject searches in engineering can be quickly and easily facilitated with this set. As a useful reference resource, it gives quick, multiple access to a wide range of timely literature for the scholar, researcher, and librarian.

R. GUY GATTIS

Directories

407. Directory of American Research and Technology 1988: Organizations Active in Product Development for Business. 22nd. New York, NY: R. R. Bowker, 1987. 763p., index. $199.95. 0-8352-2417-1. 0886-0076.

This reference volume is the latest in a series of annual publications profiling research and product development capabilities of public and private firms within the United States. Coverage is limited to nongovernmental facilities, including cooperative organizations, foundations, and universities. Of the 11,942 organizations included, 977 are new with this edition. Reference librarians should be aware that in some cases subsidiaries have been listed as separate companies at their own request. Each entry provides full address information, names of the executive staff, employment figures, and the areas of research and development in which the body is active. Indexing is provided by geographic location, personal names, and subject access, by type of industry and scientific subject area. This volume is also available as an online database on the Pergamon Infoline service. Suitable for large public libraries, college and research institutions, and corporate collections.

ROBERT B. MARKS RIDINGER

408. The History of Invention. Trevor I. Williams. New York, NY: Facts on File Publications, 1987. 352p., illus. (part col.), bibliog., index. $35.00. 0-8160-1788-3.

This is a broad encyclopedic survey of the history of technology, with half of the text devoted to historical influences such as "The Rise of Islam" and "The Arts of War" and half of the text focused on specific industrial developments in shipping and navigation, mining and metals, and agriculture. The biographical dictionary is particularly helpful in determining the origin of particular inventions. Maps and timelines add to the visual pleasures of this volume. The bibliography is not extensive but does provide an opportunity to pursue further study. Inclusion of cut-away diagrams helps the student understand how things work and gives ideas for model construction. This would be most appropriate for school and public libraries. *ANNE PIERCE*

Encyclopedias

409. G. K. Hall Encyclopedia of Modern Technology. David Blackburn; Geoffrey Holister, eds. Boston, MA: G. K. Hall and Co., 1987. 248p., illus. (part col.). $35.00. 0-8161-9056-9.

The idea of making technical information understandable to students and laypeople through primarily visual means is a good one, and it works here. With nearly 500 excellent, specially commissioned illustrations, this single-volume survey provides the student with a basic understanding of select processes, devices, and instruments of contemporary technology. Although a single-volume work surveying modern technology raises initial questions of adequacy, this work's visual support compensates for its brevity. For example, the chapter "Seeing with Sound" is short and introductory, but the treatment is rich and stimulating. "How-It-Works" diagrams and actual sound-generated images illustrate the topics of seabed-scanning sonar, seismic reflection, ultrasound, and acoustic microscopy. This, with coverage of other related technologies, occurs in just eight pages. The chapter, however, makes a vivid impression and works to provide a conceptual understanding of the processes described.

There is a clear and successful effort here to be current (compact disk, scanning tunneling microscope, the Rutan "Voyager"), yet even the newest technologies described are consistently placed in their evolutionary context. Quick reference chronologies accompany descriptions of major fields of development. Illustrations include three-dimensional cut-away sections, process flow diagrams, exploded diagrams, and pictures from a variety of imaging techniques. Text and illustrations are effectively complementary. Arrangement is by broad conceptual classes, then subdivided by specific technologies. Articles are unsigned but contributors noted. Note that this is an introductory work; however, even the well-versed may appreciate these visual elucidations. This is highly recommended for school and public libraries. *JOHN T. BUTLER*

Handbooks

410. Handbook of Engineering Design. Roy D. Cullum. Boston, MA: Butterworths, 1988. 303p., illus., bibliog., index. £155.00. 0-408-00558-0.

This book contains 16 chapters on different aspects of engineering design. Each chapter is written by an expert or experts in that specific area. An example of its content, Chapter 2 on "Engineering Materials," is 47 pages in length, includes general statements on "choice of materials," and then describes specific materials ranging from "alloy steels" to "poly-carbonates." Examples of other chapter headings included are "Stages in Design," "Stress Analysis," "Fluid Power," "Heat Exchangers," and "Preparing a Technical Specification."

The book is well illustrated with tables and diagrams throughout that enhance the descriptive information in the text. Most chapters include a bibliography, and some include the names and addresses of companies that can supply materials or products listed. There is a seven-page directory at the end of the book listing the names and addresses of manufacturers, consultants, societies and institutions, testing and inspection firms, etc. Edited and published in England, the book reflects the geographic focus of its production; most of the firms listed at the end of chapters and in the directory at the end of the book are located in England. Recommended for university and special libraries. *RAYMOND BOHLING*

411. IES Lighting Handbook: Applications Volume. John E. Kaufman; Jack F. Christensen, eds. New York, NY: Illuminating Engineering Society of North America, 20, 1987. 530p., illus. (part col.), bibliog., index. $200.00. 0-87995-024-2.

The *IES Lighting Handbook: Applications Volume*, is a revision of the 1981 *Applications Volume*. Review of content and accuracy of the current volume was done by either the Illuminating Engineering Society's technical committees, individuals knowledgeable on the subjects, or by committees of other technical societies. Contributors to the handbook are listed in the preface. The handbook is divided into 20 sections. A few of the major changes from the 1981 volume include: a new section on lighting controls; illumination criteria for visual display terminals (VDTs); and luminance criteria for design and evaluation of roadway lighting systems. References are at the end of each section; the very extensive index, which includes acronyms, is also an index for the companion *Reference Volume*. On the inside front cover the contents for both the *Applications Volume* and the *Reference Volume* are given. A list of credits for illustrations and tables precedes the index, and on the inside back cover is a "Partial list of abbreviations and acronyms used in the handbook," along with illumination charts.

This handbook covers every aspect of lighting applications. The 18 sections are: "Lighting Design"; "Lighting System Design Considerations"; "Lighting Controls"; "Energy Management"; "Office Lighting"; "Educational Facilities Lighting"; "Institution and Public Building Lighting (including Libraries)"; "Residential Lighting"; "Theatre, Television and Photographic Lighting"; "Outdoor Lighting Applications"; "Sports and Recreational Areas"; "Roadway Lighting"; "Aviation Lighting"; "Transportation Lighting"; "Lighting for Advertising"; "Underwater Lighting"; "Nonvisual Effects of Radiant Energy"; and "Searchlights." Highly recommended for those libraries serving researchers in the field of lighting.

JO BUTTERWORTH

Manuals

412. Cost Estimator's Reference Manual. Rodney D. Stewart; Richard M. Wyskida. New York, NY: John Wiley and Sons, 1987. 620p., illus., bibliog., glossary, index. $60.00. "A Wiley-Interscience Publication." 0-471-83082-8.

This "handbook is designed to permit a thorough cover-to-cover study of cost estimating or to permit quick reference to specific areas and disciplines for those who have encountered trouble spots in a cost-estimating situation. You will be told how to develop a credible and accurate cost estimate; where to get supporting information and data; what tools and techniques are available; and whom to contact about being certified, obtaining publications and information, and gaining education and training in the profession." Learning to do cost estimates takes time and talent, and far too often those who do the cost estimating are not informed enough to do the best job possible, resulting in a poor estimate that ultimately costs the user of the estimate more money and time.

Stewart and Wyskida have written an excellent manual with the help of some experts in the field, covering estimating techniques such as cost allocation, discounted cash flow analysis, learning curves, parametric estimating, risk analysis, and the use of microcomputers. After an introductory chapter that covers the fundamentals of cost estimating, this well-written book covers "Cost Allocation," "Statistical Techniques in Cost Estimation," "Discounted Cash Flow Analysis," "Learning Curves and Progress Functions," "Detailed Cost Estimating," "Parametric Estimating," "Cost/Schedule/Technical Performance Risk Analysis," "The Use of Microcomputers for Cost Estimating," "Construction Cost Estimating," "Cost Estimating in Manufacturing," "Software Cost Estimating," "Aspects Affecting Cost Estimating in Government Procurement," "Cost Estimating as a Profession," and "Artificial Intelligence in Cost Estimating."

Most chapters have an introduction or overview followed by a detailed discussion of the topic, ending with a summary and bibliography. There is good use of graphs, charts, and diagrams

and a brief but good bibliography of books from the 1980s, through 1986. Of particular note is the "Dictionary of Estimating Terms" adapted from the *National Estimating Society Dictionary*. The index is brief but adequate.

This is a recommended book for engineering libraries and for those persons who are required to make cost estimates, including engineers, accountants, manufacturing planning personnel, mathematicians, statisticians, economists, engineering managers, and other professionals.

H. ROBERT MALINOWSKY

413. Engineering Projects for Young Scientists. Peter H. Goodwin. New York, NY: Franklin Watts, 1987. 126p., illus., bibliog., index. $11.90. Projects for Young Scientists. 0-531-10339-0.

"This book outlines the proper method for finding answers to questions you ask, gives you a chance to 'do' physics, and suggests engineering problems you may want to solve." Through a well-written text with good illustrations, this manual helps the student understand the laws of physics and how they are applied to engineering. Each chapter begins with some background information on the topic, outlines preliminary investigations, and then gives some ideas for projects. All of the projects are designed for materials that are readily available. The seven chapters cover "Physics and Engineering," "Working Scientifically," "Forces and Motion," "Sound," "Light and Water Waves," "The Physics Olympics," and "Science Fairs."

All of the projects are illustrated with things one sees everyday, such as cantilever-and-truss bridges, rockets, amusement park rides, musical instruments, cameras and lenses, and water waves. This is an excellent book for teaching the student about acoustics, optics, kinetics, and other branches of physics and a recommended book for school and public libraries.

H. ROBERT MALINOWSKY

Union Lists

414. Union List of Technical Reports, Standards and Patents in Engineering Libraries. Dorothy F. Byers, comp. Washington, DC: Engineering Libraries Division, American Society for Engineering Education, 1986. 29p. $5.00pbk. Engineering Literature Guides, no. 10. 0-87823-110-2pbk.

The American Society for Engineering Education (ASEE) Engineering Libraries Division (ELD) has made available another in its series of *Engineering Literature Guides*. Number 10 is a union list of difficult-to-locate technical reports and papers; standards and specifications; and patents held by the ELD member libraries. These libraries are located at universities with the nation's major engineering schools. The avowed aims of this publication are to help engineering librarians locate engineering materials often needed, yet not always readily available or easy to find

via the traditional interlibrary loan channels, and to promote reciprocal sharing of materials and to foster more rapid loaning among participants. The introductory pages list the participating libraries by Library of Congress (LC) codes, and are followed by a summary of the lending and copying policies of each. The union list comprises the main body of the guide and is arranged alphabetically within each of the five groups of materials: society technical papers and reports (e.g., ASME, RAND reports); government agency reports (AEC, EPA reports); standards (ANSI, military standards, and specifications); patents (U. S. and foreign); and "other" (data services, manufacturers' catalogs). The entries for each title include holding libraries identified by LC code (e.g., MiU for University of Michigan), the range of holdings in each library, and the format of the materials (paper, microfiche, film).

The listing by material type, coupled with the list of participating libraries and codes, constitutes a potentially valuable source of information and assistance both to the librarian and to the informed researcher. Telephone numbers and contact personnel are included in almost all of the library entries, a service that could prove quite useful when attempting to get a copy of a certain MIL-Standard or an EPA report on short notice. The *Union List* is scheduled to be updated approximately "every two or three years." Twenty-nine other guides currently in preparation by the ELD Publications Committee, all of which will be welcome additions to the reference world of academic and large public libraries, are listed in the endpaper.   *R. GUY GATTIS*

Yearbooks/Annuals/Almanacs

415. Annual Statistics of Academic Engineering Libraries, 1984–85. Standards Committee, Engineering Libraries Division, American Society for Engineering Education., comp. Washington, DC: American Society for Engineering Education, 1987. 70p. $12.00pbk. 0-87823-112-9pbk.

The stated purpose of this publication is "to measure college library resources directly supporting engineering education and research." The statistics were gathered from a survey sent to over 250 schools with ABET-accredited engineering programs in 1985. The questionnaire is appended. Statistics are reported in two separate sections, one for engineering libraries and one for engineering collections housed in larger libraries. Statistics focus on library expenditures, personnel, circulation policies, size and type of collection, and library services offered to both primary and secondary clientele. A salary survey of librarians from the responding libraries makes up section 3 of the work. The resulting comparative statistics will benefit library administrators, library planners, and associations who determine standards of excellence for engineering education.   *CHARLENE M. BALDWIN*

★
BUILDING CONSTRUCTION
Handbooks

416. Architectural Lighting for Commercial Interiors. Prafulla C. Sorcar. New York, NY: John Wiley and Sons, 1987. 249p., illus., index. $37.95. "A Wiley-Interscience Publication." 0-471-01168-1.

"For many years the scientific side of lighting has been separated from the artistic, the utilitarian from the aesthetic, engineering from architecture. But lighting design combines these elements and more; the visual, psychological, and physiological, to name a few." The author has put together a working handbook for lighting designers, consulting engineers, and architects that discusses many of the controversial issues involved in designing lighting for commercial buildings, including level of illuminance, brightness, uniform pattern, blanket of light, nonuniform pattern, and accent lighting.

After an introduction that discusses the light story, light and vision, and seeing color, Sorcar divides the text into four parts. The first, "Engineering Tools," which covers light sources and accessories, photometrics, lighting calculations, short-cut methods of calculation, lighting layout, electrical circuiting, and principles of lighting control. This is followed by "Architectural Tools," covering incandescent luminaires, fluorescent luminaires, high-intensity discharge luminaires, lighting patterns and forms, human reaction to light, human reaction to color and application, and environmental impression. These two sections lay out the basics in engineering and architectural design of lighting. The text is easy to follow and contains many illustrations, charts, diagrams, and graphs to aid in the design of the right lighting structure.

The last two sections are the applications sections, one covering the office environment and the other the merchandising environment. Here are shown examples of task lighting, ambient lighting, task-ambient lighting, perimeter lighting, sign and show-window lighting, and store lighting. The environmental aspects are stressed in all cases, and worst cases are shown.

This is an outstanding book and one that should be of great value to architects and engineers, as well as being a good reference source in architectural engineering libraries.
      H. ROBERT MALINOWSKY

417. Building Structural Design Handbook. Richard N. White; Charles G. Salmon, eds. New York, NY: John Wiley and Sons, 1987. 1,197p., illus., bibliog., glossary, index. $74.95. "A Wiley-Interscience Publication." 0-471-08150-7.

Structural engineering is the art and science of constructing buildings. *Building Structural Design Handbook* has two primary purposes: (1) to enable practicing structural engineers to make the

best decisions in the crucial early stages of building design, when alternative schemes are being formulated and evaluated, and (2) to provide information needed for the final design which is not readily available in standard textbooks or other traditional references." Editors White and Salmon have more than adequately carried out these purposes with the help of 41 contributors. Each chapter is separately authored, but all have the same format including a table of contents and a bibliography of additional references.

The first three chapters, "Introduction to the Handbook," "Loads," and "Design Philosophies," provide an introduction to structural engineering. Chapter 4, "Mechanical and Electrical Systems," and chapter 5, "Vertical Transportation," discuss the mechanical, electrical, and transporting aspects of building design. Chapter 6, "Welding-Related Considerations in Building Design," chapter 7, "Structural Walls and Diaphragms," and chapter 8, "Structural Form," complete the total design process. Information in these chapters is basic to the rest of the handbook. Chapters 9 through 16 cover preliminary design of different kinds of buildings and foundations: "Preliminary Design of Low-Rise Buildings," "Tall Buildings," "Preliminary Design of High-Rise Buildings," "Preliminary Design of Single-Story Open-Space Buildings," "Preliminary Design of Shells and Folded Plates," "Preliminary Design of Space Trusses and Frames," "Preliminary Design of Lightweight Membrane Structures Including Air Supported and Structurally Supported Systems," and "Foundations." These are detailed chapters and ones that the architect and engineer will use in designing structures.

The remaining chapters cover "Computer-Aided Analysis and Design," "Dynamic Loading," "Steel Design," "Cold-Formed Steel Construction," "Prestressed Concrete Design," "Reinforced Concrete Design," "Composite Construction," "Masonry," "Wood Structures," "Roofs and Roofing," "Building Facades," "Aluminum Structures," "Design for Stainless Steel," and "Design with Structural Plastics."

The editors have left out nothing and have presented the information in a clear, readable format. In addition to the tables of contents for each chapter, there is a detailed index. The illustrations are clear, and the use of many charts and graphs aids in understanding the various concepts that are being presented. Engineers, architects, designers, and students will find this an invaluable handbook. Academic and special libraries will need the book for their reference shelves.

H. ROBERT MALINOWSKY

418. **Construction Foreman's Job Guide.**
James E. Clyde. New York, NY: John Wiley and Sons, 1987. 416p., illus., index. $49.95. Wiley Series of Practical Construction Guides. 0-471-81660-4.

Construction Foreman's Job Guide covers a broad range of civil engineering terms, tools, practices, and examples from on-site situations. Although the title suggests that this reference work has been primarily written for construction supervisors, it may be more appropriate to approach the work by determining the level of experience and knowledge (rather than position) of the potential reader. The author, an experienced engineer, opens the first chapter with comments and observations on what it takes for a person to become a construction supervisor and then, in general terms, outlines the professional expectations of those working as supervisors. So, although some of the terms, such as "official visitors," "labor relations," "budgets," or "merit system," may only be of interest to supervisors, most of the information will be helpful to people working in the field who want a brief dictionary on the subject.

After the first chapter, the arrangement of terms and practices is alphabetical. The degree to which the terms are explained or described in figures or photographs varies greatly. For example, under "bituminous concrete," there are four pages of text and photographs, compared to one sentence on "eaves." Neither explanation is sufficient to understand or perform the tasks. Several pages and photographs are devoted to a list of contractor's equipment. The uneven treatment of subjects may reflect the experience of the author rather than the nature of that which is being discussed. Several appendices are included which cover information on metric system conversion, average weights of materials, useful calculations, weights and measures, and temperature conversion. The author provides the Construction Specifications Institute (CSI) Index reference numbers for the appropriate terms listed in the contents. Academic and special libraries will find this a useful reference source.

CHARLES R. LORD

419. **Designs for Glued Trusses.** 4th.
Ames, IA: Midwest Plan Service, 1981. 84p., illus. $5.00pbk. "MWPS-9." 0-89373-051-3pbk.

"This book gives about 10,000 different truss designs. There are 16,443 combinations of span, lumber grade, slope, etc., but loads below 12 psf and above about 100 psf are not included." Only truss, roof, and ceiling design and construction are covered in this book. Introductory chapters cover truss selection, materials, joints, construction, erection, and roof and ceiling construction. The main part of the handbook contains the selection tables that identify the various trusses. Builders will find this a useful handbook to have, and libraries will find that it supplements more sophisticated building engineering reference sources. *H. ROBERT MALINOWSKY*

420. **Farmstead Planning Handbook.**
Ames, IA: Midwest Plan Service, 1982 (c1974). 44p., illus. $5.00pbk. "MWPS-2." 0-89373-001-7pbk.

This is a small, well-written, and practical handbook that helps the farmer plan and develop the farmstead to increase efficiency, convenience, and profit. Charts, maps, diagrams, and line drawings supplement the text covering development, planning, activity centers, and services.

Any modern farmer would want to consult this useful handbook. Libraries in rural areas will need it for consultation by local patrons.

H. ROBERT MALINOWSKY

421. Guidelines for Laboratory Design: Health and Safety Considerations. Louis J. Diberardinis; Janet Baum; Melvin W. First; Gari T. Gatwood; Edward Groden; Anand K. Seth. New York, NY: John Wiley and Sons, 1987. 285p., illus., bibliog., index. $34.95. "A Wiley-Interscience Publication." 0-471-89134-7.

Guidelines for Laboratory Design is intended for those who will be involved in the design of new or renovated laboratories. To ensure that the needs of all are met, the engineering architect would be working with those who will be using the laboratory. Therefore, the audience includes "laboratory owners, managers and occupants, architects, engineers, health and safety personnel, and risk managers." The authors, too, bring diverse backgrounds to the subject—architecture, industrial hygiene, and mechanical engineering. There are many types of laboratories. Many laboratories, however, share the same health and safety problems. Two-thirds of the book is directed at those issues that involve a wide range of laboratories. The other third of the book contains chapters with the guidelines for specific types of laboratories. Although the problems may differ, each chapter has the same layout, with these five categories covered: (1) "Guiding Concepts"; (2) "Laboratory Arrangement"; (3) "Heathing, Ventilating, and Air Conditioning (HVAC)"; (4) "Loss Prevention, Industrial Hygiene, and Personal Safety"; and (5) "Special Requirements." As an example, the HVAC sections discuss the general requirements of ventilation in the laboratory as well as specific ventilation needs with the use of chemical fume hoods.

There is a brief chapter devoted to each of 16 laboratory types. Some of these laboratories are analytical chemistry, high-toxicity, clean room, radiation, gross anatomy, and animal research. In addition, Appendix XI conveniently contains several pages displaying, in charts, which health and safety considerations are applicable to which types of laboratories. Numerous agencies and associations have issued regulations, codes, and standards that apply to the health and safety issues discussed. References to those regulations, etc., have been made throughout the book and the appendices.

Organizations often have written requirements for the facilities and procedures for their laboratories. *Guidelines for Laboratory Design: Health and Safety Considerations* is recommended as a helpful supplement to other laboratory design books in university and special libraries.

FRANK R. KELLERMAN

422. Painter's Handbook. William McElroy. Carlsbad, CA: Craftsman Book Co., 1987. 318p., illus., glossary, index. $21.25. 0-934041-28-8.

The purpose of this handbook is "to give all the information needed to handle nearly any residential, commercial, or industrial painting job—quickly, efficiently and profitably." After a brief history of paint, information is given on how to create a profitable paint contracting company. Pointers are given on an accounting system and how to carry out accounting, inventory, and management control once the business is established. This is followed with chapters on selling your service, estimating, and painting safety. The next area covered is choosing the paint and color and surface cleaning and preparation. Common paint problems and water and fire damage are discussed with suggested solutions to the problems. The chapter on tools covers brushes, pads, rollers, spray equipment, paint removal tools, ladders, and scaffolds. Procedures are given on painting new construction, spray painting, texturizing, and alternatives to paint as a wall covering. There are few handbooks of this type available that explain in simple terms the "nitty-gritties" of painting. There is also a glossary. Anyone interested in becoming a painter will need this handbook. Seasoned and professional painters will find a wealth of useful information. Public libraries will find this a good source of information on a profession that has few books written about it.

H. ROBERT MALINOWSKY

423. Structural Details for Concrete Construction. Morton Newman. New York, NY: McGraw-Hill Book Co., 1988. 248p., illus., index. $22.50pbk. 0-07-046360-3pbk.

A collection of structural drawings illustrating the most commonly used details in concrete construction, *Structural Details for Concrete Construction* is intended for architects, engineers, and contractors involved in the design and construction of buildings. Morton Newman, a civil engineer and president of Newman Engineering Company, proposes to improve the communication between the engineer and the contractor through clear and complete structural drawings. Providing a compilation of the details commonly used in the concrete construction industry, each drawing is accompanied by a description outlining materials, conditions, and construction methods. The concrete construction details included cover continuous footings, grade beams, pit walls, basement walls, reinforcement splices and bends, retaining walls, beam sections, concrete joists, caissons, spread footings, column sections and splices, slabs, steps, and precast walls. The volume includes an index and an explanation of abbreviations.

Structural Details for Concrete Construction is a reprint of the concrete section *Standard Structural Details for Building Construction* (1968), by the same author, which also covered steel, wood, and masonry construction. In this 1988 volume, there are only three changes from the original edition: (1) the introductory material has been updated to include current ASTM (American Society for Testing and Materials) specification numbers and ACI (American Concrete Institute) standard numbers; (2) each page of drawings now

faces a blank page of graph paper headed "Notes Drawings Ideas," and (3) the format is now a spiral-bound paperback which lies flat for easy use. The primary advantage of this reprinting is that a user can selectively purchase only the structural details pertaining to concrete construction.

CAMILLE WANAT

424. Structural Details for Masonry Construction. Morton Newman. New York, NY: McGraw-Hill Book Co., 1988. 152p., illus., index. $19.50pbk. 0-07-046361-1pbk.

Structural Details for Masonry Construction is a collection of structural drawings illustrating the most commonly used details in masonry construction. Intended for architects, engineers, and contractors involved in the design and construction of buildings, the book's stated purpose is to improve the communication between the engineer and the contractor through clear and complete structural drawings. This collection is intended to assist in the process by providing a compilation of the details commonly used in the masonry construction industry. Each drawing in the book is accompanied by a description outlining materials, conditions, and construction methods. The masonry construction details included cover concrete block walls; concrete block columns and pilasters; concrete block walls supporting floors and roofs, steel framing, and concrete slabs; brick walls; brick columns and pilasters; brick walls and steel columns; brick walls supporting wood floors and roofs, and steel framing; typical brick wall sections; and concrete block retaining walls. The volume includes an index and an explanation of abbreviations.

Authored by a civil engineer and president of Newman Engineering Company, *Structural Details for Masonry Construction* is a reprinting of the section concerning masonry from the 1968 publication *Standard Structural Details for Building Construction*, also written by Morton Newman. This earlier volume also covered concrete, wood, and steel construction. In this 1988 volume, there are only three changes from the masonry section in the original edition: (1) the introductory material has been updated to include current ASTM (American Society for Testing and Materials) specifications; (2) each page of drawings now faces a blank page of graph paper headed "Notes Drawings Ideas," and (3) the format is now a spiral-bound paperback which lies flat for easy use. The primary advantage of this reprinting is that a user can now selectively purchase the structural details pertaining only to masonry construction.

CAMILLE WANAT

425. Structural Details for Steel Construction. Morton Newman. New York, NY: McGraw-Hill Book Co., 1988. 226p., illus., index. $22.50pbk. 0-07-046359-Xpbk.

Structural Details for Steel Construction is a collection of structural drawings illustrating the most commonly used details in steel construction. Intended for an audience consisting of architects, engineers, and contractors involved in the design and construction of buildings, the author's stated purpose is to improve the communication between the engineer and the contractor through clear and complete structural drawings. Morton Newman, a civil engineer and president of Newman Engineering Company, provides a compilation of the details commonly used in the steel construction industry. Each drawing is accompanied by a description outlining materials, conditions, and construction methods. The steel construction details cover base plates and connections, beam and column connections, column splices, beam connections, composite steel beams and concrete slabs, fire protection, bar joists, light steel framing, steel stairs, and clevis connections. The volume includes an index and explanation of abbreviations.

Structural Details for Steel Construction is a reprint of the steel section from the 1968 publication *Standard Structural Details for Building Construction*, by the same author, also covering concrete, wood, and masonry construction. In this 1988 volume, there are only three changes from the original edition: (1) the introductory material has been updated to include current ASTM (American Society for Testing and Materials) and AISC (American Institute of Steel Construction) specifications and data; (2) each page of drawings now faces a blank page of graph paper headed "Notes Drawings Ideas," and (3) the format is now a spiral-bound paperback which lies flat for easy use. The primary advantage of this reprinting is that a user can now selectively purchase the structural details pertaining only to steel construction.

CAMILLE WANAT

426. Structural Details for Wood Construction. Morton Newman. New York, NY: McGraw-Hill Book Co., 1988. 136p., illus., index. $22.50pbk. 0-07-046358-1pbk.

Structural Details for Wood Construction is a collection of structural drawings illustrating the most commonly used details in wood construction. The intended audience consists of architects, engineers, and contractors involved in the design and construction of buildings. A civil engineer and president of Newman Engineering Company, Morton Newman's purpose is to improve the communication between the engineer and the contractor through clear and complete structural drawings. This collection is intended to assist in the process by providing a compilation of the details commonly used in the wood construction industry.

Each drawing in the book is accompanied by a description outlining materials, conditions, and construction methods. The wood construction details included cover post-base connections, floor sheathing, wood stairs, walls, floor framing, post and beam connections, beam connections, and glued, laminated wood details. The volume includes an index and an explanation of abbreviations.

Structural Details for Wood Construction is a reprinting of the section on wood from the 1968 publication *Standard Structural Details for Building Construction*, by the same author. This earlier volume also covered concrete, masonry, and steel construction. In this 1988 volume, there are only three changes from the wood section in the original edition: (1) the introductory material has been updated to include current standards and designations; (2) each page of drawings now faces a blank page of graph paper headed "Notes Drawings Ideas," and (3) the format is now a spiral-bound paperback which lies flat for easy use. The primary advantage of this reprinting is that a user can now selectively purchase the structural details pertaining only to wood construction.

CAMILLE WANAT

427. Swimming Pools: A Guide to Their Planning, Design, and Operation. 4th.
M. Alexander Gabrielsen, ed. Champaign, IL: Human Kinetics Publishers, Inc.. for the Council for National Cooperation in Aquatics, 1987. 315p., illus., bibliog. $32.00. 0-87322-075-7.

The first edition of *Swimming Pools: A Guide to Their Planning, Design, and Operation* was published in 1969. During this time, it has become the accepted handbook on swimming pool design by the Council for National Cooperation in Aquatics, an umbrella organization for over 30 national agencies and organizations that have strong interest and involvement in aquatics.

"There are basic questions that must be answered for all those who own or operate pools, whether they are homeowners, motel or hotel pool operators, school districts, municipalities, or park boards." These questions may relate to aesthetic design, reasonable cost, safe operation, maintenance and supervision, or many other considerations. This handbook should help to answer all of these questions. It is extremely well written and understandable.

The 20 chapters cover "The Community Swimming Pool"; "Voluntary Agency and School Pools"; "Residential Pools"; "Pools for Motels, Hotels and Apartment/Condominium Complexes"; "Spas, Hot Tubs, Saunas, and Steam Rooms"; "Pools of the United States"; "Pools of the World"; "Safety Guidelines for Pool Design and Operation"; "Pool Facilities for Impaired and Disabled Persons"; "Pool Equipment and Accessories"; "Pool Design"; "Water Circulation and Filtration"; "Swimming Pool Chemistry"; "Pool Design Features for Competitive Swimming"; "Pool Enclosures"; "The Pool Bathhouse"; "Solar Energy"; "Pool Modernization and Renovation"; "Pool Energy Management"; and "Innovations in Pool Design." As can be seen from these chapter headings, no aspect of swimming pools has been overlooked. Three appendices cover "Establishing a Pool Design Program," "Pool Planning Checklist," and "Organization Addresses."

This is a must book for any facility that has a swimming pool or is contemplating having one. The amount of information contained in this handbook makes it one of the best of the books covering swimming pools and as a result is highly recommended.

H. ROBERT MALINOWSKY

How-to-do-it Books

428. Home and Yard Improvements Handbook. Ames, IA: Midwest Plan Service, 1978. 100p., illus. $6.00pbk. "MWPS-21." 0-89373-034-3pbk.

This is a "do-it-yourself guide with emphasis on making maximum use of space for storage and on home improvements." It contains descriptions and instructions for making numerous projects pertaining to storage and outdoor living, such as room dividers, closets, decks, fences, and retaining walls in addition to the many storage facilities that can be built in bedrooms, bathrooms, laundries, workshops, and basements.

H. ROBERT MALINOWSKY

Manuals

429. Carpentry Estimating. W. P. Jackson. Carlsbad, CA: Craftsman Book Co., 1987. 317p., illus., index. $25.50pbk. 0-934041-17-2pbk.

The whole purpose of this book is "to help you make fast, accurate carpentry estimates." It stresses the use of the multiplying tables throughout the book to eliminate common mathematics errors. The second most important aspect is that it emphasizes the use of good record keeping, and this book will help in creating these good records. The 12 chapters cover "Estimating Construction Costs," "Estimating Rough Carpentry," "Estimating Footings and Foundations," "Estimating Floor Systems," "Estimating Exterior and Interior Walls," "Estimating Roof Systems," "Estimating Moisture Protection and Insulation Costs," "Estimating Siding and Exterior Trim," "Estimating Finish Carpentry," "Estimating Interior Trim and Stairs," "Staying Out of Trouble with Subcontractors," and "Estimating Manhours for Carpentry." The text is well written and includes numerous examples with many line drawings, charts, and tables. An additional aid is the collection of worksheets covering rough carpentry, footings and foundations, floor systems, interior and exterior walls, roof systems, siding and exterior trim, interior wall and ceiling finish, interior trim, and weekly timesheet. This is an extremely practical manual for anyone who is involved in carpentry work. It would also be useful for the architectural student who needs to figure cost estimates. School and public libraries will find it useful for its many charts and tables.

H. ROBERT MALINOWSKY

430. Carpentry for Residential Construction. Byron W. Maguire.
Carlsbad, CA: Craftsman Book Co., 1987. 388p., illus., index. $19.95. 0-934041-21-0.

"This book was written for carpenters and builders who need a good reference manual on the carpentry trade." The first four chapters cover the basic skills one needs to do any carpentry job—planning, drawing and specifications, estimating, and programming. The fifth chapter is a brief dictionary of carpentry and building terminology. Each of the remaining chapters covers a particular aspect of construction and building—cement work and foundations, wall framing, sheathing, roof framing, shingling, cornice, window-unit installation, door-unit installation, door installation and maintenance, siding, drywall installation, ceiling tile installation, trim, floor installation, and paneling. These chapters have the same format: glossary of basic terms; description of what is to be constructed; and examples that list materials needed, tools required, manhours to finish, and step-by-step procedures for accomplishing the job. The text is well written and easy to follow with numerous illustrations. There is a pictorial appendix that shows what should be in a basic carpenter's toolkit. This is a useful manual for any carpenter and a good reference book for school and public libraries. The paperback format and "tight" binding may present a problem if the book gets heavy use.   *H. ROBERT MALINOWSKY*

431. The Complete Guide to Remodeling Your Home: A Step-By-Step Manual for Homeowners and Investors. Kent Lester; Una Lamie. White Hall, VA:
Betterway Publications, Inc., 1987. 272p., illus., glossary, index. $18.95pbk. 0-932620-73-6pbk.

The purpose of this comprehensive guide is to assist people in making or saving money through remodeling by providing the essential information necessary to make maximum use of remodeling money. It is equally useful for both homeowners/home buyers and investors who are interested in remodeling as a part-time or full-time business. The book is divided into two sections—planning and implementation. Planning covers all the steps that precede the remodeling work itself: legal and financial considerations, tax laws, locating the right property, evaluating remodeling potential, estimating the cost, and working with subcontractors and material suppliers. Implementation covers the remodeling work itself and provides step-by-step instructions, sample specifications, estimated completion time and cost, and a rating of various projects regarding degree of difficulty. Many specific projects are described in detail. It also includes advice on when to "do it yourself" and when to work with a subcontractor. Many clear, detailed illustrations are provided in this section; and a detailed table of contents, an appendix of 11 reproducible forms (including a renovation estimate checklist, cost estimate summary, plan analysis checklist, subcontractor's agreement, draw schedule sheet, etc.), a glossary

of approximately 250 terms, and a detailed index are included. School, public, college, and trade school libraries will find this a useful reference source.   *JULIE BALDWIN*

432. The Home Inspection Manual: 101 Things to Know before You Buy a House. Alfred H. Daniel. Pownal, VT:
Storey Communications, Inc., 1987. 94p., illus., glossary, index. $19.95, $9.95pbk. "A Garden Way Publishing Book." 0-88266-457-3, 0-88266-481-1pbk.

"This book was born out of the need to educate the prospective home buyer." It is divided into three parts, with part 1 covering those items that are considered standard inside the home. This includes the foundation, basement, water heater, furnace, wiring, windows, floors, stairs, and fireplace among others. Part 2 covers everything that is considered standard on the outside of the home, such as roof, guttering, porch, lighting, and driveway. Finally, part 3 covers items that may or may not be found in or around the home such as septic system, burglar alarm, appliances, fencing, garden, swimming pool, and skylight. Each item is discussed, and an explanation covers those specific areas for which one must watch. There is a very good glossary and index. Probably the most useful part of the book is a 56-page checklist that one can use when inspecting a home for purchase. This book, obviously, is intended for use by individuals, but it would be useful as a reference source in public libraries.   *H. ROBERT MALINOWSKY*

433. Practical Construction Equipment Maintenance Reference Guide. Lindley R. Higgins. Tyler G. Hicks, ed. New York, NY: McGraw-Hill Book Co., 1987. 411p., illus., index. $39.50. McGraw-Hill Engineering Reference Guide Series. 0-07-028772-4.

The contents of this book are selected from the *Handbook of Construction Equipment Maintenance,* which was also edited by Lindley Higgins and published in 1979. Divided into three sections—Section 1, the largest, is on "Basic Maintenance Technology"; section 2 is on "Maintenance of Power Systems"; and section 3 is on "Maintenance of General Contact Elements," such as vehicle tires—this book's primary audience will be technical personnel who are working with or have responsibility for maintenance of equipment in the construction industry, as well as the general reader interested in this topic. Its coverage is sufficiently broad to be of value to maintenance supervisors and sufficiently detailed to be of value to the maintenance worker doing maintenance or repair. The guide is written in clean style and leads the reader through a maintenance procedure from beginning to end using well-designed figures and tables. A detailed index facilitates use of the guide.

RAYMOND BOHLING

**434. Structural Masonry Designers'
Manual.** 2nd. W. G. Curtin; G. Shaw; J. K.
Beck; W. A. Bray. Boston, MA: BSP
Professional Books, 1987. 488p., illus., index.
$98.00. 0-632-01888-7.

Although only five years have passed since the
first edition of this manual, enough new material
has appeared to merit a new edition. "Since the
issue of the first edition there has been a rapid
growth in the use of structural masonry by designers, some progressive higher educational institutions have started instruction in the subject,
and a few manufacturers have shown genuine interest." All of this interest has been brought
about through "the improved reliability and consistency of bricks and blocks by the manufacturers, the increasing availability of design guidance
by their trade associations, the development and
applications by some engineers, together with the
issue of a revised and extended Code of Practice."

This is not a general introduction to masonry
design but rather a very detailed and technical
discussion. A certain amount of mathematical
skill is assumed. After a four-page introduction
there are 15 sections covering various areas of
masonry design: "Advantages and Disadvantages
of Structural Masonry"; "Design Philosophy";
"Limit State Design"; "Basis of Design (1): Vertical Loading"; "Basis of Design (2): Lateral Load-
ing—Tensile and Shear Strength"; "Strapping,
Propping and Tying of Loadbearing Masonry";
"Stability, Accidental Damage and Progressive
Collapse"; "Structural Elements and Forms";
"Design of Masonry Elements (1): Vertically
Loaded"; "Design of Masonry Elements (2):
Combined Bending and Axial Loading"; "Design
of Single-Story Buildings"; "Fin and Diaphragm
Walls in Tall Single-Story Buildings"; "Design of
Multi-Story Structures"; "Reinforced and Post-
Tensioned Masonry;" and "Arches."

There are numerous drawings to illustrate the
discussions with the text keyed to the drawings.
There is constant reference to other parts of the
text when necessary, so that all areas are covered.
Each chapter also contains work examples that
help in understanding the discussion. Designing
masonry projects is critical when it comes to
safety. The failures of masonry buildings in the
past have prompted the industry to become much
more rigid and conscious of standards.

Although this book has a British slant, that
factor does not prohibit in any way its usefulness
in other parts of the world. The appendices include a list of materials giving such data as size,
classification, variety, strength and durability, and
testing; components; movement joints; provision
for services; and tables of design loads.

There are few books that cover in detail the
design qualities of masonry, especially brickwork.
This manual should continue to be an excellent
source of information for the architectural engineer, student, and construction engineer. It is
well written and indexed. Libraries from public to
special will find it a useful addition to their reference shelves.   *H. ROBERT MALINOWSKY*

Textbooks

**435. Building Construction: Materials
and Types of Construction.** 6th. Donald
C. Ellison; W. C. Huntington; Robert E.
Mickadeit. New York, NY: John Wiley and
Sons, 1987. 434p., illus., glossary, index.
$49.95, $32.95pbk. 0-471-85052-7,
0-471-87427-2pbk.

This sixth edition of a well-accepted textbook
on building nomenclature, methods of construction, and building materials extensively updates
the previous edition published in 1981. A general
statement that introduces the subject matter has
been added at the beginning of each chapter as
well as an objectives statement to outline the
purpose of individual chapters. In addition, as an
aid in reinforcing information learned, a summary
and review questions have been added at the end
of each chapter.

A new chapter has been added on the application of composite materials, and new chapter
subheadings on metal building systems, single
membrane roofing, and drywall construction are
also included. A glossary of construction terms, a
list of trade associations, and an extensive index
provide a helpful ending to this popular textbook.

Building Construction is intended for "students of architecture, building construction and
construction technology." Libraries may find it to
be a useful reference source, especially in the
glossary and the list of trade associations.

                                      RAYMOND BOHLING

★
CHEMICAL ENGINEERING
Abstracts

**436. EMA: Engineered Materials
Abstracts.** Metals Park, OH: American
Society for Metals, 1986-. v.1- , index.
Monthly. $860.00 per yr., $720.00 per yr
with *Metals Abstracts.* 0951-9998.

EMA is a monthly abstracting tool published
by Materials Information, which is a joint effort
of the Institute of Metals of London, England,
and the American Society for Metals of the United States. It complements *Metals Abstracts*, another of their abstracting journals, and is designed to meet the informational needs of material scientists, engineers, and others seeking information about ceramics, polymers, or composite
materials. The coverage is international. Literature dealing with metals is not abstracted unless
it is about laminated or composite structures that
contain metal flakes, fibers, etc.

The abstracts are listed in a classed arrangement. Although further subdivided, there are sev-
en main classes: fundamental characteristics, laboratory techniques and quality control, properties,
raw materials, fabrication and finishing, engineering design and applications, and other coverage.
The compositions of each subclass are defined

inside the back cover of each issue. Each subclass is divided by type of materials, i.e., cermics, composites, polymers, and combined coverage. The latter category is for articles covering more than one of the materials classes. Each abstract includes the abstract number and title in boldface type, the abstract, initials of the abstractor, author(s) names(s) and location, source of the publication, and language of the original articles. ISBNs are given if the source of the article is a published proceeding or a book. In some instances, the term "Retroactive Coverage" is used to indicate an older publication not previously indexed and abstracted in *EMA*.

The abstracts in each monthly issue can be located through subject, trade name, material, author, or corporate author indexes. The subject index uses the *Engineered Materials Thesaurus* for vocabulary control. "See" and "see also" references are used in the subject, trade name, and material indexes to lead the user from broader or synonymous terms to the specific term. The information in the monthly indexes is cumulated and published as a separate volume at the end of the year. The first two pages of each issue include "Guidelines for Optimum Use" that explain the arrangement of the abstracts and each of the five indexes. This tool is recommended for any library which serves material scientists.
LUCILLE M. WERT

Encyclopedias

437. Encyclopedia of Industrial Chemical Additives, Volume 4. Michael Ash; Irene Ash, comps. New York, NY: Chemical Publishing Co.., Inc., 1987. 400p., index. $82.50. 0-8206-0320-1.

This work supplements and updates the three-volume set published in 1984. The compilers' goal is "to present comprehensive coverage of national and international additives in industry." Additives are grouped into sections, arranged alphabetically from accelerators to waxes, based on their principal function. Within each section, the additives are listed alphabetically by trade name. A typical entry gives the trade name, manufacturer, chemical name, applications, composition, properties, and description of reference sources, e.g., data sheet or chart. The sources of this information come from the manufacturer, and there is no consistency in the information provided. It is not clear which categories of information would comprise a complete profile. Following the table of contents is a list of contributors and their addresses. These seem to be the manufacturers who have supplied the information for their products. There is a helpful explanatory list of the abbreviations used in the entries and an index by tradename. This easy-to-use book and its previous volumes would be useful additions to chemical technology collections in university or technical libraries.
SUE ANN M. JOHNSON

Handbooks

438. ASTM Fire Test Standards. 2nd. ASTM Committee E-5 on Fire Standards. Philadelphia, PA: American Society for Testing and Materials, 1988. 872p., illus., index. $49.00. "03-505088-31." 0-8031-1173-8.

This manual is divided into 19 reprinted sections from the *Annual Book of ASTM Standards* covering such general topics as fire standards, electrical insulation, textiles, plastics, and chemicals. Within each section are one or more tests relating to that general area. The tests may cover flammability, flash point, fire tests, smoke, temperature, or any of a number of other potential hazards. Each test method is divided into sections.

First is the scope of the test, which includes the purpose; then any references or applicable documents are listed. The third section is on the significance or use of the test, followed by a decription of the apparatus to be used, generally with a detailed diagram. Occasionally there will be a photograph of the apparatus, but these are few and add little to the book. The fifth section describes the criteria of the test specimen. Following that are sections unique to the individual test, and those included might be calibration, test sample, protection, or clean-up. The next to last section is a guideline to reporting, and the last section is a statement or description of the precision of the test. This compilation of tests is complete, well documented, and easy to follow for the individual familiar with fire test procedures. The diagrams are useful in constructing the necessary test apparatus, and each procedure takes the tester through a step-by-step process of what needs to be done.

For those who are in need of testing materials for burning characteristics, this text is a must. Libraries serving research and development facilities will need this on their reference shelves. However, libraries that own the *Book of ASTM Standards* may not want to purchase this volume unless they prefer the material self-contained in one volume.
JOHN NISHA

439. ASTM Standards and Literature References for Composite Materials. ASTM Committee D-30 on High Modulus Fibers and Their Composites. Philadelphia, PA: American Society for Testing and Materials, 1987. 616p., illus., bibliog., index. $44.00. 0-8031-0986-5.

This volume is a collection of standards related to composite materials as issued by the American Society for Testing and Materials (ASTM). ASTM's purpose is to develop "voluntary full-consensus standards" through its committees of users, producers, and academic and government representatives. This particular compilation, reprinted from the *Annual Book of ASTM Standards*, was developed by ASTM's

Committee D-30 on High Modulus Fibers and their composites. [High modulus fibers are those with a Young's modulus that is greater than 20 GPa (3x10 to the 6th psi).]

ASTM Standards and Literature References for Composite Materials is divided into three sections. The first contains the D-30 standards, consisting primarily of test methods for content, strength, compressive properties, fatigue, tensile properties, and density of various high modulus fibers and composites reinforced by these fibers. The next section reprints all other related ASTM standards which are referenced in the D-30 standards in the first section. The final portion of the book is a list of ASTM special technical publications on composite materials, including their tables of contents, and the tables of contents of all issues, through Summer 1987, of ASTM's *Journal of Composites Technology and Research* (formerly called *Composites Technology Review*). The book also contains an index to the standards reprinted in this volume and application for membership in Committee D-30.

The value of this compilation is that it pulls together in a single volume all the relevant ASTM standards for users and producers of high modulus fibers and their composites. For engineers involved with these composite materials, this single volume will be handier (and far less expensive) than the multivolume *Annual Book of ASTM Standards* and has the added value of the references to related ASTM publications.

CAMILLE WANAT

440. ASTM Standards on Color and Appearance Measurement. 2nd. ASTM Committee E-12 on Appearance of Materials. Roberta A. Storer, ed.

Philadelphia, PA: American Society for Testing and Materials, 1987. 341p., illus., bibliog., glossary, index. $39.00pbk. "PCN: 03-512087-14." 0-8031-0976-8pbk.

This compilation of standards "is intended to guide the individual who needs to use, select, interpret or create a numerical method for the measurement of an appearance attribute or attributes of a material or product." This includes colorimetry, color, color difference, gloss, image clarity, haze, turbidity, opacity, reflectance, retroreflectance, transmittance, transparency, yellowness, whiteness, spectrophotometry, and goniophotometry.

Forty-four ASTM standards have been reprinted from the *Annual Book of ASTM Standards*. In addition, the book includes the titles of 76 additional standards that are applicable to only one material. Each standard includes scope, definitions, sampling, application, apparatus, procedure, report, and summary with additional information when necessary to interpret the standard or to apply its test to a product. Some standards have bibliographies. Standard E 284, "Standard Definitions of Terms Relating to Appearance of Materials," is a 10-page glossary of accepted terms used in this handbook when discussing the appearance of materials. This spe-

cialized handbook of standards brings together material that is scattered throughout the *Annual*, thus making it more accessible to the user.

H. ROBERT MALINOWSKY

441. The Chemical Formulary: Collection of Commercial Formulas for Making Thousands of Products in Many Fields, Volume 27. H. Bennett, ed.

New York, NY: Chemical Publishing Co.., Inc., 1986 (c1987). 358p., index. $45.00. 0-8206-0318-X.

The twenty-seventh volume of *The Chemical Formulary* continues the tradition established in the previous volumes, listing the most recent chemical formulas for a variety of commercial products. Formulas are listed by chapter for adhesives, food and beverages, cosmetics, coatings, detergents, drugs, metal treatments, elastomers, polymers, and resins. An additional chapter listing miscellaneous formulas (emulsions, pesticides, some paints and inks, etc.) is also included. All formulas include instructions for preparation. The introductory chapter discusses the methods and techniques necessary for obtaining satisfactory results, and lists many simple formulas, both for practice purposes and because of their importance as basic formulas. Subsequent chapters list more complicated formulas, with suggested modifications for specific purposes. An appendix contains other useful information such as a review of federal laws and regulations governing food and chemical products, a list of incompatible chemicals, elementary first aid information, suppliers of trademark chemicals, and tables of miscellaneous information. *The Chemical Formulary* would be useful to both the layperson and the practicing chemist; as such it would be a significant addition to public, academic, or special library collections.

LOREN D. MENDELSOHN

442. Engineers' Guide to Composite Materials. John W. Weeton; Dean M. Peters; Karyn L. Thomas. Metals Park, OH: American Society for Metals, 1987. 364p., illus., bibliog., glossary, index. $91.00. 0-87170-226-6.

This is an excellent and timely new reference book in this important and expanding field providing a comprehensive source of information on composites. Section 1, "Introduction to Composite Materials," provides good background information for the newcomer to the field of composites and, also, is of value to the more experienced engineer looking for materials alternatives. Section 2 is on "Economic Outlook for Composites and Reinforcements," and sections 3 and 4 are intended to aid engineers with design work and include a section on equations for metals, ceramics, and polymer composites, followed by a section that reports 47 case histories of composites use. The applications section includes histories of nine fields: aerospace, automation, marine, medical, recreational, electrical, electronics, structural, and nonautomotive machinery and parts.

The largest section of the book, 225 pages in length, gives comprehensive coverage of properties, fabrication methods, and application, and testing methods for using composite materials in structural reinforcement. The guide is generously illustrated with tables, diagrams, and graphs particularly in the properties data section. To supplement the design and properties sections, a 50-page section includes a bibliography of books on composites, a directory of consultants, a directory of laboratories and information centers, a list of societies, trade associations and institutes, a directory of manufacturers, and a 20-page glossary of terms. A detailed 30-page subject index further enhances the value of this reference work. This guide will be an excellent sourcebook of information in any personal, academic or research library for which books in materials engineering are acquired. *RAYMOND BOHLING*

443. Handbook of Fillers for Plastics.

Harry S. Katz; John V. Milewski, eds. New York, NY: Van Nostrand Reinhold Co., 1987. 467p., illus., index. $69.95. 0-442-26024-5.

Fillers are used in plastics for a variety of reasons, including cost and property enhancement. The *Handbook of Fillers for Plastics* provides information that allows comparison of fillers and evaluations of which filler will work best for a given application. It covers in detail an area in which there has been very little written and should be useful to those who have applications for fillers in their work.

The handbook essentially is an expanded second edition of part of the 1979 *Handbook of Fillers and Reinforcements for Plastics*. Intended by the authors to provide information on function, cost, and processing of fillers, it includes chapters on mineral, metallic, conductive, magnetic, and flame-retardant fillers, as well as in-depth information on solid and hollow spherical fillers.

The handbook also contains a lengthy chapter on coupling agents intended for use with fillers. In each chapter, listings of major suppliers are included, and data are provided on their products to allow comparisons of the fillers. The handbook also contains property data on each type of filler; however, there is inconsistent inclusion of references to sources of the data. Also, the data often are not as precise as in other sources checked; those researchers needing very precise data may need to check other sources. This reference volume should be of primary interest to engineers or scientists utilizing plastics in their products and to the special libraries serving these patrons. *JODY KEMPF*

444. Handbook of Reinforcements for Plastics.

John V. Milewski; Harry S. Katz, eds. New York, NY: Van Nostrand Reinhold Co., 1987. 431p., illus., index. $69.95. 0-442-26475-5.

The *Handbook of Reinforcements for Plastics* is the companion volume to the *Handbook of Fillers for Plastics* and covers information on a variety of reinforcements in a manner similar to the coverage of fillers in its companion volume (see review of *Handbook of Fillers for Plastics*). This handbook greatly expands the coverage of reinforcements from the 1979 *Handbook of Fillers and Reinforcements for Plastics* by the same authors.

The compilation of such extensive information on reinforcements for plastics is special to this handbook. The handbook includes chapters on flake and fiber reinforcements, fiberglass, high modulus filaments, and a long chapter on short fiber reinforcements. It also contains a short chapter on procedures and equipment for utilizing reinforcements and includes a list of suppliers of the equipment.

Because of the extent of the information on reinforcements, this volume should be very useful to its intended audience of design engineers, materials scientists, polymer chemists, compounders, and molders, as well as to those libraries serving them. *JODY KEMPF*

445. Pigment Handbook: Volume 1— Properties and Economics.

2nd. Peter A. Lewis, ed. New York, NY: John Wiley and Sons, 1988. 945p., illus., bibliog. $412.00set. "A Wiley-Interscience Publication." 0-471-82833-5v. 1.

The *Pigment Handbook* consists of three comprehensive volumes: Volume I, "Properties and Economics;" Volume II, "Applications and Markets;" and Volume III, "Characterization and Physical Relationships." This review is concerned with volume I, which is a revision of the 1973 edition. Some of the changes which have been incorporated include expanding the statistics with recent data and information on newly developed organic and inorganic pigments. Lewis also states that "there are some pigments and classes of pigments that no longer occupy a large enough place in the market or whose usage has been totally discontinued and so are not featured in the second edition." In addition, "the majority of chapters include a new and unique section concerning the effect of pigment on health and environment." The definition of a pigment and a description of fastness properties are outlined in the introductory notes. The primary sections are divided into the following subject areas: white pigments; extender pigments; colored inorganic pigments; colored organic pigments; black pigments; metallic pigments; nacreous pigments and interference pigments; luminescent organic pigments; flourescent and phosphorescent inorganic pigments; and food, drug, and cosmetic colors.. Under each of these primary sections, there are several specific chapters authored by an impressive list of international experts. Each chapter contains relevant diagrams, photographs, charts and tables. Bibliographies and references accompany many of the chapters.

This volume does not have an index. There is a significant amount of useful information which is only accessible either to those familiar enough with the subject matter to know which chapter it might be in or by thumbing through the volume. For example, many of the chapters contain company or manufacturer information. An index to this information would have enhanced the reference value.

This technical volume is a rich resource for readers who may have interest in any of the following areas regarding pigments: history, physical and chemical aspects, international economics, producers and manufacturing processes, and applications. Academic and special libraries will want the full set for their reference collections.

CHARLES R. LORD

446. Quality Assurance of Chemical Measurements. John Keenan Taylor. Chelsea, MI: Lewis Publishers, Inc., 1987. 328p., illus., bibliog., glossary, index. $59.95. 0-87371-097-5.

"Chemical measurement data are often the basis for critical decisions on vital matters, ranging from the health of individuals, to the protection of the environment, to the production of safe, reliable, and useful products and services. Obviously, the data used for such purposes must be reliable, and there must be unequivocal evidence to prove it. The philosophy and procedures by which this is achieved and demonstrated are called quality assurance." To have a good quality assurance program, one needs to understand the basic principles of chemical measurement. This is the purpose of this handbook. The 25 chapters cover a logical discussion from a definition of quality assurance, through the philosophy of good measurements, and finally to the program itself: "Precision, Bias, and Accuracy," "Statistical Control," "Statistical Techniques," "Chemical Analysis as a System," "The Model," "Planning," "Principles of Sampling," "Principles of Measurement," "Principles of Calibration," "Principles of Quality Assurance," "Principles of Quality Control," "Blank Correction," "Control Charts," "Principles of Quality Assessment," "Evaluation Samples," "Reference Materials," "Traceability," "Quality Audits," "Quality Circles," "Validation," "Reporting Analytical Data," "An Improved Approach to Performance Testing," "Correction of Errors and/or Improving Accuracy," "Laboratory Evaluation," and "The Quality Assurance Program."

The emphasis is on chemical measurement, but the concepts can be used by any field that relies on measurement as a basis for quality control. Each chapter discusses the topic in general terms followed by the specifics. For example, in the chapter on sampling, there is an introduction followed by initial considerations with a glossary, sampling plan, tracing, management, safety, subsampling, quality assurance of sampling, sample uncertainties, statistics, acceptance testing, and other sampling considerations.

Because the concepts are the same for all fields, this book would be useful as a reference work in any college, academic, or special library. It can be used as a textbook in quality assurance; it is based on a short course given by the author.

H. ROBERT MALINOWSKY

★
CIVIL ENGINEERING
Atlases—Science

447. Civil Engineering Practice: 4/Surveying, Construction, Transportation, Energy, Economics and Government, Computers. Paul N. Cheremisinoff; Nicholas P. Cheremisinoff; Su Ling Cheng, eds. Lancaster, PA: Technomic Publishing Co.., Inc., 1988. 685p., illus., bibliog., index. $165.00. 0-87762-537-9.

Civil Engineering Practice is a five-volume encyclopedia covering the entire scope of today's civil engineering theory and practice. The volumes cover "Structures" (vol. 1), "Hydraulics/ Mechanics" (vol. 2), "Geotechnical/Ocean Engineering" (vol. 3), "Surveying/Construction/ Transportation/Energy/Economics and Government/Computers" (vol. 4), and "Water Resources" (vol. 5). It is volume 4 to which this reviewer had access; it is assumed that the other volumes are similar in quality and presentation. Coverage of major topics is at a high level of detail. This volume contains a large number of data tables, graphs, diagrams, and other schematics. Complete presentations of necessary equations and example problems illustrating how to solve real-life problems also appear. Each topic concludes with a list of pertinent literature references. The intent of the editors and contributing authors is to provide information of enduring value, supplying the practitioner with authoritative, up-to-date information in all areas of civil engineering. Because civil engineering has, in recent years, burgeoned into a variety of new subdisciplines, the volume covers many areas in which new developments have only recently come about. The range of topics appears suitably broad yet manageable enough to allow for depth of coverage. The volume is attractively and sturdily bound for heavy use and is printed on acid-free paper. The editors are all recognized authorities in their fields. This encyclopedia is unmatched by any other single reference source and should be of considerable use to all practicing engineers and engineering libraries. It is definitely recommended.

FRED O'BRYANT

Encyclopedias

448. Civil Engineering Practice: 1/ Structures. Paul N. Cheremisinoff; Nicholas P. Cheremisinoff; Su Ling Cheng, eds. Lancaster, PA: Technomic Publishing

Co.., Inc., 1987. 814p., illus., bibliog., index. $165.00. 0-87762-529-8.

Structures is the first volume in a five-volume set entitled *Civil Engineering Practice*. It contains 31 articles, some containing references, divided into six sections: "Reinforced Concrete Structures," "Structural Analysis," "Stability," "Pavement Design," "Wood Structures," and "Composites." The set is edited by Paul N. Cheremisinoff and Nicholas P. Cheremisinoff, both of whom have edited numerous other engineering handbooks, and Su Ling Cheng. Although this set is meant as a collection of the present knowledge on civil engineering, this volume concentrates, for the most part, on concrete and steel structures. Only two chapters can be found in the section on "Wood Structures." A more thorough treatment of wood structures can be found in the *Timber Designers' Manual* (Granada Technical Books, 1984). Two other recently published handbooks, *Building Structural Design Handbook* (Wiley, 1986) and *Handbook of Concrete Engineering* (Van Nostrand Reinhold, 1985), contain articles on topics similar to those covered in this book. In fact, in some cases, the bibliographies for articles in these handbooks are more current than those for corresponding articles in *Structures*. This book would be useful in libraries without a current selection of handbooks; however, its treatment of some topics on structural engineering (notably timber construction) is rather cursory. It is, therefore, recommended for libraries without current handbooks. *LINDA R. ZELLMER*

Guides to the Literature

449. Selective Guide to Literature on Engineering Geology. Cecilia P. Mullen, comp. Washington, DC: Engineering Libraries Division, American Society for Engineering Education, 1987. 24p., bibliog. $5.00pbk. Engineering Literature Guides, no. 7. 0-87823-107-2pbk.

This American Society for Engineering Education (ASEE) guide represents the latest in an on-going series of planned guides to the literature of various engineering topics. It is intended for use by undergraduate or graduate students and emphasizes the major disciplines of soil mechanics, rock mechanics, and foundations. Other subfields are mentioned only superficially or are omitted. Some of these are covered in other ASEE guides. By no means comprehensive or intended as a replacement for any of the standards in the field, this guide is largely limited to English-language sources published since 1980 (with a few exceptions), and each entry is accompanied by a brief descriptive annotation. By virtue of its inexpensive price, it is best suited for personal purchase by students wanting an overview of the field.

FRED O'BRYANT

Handbooks

450. Construction Surveying and Layout. Paul Stull. Carlsbad, CA: Craftsman Book Co., 1987. 253p., illus., index. $19.25. 0-934041-25-3.

Surveying and layout work on construction sites does not have to be difficult to learn. The purpose of this book is to assist readers in their understanding of the basic aspects of construction surveying. The author attempts to distinguish this manual from those which are too technical and detailed to be of practical use. As a result, it should be used as a companion to other more technical manuals and handbooks.

The book begins with the historical development of land surveys, emphasizing the terminology which is still used in the field. The introductory orientation also includes building site considerations and an explanation of a survey. Builders, contractors, and developers can learn how to apply the principles of surveying to their own projects including the use of a transit and the basics of leveling and plotting. The book is well illustrated with charts and figures, and the appendices including two detailed discussions on geometric and trigonometric applications in construction are added strengths of the book. Stull has done a good job in presenting the practical aspects of construction surveying without going into a great amount of technical detail. Its greatest use will be for the practicing engineer. Public and college libraries will find it a useful reference work on surveying. *CHARLES R. LORD*

451. Excavation and Grading Handbook. Rev. Nick Capachi. Carlsbad, CA: Craftsman Book Co., 1987. 380p., illus., index. $22.75. 0-934041-29-6.

Written by an experienced excavation contractor, this practical manual is organized into several small chapters, making it easier for the user to find particular procedures. It is a revision of the 1978 edition incorporating many new changes in excavation and grading, including new information on the use of contour line drawings, laser levels, and techniques in trench compaction. One of the areas where these changes has been significant is that of heavy equipment. Throughout the author expresses the belief that, for a business to remain competitive, operators must know and incorporate new technology. This handbook assists in that process.

Each chapter is well illustrated with photographs and diagrams. There is a list of abbreviations and a useful glossary with references to illustrations in the previous chapters. Filled with common-sense ideas and practical suggestions, this manual begins with the basics of understanding road survey stakes and concludes with instructions on paving with different types of materials and laying pipe for special jobs.

Although this book has been "adopted as the primary reference by many schools and in many apprentice training programs," the primary audience is really anyone who is involved with the management of excavation and compaction re-

sponsibilities at construction projects. It is recommended for private use and also a useful reference work for public, academic, and special libraries. *CHARLES R. LORD*

452. **Pipe and Excavation Contracting.**
Dave Roberts. Carlsbad, CA: Craftsman Book Co., 1987. 400p., illus., index. $23.50. 0-934041-22-9.

This manual provides a general orientation for the person who wants to develop skills and business fundamentals necessary to run a pipe and excavating contracting operation. Because the book is primarily directed to underground utility contractors, it covers "estimating and bidding, equipment and crew performance, practical work methods and techniques, equipment operation, surveying, site clearance, compaction, water systems, and sewer systems."

These topics are discussed in major areas: building a business by understanding how to obtain the jobs, knowing the specific details to accomplish the work, and selecting and using the appropriate heavy equipment. The author assumes a moderate level of knowledge and experience in the readers. For example, the emphasis in the chapter on "Operating a Backhoe" is on developing present skills and not aimed at someone who has never operated a backhoe. Each chapter contains useful diagrams, checklists, charts, and examples.

The format, however, is not clear enough and the binding not substantial enough to withstand the wear and tear of on-site use. However, the author has succeeded in his commitment to have a book that helps people understand what they need to know to have a successful and profitable venture in pipe and excavating contracting. Large public, academic, and special libraries will find this a useful source of information on pipe excavating. *CHARLES R. LORD*

453. **The Surveying Handbook.** Russell C. Brinker; Roy Minnick, eds. New York, NY: Van Nostrand Reinhold Co., 1987. 1,270p., illus., bibliog., index. $72.95. 0-442-21423-5.

Written by over 30 specialists, this excellent handbook was compiled for both the professional and others who need concise information on a variety of surveying topics. For example, whereas the first chapter includes the practical specifics of notekeeping, later chapters written by attorneys discuss the role of surveyors in land litigation and outline proper courtroom procedures. Other articles include information on such topics as electronic distance-measuring instruments (EDMJs), geodesy, inertial and satellite positioning surveys, photogrammetry, optical tooling, survey business management, and land information systems. The editors clearly state that, though the purpose of the handbook is not to be exhaustive, they have selected topics which are very appropriate to "surveyors, civil, agricultural and other engineers, foresters, architects, archaeologists, geologists, small home builders, realtors, title companies, and lawyers."

Each article is amply illustrated with drawings, tables, and photographs as appropriate. In addition, chapters are subdivided by both numeric and topical descriptions which greatly assist the reader in using information. All chapters include references, and some have attached helpful explanatory notes. Directory information includes lists of the state offices of the Bureau of Land Management, agencies in the public-land states where original surveys are housed, and the state registration boards. The volume also contains abbreviation, number, and symbol charts. Thorough indexing has been done with notations to assist the user in locating illustrations and material in tables.

Although the how-to aspect of surveying is the primary focus of each contribution, there is a wonderful continuity throughout the volume highlighting some of the historical developments of surveying. The reader can move from Roman times in the sixth century B.C. to 1831, when William Young of Philadelphia constructed what is believed to be the first American-made transit, to the 1960s and 1970s, when great strides were made in the evolution of satellite positioning systems. As stated in one of the chapters, "Surveying is an art and science. Art can stand retouching, science cannot." This handbook helps to identify the art that can be retouched and the science that needs to be understood. It is recommended for public, academic, and special libraries as well as professionals associated with the surveying industry. *CHARLES R. LORD*

Histories

454. **Landmarks in American Civil Engineering.** Daniel L. Schodek. Cambridge, MA: Massachusetts Institute of Technology, 1987. 383p. illus., bibliog. $50.00. 0-262-19256-X.

Landmarks in American Civil Engineering is a description of engineering projects in the United States which have been designated as National Historic Civil Engineering Landmarks by the American Society of Civil Engineers. Written by Daniel Schodek, a professor of architectural technology at Harvard University, it presents a short history of American civil engineering and the roads, canals, railroads, bridges, tunnels, waterworks, dams, buildings, power complexes, airports, and coastal facilities which have been designated as engineering landmarks. Notable environmental engineering, urban planning, and surveying projects are also discussed. Maps showing the location of and bibliographies of references on these landmarks are also included. Each description of a project includes a brief history and explanation of the project's design. Photographs, both historical and recent, are included to give the reader an idea of the project's appearance. Maps of individual sites are also provided to convey additional information.

Several other books on the history of civil engineering have been published. Most are dry, technical discussions of engineering and the development of the discipline in America. The brief descriptions in *Landmarks in American Civil Engineering* are easily understood by nonengineers, and additional information on each can be found using the bibliography. This book is recommended for all library collections which serve patrons who may be interested in the history of engineering and construction projects in the United States. *LINDA R. ZELLMER*

Manuals

455. Surveying Ready Reference Manual. Guy O. Stenstrom, Jr. New York, NY: McGraw-Hill Book Co., 1987. 253p., illus., index. $28.50. 0-07-061164-5.

Surveying Ready Reference Manual is a collection of surveying methods based on the practical experience of the author, a registered professional engineer and land surveyor. Meant as a desk and field reference, it summarizes surveying methods learned by the author through practical experience, rather than by methods described in textbooks. In 16 chapters and an appendix, it covers basic information, measurements, calculations, and surveying techniques associated with surveying for common construction.

Compared to the recently published ASCE *Engineering Surveying Manual* (1985), *Surveying Ready Reference Manual* is meant for use by engineers dealing with practical aspects of surveying, rather than with the theoretical aspects required of engineers who oversee projects which require surveys. Because it is meant as a field and desk manual, McGraw-Hill and the author have endeavored to keep its size small enough for field use, 14.5 by 11 cm. and its print approximately the size of the 15-point micro font available on some computer printers. If larger print had been used, the book would be about the same size as most field guides. This would be more visible in a bag or backpack in field situations or on a disorganized desk and be more convenient for library shelves. The largest type size would also be easier on the eyes of potential users. It is hoped that future printings of this useful book will take this into account. This book is recommended for college and university libraries with engineering programs and for engineers who have to supervise field surveys. *LINDA R. ZELLMER*

★
ELECTRICAL/ELECTRONICS ENGINEERING
Dictionaries

456. Dictionary of Electronics and Electrical Engineering: English-Japanese-German-Russian. 3rd. Seiichi Ishibashi. New York, NY: Plenum Press, 1987. 1773p. $195.00. 0-306-42749-4.

This is a very ambitious work that has been expanded from 25,000 entries in the 1964 edition to 42,000 in this third edition. As its title indicates, the dictionary focuses on electronics and electrical engineering, but, in addition, it also includes terms from chemistry, mathematics, and physics and from several other engineering fields and computer science. Verbs, adverbs, and adjectives that are commonly used as descriptive terms in electrical engineering and electronics are also included.

The dictionary is divided into four parts. The first part is an alphabetical listing of the 42,000 entries in English followed by the equivalent terms in Japanese, German, or Russian. The second, third, and fourth parts list the Japanese, German, and Russian terms, with a page and line number reference back to the English section. The Japanese words are arranged according to the Japanese "syllabary," and the German and Russian terms are arranged in alphabetical order. The last section of the dictionary contains a list of common abbreviations and units of technical measurements in equivalent terms for the four languages. This is an impressive work and will be a welcome reference source in any academic or research library. *RAYMOND BOHLING*

Directories

457. The RAE Table of Earth Satellites, 1957–1986. 3rd. D. G. King-Hele; D. M. C. Walker; J. A. Pilkington; A. N. Winterbottom; H. Hiller; G. E. Perry, comps. New York, NY: Stockton Press, 1987. 936p., index. $130.00. 0-935859-05-5.

This is an unusual reference book compiled at the Royal Aircraft Establishment in Farnborough, Hants, England. It is a listing of nearly 3,000 satellite launches and the resulting artificial earth satellites. The quantity of detail is astonishing. The tables list a great amount of hard factual data concerning each satellite including such details as apogee, perigee, orbital inclination, size, weight, launch date and time, etc. There is a helpful introduction and an extremely useful name index which in itself is a wonderful tool for identifying and verifying information about individual satellites.

Jane's Spaceflight Directory says that "The Royal Aircraft Establishment's *Table of Earth Satellites* is an essential and unique source of information." High praise is indeed due this remarkable volume, which is an expanded third edition now covering the years 1957 to 1986. It is highly recommended for the special library collection and the reference section of any large research library. *GEORGE M. A. CUMMING, JR.*

458. Satellites International Joseph N. Pelton; John Howkins. Janet Greco, ed. New York, NY: Stockton Press, 1987. 265p., illus., bibliog. $190.00. 0-935859-07-1.

From the title and price, one would believe that this was a comprehensive directory of satellites; unfortunately it is disappointing. The first 120 pages is an international country-by-country directory of government agencies, satellite manufacturers, communications agencies, etc. The list is impressive and includes many obscure agencies, such as the Ministry of Telecommunications in Chad. The remainder of the book is devoted to various articles covering a broad spectrum of materials, ranging from a user-friendly introduction to satellites to an article on conferencing networks. These are useful and well written, but somewhat better coverage can be found in *Jane's Spaceflight Directory*. Except for the country directory, Jane's publication contains much more. Libraries with limited budgets will probably want to purchase *Jane's Spaceflight Directory* before *Satellites International*. Large institutions which are collecting comprehensively will want to purchase this directory to have as broad a coverage as possible. *GEORGE M. A. CUMMING, JR.*

459. Telecommunications Systems and Services Directory: An International Descriptive Guide to More Than 2,000 Telecommunications Organizations, Systems, and Services. 3rd. John Krol; Janice A. DeMaggio, eds. Detroit, MI: Gale Research Co., 1988. 1,116p., index. $270.00. 0-8103-2345-1. 0738-3045.

The telecommunications industry is enormous and the variety of activity kaleidoscopic. A directory that spans a field so broad and changeable is an ambitious undertaking. *Telecommunications Systems and Services Directory* tries very hard, but it is not the one-stop tool that answers every question. It is, however, the first place to start.

The main body is an alphabetical listing of organizations, companies, and agencies covering the international as well as the domestic scene. Each entry is a model of what a good descriptive directory ought to provide. There are excellent indexes by keyword, function, geographic location, and personal name. The multiple-access points make this a good source in which to locate information.

All in all this is a fine reference source, with two caveats. It is not exhaustive, and that is no surprise. With an industry so large and chaotic it is impossible to cover the field completely. Second, it is not timely. There have been three editions published irregularly over the last five years. Although interedition supplements are available, the service is not free. Other sources are published annually; there are even monthly loose-leaf services available. This is an industry where rapid change is the rule. Updates are a necessity.

This directory will appeal to those whose budgets will not support purchase of several different sources but do need some coverage of the telecommunications industry. Large academic libraries and special libraries in the telecommunications industry will want a copy of this directory and probably will want to subscribe to the interedition updates. *GEORGE M. A. CUMMING, JR.*

460. World Satellite Survey: A Guide to Commercial Satellite Communications. Roger Stanyard. London, Great Britain: Lloyd's of London, Aviation Department, 1987. 375p., illus., glossary, index. $325.00. 1-85044-116-2.

The *World Satellite Survey: A Guide to Commercial Satellite Communications* represents an enormous cumulative effort to bring together in a single source significant information about the satellite communications industry. (Military telecommunications are specifically excluded.) Its 375 pages synthesize a wealth of commercial and technical data on 92 major satellite systems, both current and planned, as well as information about the major launch vehicles available to place these systems in orbit.

Roger Stanyard, an expert on the commercial aspects of communications technology, is also a writer, consultant, and lecturer in this field. Although a few other recent books deal with the satellite arena generally (e.g., D. M. Jansky's *World Atlas of Satellites*, Artech House, 1983), no other single source treats every commercial system with the currency, depth, and breadth of *World Satellite Survey*. The information is presented from a business rather than engineering standpoint. Thus, the work is intended for an audience more inclined toward the commercial aspects of telecommunications, including hardware manufacturers, communications carriers, regulatory agencies, consultants, economists, financiers, and insurers. However, there is also plenty of information included that would be of interest to other researchers and scholars who may be doing work in telecommunications and aerospace sciences.

The book opens with a lengthy introduction, review, and market survey. Describing current trends and developments in light of their historical background, the market is first surveyed as a whole, then by geographic region. Next, the author looks at several specific technologies, including fiber optic cable competition, fixed satellite services, direct broadcast satellite services, mobile and radio determination satellite services, very small aperture systems, and satellite news gathering. The introduction concludes with a brief overview of launch vehicle technology.

The major portion of the book consists of a detailed description of 92 existing and proposed telecommunications satellite systems worldwide. For each system operator, pertinent name and address data, launch dates, project costs, and project status are given. A project overview outlines who is using each system and for what purposes, both currently and in the future. Major technical details (e.g., orbital data, broadcast frequencies, satellite dimensions, etc.) are provided in a summary fashion easily understood by lay readers. A glossary is provided to define various technical terms and acronyms. Each summary includes illustrations of the satellite itself, one of its components, and/or the "footprint" of its signals.

The third major section of the book provides support data on the launch vehicles available to boost satellites of a commercial nature into orbit. Past launch records are included, along with scheduled future launches and general technical data on payload size, weight, and shape, and upper stage configurations. Of particular interest are the data on the Soviet Proton launch vehicle and the Chinese Long March rocket recently offered to Western businesses for commercial use.

There is no question that *World Satellite Survey* represents a significant and useful compendium of data on satellite telecommunications of a nonmilitary nature. The communications satellite industry is a mainstay of modern life and likely to grow more important in the next several decades. Locating specific information, such as *World Satellite Survey* provides, will be important to businesses and potential users of satellites everywhere. Finding this data in one convenient location should be a vast improvement over the tedious and time-consuming research otherwise required. For those who can afford the hefty price or those who need access to this data, *World Satellite Survey* will provide a welcome new reference tool. *FRED O'BRYANT*

Encyclopedias

461. 500 Electronic IC Circuits with Practical Applications. James A. Whitson. Blue Ridge Summit, PA: TAB Books, Inc., 1987. 340p., illus., index. $28.95, $18.95pbk. 08306-7920-0, 0-8306-2920-3pbk.

This book is intended to "fill the gap between the electronics circuit books and the electronics project books" by providing "information about practical electronic circuit devices and their applications." James A. Whitson has successfully carried out his intentions and has included over 500 practical electronics circuits. Most of these are accompanied by well-written and easy-to-understand descriptive text, plus the technical data needed to fully understand each of the integrated circuits that are presented. Each of the 10 chapters covers a specific application or applications of electronic circuitry: (1) "Basic Electronic Circuits"; (2) "Amplifier Circuits"; (3) "Oscillator, Timer, Counter, Clock, and Multiplier/Divider Circuits"; (4) "Interfacting Circuits"; (5) "Digital and Microprocessor-Based Circuits"; (6) "Optoelectronic Circuits"; (7) "Audio and Radio Circuits"; (8) "Alarms and Safety/Security Circuits"; (9) "Special-Purpose Circuits"; and (10) "Miscellaneous Circuits."

The schematics used in each of the chapters are clear and well labelled, and the descriptive text permits the user to build a working device that has everyday, practical applications. An excellent book for the hobbyist, electronics teacher, and technician, for school and public libraries it is a useful source of information on practical integrated circuits. *H. ROBERT MALINOWSKY*

462. McGraw-Hill Encyclopedia of Electronics and Computers. 2nd. Sybil P. Parker, ed. New York, NY: McGraw-Hill Book Co., 1988. 1,047p., illus., bibliog., index. $75.00. 0-07-045499-X.

Persons who are familiar with the *McGraw-Hill Encyclopedia of Science and Technology* will immediately recognize this new work as another spin-off from that venerable reference tool. Indeed, the information contained in the *McGraw-Hill Encyclopedia of Electronics and Computers* first appeared in the sixth edition of the parent encyclopedia and its supplementary yearbooks. This second edition of the *McGraw-Hill Encyclopedia of Electronics and Computers* contains some 520 articles relating to electricity, semiconductors, integrated circuitry, computer hardware and software, robotics, data management, communications, and consumer products employing microprocessors. Also included is information on artificial intelligence, radar, computer-aided engineering, and a variety of electronic circuits, chips, and components, as well as the interrelationships among all these topics. Forty-five new articles are included for the first time in this edition, and 120 articles have been revised from the first edition. The work is fully cross-referenced and indexed. Each article is signed, and most conclude with brief bibliographies listing further references.

Users who need brief, quick, relatively rudimentary information on any of the topics included in this source will find the encyclopedia a reasonably good starting point from which to launch further investigations. However, despite the usefulness of this encyclopedia, users should be aware that as with most such compilations the very latest developments may not be included. This reviewer checked the entry for the BASIC computer programming language, for example, and discovered that no mention is made of Kemeny and Kurtz's 1984 overhaul of the original version of the language into what they call True BASIC. Neither do the encyclopedia's few paragraphs on BASIC reflect that it has developed a host of new features and improvements relating to the microcomputer boom of the last 10 years. Although this is admittedly an isolated example, it seems an important flaw, which might cause one to wonder what other omissions lurk unseen in the text.

Those who have access to or own the present edition of the *McGraw-Hill Encyclopedia of Science and Technology* need not purchase this spin-off unless user demand or convenience dictates. Individuals or libraries wishing to own a reasonably compact and current one-volume encyclopedia of electronics and computers should consider this one—albeit with an eye to its currentness and coverage. *FRED O'BRYANT*

Exam Reviews

463. The CET Exam Book. 2nd. Ron Crow; Dick Glass. Blue Ridge Summit, PA: TAB Books, Inc., 1987. 272p., illus., bibliog.

$21.95, $13.95pbk. 0-8306-1950-X, 0-8306-2950-5pbk.

There are four levels of certified electronics technicians (CET): associate, journeyman, senior, and master. Each level requires a certain expertise to pass the exams to receive that certificate. This exam book covers the material one needs to know to pass the exam for the associate electronics technician certificate. It covers the basics needed to understand and utilize the tools and equipment all technicians must use to complete their work, but does not test knowledge about specific electronic products such as a TV or computer. Each of the chapters begins with a series of questions followed by a review that explains the answers to these questions. The chapters cover "Mathematics," "Electrical," "Series and Parallel Circuits," "Oscillators, Detectors, Comparators, Demodulators," "Test Equipment and Measurements," "Electronic Components Nomenclature," "Semiconducts," "Digital Concepts," "Computer Basics," "Communications Electronics," "Safety Precautions and Checks," "Consumer Electronics Basics," "Block Diagrams and Troubleshooting," "Antenna and Wave Propagation Theory and Signal Distribution," and "Journeyman Options." An appendix includes a directory of certification administrators including DANTES (Defense Activities Non-Traditional Education Support), who are available to military personnel. This is primarily a book for the student, but could be used as a reference book in school, public, and trade libraries.

H. ROBERT MALINOWSKY

Handbooks

464. American Electricians' Handbook.
11th. Terrell Croft; Wilford I. Summers, eds. New York, NY: McGraw-Hill Book Co., 1987. 1,664p., illus., index. $59.50. 0-07-013932-6.

The eleventh edition of the *American Electricians' Handbook* maintains the standards of previous editions. It is a professional reference volume for the practicing electrician; useful to students studying electricity or related areas; and an excellent source for those who are curious about electrical principles or seeking information on the subject.

The volume reflects current standards established by NEC, ANSI, NEMA, and NESC. The handbook is divided into 11 divisions: "Fundamentals," "Properties and Splicing of Conductors," "Circuits and Circuit Calculations," "General Electrical Equipment and Batteries," "Transformers," "Solid-State Devices and Circuits," "Generators and Motors," "Outside Distribution," "Interior Wiring," "Electric Lighting," and "Wiring and Design Tables." The most significant updates in this edition are divisions 6 and 9 covering solid-state devices, circuits, and electric lighting. The general principles, terminology, and applications of each division's subject

are thoroughly described and illustrated with the appropriate standards referenced. The extensive 44-page index in this volume allows the reader to access appropriate sections easily.

This work is a basic reference tool for anyone needing information or practical advice on the selection, installation, operation, or maintenance of any type of electrical equipment or electrical wiring. *JO BUTTERWORTH*

465. Basic House Wiring. Monte Burch.
New York, NY: Sterling Publishing Co.., Inc., 1987 (c1982). 228p., illus., index. $8.95pbk. Popular Science. 0-8069-6516-9pbk.

"This book is written primarily for the beginner who is planning to build a new home, or add on rooms, or update the wiring in an older home." The information is based on and to be used in conjunction with the 1981 *National Electrical Code* and is not to replace any codes. With easy-to-follow text and good illustrations, it covers "Electricity for the Home," "Determining Your Needs," "Plans and Blueprints," "Tools," "Materials," "Basic Wiring," "Circuits," "Service Entrance and Grounding," "Wiring a New House," "Rewiring an Old House," "Wiring for Heavy-Duty Appliances," "Shop Wiring," "Farm and Ranch Wiring," "Low-Voltage and Special Indoor Wiring," "Outdoor Wiring," and "Testing and Troubleshooting." Any school or public library will find this a useful reference source on simple electrical wiring, and for the do-it-yourself person, it is also highly recommended.

H. ROBERT MALINOWSKY

466. Evanescent Mode Microwave Components. George F. Craven; Richard F. Skedd. Norwood, MA: Artech House, Inc., 1987. 165p., illus., bibliog., index. $60.00. 0-89006-176-9.

It has been only during the last 20 years that researchers have examined evanescent waves with any serious applications in mind. Evanescent waves have been traditionally defined as being below those waveguide conditions within "which progressive wave propagation cannot exist." Thus, with no apparent use, these waves have been overlooked and researched only by those intrigued by their existence. The authors, both British, want to build their case that the "below-cutoff waveguide is in many ways an ideal medium for microwave components." It is with this statement that Craven and Skedd issue an invitation to others to contribute to this area of wave research.

The nine chapters include such topics as basic theory, waveguide obstacles, filters, additional components (such as diode mounts), and antenna elements. There is also a chapter that describes the following complete subsystems in some detail: frequency multipliers, microwave receiver, Marecs diplexer, parametric amplifier, and filter-equalizers. The final chapter highlights some of the methods that have been used for microwave impedance measurements. Each chapter is well illustrated with technical drawings and photographs.

References and an index are included. One of the primary purposes of this book is to "present a coherent account of the work to date and, consequently, to provide a sound theoretical basis for the future research." The writers, who, in fact, hold British patents on their research in this area, have achieved their goal in this thoughtful and well-documented work. Recommended for academic and special libraries with strong collections in electronic engines. *CHARLES R. LORD*

467. Farm Buildings Wiring Handbook.
Ames, IA: Midwest Plan Service, 1987 (c1986). 1v. various paging, illus., index. $5.00pbk. "MWPS-28." 0-89373-067-Xpbk.

The primary purpose of this handbook is to help farmers determine which materials and methods are needed for electrical equipment and wiring in agricultural buildings. The first section covers electrical wiring including codes, lights, outlets, motor circuits, breakers, etc. The following five chapters are specific and cover standby power, alarm systems, stray voltage, lightning protection, and example buildings. This is a very practical handbook that would be useful to anyone in agricultural areas who needs to work with new wiring or rewiring of farm buildings.
H. ROBERT MALINOWSKY

468. Guide for the Design and Use of Concrete Poles.
American Society of Civil Engineers, Structural Division, Committee on Electrical Transmission Structures, Concrete Pole Task Committee. New York, NY: American Society of Civil Engineers, 1987. 52p., illus., bibliog. $10.00ppk. 0-87262-596-6pbk.

"This guide outlines the information that is to be provided by the line designer so that the engineer designing the structure has the facts he needs. It also presents the proper procedures for the design, fabrication, inspection, Testing and installation of concrete poles." This small, but comprehensive, guide covers poles which are either spun cast or statically cast and which are prestressed, partially prestressed, or conventionally reinforced. Although the concrete poles can be used in many areas, this guide stresses their use as electric utility poles. The book covers "Initial Design Considerations," "Design," "Fabrication," "Load Testing," "Assembly and Erection," and "Quality Assurance."

This is a well-written, concise, accurate and useful guide for the industry, bringing together a great amount of information that has been spread throughout various publications. Although *Guide for the Design and Use of Concrete Poles* is most useful in an industrial and manufacturing setting, it is also an excellent reference source of information on a product that few people know anything about. *H. ROBERT MALINOWSKY*

469. Handbook of Electrical Noise Measurement and Technology. 2nd.
Charles A. Vergers. Blue Ridge Summit, PA: TAB Professional and Reference Books, 1987. 440p., illus., index. $39.95. 0-8306-2802-9.

The second edition of *Handbook of Electrical Noise Measurement and Technology* updates the first edition with new information on such topics as pulse code modulation, periodic and nonperiodic signals, and decible formulas. This work attempts to answer "virtually any question on electrical noise" and represents a text that would be an excellent coursebook for classes in this field. Each chapter in the book is filled with the related terminology, diagrams, appropriate equations, a summary of the chapter, and problems and questions to test reader comprehension. Although the book contains many facts and figures for studying this subject, the lean indexing makes a quick factual reference work somewhat cumbersome. The book is better designed for study from beginning to end rather than for random reference access. For example, acronyms are not consistently indexed. FET on page 117 and rms on page 13 are not index terms; however, PAM and PARD do appear in the index. "Femtoampere" is defined on page 279 of the text, but no reference is made to it in the index. As a textbook or comprehensive work on electrical noise, more extensive indexing would greatly enhance the usefulness of this book and would make it a much more easy-to-use reference tool. References, bibliographies, or reading lists would have been useful; and careful editorial proofing would have corrected the typographical errors.
JO BUTTERWORTH

470. Handbook of Electronics Calculations for Engineers and Technicians. 2nd.
Milton Kaufman; Arthur H. Seidman, eds. New York, NY: McGraw-Hill Book Co., 1988. 1v. various paging, illus., bibliog., index. $42.50. 0-07-033528-1.

Eleven of the 24 chapters from the first edition have been revised in this second edition of the *Handbook of Electronics Calculations for Engineers and Technicians*. The revised chapters are "Selecting Semiconductor Devices," "Audio Amplifiers," "Power Suppliers," "Battery Uses and Special Cells," "Op Amp Applications," "Video Amplifiers," "Transmission Lines," "Filters," "Antennas," "Microwaves," and "Measurements." The second edition has a new chapter on "Fiber Optic Systems."

Each of the chapters follows the same basic format—introductory material, problems, and a bibliography. The essential theory of the problem is presented, but the handbook concentrates on the solution. Worked-out problems represent "75 percent of the contents." The editors refer to their work as a "cookbook" for practicing engineers and technicians. There are 18 appendices covering such topics as the Greek alphabet, degrees and minutes expressed in radians, copper-wire tables, and SI units for electricity. The index is extensive and includes acronyms and abbreviations.

Both editors are experts in the electronics field. Kaufman has been a senior engineer in industry, worked as an instructor and radar specialist with the U.S. Army, and is currently president of Electronic Writers and Editors, Inc. He has coauthored several other engineering handbooks. Seidman is a retired professor of electrical engineering. He has also been a senior engineer in industry, done consulting work, and coauthored several engineering books. This is a recommended book for academic and special libraries, as well as for personal use.   *JO BUTTERWORTH*

471. Handbook of Microwave Integrated Circuits. Reinmut K. Hoffmann. Harlan H. Howe, Jr., ed. Norwood, MA: Artech House, Inc., 1987. 527p., illus., bibliog., glossary, index. $60.00. 0-89006-163-7.

Handbook of Microwave Integrated Circuits was translated from the German *Integrierte Mikrowellenschaltungen*, published in 1983. In addition to superb efforts of the translators, Geoffrey A. Ediss and Nigel J. Keen, the English version was edited by Harlan H. Howe, Jr. The result is a clearly written handbook on the subject of microwave technology in the frequency range of 0.5 GHz to 20 GHz. Included in the latest developments emerging from research on this frequency range are microwave integrated circuits (MICs). The first chapter, "Introduction to the Technology of Microwave Integrated Circuits," is a comprehensive overview of the development of these circuits. The following chapters focus on strip transmission lines with most of the emphasis on microstrips. The microstrip is the most common MIC. The other types of strip transmission lines that are discussed are those lines which are capable and those lines which are not capable of building up a complete circuit without the need of other transmission lines.

This handbook is well-documented by an exhaustive bibliography which is almost 100 pages in length. The bibliography is organized with references for each chapter. There is also a glossary which includes a list of symbols, mathematical constants, physical constants, instructions on the notation of equations, and an explanation of the International System of Units which is used throughout the book. Each chapter contains numerous illustrations, photographs, charts, and numerical equations. The illustrations are particularly clear and are used generously to aid the textual description of important variations and developmental stamps. The author states in his preface that the book is "primarily written for a wide range of industrial engineers" and "written for special lists from other branches of electrical engineering, who wish to undertake MIC design, and for college and university students new to the field."

This book represents a wealth of information for any person who has an interest in microwave integrated circuits and, therefore, is recommended for college, university, and special libraries.
                    CHARLES R. LORD

472. Handbook of Satellite Telecommunication and Broadcasting. L. Ya. Kantor; Donald M. Jansky, eds. Boston, MA: Artech House, 1987. 498p., illus., bibliog., index. $79.00. The Artech House Telecommunication Library. 0-89006-220-X.

The U.S.S.R. has rapidly expanded the network of earth satellite communications stations during the past years. Many experts in the Soviet Union are working on this fast-expanding field. As a result, this excellent Russian handbook has been translated into English. It is a very technical handbook and is intended "for engineering technicians, engaged in the operation, design and development of satellite Communications Systems," but it could also be used by students who are specializing in radio communications.

The handbook is divided into 18 chapters covering (1) "Design Principles and Functions of Satellite Communications Systems," (2) "The Performance Characteristics of Satellite Communications Systems," (3) "The Frequency Bands Reserved for Satellite Communications and Broadcasting Systems," (4) "Power (EIRP) Calculation of Satellite," (5) "Problems of the Electromagnetic Compatibility of Satellite Systems," (6) "On the Efficient Utilization of the Geostationary Orbit," (7) "Multiple Access and Signal Separation Methods," (8) "Design Features of Receiving Equipment of Earth Stations," (9) "Transponders for Satellite Communications and TV Broadcasting Links," (10) "Waveguide and Multiplexers," (11) "Antennas of Satellite Communications Systems," (12) "Reliability Calculation of Satellite Communications and Broadcasting Systems," (13) "Basic technical data on typical satellite communications systems," (14) "Ways and Means of Transmitting Audio Broadcasting Programs, Television Audio Signals, and Facsimile through Satellites," (15) "The Gradient, Geilikon, and Grunt Transmitting Equipment," (16) "The Orbita-2 Receivers," (17) "The Ehkran Earth Receiver Installations," and (18) "Multiple Access Equipment."

This is a recommended book for university and special libraries because of its source and the chance it gives to learn from Soviet technology.
                    H. ROBERT MALINOWSKY

473. How to Read Electronic Circuit Diagrams. 2nd. Robert M. Brown; Paul Lawrence; James A. Whitson. Blue Ridge Summit, PA: TAB Books, Inc., 1988. 214p., illus., glossary, index. $20.95, $12.95pbk. TAB Hobby Electronic Series. 0-8306-0480-4, 0-8306-2880-0pbk.

"Electronic circuit diagrams are the keys to understanding the functioning of electronic circuits and electronic equipment." This book covers the various conventions for denoting the individual electronic components and their connections in different electronic circuits. It proceeds from the basic components, such as resistors, capacitors, coils, chokes, transformers, batteries, switches, relays, fuses, and circuit breakers, to the more

complex digital circuits. Between these two extremes, the other seven chapters cover transducers, indicating components, miscellaneous components, solid-state devices, vacuum tubes, interconnecting devices, diagrams, radio and TV schematics, and specialized electronic equipment.

The chapters are well written and contain many good illustrations, diagrams, and schematics. Students and laypeople will have little difficulty using this handbook. Four appendices cover electronic schematic symbols, color codes, wire sizes, and terms and abbreviations. There is also a glossary and an index. This should be a useful book for the hobbyist as well as a good reference source for school and public libraries.

H. ROBERT MALINOWSKY

474. Maintaining and Troubleshooting Electrical Equipment. Roy Parks; Terry Wireman. New York, NY: Industrial Press, Inc., 1987. 179p., illus., index. $19.95. 0-8311-1164-X.

"The purpose of this book is to present the technical subject of maintaining and troubleshooting electrical equipment in as nontechnical language as possible." Successfully achieving this purpose by avoiding the use of advanced mathematics and unnecessary use of language not familiar to a beginning student, the authors present the material in a textbook fashion, with the suggestion that the beginning student should begin with chapter 1 and proceed through each chapter in sequence. The nine chapters cover "Basic Electricity," "Direct Current (dc) Instruments," "Devices, Symbols, and Circuits," "Three-Phase Motor Starters," "Three-Phase Motors," "Direct Current Machines," "Direct Current Motor Control," "Development of Control Circuits," and "Maintenance and Troubleshooting." Following a well-written text supplemented with good line drawings, charts, and graphs, there are three appendices: "Electrical Troubleshooting Charts," "Standard Elementary Diagram Symbols," "Resistor Color Chart," and "American Wire Guage (AWG)." This is an excellent book for apprentices in industrial training and refresher training for journeymen. Vocational and trade school libraries as well as public libraries will find it a useful reference source of information on the maintainance and troubleshooting of electrical equipment. *H. ROBERT MALINOWSKY*

475. Master Handbook of 1001 More Practical Electronic Circuits. Michael L. Fair, ed. Blue Ridge Summit, PA: TAB Books, Inc., 1987 (c1979). 698p., illus., index. $19.95pbk. 0-8306-7804-2pbk.

A supplement to *Master Handbook of 1001 Practical Electronic Circuits*, published in 1975, this handbook has a minimum of text that merely identifies the circuit with no description on how it is used. Michael L. Fair intended the book to be a reference source on those needing practical electronic circuits that have been tested in industry and are of use to a wide number of people. He includes clearly printed circuit schematics for bridge circuits, solid-state switches,

logarithmic amplifiers, integrators, FSK circuits, detectors, adders, servo motor circuits, battery chargers, data transfer circuits, smoke detectors, multivibrators, frequency doublers, biomedical, video amplifiers, gadgets, photo-activated circuits, filters, sample and hold circuits, test gear and metering circuits, OP amp circuits, AM and FM broadcast receivers, converters, power supplies, regulators, readouts, interface circuits, chopper circuits, indicator circuits, audio amplifiers, waveform generators, oscillators, math function circuits, power-controlling circuits, computer-related circuits, timers and counters, sensing circuits, multiplexers, transmitter and receiver circuits, miscellaneous circuits, and TV circuits. This is a must book for the electronic hobbyist and a very useful handbook for school and public libraries. *H. ROBERT MALINOWSKY*

476. McGraw-Hill's National Electrical Code Handbook: Based on the Current 1987 National Electrical Code. 19th. J. F. McPartland; Brian J. McPartland; Steven P. McPartland; James L. McPartland; Brendan A. McPartland. New York, NY: McGraw-Hill Book Co., 1987. 1,212p, illus., index. $42.50. 0-07-045707-7. 0277-6758.

The *National Electrical Code* (*NEC*) is the official set of rules and regulations guiding the safe installation of electrical wiring and equipment, revised and reissued every three years by the National Fire Protection Association (NFPA). By its nature, its language is quasilegal and often difficult to relate to practical situations. There are tables and almost no illustrations. Hundreds of major and minor revisions were made in the 1987 edition. Several interpretive works are published in conjunction with each *NEC* revision. In fact, the NFPA publishes its own explanatory compendium, *The National Electrical Code 1987 Handbook* (1987). It, however, is not as well done as the McGraw-Hill handbook. It is about the same size, but there is much less interpretation; it merely repeats the full text of the code. Commentary and diagrams follow each section but are printed in a small, faint, brown-toned typeface which is difficult to read. Its best feature is its introductory chapter summarizing the latest code changes; the McGraw-Hill volume does not contain such a summary. The McGraw-Hill handbook is, however, the best of the 1987 editions. It is heavily illustrated with diagrams and black-and-white photos that are very sharp and appropriate; the typeface is clear and readable; the volume is well bound and physically manageable. It is intended as a companion volume and does not include the text of the *NEC*, but follows the numbering sequence of the code. The editors have selected those areas of the code which require explanation, but in over 1,200 pages of text, the coverage is broad. An extensive index is also helpful. McGraw-Hill is much more thorough and easily understood, although it also requires the separate acquisition of the text of the code. A highly recommended handbook for academic and special libraries as well as large public libraries.

CHARLENE M. BALDWIN

477. Printed Circuits Handbook. 3rd. Clyde F. Coombs, Jr., ed. New York, NY: McGraw-Hill Book Co., 1988. 917p., illus., glossary, index. $59.50. 0-07-012609-7.

Providing a very thorough and detailed treatment of printed circuit technology, this handbook's 35 chapters cover the full range of information, from basic design to the handling of industrial waste resulting from production processes. Chapters are organized in a clear, concise manner, with numbered paragraph headings for each topic and with generous and appropriate use of explanatory figures, diagrams, and tables to illustrate important details. The index, primarily to topic headings, is excellent and provides quick and specific access to information. The book also contains a nine-page glossary of terms specific to printed circuit technology.

This handbook is suitable for reference in public, special, and academic libraries and for support of instruction in the academic setting from the technical institute to the university level. The second edition, published in 1979, has had continuous use in libraries. The handbook, of course, will also be of considerable value to the electronics technician and engineer in the industrial setting.   *RAYMOND BOHLING*

478. Protective Relaying: Principles and Applications. J. Lewis Blackburn. New York, NY: Marcel Dekker, Inc., 1987. 545p., illus., index. $85.00. Electrical Engineering and Electronics, no. 37. 0-8247-7445-0.

Protective relaying "initiates the disconnection of the trouble area while operation and service in the rest of the system continue." The IEEE defines a relay as "an electric device that is designed to interpret input conditions in a prescribed manner and after specified conditions are met to respond to cause contact operation or similar abrupt change in associated electric control circuits." They protect the equipment from abrupt changes in electrical power and can be found in heating, air conditioning, stoves, dishwashers, clothes washers and dryers, elevators, telephone networks, traffic controls, transportation vehicles, automatic process systems, robotics, and space activities, to name only a few. The function of a protective relay "is to detect defective lines or apparatus or other power system conditions of an abnormal or dangerous nature and to initiate appropriate control circuit action."

This handbook presents the fundamentals and basic technology in the use of protective relays in electric power systems. After an introductory chapter that defines relays, presents some nomenclature, and discusses in general terms the uses of relays, the following chapters cover fundamental units, symmetrical components, relay input sources, protection fundamentals, system grounding principles, generator protection, transformer, reactor, and shunt capacitor protection, bus protection, motor protection, line protection, pilot protection, stability, reclosing, and load shedding.

The text is well written, technical, and includes bibliographies at the end of each chapter. To further illustrate the concepts that have been discussed, there are problems at the end of the text that have been developed over many years of actual experience. In addition to being a good reference book for technical libraries, engineers and technicians, it is a very useful handbook for users, manufacturers and computer installers.

H. ROBERT MALINOWSKY

How-to-do-it Books

479. Amplifiers Simplified with 40 Projects. Delton T. Horn. Blue Ridge Summit, PA: TAB Books, Inc., 1987. 198p., illus., index. $16.95, $10.95pbk. 0-8306-7885-9, 0-8306-2885-1pbk.

"Virtually every electronics system includes some form of amplifier circuit. An amplifier accepts a signal at its input and reproduces a replica of the original input signal at a larger amplitude at the output." Amplifiers are mentioned in almost any text that is devoted to electronics. However, information on specific amplifiers is hard to find in one reference source. The 10 chapters cover the basics of amplifier design and specific types of amplifiers: "Amplifier Basics," "Amplifier Classes," "Increasing Gain," "Amplifier Problems," "Audio Amplifiers," "Preamplifiers," "Rf and Video Amplifiers," "Op Amps," "Amplifier ICs," and "Voltage-Controlled Amplifiers." The text is concise and easy to follow for those with some electronics background. Numerous schematics aid in the explanations, and 40 projects or experiments are interspersed throughout the book to help exemplify the various concepts. Reference collections in school, public, and college libraries will benefit from this book as a central source of information on amplifiers. Students will be able to use it as a laboratory manual in electronics to learn how to build specific amplifiers.   *H. ROBERT MALINOWSKY*

480. 44 Power Supplies for Your Electronic Projects. Robert J. Traister; Jonathan L. Mayo. Blue Ridge Summit, PA: TAB Books, Inc., 1987. 244p., illus., index. $24.95, $15.95pbk. TAB Electronic Series. 0-8306-7922-7, 0-8306-2922-Xpbk.

"This book is designed to take you from the basics of current flow, electronic components, and circuit assembly through the completion of some fairly complex and certainly exotic power supplies." "A dc power supply is any component, device, or circuit that is designed to deliver an output of direct current (dc). It can be a battery, generator, or electronic circuit driven from an alternating current source." The first 100 pages contain an introduction to electronics covering "DC Power Supplies," "The Basic DC Power Supply," "DC Power Supply Components," "Voltage Regulators," and "Obtaining and Referencing Components." An understanding of electronics is needed, but the technical jargon is kept at a

minimum. The remainder of the book describes 44 power supply projects. These projects are fully described with schematics and text that any hobbyist should be able to follow. The appendices cover schematic symbols, wire size and current rating, and resistor color codes. This book has limited reference value other than being a source of power supplies that can be easily constructed. Students will find it an interesting laboratory manual. It is recommended as a teaching text for students in schools and trade institutions and as a supplementary reference source for school and public libraries.　　　*H. ROBERT MALINOWSKY*

481. The Illustrated Home Electronics Fix-It Book. 2nd. Homer L. Davidson. Blue Ridge Summit, PA: TAB Books, Inc., 1988. 465p., illus., index. $25.95, $16.95pbk. 0-8306-7883-2, 0-8306-2883-5pbk.

Homer L. Davidson has 38 years of experience in the radio and TV repair field, is a featured columnist in *Electronic Servicing and Technology*, and has written 16 electronic books. With this expertise, he is able to write an extremely practical book that explains "how to repair everything from small pocket radios to large color television receivers." It is an easy book to use, because each chapter covers a specific kind of electronic apparatuses. Within each chapter, various problems are highlighted, followed by an explanation of what is wrong and how the problem can be corrected.

The first two chapters discuss the tools that are needed to make the repairs and some tips on what to do or not to do. The chapters that follow cover common AM/FM radios, car radios, car cassette players, common protable tape players, portable cassette players and boom-boxes, cassette decks, phonographs, compact disc players, TVs, VCRs, telephone answering machines, cordless telephones, electronic games, door bells, calculators, intercoms, and battery chargers. A final chapter gives some pointers on where and how to obtain electronic parts and a list of sources of electronic components.

This is a highly recommended book for the hobbyist and home-repair person. Its step-by-step instructions are well written, so that anyone with a minimum of electronic expertise will be able to use the book with ease. School and public libraries will find it a valuable reference source on various electronic apparatuses.

H. ROBERT MALINOWSKY

482. Major Appliances. Chicago, IL: Time-Life Books, Inc., 1987. 144p., illus., index. $12.95. Fix It Yourself. 0-8094-6204-4.

Major home appliances are not constructed as well as they were some years ago. As a result, repair bills can be rather high, unless you are a person who can do some of the repairs yourself. Most of the time the repairs are simple and easy to do, but some may require a professional. This book is designed for those who have the knack to do it themselves.

After a brief introduction on how to use the book, there is a chapter that covers the most common home emergencies—fire, electrical, gas, and water. An excellent guide, this would be most helpful in the form of a wall chart. The main part of the book has a chapter devoted to each of the major appliances—refrigerators, freezers, electric ranges, gas ranges, dishwashers, garbage disposers, clothes washers, and clothes dryers. Arrangement is the same for each chapter, with an introduction that briefly describes the appliance in terms of its lifespan and use, followed by a troubleshooting chart giving the symptom, possible cause, and procedure for correction. Finally there is a pictorial, step-by-step guide to the maintenance of the specific appliance. A final chapter covers tools and techniques with a pictorial diagram of basic tools and information on diagnosing electrical problems; working with electricity, power cords, and terminal blocks; servicing 120-volt power cords; servicing 240-power cords; repairing damaged wiring; reading wiring diagrams and timer charts, wiring diagram symbols and timer chart, and sample wiring diagram for a clothes dryer; working with gas, plumbing, and cosmetic repairs; and getting help when you need it.

This is a recommended home reference book for those who want to cut repair costs by doing the repairs and servicing themselves as well as a useful reference source for school and public libraries.　　　*H. ROBERT MALINOWSKY*

483. Oscillators Simplified with 61 Projects. Delton T. Horn. Blue Ridge Summit, PA: TAB Books, Inc., 1987. 240p., illus., index. $17.95, $11.95 pbk. 0-8306-0375-1, 0-8306-2875-4pbk.

"In electronics work, an oscillator is a circuit that generates a periodic (repeating) ac signal. Oscillation is always the result of the periodic storage and release of energy." This manual covers oscillators or signal generators of all types. Information on oscillators can be found in a variety of sources, but this is the first book to bring all that information together in a concise and readable format. It is divided into 10 chapters: "Oscillators and Signal Generators," "Sine Wave Oscillators," "Other Transistor-Based Signal Generators," "UJTS," "Op Amp Circuits," "Timer Circuits," "The VCO," "The PLL," "Digital Signal Generators," and "Dedicated Signal Generator ICs." Within each of these chapters, the fundamentals are explained using many schematics, and interspersed throughout are projects or experiments that exemplify the topics being discussed. This will be a useful general reference work on oscillators for school, public, and college libraries and a practical teaching manual for students in electronics laboratories.

H. ROBERT MALINOWSKY

484. 33 Fun-and-Easy Weekend Electronic Projects. Andres Guzman. Blue Ridge Summit, PA: TAB Books, Inc., 1987. 126p., illus., index. $14.95, $8.95pbk. 0-8306-0261-5, 0-8306-2861-4pbk.

This is a book for the electronics hobbyist. It contains no background information on electronics because that information is assumed. For each of the 33 projects, there is a schematic diagram, a brief explanation of what the project is or accomplishes, and a list of parts. Projects are varied and include car alarms, headlight minders, telephone snoopers, VLF rceivers, and PLL receivers. Four appendices are of reference value giving resistor color codes, copper wire table, electronic parts suppliers, and pinout diagrams. The book is for students and hobbyists who want to learn about electronics by doing but is also a good reference source on electronics for school and public libraries. *H. ROBERT MALINOWSKY*

Manuals

485. The Benchtop Electronics Reference Manual. Victor F. C. Veley. Blue Ridge Summit, PA: TAB Books, Inc., 1987. 598p., illus., index. $34.95, $24.95pbk. 0-8306-0285-2, 0-8306-2785-5pbk.

This reference manual contains a selected list of 160 electronic topics ranging from the general, e.g., "Units of Electromotive Force," to the specific, e.g., "Thevenin's Theorem." Each topic, selected from subject areas of ac and dc power systems, solid-state and tube circuits, communications, and microwaves, is covered briefly (generally in two to seven pages), and includes a general statement, mathematical derivations of the topic, and worked examples of the mathematical derivations. This manual will be a useful source of information for the electronics technician and for students in an educational setting. School and public libraries will find this to be a popular reference tool. *RAYMOND BOHLING*

486. Home Wiring: Improvement, Extension, Repairs. 2nd. Rudolf F. Graf; George J. Whalen. Carlsbad, CA: Craftsman Book Co., 1987. 209p., illus., glossary, index. $15.00pbk. 0-934041-20-2pbk.

This is a manual to help both the professional and the novice in doing quality electrical work in existing homes. The 12 well-written chapters cover "Electricity: What You Need to Know," "Safety: Electrical Codes, Grounds, Ground Fault Protection," "System Design: Service Entrance, Branch Circuits, Calculations," "Hardware: Wire, Cable, Conduit, Boxes, Tools," "Methods: Wire Terminations, Wiring Routes, Connection Boxes," "Service Entrance," "Branch Circuits: Adding Outlets, Adding Circuits, Grounding," "Switch Controls: Circuit Interruption, Switch Loop-Split Circuits, Three-Way Switches," "Lighting Effects: Incandescent and Fluorescent Lamp Installation," "Special Circuits: Bells, Furnace, Security, Intercom, Emergency Generator," "Outdoor Wiring: Porch, Yard and Pool Lighting, Low Voltage Systems," and "Test and Troubleshooting, Fluorescent Lamps, Motors, and Appliance Repair." The information in these chapters will enable the user to figure amperage and number of new circuits needed, decide whether or not a new service panel is needed, and determine how to deal with overloads, among the many anticipated electrical tasks. The line drawings and schematics are very good but some of the black-and-white illustrations are somewhat dark and unclear. All in all, this is a good book for the person who wants to redo or to add to the existing wiring. There is a glossary which helps make this a useful reference source on home wiring for school, public, and home libraries. *H. ROBERT MALINOWSKY*

487. How to Test Almost Everything Electronic. 2nd. Jack Darr; Delton T. Horn. Blue Ridge Summit, PA: TAB Books, Inc., 1988. 175p., illus., index. $15.95, $8.95pbk. TAB Hobby Electronic Series. 0-8306-7925-1, 0-8306-2925-4pbk.

This is a practical manual on "electronic tests and measurements, how to make them with all kinds of electronic test equipment, and how to interpret the results." The authors guide the user through the process of measuring quantity, output, and quality of electrical voltage, current, and resistance. Ten well-written chapters cover "Test Equipment," "DC Voltage Tests and Power Supplies," "Current Tests," "VOM and VTVM Tests," "Oscilloscpoe Tests," "Component Tests," "TV Tests," "Special Tests," "Signal Tracing and Alignment Tests," and "Digital Circuits." The instructions for carrying out the testing are straightforward and understandable for the layperson. Safety precautions are stressed, and the way in which to read the results accurately is of primary importance to the authors. The book does not give detailed instructions on servicing any given piece of equipment.

Jack Darr and Delton T. Horn both have many years of experience in servicing and repairing electronic equipment. Their practical experience has enabled them to write a manual that should be very useful to any hobbyist or home-fix-it person. School and public libraries will find it a useful reference source on electronic test equipment. *H. ROBERT MALINOWSKY*

488. Residential Wiring. 2nd. Jeff Markell. Carlsbad, CA: Craftsman Book Co., 1987. 344p., illus., bibliog., index. $18.25. 0-934041-19-9.

Although written for the beginning professional and apprentice electricians, this book can be a useful tool to anyone interested in tackling a residential wiring project. As an electrician, author, and teacher, Jeff Markell has drawn from more than 33 years of experience to produce this comprehensive yet concise guide with an emphasis on standards defined by the *National Electric Code*. The introductory portions of this book give some background on the basic physical properties of electricity, and subsequent chapters walk the reader through step-by-step instructions on tools, safety, materials, and techniques for installing new wiring or the rewiring of old work. Also included are sections on troubleshooting and repairs as well as supplementary systems such as

signaling and warning systems, communications, and entertainment systems. The presentation of the material is systematic and straightforward with ample illustrations and diagrams. There are two appendices covering electrical code tables most likely to be applied in residential wiring situations and building and floor plans. This book would be appropriate for public libraries as well as for trade school libraries that offer courses or apprenticeships for electricians.

DOUGLAS HINES

489. Solid-State Microwave Devices.
Thomas S. Laverghetta. Boston, MA: Artech House, 1987. 196p., illus., bibliog., glossary, index. $60.00. Artech House Microwave Library. 0-89006-216-1.

Written in straightforward and understandable language, this technical manual brings together widely scattered information for the design engineer to use in developing microwave solid-state devices. The first chapter covers some terminology associated with microwaves and compares those that were developed in the 1970s with those available today. Chapter 2 discusses the materials that are used to fabricate microwave diodes and transistors including germanium, silicon, gallium arsenide, and indium phosphide; chapter 3 covers PN and Schottky junctions that are used in solid-state devices. Chapter 4 covers in detail six types of microwave diodes—Schottky, PIN, Varactor, Gunn, IMPATT, and TRAPATT—and chapter 5 covers microwave transistors—bipolar, field effect, and high electron mobility. The final chapter brings together all the discussions by presenting fabricated components such as amplifiers, oscillators, attenuators, switches, detectors, mixers, and phase shifters.

With this manual, individuals will be able to design devices and components more effeciently. The charts, diagrams, and illustrations are clear and supplement the text. There is a brief glossary of acronyms and a brief index. Libraries serving electrical and electronics engineering researchers will find this a useful book to have on the reference shelves. *H. ROBERT MALINOWSKY*

490. Troubleshooting and Repairing Audio Equipment.
Homer L. Davidson. Blue Ridge Summit, PA: TAB Books, Inc., 1987. 325p., illus., glossary, index. $24.95, $16.60pbk. 0-8306-7167-6, 0-8306-2867-3pbk.

Troubleshooting and Repairing Audio Equipment is "to acquaint the experimenter, tinkerer, home owner, or hobbyist, whether beginner or intermediate, with how each unit works and how to make simple repairs." The book is written in an informational style with step-by-step instructions for testing and repairing audio equipment. The author, an owner and operator of a radio and television repair business for 30 years, also emphasizes when equipment should be taken to a "trained technician." In many of the chapters, the author gives actual "case histories" to demonstrate problems and solutions. The types of equipment covered include compact cassette portable stereo players, compact cassette tape decks, boom-box cassette players, deluxe AM-FM-MPX tuner tests, deluxe amplifiers, auto stereo cassettes, compact disc players, telephone answering machines, and stereo speakers.

There are many black-and-white pictures of the equipment, as well as illustrations, troubleshooting/service charts, and schematics. Panasonic Consumer Electronics, RCA Corp., and Radio Shack are all acknowledged in the introduction for their contributions of service literature and schematics. There is a glossary, an index, and an excellent appendix that lists "Manufacturers of Consumer Electronics Equipment" with their addresses. The book would be very useful to anyone interested in simple home repairs of audio equipment and will be useful in school and public libraries. *JO BUTTERWORTH*

Textbooks

491. Communications Receivers: Principles and Design.
Ulrich L. Rohde; T. T. N. Bucher. New York, NY: McGraw-Hill Book Co., 1988. 584p., illus., bibliog., index. $59.50. 0-07-053570-1.

Communications Receivers: Principles and Design is an extensive coverage of the subject. The 10 chapters are "Basic Radio Considerations," "Radio Receiver Characteristics," "Receiver System Planning," "Antennas and Antennal Coupling," "Amplifiers and Gain Control," "Mixers," "Frequency Control and Local Oscillators," "Demodulation and Demodulators," "Other Receiver Circuits," and "Receiver Design Trends."

Although the work is more textbook oriented, the index and the list of abbreviations (Appendix B) make it useful as a reference tool. In the preface, the authors state that their "effort has concentrated on single channel communications receivers at frequencies where lumped circuit elements (or some simple transmission-line elements) would serve for RF and oscillator circuits." References are given at the end of each chapter for those who seek more information or analysis

Both authors have worked for RCA, and contributions from other RCA engineers are acknowledged. Dr. Rohde has taught radio courses and been involved with RCA's military HF and UHF communications. Dr. Bucher spent 40 years in the design and analysis of radio receivers, transceivers, and systems at RCA. This is a recommended book for academic and special libraries. *JO BUTTERWORTH*

492. Converter Circuits for Superconductive Magnetic Energy Storage.
Mehrdad Ehsani; Robert L. Kustom. College Station, TX: Texas A & M University Press for the Texas Engineering Experiment Station, 1988. 246p., illus., bibliog., index. $68.50. The Texas Engineering Experiment Station Monograph Series, no. 4. 0-89096-257-X.

"Superconductive magnet coils are increasingly being used for energy storage and generation of high magnetic fields in research and industry." These coils have zero resistance to current and, thus, are very attractive energy storage sources. This text analyzes some of the recent work on the development of switching power converters for superconductive magnets. Because this is becoming an important new area of research, this text could well be used as a reference source in any technical library.

After an introductory chapter that gives the background needed to understand the rest of the book, the authors cover "Nonuniform Switching Circuits," "Analysis of Nonuniform Switching Circuits: Flying Capacitor Bridgees," "Uniform Switching Energy Transfer Circuits: Inductor-Converter Bridgees," "Analysis of the Inductor-Converter Bridge Circuits," "Methods of Phase and Frequency Shifting in the Inductor-Converter Bridge Circuits," "Feedback Control of the Inductor- Converter Bridge," "Some Experimental Results from Model Energy Transfer Systems," and "Converters for Diurnal Superconductive Magnetic Energy Storage."

This is a highly technical book and not one that the general reader will be able to understand. For the technical researcher, however, it is very well written, with numerous diagrams, illustrations, and mathematical equations that adequately describe the topic being discussed. University students involved in research pertaining to magnetic energy storage will want to be aware of this book and will expect it to be in their academic libraries. *H. ROBERT MALINOWSKY*

493. Electronic Communications: Modulation and Transmission. Robert J. Schoenbeck. Columbus, OH: Merrill Publishing Co, A Bell and Howell Information Co, 1988. 628p., illus., index. $37.95. 0-675-20473-9.

Electronic Communications: Modulation and Transmission is clearly a textbook, not a reference book. The author acknowledges this in the preface of the work and, as a result, has "written in an informal, conversational tone." It is divided into 16 chapters ranging from "AM Transmitters and Receivers" and "FM Transmitters and Receivers" to the "Communication Satellite." Each chapter concludes with review questions that thoroughly test the reader's comprehension. There are between 27 and 108 review questions per chapter. Extensive illustrations and schematics with precise descriptions fill the chapters. For example, the "American parking area in the Clarke orbit," Fig. 16.1, shows the placement, by international agreement, of U.S. satellites in the "equatorial geosynchronous orbit, called the Clarke orbit." Formulas are given, however the author has kept mathematics to a minimum. There is one appendix, "FM Vector Analysis," and an index. The index does not include acronyms, and this would hinder reference work, but as stated previously this volume is intended to be a textbook not a reference book; it should be part of the circulating collection of academic libraries.

The table of contents is very precise and identifies acronyms in parenthesis after appropriate terms. There are no bibliographies or reading lists. *JO BUTTERWORTH*

494. Electronic Test Equipment: Theory and Applications. T. J. Byers. New York, NY: McGraw-Hill Book Co., 1987. 335p., illus., bibliog., glossary, index. $39.95. "Intertext Publications, Inc.." 0-07-009522-1.

This work describes, in a simple, straightforward fashion, various examples of electronic test equipment. Its purpose is to create an understanding of electronics as "the science of measurement" to people in all disciplines who utilize electronic measuring devices and want to know about how they work. The first chapter concerns simple analog instruments such as voltmeters; the final chapters deal with modern computer-based test instruments. Emphasis is on oscilloscopes, for which the author devotes three of eight chapters. Other chapters treat digital instruments and signal generators. The final chapter focuses on the GPIB (General Purpose Interface Bus), the IEEE-408 protocol for microcomputer interface with PC-based test equipment. A glossary, two pages of references, and a short index complete this book. A paragraph or two serve as a general introduction to each chapter, but there is no overview chapter to introduce the work. (The preface is not well written and does not substitute). The book is heavily illustrated with diagrams and photographs of almost every piece of equipment described. Academic libraries with engineering programs will want a copy of this text. *CHARLENE M. BALDWIN*

495. Principles of Electronic Instrumentation and Measurement. Howard M. Berlin; Frank C. Getz, Jr. Columbus, OH: Merrill Publishing Co., A Bell and Howell Information Co., 1988. 466p. illus., glossary, index. $36.95. International Series in Electrical and Electronics Technology. 0-675-20449-6.

The 16 chapters of this textbook are written to be self-contained study units covering the principles of instrumentation and measurement. Topics include measurement fundamentals and units, standards, grounding, shielding, noise, AC and DC indicating meters, analog and digital meters, potentiometers and analog recorders, AC and DC bridges, transducers, oscilloscopes, and RF and fiber-optic measurements. The authors state in the preface that their work "is primarily intended for a single-term course in instrumentation and measurement at either a two-year technology school or a standard electrical engineering technology program." Trade school libraries will want a circulating copy. They also mention that "it is assumed that the student has had previous training in basic AC and DC circuits, as well as fundamentals of solid-state devices such as diodes, transistors, and operational amplifiers."

The chapters are clearly formatted with the teaching environment in mind. Each chapter begins by outlining the instructional objectives including the important terms which will be defined in the text. This extra effort to alert the reader sets this text apart from many in the field. In addition, there are summaries, glossaries, and problems at the end of each chapter. As a result, the student and instructor do not have to wander to the back of the volume to define important terms. The answers to the odd-numbered problems are located in the back, just before the volume's index. There are numerical examples with solutions, diagrams, and figures woven throughout the text. *Principles of Electronic and Instrumentation and Measurement* can also be used as a reference book. There are several appendices listing symbols and conversion factors. Some readers may be interested in knowing that a laboratory workbook has been written by the authors, Berlin and Getz, which can be used along with this textbook.   *CHARLES R. LORD*

★
ENERGY
Dictionaries

496. **A Dictionary of Nuclear Power and Waste Management with Abbreviations and Acronyms.** Foo-Sun Lau. Letchworth, Great Britain: Research Studies Press, Ltd., Dist. by John Wiley and Sons, New York, NY, 1987. 396p., bibliog. $170.00. Research Studies in Nuclear Technology, 1. 0-86380-051-3.

The most recent attempt to provide access to the terminology of nuclear energy, engineering, and waste disposal since Ralf Sube's *Dictionary of Nuclear Engineering in Four Languages* (Elsevier, 1985) and Michael Stephenson's *The Nuclear Dictionary* (Longman, 1985), this volume presents both the language of the field and 65 pages of commonly used abbreviations. These last range from symbols for chemical elements and compounds to acronyms for such international bodies as the International Radiation Protection Association. Six appendices provide listings of the atomic masses of elements, half-lives of nuclides, physical constants, a geological time scale, tables of scientific units, and equivalencies. A bibliography of sources consulted completes the work.

Although this is a somewhat expensive book, it is one that would be useful in large public libraries with a strong emphasis on science and technology; in college and university libraries with research conducted in the areas of nuclear energy and waste disposal; and in research libraries supporting research in nuclear power, waste management, and nuclear engineering.

ROBERT B. MARKS RIDINGER

Directories

497. **Wind Energy 1987: Wind Turbine Shipments and Applications.** Thomas F. Jaras. Great Falls, VA: Wind Data Center, Stadia, Inc., 1987. 316p., illus., index. $300.00. 0-944038-00-X.

Wind turbines, their manufacture, applications, and marketing is the focus of this work, "the first in a series of publications containing statistical information and analyses on the wind energy industry." It is international in scope, covers the years 1981–1986, and presents data in pictorial graphs and tables. The statistics come from a survey of wind turbine manufacturers. An explanation of the survey method and the statistical database is in the introductions. The book is arranged in six sections: (1) introduction and description; (2) wind markets and trends; (3) wind turbine market statistics by region; (4) statistics by rotor diameters; (5) statistics by nine applications, "ranging from water heating and water pumping to grid-connected windfarm"; and (6) turbine manufacturing and export/import activity. Approximately half of this book is an appendix listing wind turbine manufacturers. "Each manufacturer's description consists of four parts: range and address; wind industry activities; other corporate activities; and wind turbine products."

This book has a table of contents, list of illustrations, and an index for the appendix. The text is in large, easy-to-read print, and the graphic information is clearly presented. This book would be useful for large public libraries or academic libraries with energy or business collections, and special libraries with an interest in energy resources.   *SUE ANN M. JOHNSON*

Handbooks

498. **Pipe Line Rules of Thumb Handbook: A Manual of Quick, Accurate Solutions to Everyday Pipe Line Problems.** 2nd R04 E. W. McAllister. Houston, TX: Gulf Publishing Co., 1988. 444p., illus., bibliog., index. $45.00. 0-87201-695-1.

Pipe Line Rules of Thumb is a new edition of a 1978 handbook. The subtitle, "A Manual of Quick, Accurate Solutions to Everyday Pipe Line Problems," defines its scope and purpose. The author, E. W. McAllister, has had more than 30 years of experience in the pipeline industry as an engineer, administrator, and project manager. Divided into 21 sections, this book has numerous charts, tables, and other illustrations. Many of the sections have bibliographies. In addition to an overall table of contents, each section has its own detailed contents page. This book is clearly written, well organized, and easy to use and understand. It would be an excellent addition to any reference collection that needs petroleum engineering information.   *SUE ANN M. JOHNSON*

499. The Storage and Handling of Petroleum Liquids. 3rd. John R. Hughes; Norman S. Swindells. London, Great Britain: Charles Griffin and Co.., Ltd., 1987. 332p., illus., index. $57.50. 0-471-62966-9.

The literature of technology has its own classics; these are books that are extremely useful and that, with revisions, are useful over a period of decades. *Storage and Handling of Petroleum Fluids* is a minor classic of this sort and is made so by being a rather complete summary of its subject. A search of the available literature shows that it has little competition in the field of petroleum safety. A British book, it refers to various British standards and agencies; it also uses the British spelling of words (vapour instead of vapor), but the book remains very useful in the United States. Measurements are in metrics, but in most cases, the equivalents in the United States system are given.

The accent of this book is on practical matters, starting with an explanation of how a flammable or explosive situation develops and can be ignited. Then it discusses safety considerations in different phases of petroleum storage and transportation, such as at jetties or while cleaning storage facilities. A section on fighting oil fires concludes the text.

Hughes's book is indispensable in any library in the petroleum and petrochemical industries; universities with programs in petroleum engineering would also find it a necessary item. Chemical engineering reference collections would also be enriched by having this handbook. Academic libraries will probably find it more appropriate for the circulating collection than as a reference item.

KENNETH QUINN

Manuals

500. Boilers Simplified: A Guide to Basic Boiler Operation. Lionel Edward LaRocque. Troy, MI: Business News Publishing Co., 1987. 227p., illus., index. $27.95. 0-912524-40-5.

"Simpley stated, a boiler is an apparatus designed to convert fuel energy into a working medium (steam or hot water). The working medium, in turn, transfers heat." This is an instructional manual on boilers. Each chapter covers a particular aspect of boilers in simple language with little or no technical jargon. At the end of the chapter are a series of review questions. There are good drawings and photographs to aid in the interpretation of what is being discussed. The six sections cover hydronic systems, types of boilers, systems components, fuels and combustion, boiler operation, and water treatment. Appendix A contains information on boiler room safety; appendix B is a sample of the boiler operator's preparatory licensing examination; and appendix C contains the solutions to the review questions. This is not a technical book and is not intended for in-depth study of boilers. It is, rather, a general introduc-

tion for individuals whose work requires them to be around boilers. Special libraries serving these individuals will want a copy for their circulating reference shelves. *H. ROBERT MALINOWSKY*

501. Energy Projects for Young Scientists. Robert Gardner. New York, NY: Franklin Watts, 1987. 127p., illus., bibliog., index. $11.90. Projects for Young Scientists. 0-531-10338-2.

"An understanding of energy, its sources, and ways to reduce its use or improve the efficiency with which we use it will become increasingly important as some of our fossil-fuel energy resources approach depletion in the years to come." With this in mind, the author has outlined, with a well-written text and good illustrations, some ideas that will help the student understand energy in its various forms. Each chapter begins with a general discussion of the topic and then presents projects that help to illustrate what has been discussed. The projects vary in difficulty, but most can be done with a minimum of preparation, and all illustrate things that are around us on a daily basis. The first chapter is devoted to "Science Projects and Fairs." A bibliography of books pertaining to science fairs is included, as is a list of national competitions. The following six chapters cover "Energy, Work, Power, and Efficiency," "Heat and Thermal Energy," "Electrical Energy," "Energy of Motion and Position," "Solar Energy," and "Saving Energy at Home and School." Two appendices cover "R Values of Various Materials" and "Specific Heats of Some Common Substances." Through the introductory materials and the various projects, the student will have an "understanding of the basic concepts of energy, work, and power" and "of heat and thermal energy, electrical energy, energy of motion and position, solar energy, and energy conservation." School and public libraries will want to have a copy of this book.

H. ROBERT MALINOWSKY

Tables

502. ASTM Standards on ASTM Viscosity Tables for Kinematic Viscosity Conversions. 4th. ASTM Subcommittee D02.07 on Flow Properties. Roberta A. Storer, ed. Philadelphia, PA: American Society for Testing and Materials, 1987. 36p. $18.00. "PCN:03-043040-12." 0-8031-0990-3.

This, the fourth edition of the *ASTM Viscosity Tables for Kinematic Viscosity Conversions*, has been reprinted from the *Annual Book of ASTM Standards*. The standards were sponsored by the ASTM Subcommittee D02.07 on Flow Properties of the ASTM Committee D-2 on Petroleum Products and Lubricants. The pamphlet was designed to provide a quick source of information to the D 2161-87, ASTM Practice for Conversion of Kinematic Viscosity to Saybolt

Universal Viscosity or to Saybolt Fural Viscosity, and to the D 2270-86, ASTM Practice for Calculating Viscosity Index Kinematic Viscosity at 400 and 1000C.

Two appendices are included. The first is a reference to an article which gives equations for the conversion of different viscosity units not discussed in this pamphlet, and the second is a list of viscosity related conversion factors. This pamphlet will be useful for researchers working on petroleum products and lubricants.

LUCILLE M. WERT

★
ENVIRONMENTAL ENGINEERING
Handbooks

503. Building Services Materials Handbook: Heating, Sanitation and Fire Protection. Building Services Research and Information Association. New York, NY: F. N. Spon, 1987. 682p., bibliog., index. $110.00. 0-419-14310-6.

Building Services Materials Handbook is intended to provide information on materials and components, e.g., controls, heat generators, etc., for a wide variety of applications under normal environmental condtions. It is divided into four sections: Secton 1, General; Section 2, Services; Section 3, Components; and Section 4, Materials.

Section 1 gives information on units of measurement, different types of environment, and provides a framework for use of the other three sections. Section 2 has service datasheets for components of building services systems, such as space heating, that supply details important for selecting component parts. Relevant specifications in British Standards notation are identified on component data sheets as well as descriptive detail in Section 3 for standards to be followed in specifying building components such as "tanks." Data sheets in Section 4 supply a wealth of information on physical characteristics of materials including suitability for different environmental conditions in forming techniques, joining, and typical applications.

This handbook will be useful to the builiding and systems design engineer and to the technical and professional architectural design student. Academic libraries will especially want a copy for their reference collections. *RAYMOND BOHLING*

504. Red Book on Transportation of Hazardous Materials. 2nd. Lawrence W. Bierlein. New York, NY: Van Nostrand Reinhold Co., 1988. 1,203p., index. $96.50. 0-442-21044-2.

This handbook is an essential guide to the regulations governing the transport of hazardous materials. Written with the user in mind, the *Red Book* serves "the full range of people dealing with hazardous materials regulation, including shippers, carriers, container manufacturers, freight forwarders, lawyers, educators, legislators, and regulators." Although not a listing of the regulations themselves, it is an outline and summary of their scope, a guide to the manner in which they are applied, and a reference resource for compliance with the often bewildering range of Department of Transportation (DOT), Environmental Protection Agency (EPA), and related state and international agencies. This is the second edition of the *Red Book* and is a welcome replacement for the first edition, orginally published in 1976. Even so, with the pace of change in legislation, it still must be carefully read in conjunction with the current regulations and not viewed as a substitute for them. Indeed, the author points out in the introduction that it is recommended that the relevant sections be copied and maintained for joint use with the actual text of the regulations. Although it could be profitably used by many people, its primary beneficiary is the shipper of those materials that fall under the hazardous classification. Not only are all statutory bases for regulations reviewed, but all of the relevant regulatory sources are referenced and explained in a language devoid of the legal phrasing and format characteristic of the Code of Federal Regulations (CFR). The format of the volume addresses, each in turn, the many responsibilities of the shipper of hazardous materials: identification, classification, naming, packing, marking, labelling, placarding, and the proper documentation and shipping papers to certify compliance with the regulations. Chapter 3, "The What, Who, Where, When, and Why of Hazardous Materials Transportation," outlines what the regulatory standards are, to whom they apply, and the circumstances of where and when they are in effect. All pertinent CFR titles are treated, and the appropriate chapters, subchapters, and parts are referenced. Useful appendices include tables listing "Vapor Pressure Curves for Selected Flammable Liquids." Another appendix of interest is Chapter 9 of the United Nations "Orange Book," the UN recommendations in the transport of dangerous goods. This chapter has been proposed by regulatory bodies as a replacement for the 49 CFR Part 178 in DOT's rulemaking Docket No. HM-181.

The final appendix, and a particularly valuable one, contains "Selected DOT Rule-Making Notices and Ammendments" Docket Nos. HM-1 to HM-198. "Dockets" are the files established by DOT when a specific subject is to be considered in the rulemaking process. Each docket has its own number and is assigned sequentially from the initial year of DOT in 1967, so that the first was HM-1 and the latest in the appendix is HM-198. These are especially valuable because they contain the preambles to the notices of proposed rules and outline the basis for and the intent of proposed changes in rules. Information not part of the final rules as published in the CFR is contained in each docket, including historical and background information summarizing the consideration of possible solutions leading to the issuance of a final rule, the comments submitted by interested parties and the public, and all references to publications in the *Federal Register*. This is a rich resource to have available in a single place. This volume, like the regulations

they seek to explain, does not make light reading. But it does an admirable job of interpreting the rules and plainly stating how they apply. Used as intended, in conjunction with the CFR regulations themselves—and particularly with the helpful addition of the docket material—the new *Red Book* serves as a useful tool in both practical and research applications and highly recommended for academic and special libraries.

R. GUY GATTIS

505. Septic Systems Handbook. O. Benjamin Kaplan. Chelsea, MI: Lewis Publishers, Inc., 1987. 290p., illus., bibliog., glossary, index. $44.95. 0-87371-095-9.

The proper control of rural sewage is an environmental concern that is too often overlooked. Unfortunately all civil engineers, geologists, sanitarians, and architects do not have the proper background knowledge to design suitable rural sewage systems, which are called septic tank systems. This handbook is intended to give these professionals the basic theory necessary for designing septic tank systems. It covers the disposal of rural sewage, which is "any domestic wastewater, be it blackwater, which drains down the toilet, or greywater, which drains from all other plumbing fixtures." The septic tank is just that—a tank that collects sewage. It separates the floatable sewage from the sludge, leaving the clarified sewage that drains into a leachfield. The sludge and floatables are periodically siphoned off by a licensed tank pumper.

Dr. Kaplan covers all aspects of a septic tank system—the septic tank, leachfield, soil water movement, percolation test, factors affecting the failure of leachlines, size of seepage pits, onsite sewage disposal tehnologies, degradation of groundwater by septic systems, nitrate in groundwater, mounding, land use and septic systems, uniform plumbing code, and ethics. The text is understandable for a generalist yet detailed enough for the professional. Twelve appendices give specific information on sewage organisms, environmental problems, sizing, and codes. A three-page glossary covers the terms commonly encountered in reading about septic tank systems. A very informative book for school, public, academic, and special libraries as well as for individuals who want to install their own septic tank.

H. ROBERT MALINOWSKY

★
HANDICRAFTS
Encyclopedias

506. Encyclopedia of Furniture Making. Rev and Exp. Ernest Joyce; Alan Peters. New York, NY: Sterling Publishing Co.., Inc., 1987. 510p., illus. (part col.), index. $29.95. 0-8069-6440-5.

This encyclopedia, first published in 1970, has become "the woodworker's bible." Alan Peters has revised and expanded this bible to include the current knowledge and art of furniture making (which has truly become an art, rather than just a passing hobby), making the encyclopedia even more important as a reference source.

This is not an encyclopedia arranged with words "a" at the beginning and "z" at the end. It is, instead, arranged by broad topics. Within each topic, the discussions are comprehensive, technically accurate, and easy and enjoyable to read. Part 1 covers the basic materials of furniture making—woods, veneers, manufactured boards, plastics, leathers, metals, adhesives, and abrasives. Part 2 discusses the tools and equipment of the trade. In parts 3 and 4, the basic techniques of joint construction and advanced construction are covered including edge jointing, dovetailing, leg and frame construction, and drawer construction. Metal fittings and fasteners are covered in part 5, with advanced techniques of veneering, marquetry, inlay, lining, and curved work covered in part 6. Part 7 gives scme pointers on running a professional workshop, and part 8 covers draughtsmanship and workshop geometry. Part 9 is a detailed section that describes furniture designs and constructional details. This section includes numerous color and black-and-white photographs of striking furniture designs. Finally, part 10 describes the procedures for restoration, repairs, and wood finishes.

This is an excellent encyclopedic textbook with accurate, up-to-date information on furniture making. Any type of library should have a copy, and those who design and build furniture will certainly need this for their library.

H. ROBERT MALINOWSKY

How-to-do-it Books

507. The Backyard Builder's Book of Outdoor Building Projects. Rodale's Practical Homeowner Magazine. Emmaus, PA: Rodale Press, 1987. 499p., illus., index. $24.95. 0-87857-696-7.

Rodale Press (*Rodale's Practical Homeowner* magazine) continues its tradition of publishing popular nonfiction with this book of outdoor building projects. Simple, clear diagrams and photographs accompany easy-to-follow instructions for over 125 backyard projects ranging from children's lawn furniture to gazebos and hot tub installations. Some projects could easily be completed in a day, and most can be adapted to different needs and levels of expertise.

The arrangement is straightforward. The clear type and bold project headings are visually appealing. The binding is good, and the book lays open flat. It is questionable whether the pasteboard cover can take the abuse it may receive in a workshop.

Every project has a photograph of the finished piece, plus a detailed construction drawing to show how the pieces go together. Each includes a materials list and explanations of procedures, but

the writing tends to be unimaginative, and the introductory chapter explaining basic techniques and materials is somewhat incomplete. Despite these two negative points, it is still a good how-to book for school and public libraries.

LINDA VINCENT

508. How Things Work in Your Home (and What to Do When They Don't). Time-Life Books. New York, NY: Henry Holt and Co., 1987 (c1985). 368p., illus., bibliog., index. $12.95pbk. "An Owl Book." 0-8050-0126-3pbk.

"The purpose of this book is to help you achieve more comfort and convenience, at less cost, by taking over household maintenance and repair jobs that anyone can do safely and efficiently." The text first describes how the machines work, then how to disassemble, replace parts, and make adjustments, and, finally, troubleshooting. The first part covers "buying, using, and maintaining tools," followed by seven sections that cover plumbing, electricity, small appliances, large appliances, heating, cooling, and outdoors. Specific topics, machines, or appliances covered include rewiring lamps, replacing light fixtures, can openers, cordless appliances, humidifiers, vacuum cleaners, shavers, toasters, coffee makers, hair dryers, washers, dryers, dishwashers, refrigerators, cooking ranges, heating systems, air conditioners, lawn mowers, and chain saws. Excellent diagrams, line drawings, and charts supplement a text that is understandable for the home-repair person. This is an excellent manual for school and public, as well as home, libraries.

H. ROBERT MALINOWSKY

509. Projects from Pine: 33 Plans for the Beginning Woodworker. James A. Jacobson. Blue Ridge Summit, PA: TAB Books, Inc., 1987. 177p., illus., bibliog., index. $17.95, $10.95pbk. 0-8306-7871-9, 0-8306-2871-1pbk.

"The intention of do-it-yourself projects is to get something done rather quickly with a minimum of investment and, hopefully, some success and a large degree of satisfaction." This book should help to bring about that satisfaction. Pine is a wood that is easy to work with, can be decorated in a number of ways, and has its special rewards in the finished product. For any school that teaches shop practices, this book would be an excellent reference source; it not only details 33 projects but gives the user pointers on how to design the projects. Of special reference value is the chapter on tools and crafting supplies, which discusses such implements as saws and hand tools. There is also an appendix of suppliers and magazines. This would be a useful reference work for school and public libraries as well as for personal libraries.

H. ROBERT MALINOWSKY

510. 24 Table Saw Projects. Percy W. Blandford. Blue Ridge Summit, PA: TAB Books, Inc., 1988. 120p., illus., index. $6.95pbk. 0-8306-2964-5pbk.

The table saw is one of the most useful and versatile power tools in any woodworking shop and can, when used properly, produce some beautifully cut materials for any project, large or small. This is a book of 24 projects: glass-fronted display board, laundry box, double-door cabinet, checker-top magazine table, display shelves, kitchen boards, blanket racks, paneled chest, carry-all, dresser stool, stacking trays, classic table, wall tables, yard seat, jewelry box, garden gates and fences, paneled cabinet, plant stands, matching table and benches, bed ends, desk, bar or counter, toolbox, and sawing aids. It is not a book on techniques, although some pointers are given in the detailed instructions for cutting and assembling the item. School, public, and personal libraries will find this a useful reference source of table saw projects. *H. ROBERT MALINOWSKY*

★
HYDRAULIC ENGINEERING
Handbooks

511. Private Water Systems Handbook. 4th. Ames, IA: Midwest Plan Service, 1987 (c1979). 72p., illus., glossary, index. $6.00pbk. "MWPS-14." 0-89373-045-9pbk.

This handbook is for those people who are not privileged to have community water. "It is for rural and farm water system owners and the well contractors, pump installers, plumbers and others who provide supplies and services." It is clearly written with as little technical jargon as possible and has numerous charts and illustrations. It presents practical information on water quantity, water sources, pumps, pressure tanks and controls, piping, and water treatment. Any rural family with its own water system would find this a useful handbook, and school and public libraries will find much useful information on private water systems. *H. ROBERT MALINOWSKY*

512. The Sextant Handbook: Adjustment, Repair, Use and History. Bruce A. Bauer. Camden, ME: International Marine Publishing Co., 1987 (c1986). 191p., illus., bibliog. $19.95. 0-87742-956-1.

Relatively few texts have focused on the marine sextant, the navigation instrument used to measure altitudes of celestial bodies. The *Sextant Handbook*, serving as a generic technical manual to the instrument for both beginning and experienced navigators, helps to fill that void. As clearly intended, the work is about the instrument itself, and not about navigating. "Ownership of sextants is widespread, but proficiency in their use and adjustment is not." The author observes that, although the sextant is viewed as an essential seafaring possession, many fear using it be-

cause of its reputed fragility and, therefore, do not use it effectively or at all. Through a well-organized text of practical information and a friendly, informal presentation, the author mitigates those anxieties.

A concise, yet fascinating, history of the development of the sextant and other celestial navigation devices precedes the main body of the work. From the latitude hooks used by the Polynesians, to the astrolabe, to the simultaneously invented double-reflecting instruments of Godfrey and Hadley, history is conveyed while the functional principles of the instrument are explained. Main chapters focus on the anatomy of the sextant, assembly, step-by-step adjustment procedures, care and maintenance, repair, attachments and accessories, sighting techniques (including rough weather), timing, and correction. The down-to-earth usefulness of this book is evidenced by the chapter on how to buy a sextant, new and used. A list of sextant manufacturers, distributors, and dealers is provided in the appendices. Illustrations effectively support the text.

Disguised as a personal conversation with an expert, the *Sextant Handbook* succeeds as a technical manual. Although the author's informal writing style may not appeal to everyone, the practicality and accessibility of his knowledge should. The work is relatively short; a logical and welcome expansion of the book would offer more navigational theory and techniques for the use of the instrument. It is recommended for public libraries and collections supporting navigation sciences.   *JOHN T. BUTLER*

★
INDUSTRIAL ENGINEERING
Handbooks

513. **Engineering Design for the Control of Workplace Hazards.** Richard A. Wadden; Peter A. Scheff. New York, NY: McGraw-Hill, Inc., 1987. 735p., illus., bibliog., index. $69.50. 0-07-067664-X.

Occupational safety and health are basic necessities in industry, thanks to requirements by OSHA, increased employer liability, and heightened worker awareness. The aim of this handbook is to provide ventilation and safety engineers and industrial hygienists "with guidelines with which to evaluate the problem, and design tools with which to make at least initial estimates of the control technology required." The book is extremely well written and contains a wealth of information in the form of graphs and charts. Each chapter covers a special area: workplace hazard control, occupational injury and disease, hazard evaluation and control, design criteria, ventilation, air cleaning, process alteration, worker-machine interactions, guarding, ionizing radiation, nonionizing radiation, heat, and noise and vibration. Chapters are authored by authorities, most of whom work in industry.

The information contained in the charts is very useful. Examples of such charts include "Relative Mortality Risks for Steelworkers," "Design Criteria for Air Contaminants," "Characteristics of Flammable Gases and Vapors," "Explosion Characteristics of Various Dusts," "Comparison of Human Capabilities with Machine Alternatives," and "Maximum Permissible Body Burdens and Maximum Permissible Concentrations of Radionuclides in Air and in Water for Occupational Exposure." Unfortunately there is no easy access to these charts and tables because they are not listed or indexed separately. Although the text is easy to comprehend, there are times when some advanced mathematics is necessary. This, however, does not create any major problems because most design engineers will be capable of handling the mathematics. There are extensive bibliographies at the end of each chapter, but the index is very brief. Both authors are environmental engineers in Illinois universities. The book is highly recommended for university and special libraries.   *H. ROBERT MALINOWSKY*

514. **Industrial Refrigeration.** W. F. Stoecker. Troy, MI: Business News Publishing Co., 1988. 386p., illus., bibliog., index. $39.95. 0-912524-42-1.

In the field of refrigeration, there is a considerable body of literature on air conditioning; but in the field of industrial engineering, individual publications, especially recent publications, are scarce. This book does a very creditable job in helping to close that gap. It covers the topic in twelve chapters from the general in chapter 1, "The Field of Industrial Refrigeration," through the specific such as chapter 6, "Evaporators-Air Coils and Liquid Chillers."

It is not a do-it-yourself book on installation and maintenance of refrigeration systems but focuses, instead, on principles of operation and applications to provide an understanding of different refrigeration systems. Written for the individual interested in design, engineering, sales, etc., of refrigeration systems, this book should provide a solid understanding of the industrial refrigeration field. Special libraries as well as trade school libraries will want a copy of this handbook.   *RAYMOND BOHLING*

515. **Piping and Pipe Support Systems: Design and Engineering.** Paul R. Smith; Thomas J. Van Laan. New York, NY: McGraw-Hill Book Co., 1987. 334p., illus., index. $44.95. 0-07-058931-3.

The purpose of this handbook is to help the reader in the design and analysis of piping systems. It includes applications in power, industrial, and chemical fields. The first three chapters, "Piping Systems and Power Plant Evolution," "Codes, Standards, and Regulations," and "Technical Piping Documentation," serve as an introduction. The chapter on codes, standards, and regulations is especially important in the power industry. Chapter 4, "Overview of Pipe Stress Requirements," provides additional code information for power, chemical, petroleum refinery, nu-

clear power, and gas transmission and distribution. Chapter 5, "Piping Design Loads," presents the facts and figures on how to calculate the various loads that may be encountered, and the final four chapters cover the actual details of piping and support design: "Pipe Support Hardware," "Piping Support Design Process," "Manual Calculation Methods," and "Computer Applications for Design and Analysis."

The text is concise and includes numerous charts, graphs, schematics, diagrams, and illustrations. The authors have written this book based on a set of lecture notes for a piping design and analysis workshop. This makes it an even more practical and up-to-date presentation. Piping engineers will find this a valuable handbook, and libraries will want it for their reference collections in industrial engineering.

H. ROBERT MALINOWSKY

★
MANUFACTURING
Catalogs

516. **Handloader's Digest.** 11th. Ken Warner, ed. Northbook, IL: DBI Books, Inc., 1987. 352p., illus., bibliog., index. $15.95pbk. 0-87349-009-6pbk.

For the serious firearms owner, handloading one's own cartridges is a way of life. Improvements on methods and equipment are constantly being made, resulting in an ever-changing dialogue between one shooter and another. This 11th edition of the *Handloader's Digest* has 31 signed articles covering actual experiences, testing, new methods, and descriptions of handloading with clear black-and-white illustrations. The second part of this comprehensive manual is a directory/catalog of anything and everything that a handloader may need: "Tools and Accessories for Metallic Cartridges" (part 1), "Tools and Accessories for Shotshells" (part 2), "Tools and Accessories for Bullet Swagging" (part 3), "Tools and Accessories for Bullet Castings" (part 4), "Components" (part 5). A sixth part lists books, manuals, videos, computer software, chronographs, boxes, and labels. This is a must for any serious shooter's library and a good reference work for public and special libraries interested in ballistics.

H. ROBERT MALINOWSKY

Directories

517. **Best's Safety Directory: Industrial Safety, Hygiene, Security: Featuring OSHA Summaries, OSHA Self-Inspection Checklists, Safety Guidelines, Training Series, Monthly Safety Training Topics, Technology Series, Product Applications, Purchase Sources.** Maryann M. Grace; Linda L. Bell,

eds. Oldwick, NJ: A. M. Best Co., 1988. 2v. in 1,836p., illus., index. $25.00. 0090-7480.

This directory in two volumes is organized into six main areas: "Administration," "Apparel," "Noise," "Operational Safety," "Pre-operational Plant Safety," and "Security." Areas are further subdivided when appropriate. For example, "Operational Safety" is divided into three chapters: machine guards, process guidelines, tool handling; ordinary materials handling; and hazardous materials handling. OSHA standards and requirements are included in each chapter; articles, under the title of "Training Series," and short summaries, under the heading of "Training Topics," are also presented. Additionally there is a buyers' guide of product applications and sources.

The OSHA requirements are a necessary inclusion in any safety directory and are standard. The "Training Series" articles are informative and well written and cover a topic relevant to that particular chapter. They are available as reprints for companies wishing to use them for education. The training topics immediately following the training series article summarize the article, with the important ideas in bullet form. These are also available as reprints and can be used as reminders or in groups to stimulate discussion on a particular topic. Following each section of guidelines and training series is the buyers' guide. Products are listed alphabetically and described in reasonable detail. Major sources of the product are then listed, including mailing address and phone number. These are followed by minor sources where only the company name is listed. Their addresses are in the index at the end of volume 2. Ads generally range from a quarter page to a full page and may be in color. The pages of the book have chapter indicators on the outside edges, so that finding a chapter is made easier. The index of products and index of advertisers are both easy to use. This book is a must for O & M people in all areas and for special libraries as well as large public and academic libraries.

ANNE PIERCE

Encyclopedias

518. **Great Combat Handguns: A Guide to Using, Collecting and Training with Handguns.** Leroy Thompson; Rene Smeets. John Walter, ed. Poole, Great Britain: Blandford Press, Dist. by Sterling Publishing Co., New York, NY, 1987. 224p., illus., index. $24.95. 0-7127-1444-4.

"The guns evaluated for this book include some classics, dating back to the turn of the century, and many brand-new designs. Each has been chosen because it offers features making it specially attractive for combat use, or is so widely used by police, civilians and military agencies that it could not be overlooked." This is an encyclopedia of handguns for those interested in combat handguns. The first chapter is a history of handguns beginning with Europe up to the modern era. The second chapter is a short discussion

on handgun shooting, and the third chapter helps in choosing a handgun. The text in these first chapters is interestingly written and illustrated with many good photographs and pictures.

The bulk of the book is "The Directory," a listing of 50 recommended handguns from around the world. For each gun, a descriptive list gives type, maker, caliber, overall length, barrel length, weight, cylinder capacity, and construction. (This information is boxed at the beginning of the discussion.) Following is a narrative that discusses the history, good points, and uses of the handgun. There are over 300 photographs, drawings, and exploded diagrams of the various handguns.

Some individuals will have mixed feelings about this book because it makes combat handguns look inviting to the casual reader. It should be emphasized that these guns are for combat use and are not intended for the normal recreational use. It contains, however, an excellent discussion of handguns, and for the collector of such guns this book would be invaluable.

H. ROBERT MALINOWSKY

519. **The Luger Book, Luger: The Encyclopedia of the Borchardt and Borchardt-Luger Handguns, 1885–1985.** John Walter. London, Great Britain: Arms and Armour Press, Dist. by Sterling Publishing Co., New York, NY, 1986. 287p., illus., bibliog., glossary. $40.00. 0-85368-886-9.

Much new information has been brought to light since a former book, *Luger,* was published in 1977. Completely rewritten, *The Luger Book* is a synthesis of the old 'all-in-one volume' approach with the most important new material." A bibliography, glossary, and nomenclature table precede the introduction, which is actually a history of the luger covering "From Maxim to Borchardt," "The Perfection of Automaticity," "The Early Automatics," "From Borchardt to Luger," "The Borchardt-Luger and the Military," "The German Trials and Adoptions," "The First World War," "The Armistice, Versailles and beyond," "Mauser and the Third Reich," "The Second World War," and "The Replacement of the Pistol."

The main part of the book is an alphabetically arranged section covering all aspects of the luger, including names, countries, personalities, manufacturers, and individual parts. Excellent photographs and sketches are included. There is a great amount of effort to include as much biographical information as possible, as well as specific information on individual models of the luger. Because the luger was so prominent in Germany, there is a great amount of information relating to its use during the wars, much of this being translated from German documents.

Although this is an encyclopedia, it is actually a history of the luger tracing its development from the early days of the Borchardt to the last Mauser-Parabellum. This is a highly recommended book for anyone who is collecting lugers or has an interest in this type of handgun. With very

accurate information and excellent illustrations, this is a good one-place location for information on manufacturers, distributors, guns, ammunition, leatherware, accessories, proof and inspection, and unit marks.

H. ROBERT MALINOWSKY

Field Books/Guides

520. **Robotics Careers.** Ellen Thro. New York, NY: Franklin Watts, 1987. 111p., illus., bibliog., index. $11.90. High-Tech Careers. 0-531-10425-7.

Robotics Careers covers one of the faster growing careers in the world today. It details job opportunities, job descriptions, educational requirements, salaries, promotion possibilities, college and training programs, and other information about the industry. The 11 chapters cover "What Is Robotics?," "Opportunities in the Field," "Designing New Robots," "Hiring Robots for a Factory," "Planning the Use of Robots," "Programming Robots," "Servicing Robots," "Selling Robots," "Inventing Robots of the Future," "Running Your Own Business," and "Robotics in Oceanography and Space Science."

This is an excellent survey of the field and a good reference source for any school or public library career section.

H. ROBERT MALINOWSKY

Handbooks

521. **Bulk Materials Handling Handbook.** Jacob Fruchtbaum. New York, NY: Van Nostrand Reinhold Co., 1988. 488p., illus., bibliog., index. $79.95. 0-442-22684-5.

This handbook orginally was written as a training manual for the staff of the author's firm. Both the content and the nomenclature of the present book have been updated and revised in consultation with specialists in the field. The author, Jacob Fruchtbaum, has 60 years of experience as a consulting engineer, has authored other books in the area of industrial engineering, and has lectured at Syracuse University. Because the curricula of most engineering schools and colleges do not cover the topic of bulk materials-handling, the book was designed as a training manual for practicing engineers and as a reference tool for engineering students and others who need information on this topic. It will be of special interest to industrial and chemical engineers.

The volume emphasizes the design aspects of equipment used for bulk materials handling systems and for dealing with problems related to bulk materials handling, e.g., dust collection, explosives, and ventilation. In addition, the chapters discuss the various types of conveyors, drives, crushers and screens, hoists, car shakers, gates, chutes, and spouts. One chapter covers the handling of specific materials. The list of 35 separate materials discussed in this chapter is not

intended to be all-inclusive but is instead a guide for similar products. The coverage of the materials included ranges from flour and rice to ashes, pitch, and zirconium. The final chapter discusses the layout of a materials-handling plant.

There are approximately 300 illustrations throughout the text. These are line drawings of "successful" installations of materials handling equipment complete with specifications. A separate section of the table of contents lists all of the illustrations by chapter and page. Also, there are approximately 140 tables throughout the text. These are listed in a third section of the table of contents. In addition to the table of contents, the volume is well indexed with references to materials found in the figures and tables, but the index does not indicate that the information is not in the text. The volume has a few references to additional sources of information scattered throughout the text but does not have a bibliography or lists of references at the ends of the chapters. Because this is a field with little published information, a bibliography of available sources would be useful. The book is recommended for industrial and chemical engineers and engineering library collections in colleges and industries.   *LUCILLE M. WERT*

522. Computer-Integrated Manufacturing Handbook. Eric Teicholz; Joel N. Orr. New York, NY: McGraw-Hill Book Co., 1987. 466p., illus., bibliog., index. $59.95. The McGraw-Hill Designing with Systems Series. 0-07-047774-4.

Computer-integrated manufacturing (CIM) is defined as the "complete automation of the factory, with all the processes functioning under computer control and only digital information tying them together." As the authors note, CIM will not be common before the late 1990s, and only isolated pockets of factory automation exist now. The *Computer-Integrated Manufacturing Handbook* is primarily aimed at manufacturers considering implementation of CIM.

This book is divided into four parts. The introductory section provides an overview of the industry, including case studies, future trends, and a review of the obstacles to implementation. The second section deals with the various component facets of CIM, such as CAD/CAM systems, group technology, process planning, numerical control systems, robotics, inventory control, and communications. The third part of the book outlines the planning process for computer-integrated manufacturing, and the final section deals with the implementation and management of the CIM environment.

Chapters are individually authored, and only a few include brief bibliographies; the only chapter with an extensive bibliography is the one on the economics of computer-integrated manufacturing. An index is included. The 21 authors are drawn primarily from industrial and consulting backgrounds, though a couple are academics. The two editors, Eric Teicholz and Joel Orr, are both presidents of consulting firms which specialize in

computer-assisted design and drafting (CADD) and in industrial automation, respectively. Recommended for academic and special libraries.
                    CAMILLE WANAT

523. Flayderman's Guide to Antique American Firearms and Their Values. 4th. Norm Flayderman. Northbrook, IL: DBI Books, Inc., 1987. 624p., illus., bibliog., index. $22.95pbk. 0-87349-016-9pbk.

The cover of this very comprehensive handbook states its claim to be "The Complete Handbook of American Gun Collecting." Norm Flayderman is the world's best-known antique arms dealer and authority on historic weapons of all types. This handbook lists over 3,400 firearms. The first three chapters cover collecting, values and condition, restoration and fakes, and the arms library. The rest of the book is divided by type of firearm with each entry fully described, giving date manufactured, number produced, pertinent information on calibers, serial numbers, and special makes plus values in dollars. The 13 chapters cover "Major American Manufacturers," "American Military Single Shot Pistols," "American Percussion Pistols," "American Metallic Cartridge Pistols," "American Military Longarms," "Confederate Firearms," "Kentucky Rifles and Pistols," "Percussion Sporting and Target Rifles, Plains Rifle, Flintlocks, New England Rifles," "Lever Action and Other Repeating Rifles," "Single Shot Breech-Loading Cartridge Rifles," "Revolving Rifles," "American Colonial and Revolutionary War Firearms," and "American Shotguns and Fowling Pieces." There are good black-and-white photographs for over 1,400 entries. This is an excellent book for the collector and a good reference book for public libraries interested in firearms.   *H. ROBERT MALINOWSKY*

524. Gun Digest, 1988. 42nd. Ken Warner, ed. Northbrook, IL: DBI Books, Inc., 1987. 480p., illus., bibliog., index. $16.95pbk. 0-87349-010-Xpbk.

This annual is by far "The World's Greatest Gun Book." In addition to being a catalog of over 1,200 different guns and accessories available on the market, it contains a digest of 44 articles written by gun enthusiasts, specialists, and historians. The articles cover a wide range of topics including "The Lever and Why It Lasted," "A Second Look at the Glock 17," "Handguns Today: Sixguns and Others," "Rifle Review," "Coming to Terms with a Muzzle-Loading Deer Rifle," "Black Powder Review," "Getting Hooked on Air-Soft Guns," "Shotgun Review," "Scopes and Mounts," "The Art of the Engraver," "Custom Guns," and "Black Powder through the Centuries."

The main part of this catalog is the "Guns Illustrated Catalog," which contains over 1,200 entries of all the current guns and accessories that are on the market. Each entry includes full name, caliber, barrel, weight, stocks, sights, features, and price, plus for most entries, a black-

and-white photograph. The final portion of the book is a manufacturers' directory of every possible item associated with firearms, from ammunition, to decoys, to guns, to sights, to stocks.

This annual is a must for any gun enthusiast as a source of information on current guns and accessories that are available on the market. Because it is paperbound, it will become somewhat torn through heavy use, and libraries may want to have it permabound. However, this is an annual, so this may not be necessary. Public and special libraries with an interest in firearms will need this publication as a standard reference source. *H. ROBERT MALINOWSKY*

525. **Guns Illustrated, 1988.** 20th. Harold A. Murtz, ed. Northbrook, IL: DBI Books, Inc., 1987. 320p., illus., bibliog., index. $14.95pbk. 0-87349-011-8pbk.

Guns Illustrated contains 20 separately authored articles on many different aspects of firearms, including "Guns that Shoot Electronic Bullets," "The 22 Winchester Magnum Rimfire," "The Revolutionary Sauer Model 200," "Famous Guns of the Old West—The Derringer," "Sniper Garand," "Elbit Falcon Optical Gunsight," and "The Indestructible 25-06." Each article is written by a specialist or historian and includes many black-and-white illustrations. The main part of this catalog is the "Guns Illustrated Catalog," which contains over 1,200 entries of all the current guns and accessories that are on the market. Each entry includes full name, caliber, barrel, weight, stocks, sights, features, and price, plus for most entries, a black-and-white photograph. The final portion of the book is a manufacturers' directory of every possible item associated with firearms, from ammunition to decoys to guns to sights to stocks.

This publication is very similar to the *Gun Digest*, an annual by the same publisher. The main difference is that the *Gun Digest* has twice as many articles. The catalog portions are identical as are the directory sections. However, even though these sections are identical, the gun enthusiasts will want both books for the articles. Public libraries will want both for the specialized information included in the articles, but if budgets will not allow both, *Gun Digest* would be the recommended choice. *H. ROBERT MALINOWSKY*

526. **Handbook of Product Design for Manufacturing: A Practical Guide to Low-Cost Production.** James G. Bralla, ed. New York, NY: McGraw-Hill Book Co., 1986. 1v. various paging, illus., index. $98.00. McGraw-Hill Handbooks in Mechanical and Industrial Engineering. 0-07-007130-6.

"The prime objective of the *Handbook* is to aid those involved in the manufacture of commercial products to design them to be easily made...to enable designers to take advantage of all the inherent cost benefits available in the manufacturing process which will be used." The subject matter of this handbook includes both product engineering and manufacturing engineering information. It is a complete summary of all of the workings and capabilities of the various significant manufacturing processes. To this end, the design engineer will find useful information on design, and the manufacturing engineer will find information on tooling, equipment, operation sequence, and other technical data.

There are 54 contributors from both industry and academia. The editor, James G. Bralls, has over 30 years' experience in manufacturing as a line manager, consultant, and industrial and project engineer and is currently vice-president for manufacturing of Alpha Metals, Inc. These experts have produced a handbook that brings together information that has been scattered in many sources, and some which has never been in print. It is divided into eight sections. The first is an introduction that gives background on the purpose, contents, and use of the book plus a chapter on "Economics of Process Selection." Section 2, "Economical Use of Raw Materials," covers hot-rolled steel, cold-finished steel, stainless steel, aluminum, copper, brass, magnesium, and other ferrous and nonferrous metals. Section 3, "Formed Metal Components," covers such topics as metal extrusions, metal stampings, spun-metal parts, rotary-swaged parts, and electroformed parts. Section 4, "Machined Components," includes detailed information on the machined production of the various products. The remaining sections cover "Castings," "Nonmetallic Parts," "Assemblies," and "Finishes."

Each chapter begins with a summary of how each manufacturing process produces its result and usually includes a schematic representation of the operation plus photographs or drawings. It "tells readers how large, small, thick, thin, hard, soft, simple, or intricate the typical part will be, what it looks like, and what material it is apt to be made from." Within the discussion are tables of various sorts to help in the design process. Finally, each chapter contains recommendations for more economical product design. The handbook is useful at various levels of design from the new product to the retooling or replacing of current tooling. The text is well written, but the typeface is somewhat small. The schematics and line drawings are excellent, but the photographs tend to be dark and in some cases too small to see the detail that is being illustrated. The detailed index is well organized and makes accessing the information quick and easy.

This is a must reference book for any university or special library that serves people involved in manufacturing and design. Its greatest use, however, will be in the manufacturing areas, where it will be used by product designers and manufacturing engineers as well as by operation sheet writers, drafters, and value, tool, process, production, cost reduction, research and development, and industrial engineers. *H. ROBERT MALINOWSKY*

527. **Handbook of Reliability Engineering and Management.** W. Grant Ireson; Clyde F. Coombs, Jr, eds. New York,

NY: McGraw-Hill Book Co, 1988. 1v.
various paging, illus., bibliog., index. $59.50.
0-07-032039-X.

This revised edition of the 1966 *Reliability Handbook* has been reorganized, and the 15 authors of the chapters are mostly new and all have impressive credentials from a lengthy list of major companies. W. Gran Ireson, an editor for both editions, is a professor emeritus at Stanford University. His coeditor for this book is Clyde F. Coombs, Jr., who is associated with Hewlett-Packard.

Product reliability has become an important concern of manufacturing companies in recent years, and this handbook provides guidelines for achieving maximum product integrity along with cost-effective management. There are four major sections, covering the introduction, management, engineering, and mathematics of reliability. Nineteen chapters include such aspects as cost accounting, collecting reliability information, production problems, product testing, designing for people, failure analysis, and software reliability. There is a good balance of the many aspects, from planning and management to applying statistical analysis to producing a reliable product that will achieve acceptable performance, will survive a minimum length of time, and can be tested to determine the success of the manufactured item. The detailed organization presented in the table of contents, the section headings in the chapters, and the index provide good access to the subject material when particular information is needed. A breif bibliography concludes each chapter.

This up-to-date handbook will be a valuable resource for the engineer and managers of companies that produce products which must compete in the marketplace. Recommended for the technical collection, engineering library, and the manufacturing community.   *JANICE SIEBURTH*

528. **Handgun Digest.** Dean A. Grennell. Northbrook, IL: DBI Books, Inc., 1987. 256p., illus., bibliog., index. $12.95pbk. 0-87349-013-4pbk.

This handbook, for those interested in handguns, contains a wealth of interesting and useful information. It covers equipment, law enforcement, hunting, cartridges, competition, and reloading in 21 review articles—"Handguns, an Overview," "Handgun Safety," "Handguns in Competition," "Handgun Big Game Hunting," "Handgun Collecting," "Handguns for Law Enforcement," "Ancestry and Development of the Detective Special," "The Glaser's a Hell-Raiser," "Birth of a Wildcat," "Handgun Cartridges," "Homebrewed Handgun Ballistics," "Basic Handgun Care and Maintenance," "Handgun Bulletmaking," "Reloading Handgun Ammunition," "Handgun Holsters," "FBI Firearms Training," "The Handgun Hunter's Contender," "The Last ASP," "Instructing Beginning Handgunners," "Test Firing Techniques and Equipment," and "I Can't Collect Handguns." A manufacturing directory lists the source for everything a handgun enthusiast would need. This is a

paperbound book, so especially heavy use may be a problem unless some type of permabound cover is used. Public and special libraries who serve those interested in firearms will want this handbook, as will any handgun enthusiast.

H. ROBERT MALINOWSKY

529. **Maintenance Engineering Handbook.** 4th. Lindley R. Higgins, ed. New York, NY: McGraw-Hill Book Co., 1988. 1v. various paging, illus., bibliog., index. $82.50. 0-07-028766-X.

It has been 33 years since the first edition of this well-accepted handbook appeared. During that time, there have been a great many changes in the field of maintenance engineering, due to the advances in technology. "Maintenance has grown from an activity that could not even be graced by the title of an art into a precise, technical engineering science." The editor hopes that this special handbook will "pass along invention, ingenuity, and a large dose of pure basic science." This he has done in an excellent manner with the help of 63 contributing authors who have written 61 chapters in 12 sections. These 12 sections divide the book into the separate areas of maintenance: "Organization and Management of the Maintenance Function," "Establishing the Costs and Controls of Maintenance," "Applying the Computer to Maintenance Management and Control," "Maintenance of Plant Facilities," "Sanitation and Housekeeping," "Maintenance of Mechanical Equipment," "Maintenance of Electrical Equipment," "Maintenance of Service Equipment," "Lubrication," "Instruments and Vibration," "Maintenance Welding," and "Chemical Corrosion Control and Cleaning."

The text is well written and keeps the engineer and the goal of excellent maintenance in mind. Each chapter has a general overview of the chapter followed with a systematic discussion of how the maintenance should be performed. For example, the chapter on electric motors begins with a general introduction that points out what is needed in a maintenance plan. This is followed by precise and detailed information on various types of motors, including design characteristics, application data, and special types. A scheduled routine motor care maintenance program is outlined covering such areas as lubrication, heat, noise, vibration, winding insulation, brushes and commutators, and collector rings. Finally, a series of flow diagrams pinpoints 14 troubleshooting areas. The text is supplemented with numerous charts, tables, graphs, sketches, and photographs. Many of the photographs, however, are dark and hard to see. For the maintenance engineer, however, they should be adequate. There is a minimum of highly technical mathematics.

Lindley R. Higgins has done an excellent job in bringing all of this information together. This latest edition contains 40 percent completely new material and another 30 percent revised material. Higgins is a registered professional engineer and was for many years a plant maintenance engineer at the Colgate-Palmolive Company, where he designed maintenance-free machinery and equipment. Every maintenance engineer should have a

copy of this handbook on hand at all times. Academic and special libraries will need a copy for reference purposes.

H. ROBERT MALINOWSKY

530. **Performance Testing of Shipping Containers.** 2nd. ASTM Committee D-10 on Packaging. Philadelphia, PA: American Society for Testing and Materials, 1987. 255p., illus., bibliog., index. $19.00. 0-8031-0463-4.

Through the work of its various committees, the American Society for Testing and Materials (ASTM) develops and issues voluntary and agreed-upon standards for products, systems, materials, and services. ASTM Practice D4169, "Performance Testing of Shipping Containers and Systems," was developed by ASTM's Subcommittee on Shipping Container Environment (a subgroup of the Committee on Packaging). Now in its second edition, this volume reprints this standard practice D4169 from the *Annual Book of ASTM Standards*, along with related standards, documents, and references. *Performance Testing of Shipping Containers* consists of two parts.

The first section is a reprinting of the ASTM standard (D4169) on performance testing of shipping containers, as well as all related ASTM documents on topics such as conditioning containers for testing, terminology for packaging, and various testing methods used for boxes, bags, crates, and other containers. These test methods include compression, drop, impact, and vibration tests. The second part of the collection lists all the government and industry (i.e., non-ASTM) documents cited in D4169. About half of these documents are reprinted in this volume; complete references are given for the remainder (full abstracts and tables of contents are provided for half of the documents not reprinted here). An index covering only the reprinted ASTM standards in the book is included. For special libraries involved in container performance testing, this volume is a handy and compact reference.

CAMILLE WANAT

531. **Plastics Mold Engineering Handbook.** 4th. J. Harry DuBois; Wayne I. Pribble, eds. New York, NY: Van Nostrand Reinhold Co., 1987. 736p., illus., bibliog., index. $59.95. 0-442-21897-4.

The third edition of this handbook was published in 1978. The basic principles of plastics molding have not changed much since that edition, and as a result, this fourth edition has little revision in the first six chapters; these cover "Introduction to Plastics Processing," "Basic Mold Types and Features," "Tool-Making Processes, Equipment and Methods," "Materials for Mold Making," "Design, Drafting and Engineering Practice," and "Compression Molds." The remaining eight chapters have been completely revised, and one, "Blow Mold Construction and Design," is new to this edition. The remaining chapters cover "Transfer and Injection Molds for

Thermosets," "Injection Molds for Thermoplastics," "Cold Mold Design," "Extrusion Dies and Tools for Thermoplastics," "Molds for Reaction Injection, Structural Foam and Expandable Styrene Molding," "Care, Maintenance and Repair of Plastics Molds," and "Thermodynamic Analysis of Molds."

Each chapter is authored by a specialist in a particular field of plastics molding. The chapters are well written with good photographs, excellent schematics, and line drawings. Most of the illustrations are adjacent to the relevant text. The text is intended to give the user all of the information needed to effect a good design for a plastics mold. There are explicit instructions on the use of the materials, and problems encountered in designing plastics molds are discussed. As a safety measure, extensive checklists are given so that potential hazards, weaknesses, and misunderstandings can be foreseen. DuBois published the first edition in 1946 with the idea that plastics molding was a science as well as an art. Before his death in 1986, he advised Pribble one what should be included in this edition. Pribble, a top consultant in the plastics industry, received the 1985 Designer of the Year Award from the Society of Plastics Engineers. This is a recommended book for all university engineering libraries and for special libraries that support the field.

H. ROBERT MALINOWSKY

532. **Production and Inventory Control Handbook.** 2nd. James H. Greene, ed. New York, NY: McGraw-Hill Book Co., 1987. 1v. various paging, illus., bibliog., index. $74.95. 0-07-024321-1.

This handbook, prepared under the supervision of the American Production and Inventory Control Society (APICS), is the second edition of a work originally published in 1970. It is a major work in the field, presenting 31 signed chapters by over 85 contributors in nine major sections. The handbook presents current thinking on the subject of production and control: production and inventory management; strategic business planning; information requirements; tactical planning and control; manufacturing information systems; warehouse and distribution operations; just-in-time concepts and applications; managing the function; and techniques and models. References conclude each chapter. This edition presents the "big picture" integration of manufacturing functions, called the "closed loop" concept of manufacturing, MRP II. The chapters move from strategic planning and predictions about human resource and facility requirements through scheduling and capacity plans for the factory itself, to the manufacturing process, to distribution. New areas of this second edition include CAD/CAM, quality control, and efficiency techniques such as quality circles, and kanban, or just-in-time methods. This edition also adds major new sections on strategic planning, master production scheduling, capacity planning, robots, and computers. This handbook would be useful for the engineer in training, as well as for the practicing manufactur-

ing engineer or supervisor. It is recommended for anyone in preparation for APICS certification and special or academic libraries serving those individuals. *CHARLENE M. BALDWIN*

533. Productivity and Quality Improvement: A Practical Guide to Implementing Statistical Process Control. John L. Hradesky. New York, NY: McGraw-Hill Book Co.., 1988, 1988. 243p., illus., bibliog., glossary, index. $39.95. 0-07-030499-8.

Productivity and quality are key words in industry today as companies strive to compete in world markets. This handbook presents a plan that will enable a company to examine its organization and processes, analyze each step and quantify it, then decide how to improve productivity and yet keep costs under control. The author is president of a consulting company and has had 25 years of experience in various major companies.

The 12 chapters are each devoted to one of the 12 steps for PQI, the Productivity and Quality Improvement process. A major objective is to provide understanding of the statistical process control for management and staff. Starting with an examination of a problem area and the development of a goal for improvement, a plan is formulated, with a timetable. There are specific performance measurements, evaluation techniques, assignment of responsibilities for each step, as well as anticipated problems, and the establishment of accountability.

A key to success is the support of management and developing a spirit of enthusiasm for change. Participants must work together in teams, and strong leadership is required. Each project is broken down into very specific steps which can be described in quantifiable terms, and goals are in percent gains or improvement. Flow charts and figures illustrate processes and the effects of changes. Sample forms assist in planning and the collection of data, and worksheets show the use of statistical analysis and assist in the calculation of potential errors, a percent tolerance or a range in variation of measurements. Numerical criteria are developed as specific steps are analyzed in an industrial process. Calculations are clearly illustrated with examples and explanations. A glossary is included, as are a brief bibliography of seven books published between 1979 and 1984 and an index. Recommended for a company library and for academic business and engineering collections. *JANICE SIEBURTH*

534. Robot Design Handbook: SRI International. Gerry B. Andeen, ed. New York, NY: McGraw-Hill Book Co., 1988. 384p., illus., bibliog., index. $42.50. 0-07-060777-X.

The *Robot Design Handbook* is an updated, revised version of *Robot Handbook* which was published privately by SRI in 1983. SRI international, formerly Stanford Research Institute, is a basic and applied research facility headquartered in Menlo Park, California. The editor, Gerry B. Andeen of SRI International's Mechanical Research Department, is a robotic designer. The emphasis of this book is on robotic manipulators, i.e., "devices that are attached to foundations (which may be mobile) and reach out to manipulate objects." According to the introduction, the book's threefold purpose is to provide basic knowledge for manipulator development, promote mutual understanding among participants on a robot design team, and highlight areas for development and future research. Following the introduction, there are chapters on design, tools, components, mathematics, and research. Throughout the book are many excellent illustrations. A bibliography and index follow the work's 15 chapters. Written for robotic engineers and industrial managers, *Robot Design Handbook* is not a comprehensive reference source for robotics information. It would, however, be a valuable addition to a university library's engineering collection. *SUE ANN M. JOHNSON*

535. Selected ASTM Standards on Packaging. 2nd. Philadelphia, PA: American Society for Testing and Materials, 1987. 217p., illus. $39.20pbk. "PCN:03-410187-11." 0-8031-0964-4.

The material in this handbook has been reprinted from the *Annual Book of ASTM Standards*, the accepted source of information on standards set by ASTM. It includes all of the standards that relate to packaging: C 149, D 528, D 585, D 642, D 644, D 685, D 774, D 775, D 828, D 880, D 882, D 895, D 959, D 996, D 997, D 999, D 1185, D 1251, D 1596, D 2176, D 2221, D 2808, D 2911, D 3078, D 3079, D 3199, D 3332, D 3580, D 4003, D 4168, D 4169, E 96, and E 685. These standards cover everything from thermal shock test on glass containers to bursting strength of paper, drop test for loaded boxes, drop test for filled bags, and folding endurance of paper. Every type of packaging product is included with the corresponding recommended ASTM standards for durability and quality. Discussion of each standard typically includes sections on scope, applicable documents, significance and use, definitions, apparatus, test specimens, conditioning, number of test specimens, speed of testing, procedure, calculations, report, and precision and bias. There are numerous charts, diagrams, and calculations which help in describing the standard. Standard D 996 is a 13-page glossary called "Standard Terminology of Packaging and Distribution Environments." Any library that serves manufacturing industries involved in packaging will want this book.

H. ROBERT MALINOWSKY

536. The Warehouse Management Handbook. James A. Tompkins; Jerry D. Smith, eds. New York, NY: McGraw-Hill Book Co., 1988. 702p., illus., bibliog., index. $69.95. 0-07-064952-9.

The Warehouse Management Handbook is a comprehensive source of information on warehousing, consisting of 29 chapters written by 36 contributors representing a variety of businesses. It is divided into four parts: "Introduction to Warehousing," "Warehouse Planning," "Warehouse Equipment," and "Functions of a Warehouse Operating System." There is an index. Many chapters have a bibliography, and there are extensive illustrations throughout.

The editors are James A. Tompkins, founder and president of Tompkins Associates, Inc., and Jerry D. Smith, executive vice-president of the same consulting firm, which specializes in facilities planning, materials handling, warehousing, and integrated automation. In the mid-1970s, these men set out to make warehousing a profession. Although their first efforts were not met with enthusiasm, interest in warehousing information is now in demand. Warehousing consulting, publications, and conferences have all proliferated since these editors began their crusade. This book would be useful for libraries with business or industrial engineering collections.

SUE ANN M. JOHNSON

Manuals

537. Gunsmithing with Simple Hand Tools: A Basic Manual for the Advanced Amateur on Use of Hand Tools for Minor Alterations, Improvements, and Reconstruction of Firearms. A. D. Dubino. Harrisburg, PA: Stackpole Books, 1987. 205p., illus., index. $19.95. 0-8117-0784-9.

Gunsmithing with Simple Hand Tools is a well-written manual for those who desire to make minor alterations, improvements, and reconstruction of firearms that are currently owned. The author stresses that anything that can be done with the massive machine tools and mass production techniques can be done with simple hand tools—files, hacksaws, drills, hammers, and chisels.

The detailed text covers a variety of techniques and applications that the typical gun owner can handle. After an introductory chapter, five chapters cover basic hand tools, the workshop, metals, the file, and abrasives. These chapters give the background information that is needed to carry out the applications in the following chapters; these cover polishing the gun in preparation for blueing, blueing the gun, small parts and miscellaneous work, action tune-up, machined trigger guard, side lock, lock plate, tumbler, bridle, sear, bridle and sear, and lock completion. Clear line drawings are used throughout the book, as are some adequate black-and-white photographs. There is a brief index.

This is a must book for any gun owner, regardless of whether or not the actual techniques are attempted. Public libraries will want a copy for their circulating reference collection.

H. ROBERT MALINOWSKY

538. Injection Molds and Molding: A Practical Manual. 2nd. Joseph B. Dym. New York, NY: Van Nostrand Reinhold Co., 1987. 395p., illus. (part col.), index. $44.95. 0-442-21785-4.

In the introduction to this second edition, the author cites 50 years of practical experience that provided the background of knowledge for this book. His thorough knowledge of mold design and operation in the manufacture of plastic products is evident in the detail and clarity of the text and the step-by-step explanation of mold making and the molding process. The book covers the subject of molding of plastics, from basic design of molds, including the nature of plastics, to troubleshooting of possible production problems in the operation of injection molding equipment. Thermoplastic molding is emphasized because the greatest activity in the moldings field occurs in this area. Illustrations are clear and of sufficient number to clarify key details in equipment design and use. This book will be of value to anyone needing information on mold making and plastics molding, particularly the designer and engineer directly involved in plastics molding processes.

RAYMOND BOHLING

★

MECHANICAL ENGINEERING
Abstracts

539. Elevator Abstracts: Including Escalators. G. C. Barney, ed. Chichester, Great Britain: Ellis Horwood, Ltd.. for International Association of Elevator Engineers, Dist. by John Wiley and Sons, New York, NY, 1987. 178p., index. $69.95. Ellis Horwood Series in Transportation. 0-7458-0180-3.

This reference work covers over 25 years of articles, books, standards, patents, papers, and dissertations covering the field of international research on elevators and escalators. Most entries include a brief bibliographic citation and an abstract. The primary arrangement of the material is by individual author, although some corporate entries are included. An additional author and subject keyword index are located in the back of the volume. Each of the about 80 index terms is tagged with the number of the appropriate abstract. This allows the reader to correlate an author with a specific subject.

It is important to note, though, that at the end of the subject keyword index is a list of abstracts not indexed by subject. Access to these abstracts is only by the primary arrangement of authors. The abstracts vary in length. Although several of the bibliographic citations include the country of origin, it would also have been helpful to have included a list of the complete titles of the cited material. This is critical when such a

broad range of materials is covered. Academic and special libraries will find this a useful reference source on a subject in which information is widely dispersed. *CHARLES R. LORD*

Dictionaries

540. Boiler Operator's Dictionary: A Quick Reference of Boiler Operation Terminology. Phil Roman, ed. Lionel Edward LaRocque, comp. Troy, MI: Business News Publishing Co., 1988. 149p. $6.96pbk. 0-912524-41-3pbk.

The *Boiler Operator's Dictionary* is a handy pocket guide to boiler operation. This small dictionary conveniently fits into the shirt pocket for quick and easy reference and is, therefore, a dictionary for the practicing engineer. The definitions are clear, concise, and are written in easy-to-understand sentences. Where there are more than one definition associated with a term; depending on a preceding adjective, the author bullets each individual definition. When there are synonyms, the author refers the reader to one main term, however, this is minimal. Because of the concise nature of the definitions, the author could have used the same definition with a cross-reference to the second term to avoid having to look up the other term. Generally, this is a very useful dictionary for the engineer and would have some value on the reference shelves of a special library. *JOHN NISHA*

Guides to the Literature

541. Selective Guide to Literature on Mechanical Engineering. Hugh Lockwood Franklin, comp. Washington, DC: Engineering Libraries Division, American Society for Engineering Education, 1985. 22p. $5.00pbk. Engineering Literature Guides, no. 2. 0-87823-102-1pbk.

This bibliography cites print and nonprint sources of information on the subject of mechanical engineering. About 150 unnumbered sources are arranged by type. There are 10 sections: "Guides to the Literature"; "Bibliographies"; "Indexing and Abstracting Services"; "Computers and Databases"; "Dictionaries and Encyclopedias"; "Yearbooks and Review Series"; "Handbooks and Manuals"; "Directories of Organizations, Manufacturers, and Products"; "Biographical Sources"; and "Standards and Specifications." Over one-third of the listings fall into the "Handbooks and Manuals" section. The section on computer databases lists those which cover mechanical engineering topics. Most of the citations have evaluative annotations, but some have no descriptive information. The guide was published in 1985, so it lacks references to some major new and revised works in the field (Marks, Kutz, and *Machinery's Handbook*, for example). There is no author or title index, a limitation mitigated only by the small size of the guide. The bibliography

updates, but is not strictly a supplement to, an earlier (1970) guide to the literature of mechanical engineering, which was also published by the ASEE (*Guide to Literature on Mechanical Engineering*, compiled by James K. K. Ho). The new edition adds chapters on computers and yearbooks and drops the chapters on periodicals and research centers. This is a budget production, photocopied from typescript and velo-bound, but at its modest price, it is a useful compendium for mechanical engineers and a concise guide for the libraries which serve them.

CHARLENE M. BALDWIN

Handbooks

542. How to Design Heating-Cooling Comfort System. 4th. Joseph B. Olivieri. Troy, MI: Business News Publishing Co., 1987. 328p., illus., index. $29.95. 0-912524-36-7.

"A building without a good environmental system is nothing more than a mausoleum." Joseph B. Olivieri firmly believes this statement and has written the fourth edition of *How to Design Heating-Cooling Comfort Systems* so that users of the book can design a building that includes an "environmental control system which will comfort the body as the beautiful building comforts the soul."

After a brief introduction in which terms are defined, 14 chapters cover comfort, heating and cooling loads, anatomy of heating, ventilation, refrigeration, psychrometrics, air and how to move it, water and environmental control, comfort, temperature controls, moisture in construction, sound, air cleaners, and design and energy conservation. Text in each chapter is kept to a minimum. Instead, there are numerous tables, charts, and diagrams that outline the topics under discussion. Tables cover such areas as "Thermal Resistances of Plane Air Spaces," "Climatic Conditions for United States and Canada," and "Pressure Drop in Divided Flow Fittings."

The text, well written and understandable for both the layperson and the technician or architect, makes for an excellent book with a great amount of information that could be found in a number of other books. However, its greatest value comes from the fact that the information is all contained in one volume rather than in many. It is meant to be used by the designer, architect, and the person actually doing the work, but students will find this a valuable source of information in class work that requires them to design heating and air conditioning systems. Libraries will want this as a reference source for anyone involved in the design of heating and air conditioning. *H. ROBERT MALINOWSKY*

543. Industrial Control Handbook. E. A. Parr. New York, NY: Industrial Press, 1987. 2v., illus., index. $29.95 per vol. 0-8311-1175-5 v. 1, 0-8311-1178-X v. 2.

Written in two volumes, the *Industrial Control Handbook* is a very thorough treatment of industrial controls. Volume 1, "Transducers," covers differences between sensors and transducers, types of sensors and transducers, and methods of measurement. Definite strengths of this text are the author's clear definitions of terms and liberal use of examples and diagrams to illustrate his narrative. In volume 2, "Techniques," topics include amplifiers, rotating machines, circuits, and hydraulics. This is an excellent text and reference source.

Although these books are not written for the layperson, an individual with minimal background in this area certainly will gain a good general knowledge and understanding of the topics. However, to gain the greatest benefit from the author's work, one must have a background in calculus and physics. Regardless of the reason, these books are for those who want a better understanding of industrial control. Libraries with engineering collections will want this for their reference collections. *JOHN NISHA*

544. Machinery's Handbook: A Reference Book for the Mechanical Engineer, Designer, Manufacturing Engineer, Draftsman, Toolmaker, and Machinist. 23rd. Erik Oberg; Franklin D. Jones; Holbrook L. Horton. Henry H. Ryffel; Robert E. Green; James H. Geronimo, eds. New York, NY: Industrial Press, Inc., 1988. 2,511p., illus., index. $55.00. 0-8311-1200-X.

Machinery's Handbook is a practical handbook whose audience is primarily design and production departments of large and small manufacturing plants; jobbing shops; and trade, technical, and engineering schools which teach the design and fabrication of machinery.

This 23rd edition takes the same practical approach as the 22nd (1984). It is about the same length, looks the same (sits well in the hand, opens well), and has the same effective index. A few new sections have been added, notably numerical control and jig boring, but it essentially contains the same information, updated when appropriate, referring to revised standards and specifications.

The 23rd edition, however, has been completely reorganized. Formerly there were 58 separate chapters; now there are over 200 chapters arranged under 13 major sections: mathematics; mechanics; strength of materials; properties, treatment, testing; dimensions; tooling and toolmaking; machinery operation; manufacturing processes; fasteners; threads; gears; bearings; and measuring units. This new arrangement greatly facilitates the use of this handbook.

This is not a handbook for information on the theory and principles of mechanical engineering; it is not a handbook of management, or systems and control, or of computerization. (The term "computers" does not even occur in the index.) It is a handbook which stresses the metal-working industries and tool design. It includes step-by-step calculations of formulas and an increased

inclusion of metric data. A companion guide has been published, *Machinery's Handbook Practical Companion: Machinery's Handbook Guide* (23rd ed., 1988), to increase the applicability of information in the handbook by providing examples, test questions, and illustrated applications. This is a highly recommended handbook for academic and special libraries. *CHARLENE M. BALDWIN*

545. Marks' Standard Handbook for Mechanical Engineers. 9th. Eugene A. Avallone; Theodore Baumeister, III, eds. New York, NY: McGraw-Hill, 1987. 1v. various paging, illus., bibliog., index. $89.00. 0-07-004127-X.

For over 70 years, *Marks'*, as this handbook is known, has been recognized as the classic in its field. The time between the eighth and ninth editions was too long, however, and much has changed since 1978, mostly related to computerization and robotics. In fact, a competitive volume, *Mechanical Engineer's Handbook*, edited by Myer Kutz (Wiley, 1986), was published to fill the gap. The Kutz book and *Marks'* are similar. Kutz offers 78 authored chapters in 2,316 pages but entirely devotes its first nine chapters to computers. *Marks'* (9th ed.), arranges its signed contributions into 19 separately paged sections (1,936 pages). There are large sections on mathematics, units of measure, fluid and solid mechanics, heat, strength of materials, materials of engineering, power generation, and transportation, as well as chapters on building construction, industrial engineering, and environmental control but little coverage on computers. Both have excellent indexes. Both are bound in a single volume, but *Marks'* is much easier to handle because of its thinner pages and choice of cover.

Marks' may not be as practical as other handbooks, and is less theoretical than Kutz's, but it is just as useful with its heavy emphasis on illustrations—tables, drawings, graphs, charts—and specific chapters dealing with shop processes, fans, fuels, furnaces, and welding. This latest edition of *Marks'* presents current practice, latest methods, and techniques in an easy-to-read and easy-to-use format. It is recommended not only for college, university, and research centers but also for public libraries. *CHARLENE M. BALDWIN*

546. Reciprocating Pumps. Terry L. Henshaw. New York, NY: Van Nostrand Reinhold Co., 1987. 329p., illus., bibliog., index. $48.95. 0-442-23251-9.

"Reciprocating pumps have been in use for over 2,000 years" and one of the earliest pumps of this type was built in Egypt about 150 BC. It was the workhorse of the pumping industry until the development of the centrifugal pump in the early 1900s and was largely displaced in many applications by the new pump design. The technical literature reflected this displacement, and, consequently, little has been written about recip-

rocating pumps for many years. The gap in the literature has become more evident in recent years because of increasing interest in the energy efficient design of the reciprocating pump.

This book fills this major gap in the technical literature on reciprocating pumps and, as a result, would be useful to public, academic, and special libraries. It updates information on newer designs and describes the operation, primarily, of power pumps that operate at two or three times the speed of earlier designs in use since the 1940s. It answers questions relating to selection, application, installation, maintenance, and operation of the pump for the designer, installer, maintenance personnel, and operator. The detailed descriptions will also be of value in the academic setting for the engineering student or instructor for an operations or equipment design project.

RAYMOND BOHLING

547. The Service Hot Line Handbook: A Compendium of Highlights from the Manufacturers' Service Advisory Council (MSAC) Hot Lines, Volume III: Electrical Components and Service, Non-Electrical Components and Systems. Phil Roman, comp. Troy, MI: Business News Publishing Co., 1987. 157p., illus., index. $12.95pbk. 0-912524-35-9pbk.

This handbook is a continuation of the *MSAC Hot Line Handbook* of the Manufacturers' Service Advisory Council (MSAC). It is a guide to the maintenance and repair of air conditioning, refrigeration, and heating equipment. This volume covers "Electrical Components and Service" and "Non-Electrical Components and Systems." The entire book is in the form of questions and answers about specific maintenance and repair problems. This book is of primary use in the shop but has some limited reference value for general public library reference collections.

H. ROBERT MALINOWSKY

548. Shock and Vibration Handbook. 3rd. Cyril M. Harris, ed. New York, NY: McGraw-Hill Book Co., 1988. 1,312p., illus., bibliog., index. $76.50. 0-07-026801-0.

Now in its third edition, the *Shock and Vibration Handbook* has long been the standard reference book in this field. Intended to be a working and comprehensive handbook for engineers and scientists, it could also be useful as supplementary material for university courses on the topic. The handbook brings together both the theoretical foundations of vibration and shock and the current applications of the theory to engineering practice. About 50 percent of this new edition is either completely new or extensively revised since the 1976 edition. New topics addressed include modal analysis, fluid flow-induced vibration, and seismic qualification and condition monitoring of equipment. Many other chapters were extensively revised to cover current developments, including, among others, chapters dealing with transducers, measurement techniques, vibration standards, isolators, data analysis, and ground-motion-in-duced vibration of structures. The 44 chapters were written by 60 authors drawn from universities, industry, and government laboratories. The editor, Cyril M. Harris, is an electrical engineering and architecture professor at Columbia University and a member of both the National Academy of Sciences and the National Academy of Engineering.

The *Shock and Vibration Handbook* is arranged by broad groups of topics. The first group of chapters deals with theory, followed by chapters on measurement and instrumentation. Vibration standards are covered in a single chapter, followed by chapters dealing with analysis and testing, such as modal analysis and shock and vibration data analysis and testing machines. Although not grouped together as one might expect (and as the preface implies), there are chapters on structural vibration caused by ground motion, fluid flow, and wind. Next there is a selection of chapters dealing with the control of vibration, including the use of isolators, damping treatment, and balancing procedures. The last few chapters deal with equipment and packaging design and with the effects of shock and vibration on humans. The handbook includes an extensive index. It is highly recommended for academic and special libraries.

CAMILLE WANAT

Manuals

549. Small Air-Cooled Engines Service Manual. 15th. Overland Park, KS: Intertec Publishing Corp, 1987 (c1986). 392p., illus. $14.95pbk. 0-87288-223-3pbk.

Small engines are used in much equipment that the layperson has around the house, including lawn mowers, garden tractors, and generators. All are known as "internal combustion reciprocating engines." Servicing these engines is usually fairly easy, especially when a service manual is available that guides the person through the procedures.

The first part of this service manual covers the general information on principles and fundamentals of operation of these small engines, troubleshooting, and general service information. The main part of the book covers specific makes of engines: Acme, Advanced Engine Products, Briggs and Stratton, Chrysler, Clinton, Cox, Continental, Craftsman, Deco, Grand, Homelite, Honda, Jacobsen, Kawasaki, KioritzEcho, Kohler, Lauson, McCulloch, O&R, Onan, Power Products, Sachs, Solo, Tecumseh, West Bend, Wisconsin, and Wisconsin Robin. Each chapter has the same format: list of models, engine information, maintenance, and repairs. Excellent schematics, fully labelled, supplement the text and the step-by-step instructions on how to do the maintenance and repairs. At the end of the manual is a "Service Shop Tool Buyer's Guide" that gives the names and addresses of tool and equipment companies and manufacturers.

This is considered the best manual covering small engines; the fact that it is in its 15th edition is sufficient proof. Schools with shop mechanics classes will find it very useful, and public libraries will want it with the other service manuals. *H. ROBERT MALINOWSKY*

550. Small Diesel Engine Service Manual. 2nd. Overland Park, KS: Intertec Publishing Corp, 1987. 288p., illus. $12.95pbk. 0-87288-238-1pbk.

This service manual covers the following makes of small diesel engines: Continental, Deutz, Farymann (Briggs and Stratton), Kirloskar, Kubota, Lister, Lombardini, MWM, Onan, Perkins, Petter, Peugeot, Slanzi, Volkswagen, and Wisconsin. Each entry is divided into two parts covering maintenance and repairs. Prior to the actual service information on each make, the various model numbers for that particular manufacturer are listed with the number of cylinders, bore, stroke, and displacement. There are numerous schematics and photographs plus "exploded views" of various parts of the engine, and the views are numbered, with each part identified. Most of the photographs are clear and easily show what is being illustrated. This is an excellent service manual covering engines that may be found on compressors, generators, construction equipment, farm equipment, industrial equipment, mixers, pumps, and welders. It is an essential manual for persons who service diesel engines and would be a useful reference manual for school libraries with vocational training classes and for public libraries.

H. ROBERT MALINOWSKY

551. Valve Selection and Service Guide. Matthew McCann, ed. Troy, MI: 1986, 1986. 202p., illus., index. $39.95. 0-912524-25-1.

The title, *Valve Selection and Service Guide*, identifies in brief terms the contents of this book. It is intended for technicians and service or maintenance personnel who are responsible for proper valve operation in their daily work. Chapter 1 opens with a brief introductory statement on the history and function of valves in modern society. Chapter 2 provides a listing of the large variety of valve designs that are available in the valve industry. Chapter 5 presents an interesting and thorough description of pressure and temperature regulators which is a type of valve through which flow is regulated internally rather than externally as in a control valve. The book is illustrated generously throughout with diagrams, charts, tables, and photographs that complement the overall high quality of the text. This is an excellent book for the technician and would be a useful reference source in special libraries where information on valves is necessary.

RAYMOND BOHLING

Textbooks

552. Foundations for Machines: Analysis and Design. Shamsher Prakash; Vijay K. Puri. New York, NY: John Wiley and Sons, 1988. 656p., illus., bibliog., index. $69.95. "A Wiley-Interscience Publication"; Wiley Series in Geotechnical Engineering. 0-471-84686-4.

A thorough and well-balanced treatment of an important geotechnical engineering topic sparsely represented in the literature is covered in *Foundations for Machines: Analysis and Design*. As the authors assert, machine-foundation engineering requires coordinating knowledge of machine forces and vibrational theory (mechanical engineering) with that of natural and man-made foundations (civil engineering). Few texts recently published provide such an extensive, coordinated coverage of the topic.

This text is usefully organized around a strong theoretical base which includes chapters on vibrational theory, wave propagation in elastic medium, and an extensive discussion on dynamic soil properties, methods for their determination, and evaluation of data. Other major components are chapters on foundations for specific machine types—reciprocating, impact, high-speed rotary, others—and chapters on specific foundation types for dynamic loads—piles and embedded block using elastic half-space and linear elastic weightless spring analyses. Extensive bibliographies are included with each chapter.

The text has strong instructional objectives as are indicated by chapters that are complete with examples and overview, and some with problem exercises (solutions included). Methods of testing and analysis are carefully delineated, and two detailed case studies are presented. Although the authors note rapid movement in areas of research, such as the analyses of embedded foundations and piles under dynamic loads, it is interesting to note that the average age of a reference cited in various chapters on foundation engineering for certain machine and foundation types ranges from 12 to 24 years.

The time for a state-of- the-art summation of the theory and practice of foundation engineering for machines is ripe, and here is a well-delivered response. This is a strongly recommended book for university libraries. *JOHN T. BUTLER*

★

MINING/METALLURGY ENGINEERING
Guides to the Literature

553. Selective Guide to Literature on Mining Engineering. Charlotte A. Erdmann, ed. Washington, DC: Engineering Libraries Division, American Society for Engineering Education, 1985. 32p. $5.00pbk.

Engineering Literature Guides, no. 6.
0-87823-106-4.

This addition to the series of *Engineering Literature Guides,* published by the American Society for Engineering Education, includes references to nearly 200 sources, both print and online, which provide access to the literature of mining engineering. Most of the sources are in the English language and have been published after 1970. Many of the citations include succinct annotations which define accurately the scope of the referenced item. A useful reference source for academic libraries. *EARL MOUNTS*

Handbooks

554. Engineered Materials Handbook: Volume 1—Composites. Theodore J. Reinhart; Cyril A. Dostal, eds. Metals Park, OH: ASM International, 1987. 983p., illus., index. $104.00. 0-87170-279-7.

Sixty years ago the American Society for Metals (ASM) published the first edition of its well-known *Metals Handbook,* a reference tool for sources of information on metals and the metal working industry. The publication of the ninth edition of that handbook is now in process. This new title, *Engineered Materials Handbook,* supplements but does not replace the older title. It is the result of the ASM International's commitment to expand its scope and to provide information on all materials of interest to the industries the society serves. As in the *Metals Handbook,* each volume of the new title will discuss a different aspect of engineered materials. Volume 1 is a detailed discussion of composites. Future volumes will cover engineering plastics, ceramics, and other high technology materials.

With this new title, the society continues to use the same pattern of production of the handbook as it did in its earlier one. Each volume will be the work of a large number of contributors under the direction of the Handbook Committee of the society. Approximately 520 authors and reviewers were involved in the preparation of volume 1. These include scientists, engineers, and technicians selected from the worldwide community working in materials and are listed in the front of the volume along with their places of employment.

Volume 1 covers "all phases of advanced composite technology" with "emphasis on engineering properties and manufacturing." The 174 chapters are organized into 13 sections that cover the properties and forms of the basic fibers and matrix materials from which composites are made; the analysis and design, testing, forms, and properties of composites; the manufacturing process; machining; quality control; failure analysis; and applications.

Section 1 serves as a general introduction to the subject of composite materials. It includes a glossary of approximately 1,200 terms used in the field and a guide to sources of information. The majority of the remaining sections begin with a short one-page introduction to define the scope of the chapters which are included. The volume does not include a bibliography but has extensive lists of references at the end of each chapter. The chapter on adhesive specification summarizes various specifications and standards which apply to the adhesives. A source list of these is given at the end of the chapter.

It is the policy of the ASM to publish data in both metric and U.S. units of measure in its publications. For that reason, the primary system of units of measure in this volume is metric units based upon the *Système International d' Unites* (SI), with secondary mention of the values in U.S. units. A metric conversion guide is found at the end of the volume. Also, a list of abbreviations and symbols with their definitions and the Greek alphabet are found at the end of the volume. The index consists of an alphabetized list of subjects with "see" and "see also" references. The main headings are in boldface type, and the subheadings are in regular type. Many of the entries indicate the page on which a definition of the term can be found. This new reference tool for new materials is the first major source in this subject area. It is a welcome addition and is recommended for university, research, and special libraries. *LUCILLE M. WERT*

555. High-Temperature Property Data: Ferrous Alloys. M. F. Rothman, ed. Metlas Park, OH: ASM International, 1988. 489p.,illus., bibliog. $120.00. 0-87170-243-6.

High-Temperature Property Data: Ferrous Alloys is an excellent new reference work in metals that was made possible by the availability of new production and printing technology using computer manipulation of data and laser jet printing. This technology significantly reduced both the time and cost of production. A spin-off is the availability of the data in electronic form, which will facilitate updating of the publication and could, conceivably, be made available for electronic searching. This is the first of the two-volume pair on high-temperature applications of alloys that together will provide comprehensive and comparative data for discriminating judgments on the selection of alloys for high temperature applications. The book is divided into sections representing 12 families of ferrous-alloy materials. Each section has an introduction that describes the general characteristics of the family, including the nature of the composition described, the metallurgical characteristics, and the typical temperature range of applications. The introductory statement is followed by summaries of the most common alloys in the section; these summaries include the alloy name, alternative designations, specifications, and composition. Following this information, mechanical and physical property data are then described in tables and graphs for each alloy. The information on physical data is especially complete. This book will be a valuable reference source for the design and materials engineer and for any academic or research library collecting information in materials technology. *RAYMOND BOHLING*

556. Metals Handbook: Volume 13—Corrosion. 9th. Lawrence J. Korb; David L. Olson; Joseph R. Davis, eds. Metals Park, OH: ASM International, 1987. 1,415p., illus., bibliog., glossary, index. $98.00. 0-87170-007-7 v. 1.

"Cost of corrosion to U.S. industries and the American public is currently estimated at 170 billion dollars per year." Avoiding the consequences of corrosion involves the engineering disciplines of chemistry, design, and metallurgy. For this reason, the editors have devoted an entire volume of the *Metals Handbook*, 9th edition, to corrosion and intend it to be a definitive work on the subject, a reference source engineers will turn to first for corrosion information. This volume is arranged in eight major sections: (1) Glossary of Terms; (2) Fundamentals; (3) Forms of Corrosion; (4) Testing and Evaluation; (5) Design; (6) Protection Methods; (7) Specific Alloy Systems; and (8) Specific Industries and Environments. With the exception of the glossary, each section contains signed, illustrated articles, with bibliographies. Following the text is a "Metric Conversion Guide" for the expression of weights and measures in SI units, a list of "Abbreviations and Symbols," and a detailed comprehensive index. Without a doubt this volume belongs in all university and technical libraries with engineering collections. *SUE ANN M. JOHNSON*

557. Metals Handbook Comprehensive Index. 2nd. Metals Park, OH: ASM International, 1988. 657p., bibliog. $98.00. 0-87170-298-3.

This is the second edition of the comprehensive index of the *Metals Handbook*. It was designed to be a convenient single-volume index for the 16 volumes of the *Handbook*. It indexes 12 volumes of the ninth edition and four volumes of the eighth edition. It was prepared under the auspices of the American Society of Metals, Handbook Committee, and published by the society. The basis of the first edition of the comprehensive index was the individual indexes of the 16 original volumes of the *Handbook*. These indexes were rekeyboarded, edited, corrected, and merged. When terms could not be merged, cross-references were established to assist users in locating the information.

The database for the first edition was used to produce the second edition. The indexes of the four new volumes of the ninth edition were added to the database, and entries for superseded materials in the eighth edition were removed. This same procedure will be followed to keep the *Comprehensive Index* up to date as new volumes of the *Handbook* are published in the future. The second edition of the *Comprehensive Index* contains approximately 80,000 entries referring to 12,000 pages of text in the 16 volumes of the *Handbook*. Except for a short list of numbered alloys, the index is alphabetic by topic. The main entries are in boldface type, with subdivisions listed alphabetically under the main entry in regular type. The numbers following the entries referring to the edition and volume are in boldface

type and pages in regular type. To prevent confusion resulting from indexing volumes from both the eighth and ninth editions of the *Handbook*, a footnote appears on every even-numbered page listing the specific volumes indexed from each edition. The index has cross-references between related topics. However, the preface cautions users to check both general and specific entries, since every variance could not be handled by cross-references This index is recommended for all libraries that have the current set of *Metals Handbook*. *LUCILLE M. WERT*

558. Worldwide Guide to Equivalent Nonferrous Metals and Alloys. 2nd. Paul M. Unterweiser; Harold M. Cobb, eds. Metals Park, OH: ASM International, 1987. 465p., illus., index. $120.00. 0-87170-306-8.

University libraries with engineering programs and special libraries with a need for materials information will welcome this completely revised, expanded, and reorganized second edition. "This current volume is specifically designed to support the task of identifying, relating, and...selecting nonferrous material from a variety of national origins" and is divided into 11 separately paged sections: (1) Introduction; (2) Wrought Aluminum; (3) Cast Aluminum; (4) Wrought Copper; (5) Cast Copper; (6) Wrought and Cast Lead; (7) Wrought and Cast Magnesium; (8) Wrought and Cast Nickel; (9) Wrought and Cast Tin; (10) Wrought and Cast Titanium; (11) Wrought and Cast Zinc. Each material section is arranged alphabetically by country or standards-issuing organization. Information on each metal or alloy is presented in columns: designation, UNS number, composition, specification, form, condition, and mechanical properties. A detailed, clearly written introduction explains the book's arrangement and use. Tables outlining the Unified Number System and a directory of standards organizations are also included. Appendix A is a "Metric-to-English Conversion" table, and appendix B contains indexes for UNS number, specification, and designation. *SUE ANN M. JOHNSON*

Tables

559. World Survey of Nonferrous Smelters: Proceedings of Symposium on World Survey of Nonferrous Smelters, sponsored by the Pyrometallurgical Committee of TMS at the TMS-AIME Annual Meeting in Phoenix, January 25–29, 1988. John C. Taylor; Heinrich R. Traulsen, eds. Warrendale, PA: Metallurgical Society, Inc., 1988. 399p., illus., bibliog., index. $90.00. 0-87339-026-1.

World Survey of Nonferrous Smelters includes papers presented at a symposium held January 25–29, 1988, in Phoenix, Arizona, and sponsored by the Pyrometallurgical Committee of the Metallurgical Society. The papers give the results of a questionnaire sent to selected nonferrous smelters

in 1985 and 1986. Data from more than 90 smelters are included, along with simplified process flow diagrams. For each metal (copper, lead, and zinc) an overview of the process is given, and detailed tables are listed arranged by continent. Data in each table are annual production, feed analysis, feed preparation system, and specifics of the smelting process such as furnace type, fuel, slag, and offgas.

Also included are converting data, fire refining, product casting, and sulfur fixation. For selected plants, a detailed textual analysis is given. Additional material includes a list of survey respondents with addresses, a short bibliography, and subject and author indexes.

Although the data are now two to three years old, this is still a recommended source of information on nonferrous smelters. It is a unique book and one that all libraries serving metallurgical researchers should have. *EARL MOUNTS*

Treatises

560. **Mineral Economics of Africa.** N. de Kun. New York, NY: Elsevier Science Publishing Co.., Inc., 1987. 345p., illus., bibliog., index. $86.75. Developments in Economic Geology, no. 22. 0-444-42795-3.

This is a book about the economics of the mineral industry in Africa. It is not a detailed discussion of the minerals themselves but rather a treatise on the African share of world resources; the dependence on Africa by the rest of the world; the distribution of those resources within Africa; and a historical overview of the development of the industry. All of this is spelled out in the introduction, along with a list of abbreviations, a legend, and a list of symbols used throughout the book. The rest of the book is divided by geographical regions of Africa: the Maghreb, Northeast Africa, the Sahel, West Africa, Central Africa, East Africa, Southern Africa, South Africa, and islands and the sea.

Within each of these geographical areas, each country is discussed in terms of what minerals are available, the economics for that country, geological information, and statistics on what is processed and sold. There is much use of tables to show most of the information, but the text is well written for even the layperson to understand. There are maps that show the locations of various minerals within each of the countries; however, the maps are very brief, and one needs to know African geography to be able to read them. For each country, the major minerals are given, followed by the mineral, energy, and natural resource goals of that country. There is a brief index plus a list of acronyms and a bibliography of books and periodicals.

This is an excellent book for university and special libraries that analyses the strategic importance of African commodities and describes the distribution of deposits, oilfields, and mines. Researchers, economists, and mineralogists will find this a very useful book.

H. ROBERT MALINOWSKY

Yearbooks/Annuals/Almanacs

561. **Metal Bulletin's Prices and Data 1987.** Paul Millbank, ed. Ruby Packard, comp. New York, NY: Metal Bulletin, Inc., 1987. 352p., glossary. $55.20. 0-947671-07-2. 0269-1698.

An annual publication of the journal *Metal Bulletin*, this source is divided into three parts— "Prices," "Statistics," and "Memoranda." The prices section includes daily, weekly, or monthly prices for metals and also a detailed explanation of the different types of prices. Also included are detailed descriptions of the various grades and forms of metals and sources of prices. An important and useful feature of this section is the large number of line and bar graphs of prices computed over a four-year period. The statistics section includes annual statistics on production and consumption for the years 1965 to 1986 and the U.S. stockpile status as of December 31, 1986. The memoranda section includes the names, addresses, and telephone and telex numbers for metals associations worldwide, arranged by country. Also included is a list of abbreviations, a table of physical constants for pure metals, a glossary in six languages of steel industry terms, a list of brand names, and a list of tariffs. This source represents the most comprehensive publication available on the prices of metals. Large public libraries and academic and special libraries will need a copy of this databook. *EARL MOUNTS*

★
TRANSPORTATION/HIGHWAY ENGINEERING
Handbooks

562. **Analysis and Design of Curved Steel Bridges.** Hiroshi Nakai; Chai Hong Yoo. New York, NY: McGraw-Hill Book Co., 1988. 673p., illus., bibliog., index. $59.95. 0-07-045866-9.

In *Analysis and Design of Curved Steel Bridges*, Nakai and Yoo present current research and techniques, as well as substantial background information on horizontally curved bridges. The book is oriented toward Japanese designs and specifications. The Japanese Specification for Highway Bridges (JSHB) design code and the design code of curved steel girder bridges established by the Hanshin Expressway Public Corporation (HEPC) are used to support design examples throughout the book.

Both authors are experts in their field; both have authored many technical articles dealing with applied mechanics and bridge engineering. Professor Nakai is a member of the Department of Civil Engineering, Osaka City University, and Professor Yoo is a faculty member of Auburn University. They have divided the work into eight chapters: "Introduction," "Basic Theory of Thin-Walled Beams," "Fundamental Theory of Curved

Girders for Analyzing Static and Dynamic Behavior," "Analysis of Flexural and Torsional Stress Resultants and Displacements in Curved Girders," "Buckling Stability and Strength of Curved Girders," "Design Codes and Specifications," "Fabrication, Details, Painting, and Erection of Curved Bridges," and "Design Examples." Each chapter has introductory material and ends with an extensive reference list.

The book is "intended for practicing engineers"; however, the reference lists at the end of each chapter and the three appendices—(A) "Application of Matric Calculus to Estimate Cross-Sectional Quantities," (B) "Vibration Parameters," and (C) "Computer Programs for Designing Curved Steel Bridges"—make it a useful reference tool for academic and special libraries. There is an author and subject index, and a conversion table is given at the beginning of the work.

JO BUTTERWORTH

Histories

563. **Great American Bridges and Dams.** Donald C. Jackson. Washington, DC: The Preservation Press, National Trust for Historic Preservation, 1988. 360p., illus., bibliog., index. $16.95. "A National Trust Guide." 0-89133-129-8.

Great American Bridges and Dams is a guide to historic bridges and dams in the United States. It contains historical and technological descriptions of bridges and dams which are accessible to the public and are "historically significant." Many of these bridges and dams have unusual designs or building materials (such as wooden covered bridges), are well-known landmarks, or deserve mention as great works in the history of civil engineering. The 360-page book begins with two chapters on bridges and dams, which include descriptive information on the design of various types.

The book continues with chapters on preservation and a "Guide to the Guide," information on how to use the book. This is followed by a region-by-region description of notable bridges and dams in various parts of the United States. This section, which comprises the largest part of the book, is organized by geographic region (New England, South, West, etc.). Descriptions of bridges and dams in each region are listed alphabetically by state and by town or city within that state. The regional breakdown makes the book difficult to use without consulting the index, which was not available in the review copy. A strict alphabetical state/location arrangement would make the book easier to use. The finished copy will also contain over 500 illustrations, although these, too, were largely absent in the review copy. *Great American Bridges and Dams* also contains an epilogue with descriptions of bridges and dams notable in their failure, such as the Tacoma Narrows Bridge ("Galloping Gertie") and the Saint Francis Dam. A brief bibliography and index complete the book.

Great American Bridges and Dams is an informative book which would be useful to the many people who enjoy traveling the back roads rather than on the major interstates. Because the descriptions are simple, the book also could be useful in public or school libraries as a resource for reports. Therefore, it is recommended for public and school library collections as well as for academic collections that are collecting comprehensively.

LINDA R. ZELLMER

★

VEHICULAR ENGINEERING
Abstracts

564. **International Aerospace Abstracts.** New York, NY: Technical Information Service, American Institute of Aeronautics and Astronautics, Inc.., for the Institute and the National Aeronautics and Space Administration, 1962-. 1- , index. Biweekly. $700.00 per year U.S., $975.00 per year foreign, $400.00 per year cumulative index, $550.00 per year cumulative index foreign, $950.00 per year U.S. for both, $1,350.00 per year foreign for both. 0020-5842.

International Aerospace Abstracts (IAA) forms one of the two essential indexing and abstracting tools for the world's aeronautics and space science and technology literature. The other tool is *Scientific and Technical Aerospace Reports (STAR)*, with which *IAA* is closely allied. Whereas *STAR* covers unpublished technical reports literature, *IAA* concentrates on periodicals, books, meetings papers, conference proceedings, and translations of journals and journal articles. Thus, *IAA* and *STAR* complement each other and cover between them virtually all the world's aerospace literature. *IAA* appears on a biweekly basis in alternating weeks with *STAR*. Both use the same subject category arrangement and similar indexing and accession numbering schemes. *IAA* lists the 75 subject categories and their scope as the table of contents for each issue. Approximately two-thirds of each issue is made up of the abstracts section. Each abstract runs from 60 to 100 words in length. Each issue also includes five indexes: subject, personal author, contract number, meeting paper and report number, and accession number. *IAA* and its indexes are cumulated annually. *IAA* is also available online via the NASA database system. *IAA* is an essential purchase for any library or organization serving researchers and others involved in aeronautics or astronautics.

FRED O'BRYANT

Biographical Sources

565. Who's Who in Space: The First 25 Years. Michael Cassutt. Boston, MA: G. K. Hall and Co., 1987. 326p., illus., index. $35.00. 0-8161-8801-7.

It is hard to imagine how a more thorough, engrossing, and potentially useful reference book on manned spaceflight could be produced. *Who's Who in Space: The First 25 Years* has it all. Cogent, introductory essays on the American, Soviet, and international space programs; biographical data on over 380 astronauts, cosmonauts, back-up crew members, payload specialists, and X-15 and X-20 test pilots; 360 individual photographs plus other group photos and equipment photos; all the NASA mission patches; a list of space-related acronyms and abbreviations with their meanings; and, chronological logs of space missions and X-plane test flights—author Michael Cassutt has packed every bit of this information and more into this excellent and useful reference work. Each of the biographical sketches was compiled from official sources and augmented with data from news accounts and other media. Along with the usual standard biographical data, the reader is constantly delighted by dozens of fascinating bits of "trivia," e.g., Who was the first mother to fly in space? Who was the first American astronaut to go to sleep in space? Which Russian cosmonaut began his working career as a lumberjack? The list goes on. This is a fascinating book to browse through and a well-organized, easy-to-use reference source in which to locate material that is otherwise scattered in a great many other sources, some almost impossible to locate even in the largest American library. This is obviously a labor of love, some seven years in the making, and there is no other spaceflight book quite like it. Other books may treat the history of spaceflight more thoroughly and in greater detail, but for such a modest price, you could not hope to find a better single source of information about manned spaceflight pioneers. Highly recommended for libraries at all levels, as well as for individual personal purchase.

FRED O'BRYANT

Directories

566. North America's Maritime Museums: An Annotated Guide. Hartley Edward Howe. New York, NY: Hartley Edward Howe, Dist. by Facts on File, New York, NY, 1987. 371p., illus., bibliog., index. $35.00. 0-8160-1001-3.

North America's maritime history and heritage is preserved in the ships that have survived the ages of sail and steam, and in the many small and often out-of-the way museums preserving both ships and the artifacts associated with them. Indeed, most would be difficult to locate or learn about without this guide to their existence and resources. The author's abiding interest in the subject has resulted in an extremely useful tool for those wishing to visit these ships and museums.

North America's Maritime Museums contains 261 entries and is organized on a geographical basis. The three major divisions are the Atlantic and Gulf Coasts, Middle America's rivers and lakes, and the Pacific Coast and Hawaii. These, in turn, are further subdivided into regions sharing common maritime backgrounds, and then into sections of the bays, coasts, rivers, and lakes. The Atlantic and Gulf Coasts section, for example, contains seven broad regions ranging from "New England (And Its Stepchild, Long Island)" to "The South Facing the Atlantic" and "The Gulf Coast." "New England" is divided into ten additional areas. The sequence is clearly laid out in the table of contents, much in the manner of reference grammars, and this is the easiest and best way to quickly locate geographic areas of interest to the user.

Each entry lists the museum location complete with travelers' directions, dates and hours of operation, admission charges, addresses, a good description of the facilities, and telephone numbers. Additional information, for example on, library hours and conditions of use ("open to researchers by appointment"), is extremely helpful. The description of the museum and its resources is in the form of a narrative discussion of the materials of significant interest and is not intended as a critical analysis or collection inventory. Helpful background notes on the early maritime history of various regions, rivers, lakes, and ports precede the entries for each section. An index of personal and place names, ship names, and subjects facilitates checking for other references in the volume.

For those interested in learning more about these places, or for preparing to explore one of the regions outlined in the text, a useful bibliography is appended to the book. Like the guide itself, it is a narrative, not a simple listing of titles, and is arranged in the same geographical manner by region. It alone constitutes an important introduction to the literature dealing with the history of North America's maritime past. This annotated guide will constitute a useful and long-lasting public and academic library reference tool both for the history buff as well as for the serious researcher for many years to come. It allows both casual and longer term interests in this important and neglected part of North America's history to be satisfied in a single source.

R. GUY GATTIS

567. Space Industry International: Markets, Companies, Statistics and Personnel. Geoffrey K. C. Pardoe, ed. Harlow, Great Britain: Longman Group UK, Ltd., 1987. 353p., bibliog., tables, glossary, index. $155.00. 0-582-00314-8.

Space Industry International has gathered together the bits and pieces of this industry's organizations, policies, goals, manufacturers, and statistics worldwide. It is arranged very conveniently for reference work, consisting of 12 country/continent chapters, a glossary, reference

sources and further reading, an international synopsis, a company and organizations index, a product area index, and a personal name index. The 12 regional chapters that cover the world include detailed information on the organization of each country's space programs and policies at a national and international level. Directory type data on firms, associations, and institutes include address, telephone, telex, facsimile, parent company, product, future plans, annual turnover over the past three years, and publications.

This directory represents an extremely useful reference tool for university and special libraries. However, it cannot be described as comprehensive. For example, McDonnell Douglas (a large aerospace company in the United States) is not listed in the industries section of chapter 12 (United States). Yet, within the discussion portion of chapter 12, "Industry and the Space Market," there is a reference to the USAF ordering "7 DELTA IIs with options for a further 13 from McDonnell Douglas." The inclusion and/or exclusion of companies may be in direct relationship to "organizations who completed the form" and returned it to the editor as mentioned in the "How to Use this Directory" section. A companion source that makes an attempt to be comprehensive—i.e., *World Aviation Directory* or *Intervia Aerospace Directory*—would be necessary for complete reference work in this subject area.

JO BUTTERWORTH

Encyclopedias

568. Album of Spaceflight. Rev and updated. Tom McGowen. New York, NY: Checkerboard Press, 1987 (c1983). 61p., illus. (part col.), index. $4.95pbk. 0-02-688502-7pbk.

Album of Spaceflight is an encyclopedic history of spaceflight, beginning with the early rockets, satellites, and unmanned flights, including the many disappointments that were encountered along the way. This is followed with accounts of the moon exploration, study of planets, and the new technologies of Skylab and the space shuttle. Finally, it gives some insight into what the future holds for spacetravel.

The text is interestingly written, accurate, and up to date. Lee Brubaker has created some outstanding illustrations that add much to the book. Of particular use for reference purposes is the chronology of spaceflight highlights beginning with *Sputnik* in 1957 and ending with the *Challenger* disaster in 1986. Each entry in this chronology gives the spacecraft, launch date, crew, and highlights of the flight. This is a recommended book for school and public libraries.

H. ROBERT MALINOWSKY

569. Land-Based Fighters. David Baker. Vero Beach, FL: Rourke Enterprises, Inc., 1987. 48p., illus. (col.), glossary, index. $12.66. Military Aircraft Library. 0-86592-351-5.

A general overview of fighter aircraft operated by the United States Air Force, this book is visually appealing with an average of one color photograph per page. Its text is clear and easy to read and understand, well within the capabilities of its intended audience of juvenile readers. This is not a detailed description of all modern fighter aircraft but rather an outline of the general characteristics and capabilities of major American military aircraft and the types of missions they might expect to fly. A list of abbreviations, acronyms, and a brief glossary and index round out the volume. This book is part of the six-volume *Military Aircraft Library*, which includes works on spy planes, helicopters, navy fighters, bombers, and research planes. It is suitable for purchase by school and public libraries, as well as interested individuals.

FRED O'BRYANT

570. Navy Fighters. David Baker. Vero Beach, FL: Rourke Enterprises, Inc., 1987. 48p., illus. (col.), glossary, index. $12.66. Military Aircraft Library. 0-86592-352-3.

This book provides a general overview of fighter aircraft associated with the modern naval aircraft carrier. Using one color photograph per page, the book is visually appealing. Clear and easy to read and understand, well within the capabilities of its intended audience of juvenile readers, this book offers descriptions of the types and capabilities of many of the aircraft currently in use by the United States Navy. A list of abbreviations, acronyms, and a brief glossary and index round out the volume. This book is part of the six-volume *Military Aircraft Library*, which includes works on land-based fighters, spy planes, helicopters, bombers, and research planes. It is appropriate for purchase by school and public libraries, as well as by interested individuals.

FRED O'BRYANT

571. Standard Catalog of American Light Duty Trucks: Pickups, Panels, Vans, All Models 1896–1986. John A. Gunnell, ed. Iola, WI: Krause Publications, 1987. $24.95pbk. 0-87341-091-2pbk.

This comprehensive encyclopedia is also a history, guide to identification, and price guide to over 700 trucks and companies from 1896 to 1986. Krause Publications is known for its outstanding guides to cars and trucks, and this new guide is no disappointment. It is a storehouse of information about trucks manufactured throughout the world. The book is arranged alphabetically by make, each section begins with a history of that particular make. This is followed by a model-by-model description that includes such information as body types, prices at time of manufacture, weight, number produced, engine, chassis, options, historical notes, illustrations of most models, and current values, based on five levels from poor to excellent. The photographs are reproductions from manufacturers brochures or other publications and, as a result, are sometimes not of the best quality. However, this is a thoroughly fascinating book to browse through. The names and descriptions of trucks that are no

longer produced would be interesting to anyone with an interest in vehicular history. This is a must book for any hobbyist, collector, dealer, or investor. Public libraries will find it very useful in their reference collections, and special libraries covering vehicles in general will need this as a historical reference book.

H. ROBERT MALINOWSKY

Exam Reviews

572. ATP: Airline Transport Pilot: A Comprehensive Text and Workbook for the ATP Written Exam. 3rd. K. T. Boyd. Ames, IA: Iowa State University Press, 1988. 160p., illus., index. $19.95pbk. 0-8138-0074-9pbk.

"The material in this book is designed to lead you in logical progression through the steps and knowledge necessary to carry out a flight under Part 121 of the Federal Aviation Regulations." The 10 chapters cover "Review of Basic Computer Functions," "Advanced Computer Problems," "Weight and Balance," "Preflight Regulations," "Flight Planning," "Review of NAVAIDS," "Enroute Regulations," "Meteorology," "Weather Reports and Depictions," and "Terminal Procedures and Regulations." Each chapter begins with a brief review of the subject and then ends with a series of problems that pertain to that chapter. The text is well written and concise. The solutions to the problems are found in appendix A. Appendixes B and C are the ATP Examination and answers. This text is essential for anyone seeking an airline transport pilot certificate. The author is a pilot with 19 years of professional experience and holds the certificates of airline transport pilot—single engine and multiengine (land), single engine (sea); flight instructor—airplane and instruments; ground instructor—advanced and instruments; aircraft dispatcher; and FAA-designated examiner. This book was formerly entitled *Airline Transport Pilot.* Recommended for libraries serving pilots.

H. ROBERT MALINOWSKY

Field Books/Guides

573. Aerospace Careers. James L. Schefter. New York, NY: Franklin Watts, 1987. 112p., illus., bibliog., index. $11.90. High-Tech Careers. 0-531-10422-2.

Aerospace Careers is a well-written guide to the aerospace industry giving job opportunities, job descriptions, educational requirements, salaries, promotion possibilities, college and training programs, and information on the industry. The eight chapters are titled: "The Future Is Now," "Building the Foundation," "Designers and Dreamers," "The Developers and Analysts," "The Builders," "Specialty Careers in the Shop," "Working for NASA," and "The Scientists."

Written in an interesting style, this guide is a must for all school and public library reference shelves covering careers.

H. ROBERT MALINOWSKY

574. Colors and Markings of the U. S. Navy Adversary Aircraft. Bert Kinzey; Ray Leader. Blue Ridge Summit, PA: TAB Books, Inc., 1987. 80p., illus. (part col.). $14.95pbk. Colors and Markings, v. 6. 0-8306-8530-8pbk.

The Colors and Markings series seeks to provide an on-going affordable group of books detailing the paint schemes, squadron markings, special insignias, and nose art carried by many of the most important aircraft in history. The volume in hand is a special edition, devoted to the Navy Fighter Weapons School, the so-called "Top Gun" training facility.

Almost 300 photographs show the various markings and camouflage schemes employed by aircraft attached to Top Gun. Many of these schemes are quite original and difficult to document or identify in other publications. Included are photos of aircraft painted to resemble "the enemy's" aircraft and used to provide realistic training for pilots sent to Top Gun. This volume does not intend to provide detailed operating specifications, which are available in other references works such as the Jane's series. Rather, it seeks only to convey the many visual details and color schemes used at the school. In this mission, it succeeds very well at a cost that is very affordable.

The book should be of interest to most military aviation enthusiasts and armchair fighter pilots. A promotional tie-in with the 1986 movie *Top Gun* should boost sales and circulation at libraries that own the volume. Libraries seeking a low-cost alternative to the Jane's volumes and willing to sacrifice detailed specifications for lots of actual photos should strongly consider this series. However, the flimsy binding definitely will not stand up to heavy use.

FRED O'BRYANT

Glossaries

575. Automotive Reference: A New Approach to the World of Auto/Related Information. G. J. Davis. Boise, ID: Whitehorse, 1987. 460p. $29.95, $19.95pbk. 0-937591-01-7, 0-937591-00-9pbk.

Automotive Reference: A New Approach to the World of Auto/Related Information is a thorough glossary of automobile-related terms. The product of computer publishing, the book contains definitions (and no diagrams) of numerous auto-related terms from "AAMCO," to "con game" (eight pages of shady practices which could be encountered by almost any driver), to "Z-car." Although some of the terms included ("adverse conditions" and "roadbed") are not directly related to cars, all are clearly defined. Organization is sometimes difficult to comprehend; the section on book organization and use is located in appendix B (on

page 453). Letter-by-letter alphabetical order is used, ignoring spaces between words. Finally, a hierarchical structure is used to group some terms together ("con game" and "engine" are followed by types of cons and terms related to engines, respectively). This book would be very useful to anyone who owns a car or is considering buying one, because it puts the technical terms associated with automobiles in plain English. Considering that no recent automotive dictionaries exist, this book is recommended for all libraries with any reference materials on cars.

LINDA R. ZELLMER

Handbooks

576. 1988 SAE Handbook: v. 1—Materials, v. 2—Parts and Components, v. 3—Engines, Fuels, Lubricants, Emissions, and Noise, v. 4—On-Highway Vehicles and Off-Highway Machinery. Warrendale, PA: Society of Automotive Engineers, Inc., 1988. 4v., illus., index. $45.00 per volume. 0-89883-881-9 set.

The Society of Automotive Engineers' (SAE) *1988 SAE Handbook* is the one indispensable reference resource for the engineering/technical library specializing in vehicle technologies. SAE is a nonprofit educational and scientific organization concerned with the advancement of automotive technology. Through its technical committees, SAE develops the standards and engineering practices of automotive technology. Its technical coverage includes automobiles; trucks and buses; off-road equipment; recreational vehicles; aerospace vehicles; mass transit systems; and marine equipment.

The *SAE Handbook* is published each year and comprises the standards and recommended engineering practices which are the benchmarks with which to measure products and performance. They are the basis for many federal safety and engineering standards, and may relate to a material, a product, process, procedure, or test method.

Published in four separate volumes, each is available separately and each contains a complete subject index and a numerical index to the standard number. Each vehicle standard, recommended practice, or information report has a designation consisting of the letter "J" combined with a number (in boldface type) and, in many cases, with a date indicating the month and year of revision. SAE J374 JUN80 refers to the standard number J374, "passenger car roof crush test procedure," revised in June of 1980. Citations in the indexes are to the volume number, section, and page. In the subject index, for example, under "roofs" is the entry "passenger car roof crush test procedure—SAE J374 JUN80.....4:34.194." Beginning with this edition, each volume contains the full indexes, allowing a single-volume purchase without a separate index volume. Previous editions comprised four volumes plus a separate index volume. In addition, each volume also contains a list of the SAE technical reports

referenced in government regulations (primarily Federal Motor Vehicle Safety Standards); a list of related technical reports, not included in the *Handbook* because of their size; and lists of materials and engineering aids (drafting templates, machines, and devices) used for testing for compliance with the standards and available from SAE. Finally, an appendix is included with the technical report procedures and other guidelines for the development of the standards and practices. Many federal, ASTM, and ANSI standards refer to SAE standards and cite them with the appropriate "J" number.

This set of reference materials, kept at hand, complete with procedures, tables, and diagrams, is an invaluable resource for engineering information. SAE maintains an easy-to-follow listing of standards by number as well as provides full coverage in subject term access. Now all four volumes can quickly and easily be navigated by the librarian, researcher, and engineer. Standards, practices, and information on testing procedures are readily available. The *1988 SAE Handbook* is a reliable research support tool, and a rich reference resource for any library.   *R. GUY GATTIS*

577. The Used Car Book: An Easy-to-Use Guide to Buying a Safe, Reliable, and Economical Used Car. 1988. Jack Gillis. New York, NY: Harper and Row, Publishers, Inc., 1987. 160p., tables. $9.95pbk. "Perennial Library." 0-06-096225-9pbk.

This edition of a very practical consumer's book covers 1981-1987 domestic and import models of used cars. Jack Gillis states that used cars are "difficult to find, hard to evaluate and carry one of the worst reputations of any product on the market." As a result, he has put together this handbook to help prospective buyers of used cars evaluate the car before actually putting any money down.

The first part, "Finding Them, Checking Them Out and Getting the Best Price," covers such topics as understanding the classifieds, odometer fraud, checking out the used car, warranties, negotiating tips for getting the best price, and strategies for selling your car. Of particular use are the various checklists for inside, outside, under the hood, test driving, and mechanic's checklist. The second part, "Keeping Them Going," gives practical information on insurance, safety, maintenance, tires, and complaints. This section includes a state-by-state listing of the various safety belt laws.

Finally, part 3, "How They Rate," points out the various "good choices" in used cars. It lists, in the preliminary text, four to five cars for basic transportation, comfortable cruising, sporty, and luxury. The bulk of this section and the biggest part of the book is the various tables that list all of the models. Preceding the tables are two lists, "Good Choices under $5,000" and "Cars to Stay Away From." The tables rate each of the cars for

every model year and give size, drive, crash test, parts cost, insurance cost, fuel economy, theft rating, bumpers, recalls, turning circle, weight, wheel base, and price range.

This is an excellent book for anyone contemplating the purchase of a used car. Public libraries will find it a valuable source of information along with other used and new car buying guides. Jack Gillis is an expert on consumer issues and has written numerous other books concerned with cars. *H. ROBERT MALINOWSKY*

Histories

578. The Aces. Christopher Maynard; David Jefferis. New York, NY: Franklin Watts, 1987. 32p., illus. (col.), glossary, index. $10.90. Wings: The Conquest of the Air. 0-531-10367-6.

This is an encyclopedia of World War I air battles and the pilots who became aces during the war. It is written for young people in clear, concise format with descriptive and colorful illustrations. Included is a description of national markings used on airplanes, an account of the first dogfight, shooting through the propellers, Immelmann's turn (half-loops), the first bombers, balloons and war, the Red Baron, and fighting over water. Three sections of reference interest are the pictorial representations of aircraft that were used in World War I; brief biosketches of 26 top aces; and a glossary of terms. Recommended for school libraries. *H. ROBERT MALINOWSKY*

579. American Aviation: An Illustrated History. Joe Christy. Blue Ridge Summit, PA: TAB Books, Inc., 1987. 394p., illus., index. $24.60. 0-8306-2497-X.

This book aims to provide a comprehensive history of American aviation. It contains descriptive, historical facts about people, events, and machines that have influenced the technological advances of today's civil and military aviation. The 18 chronologically arranged chapters begin with the first flying pioneers and end with the space shuttles. Special emphasis is given to the world wars and the birth of the air-mail era, which, together, led to the development of the airline industry. Joe Christy, member of the Aviation/Space Writers Association, longtime pilot, and aviation historian, intended this work to be a textbook for college students; however, the informal and entertaining style makes it attractive to anyone interested in the topic. Each chapter begins with an outline and concludes with review questions for the important points. The book includes numerous informative illustrations with rare photographs, plus an appendix that covers important questions and answers about American aviation. This comprehensive and important history would be more useful if it contained references or a bibliography. In spite of this, it is a good general history for school, public, and college libraries. *AZIZ RHAZAOUI*

580. The Cessna 172. Bill Clarke. Blue Ridge Summit, PA: TAB Books, Inc., 1987. 293p., illus. $12.95pbk. 0-8306-2412-0pbk.

Bill Clarke's *The Cessna 172* tells you just about anything you might want to know about this highly popular and successful aircraft except how to actually fly it. More Cessna 172s have been manufactured than any other model of airplane—over 37,000—and a great number are still in the air today. This handbook will certainly be of value to every 172 owner, helping to keep them flying happily and safely for years to come. The book's contents include a brief history of the Cessna Company and the Model 172, year-by-year specifications for the 172 and all its variants, engine specifications and topics, lists of known airworthiness directives pertaining to the 172, advice on how to locate and select a used 172 (including examples of required legal paperwork), how to inspect and care for the plane itself, information about new avionics for making the plane safer to fly, how to paint or repaint the plane, and even how to go about having it refitted for use on the water as a floatplane. The book also includes a buyer's price guide, listing current prices for all models of the 172. Clear and easy to read and understand, the text is liberally spiked with clear black-and-white photographs and exploded-view equipment diagrams illustrating maintenance instructions. A good buy for the money. Every Cessna 172 owner should pick up a copy, and libraries serving flying enthusiasts will want one for reference. *FRED O'BRYANT*

581. The Complete Guide to Rutan Aircraft. Don Downie; Julia Downie. Blue Ridge Summit, PA: TAB Books, Inc., 1987. 264p., illus., index. $14.95pbk. 0-8306-2420-1pbk.

This third edition of *The Complete Guide to Rutan Aircraft* brings the reader up to date on designer and builder Burt Rutan's activities since 1984. Rutan's influence on small aircraft design, and in particular the use of canards, looms large. This volume takes a model-by-model look at his major achievements and how they were developed, tested, and received by the public. The text is thorough and easy to understand. This is not a treatise on theoretical design concepts but rather a people-oriented narrative interspersed with clear photos and occasional illustrative line drawings. A brief biography of Rutan and his test pilot brother, Richard, sets the scene for individual chapters telling the story of each of Rutan's aircraft, including the record-breaking nonstop around-the-world *Voyager*. Each chapter offers homey dialog and useful pointers about the airplane being discussed and its use. There are a few typos throughout the book, as well as occasional editorial oversights, but these do not overly detract from the book's usefulness and readability.

Readers looking for a fuller account of the epic *Voyager* flight may wish to purchase Dick Rutan and Jeana Yeager's account *Voyager*, published by Knopf and distributed by Random House (1987). *Rutan Aircraft* is recommended for nonspecialist libraries where there is interest in

experimental small aircraft design or the work of Richard Rutan. Individuals interested in Rutan aircraft will also enjoy the book.

FRED O'BRYANT

582. De Havilland Aircraft Since 1909. 3rd. A. J. Jackson. R. T. Jackson, ed. Annapolis, MD: Naval Institute Press, 1987. 544p., illus., index. $29.95. 0-87021-896-4.

The de Havilland airplanes are probably the most prolific of all British aircraft, the first being the de Havilland biplane no. 1 built in 1909. The firm itself was founded in 1920, and the last airplane to be manufactured under the name of de Havilland was in 1962. This is "the story of classic First World War biplanes, of little-known military and civil prototypes between the wars, of the birth of the British light aeroplane movement, of pioneer pilots and historic record-breaking flights." Anyone interested in airplane history will find this a fascinating book.

After a brief history of the company, each of the planes is described in detail with excellent photographs and sketches. There are notes on designers, pilots, flights, battles, and personalities associated with the planes themselves. Each includes a complete list of specifications and data including power plants, dimensions, weights, performance, and number produced. The five appendices cover "de Havilland Projects," "The Cierva C.24 Autogiro," "Aircraft Designed by the de Havilland Technical School," "Aircraft Designed by de Havilland Aircraft Pty. Ltd.," and "Aircraft Designed by de Havilland Aircraft of Canada Ltd." The index is in three parts with a general list of companies, organizations, personnel, sites, and locations; a list of aircraft mentioned outside their own chapter, type names not listed in the contents, and aircraft built by other manufacturers; and a list of the specific names of the engines used in the aircraft. Recommended for public and special libraries. *H. ROBERT MALINOWSKY*

583. Flying Boats and Amphibians Since 1945. David Oliver. Annapolis, MD: Naval Institute Press, 1987. 144p., illus. (part col.). $19.95. 0-87021-898-0.

The author presents the recent history of the flying boat and its continuing development in an interesting, narrative style enlivened with pertinent anecdotes. The color reproduction is exceptionally brilliant, and the contrast of the black-and-white photographs is generally excellent, making this a beautifully executed book. In addition, there are drawings and charts of specifications that give clarity to the information in the text describing the various amphibious planes and flying boats. The format of the book is simple; each main chapter describes one of the "twelve major multi-engined flying boats and amphibians that have been produced...since 1945"; and a final chapter summarizes the current research and technological developments in the field.

Mr. Oliver, without criticism, describes the decline of interest in the flying boat in the years following World War II. "Due in part to the fact...planes had improved dramatically during the war years...an abundance of airports with long paved runways...products of wartime expansion, the commercial flying boat,...became regarded as relics of a bygone age." However, his presentation of the histories of the various aircraft, although containing technological and aeronautical information, does not let the reader's interest wane. He presents interesting bits of information such as Steven Spielberg's use of "The City of Cardiff" in *Raiders of the Lost Ark*, and the experiences of the owners, who lived in it for two years. The final chapter discusses both the military and commercial development concepts and the potential uses of the aircraft. "The three key markets...remain civil and environmental protection, which includes fire-fighting, aerial spraying, SAR and maritime reconnaissance; defense, for ASV, ASW and special forces transport; and transport, commercial passenger and cargo transport." The feasibility of using amphibious aircraft to replace the more expensive helicopter for many purposes is being studied, and the use of flying boats for access to remote areas without airfields is expanding. Mr. Oliver concludes, "The flying boat has survived, if not flourished, but its true worth is only now being recognized and the type's development over the next forty years will ensure the preservation of the species." Recommended for history collections with an emphasis in transportation, as well as for the general reader.

JUANITA A. CUTLER

584. Ghosts: Vintage Aircraft of World War II. Philip Makanna. Charlottesville, VA: Thomasson-Grant, 1987. 120p., illus. (part col.). $36.00. 0-934738-29-7.

At some time, every youngster probably dreams of putting on flight jacket and goggles (these days, pressure suit and helmet) and taking to the skies to do battle with the Forces of Evil. Although few persons ever really become fighter pilots, the dream lingers on—and for those who flew fighters and bombers in World War II, memories remain strong and poignant. Yet, of the 750,000 aircraft manufactured by all combatants in WWII, only a tiny fraction are still flying. Philip Makanna has assembled a brilliant, stunning, and nostalgic collection of photographs of these "ghosts of the skies," planes with memorable names like Wildcat, Mustang, Spitfire, Messerschmitt, Flying Fortress, Zero, and Kate. The large, full-color photos are vivid, evocative, haunting. These are real planes, many vintage, though some are replicas lovingly outfitted to original specifications. They are flown now by weekend pilots and enthusiasts, flown for fun and demonstration rather than for destruction and death. But the past still haunts these machines. Interspersed with Makanna's photos are archival black-and-whites from the war years, showing the same types of planes and their pilots as they were then. Brief reminiscences accompany these photos, thoughts and words from the men who actually flew the planes under fire. A nice touch is that these anecdotes come from pilots of all nations involved, proving that there were unsung heroes on all sides. All these men and women had "the right stuff," the courage and fortitude to slug

it out at 10,000 feet and may the best plane and pilot win. These vignettes add immeasurably to what otherwise might be just an interesting picture book. Those who flew these planes know, and we should all remember, that a human being was at the controls, that death was always a split second away. These planes were superb machines, as this book makes abundantly clear. They were flown by superb pilots, which this book also makes abundantly and fittingly clear. *Ghosts* is a fitting tribute to the planes and pilots of World War II. It is recommended to anyone with an interest in aviation or history or both and libraries with an interest in aviation history.

FRED O'BRYANT

585. Guardians of the Sea: History of the United States Coast Guard, 1915 to the Present. Robert Erwin Johnson.

Annapolis, MD: Naval Institute Press, 1987. 412p., illus., bibliog., index. $23.95. 0-87021-720-8.

Guardians of the Sea is a sequel to Stephen Hadley Evans' *United States Coast Guard, 1790–1915: A Definitive History* published in 1949. The first two chapters of *Guardians* traces the history of the U.S. Revenue-Cutter and Life-Saving services, which were merged in 1915 to form the U.S. Coast Guard. Johnson is an excellent writer, making this an interesting history. Unusual events, catastrophes, and wars are all presented in such a way that, once the reader begins, it is difficult to lay aside the book. Although there are few illustrations, those that are included are good selections.

Each chapter covers a specific era such as World War I, Prohibition enforcement, Depression years, World War II, Korean War, last years under the Treasury Department, and the coast guard since 1967 tracing to about 1982. The layout of the book is beautiful with wide outer margins, soft white paper, and very pleasing typeface. There is an extensive bibliography and a detailed index.

Robert Erwin Johnson is a professor of history at the University of Alabama, has published other books on naval history, and served in the coast guard. This book is highly recommended for those who have an interest in naval history.

H. ROBERT MALINOWSKY

586. Honda Motor: The Men, the Management, the Machines. Tetsuo Sakiya. Tokyo, Japan: Kodansha International, Ltd., Dist. by Harper and Row, New York, NY, 1987 (c1982). 242p., illus., index. $15.95. 0-87011-522-7.

Honda Motor was translated by Kiyoshi Ikemi and adapted by Timothy Porter from *Honda chohasso keiei*. The history of the Honda Motor Company is a remarkable story of how Soichiro Honda began building small engines and motorbikes after the war in 1946 and built his

company into an internationally known motorcycle manufacturer. With Takeo Fujisawa, Honda was able to expand and eventually began building automobiles.

This excellent, readable history has several purposes. Its first is to present a detailed description of these two individuals and "the forces of history that shaped their lives." Second, it analyzes Honda Motor's corporate management and strategies. It gives some insight into the history of the company since the retirement of the founders, and finally it describes some of the products that have been produced by Honda Motor Co. Of particular interest is the chronology, from events prior to 1946 through 1982.

This is a fascinating history of interest to any car enthusiast. Libraries will want to have a copy for their history of transportation collections.

H. ROBERT MALINOWSKY

587. The Huey and Huey Cobra. Bill Siuru. Blue Ridge Summit, PA: TAB Books, Inc., 1987. 181p., illus., index. $13.95pbk. 0-8306-8393-3pbk.

This commemorative volume takes a thorough, detailed look at the development and use of perhaps the most familiar of all helicopters, the Bell Huey. Over 25,000 of these machines have entered service during the past 30 years in over 75 model variations, and this book takes a close look at them all. Especially memorable were the Hueys used during the Korean and Vietnam wars. These models are described in considerable detail, including armament.

The book begins with a historical sketch, followed by chapters treating the use of Hueys by each of the four armed services. Next come chapters on the commercial use of Hueys, their adoption by international military and civilian forces, one-of-a-kind record setters, and future developments. Two appendices list systematically all Huey variants and their specifications. An index completes the book, which also is heavily illustrated with clear black-and-white photographs. This typical TAB Aero publication will excite and enlighten helicopter buffs of every description and will be useful in military history collections as well.

FRED O'BRYANT

588. Lockheed Aircraft Since 1913. Rene J. Francillon. Annapolis, MD: Naval Institute Press, 1987. 566p., illus., index. $29.95. 0-87021-897-2.

This is a fascinating history of Lockheed Aircraft Corp., one of the major airplane manufacturers in the United States. "The basic work includes an historical survey, tracing the development of Lockheed from the original work of the Loughead brothers, through the successive companies, up to its reorganisation as the present Lockheed Corporation in 1977." The book is divided into two parts, the first covering the origin and history of what is now the Lockheed Corporation. The Loughead brothers, whose first plane was the Model G floatplane built in 1912, named their business Lockheed in 1912. During

Second World War, Lockheed was one of the major manufacturers of airplanes for the United States and today is a major government contractor in the area of advanced tactical fighters.

The second part of the book, the biggest and the most important, is a complete description of the 58 models that have been developed through the corporation's history, including such names as Big Dipper, Starfire, Starfighter, Orion, Hummingbird, Viking, and the 1986 Lockheed X-30A. Each model is fully described including notes on the designers, engineers, performance, and successfulness of the airplane. Photographs and diagrams aid in the description. There are, also, notes on which countries purchased the planes, variant models, and, for most, complete specifications. Appendix A lists "Lockheed Aircraft Model Designations and Projects"; appendices B and C give the "Production Details and Constructor's Numbers" through 86; and appendix D lists the "Lockheed Missiles and Space since 1953."

This is an excellent history for libraries interested in aviation history and for anyone doing research on airplanes. It is well written, well documented, and well illustrated. Although mentioned only on the jacket, this is actually a second edition of a book that was originally published in 1913. It is a companion volume to the author's *McDonnel Douglas Aircraft since 1920.*

H. ROBERT MALINOWSKY

589. Shipwrecks in the Americas. Robert F. Marx. New York, NY: Dover Publications, Inc., 1987. 482p., illus., bibliog., index. $12.95pbk. 0-486-25514-Xpbk.

This is a republication of a Bonanza Books edition which was an unabridged republication of a World Publishing Co. title *Shipwrecks of the Western Hemisphere, 1491–1825.* The original edition was published in 1971, sold all of its 7,500 copies within three weeks, and was not reprinted until 1975. It is considered "the Bible for treasure divers." The first part of the book is devoted to a history of early sailing and techniques of treasure hunting with chapters on "The Ships," "Early Salvors, Treasure Hunters, and Marine Archaeology," "Locating Shipwrecks," "Surveying, Mapping, and Excavating a Site," "Identification and Dating of Shipwrecks and Their Cargoes," and "Preservation of Artifacts."

The second and largest part is a country-by-country account of some 4,000 shipwrecks in the ocean and coastal waters of Canada, United States, Florida, Mexico, the Lesser Antilles, Bermuda, the Bahamas, Cuba, Hispaniola, Jamaica and the Cayman Islands, Puerto Rico and the Virgin Islands, Central America and off-lying areas, and South America. A chapter is devoted to each of these countries or areas with the shipwrecks arranged in chronological order and giving the year, the ship's name, the ship's cargo, and the approximate location of the shipreck. In some cases, there is specific information about the cargo and crew. If the wreck has been salvaged, that is indicated. The major part of the index comprises the names of the ships. It is the index and the account of the wreck that makes this a recommended reference book for public, college, and university libraries, but by far, the primary use has been and will be a guide for the treasure hunter. H. ROBERT MALINOWSKY

590. U. S. Army Ships and Watercraft of World War II. David H. Grover. Annapolis, MD: Naval Institute Press, 1987. 280p., illus., bibliog., index. $44.95. 0-87021-766-6.

Most people do not realize that, during World War II, the U.S. Army had a sizable fleet which was very active in the war and a key in the strength of the U.S. military at that time. David H. Grover has made an effort to educate the general public with this book that "is intended to describe in one volume the size and nature of the Army's waterborne activities during the years 1941 through 1945." It further "sets out to delineate each class of vessels operated by various commands within the army, and to identify the individual vessels within those classes by their characteristics and specifications." The 15 chapters cover the various types of ships and vessels that the army used during the war: transports, hospital ships, coastal freighters and passenger vessels, tankers, tugs, minecraft, port rehabilitation vessels, repair and spare parts depot ships, communication ships, operational support vessels, waterway maintenance vessels, launches and small craft, barges, sailing ships, and envoy.

Each of the chapters begins with a general introduction of the ship in question giving statistics, uses, and interesting historical data. The main part of each chapter is the listing of various ships in that particular category with pictures of a representative number. Most of the lists give specifications and builder and, for some, the disposition of the ship either during the war or after the war. The black-and-white photographs are for the most part clear, with the exception of some that were taken during the war. A rather extensive bibliography is included, as is an index to the individual names of the ships mentioned in the various lists. An excellent reference source of information that is hard to locate, any military historian will want to consult this book, and those who served in the Army during World War II will find it an interesting book through which to browse.

H. ROBERT MALINOWSKY

591. U. S. Small Combatants: Including PT-Boats, Subchasers, and the Brown-Water Navy: An Illustrated Design History. Norman Friedman. Annapolis, MD: Naval Institute Press, 1987. 529p., illus., index. $46.95. 0-87021-713-5.

Small combatants are those ships and crafts that are basically small coastal craft "capable perhaps of crossing an ocean but not of fighting very far from shore." They are those ships that were used to protect U.S. coasts and to operate in overseas coastal or inland waters. Virtually all of these crafts are high-speed with limited size and weight. The introduction to this encyclopedic his-

tory discusses the development of the small combatant, its uses, and its future. With excellent text, clear black-and-white photographs, and numerous charts and tables, the remaining 15 chapters cover every type of small combatant that the U.S. has developed: "Subchasers and Eagle Boats," "Second Generation Subchasers," "Small Fast Craft and PT Origins," "The Early PT Boats," "The Wartime PT," "The Postwar PT's," "Postwar ASW Craft," "Small Combatants for Counter-insurgency," "Motor Gunboats and PTF's," "Vietnam: Beginnings," "Vietnam: Market Time and Game Warden," "Vietnam: The Riverine Force," "Vietnam: SEALs and STABS," "The PHM," and "Post-Vietnam Small Combatants."

Six appendices cover "Gunboats," "Minor Acquired Patrol Craft: SP, YP, PU, PUc," "Crash Boats and Pickets," "U.S. Small Craft Weapons: Guns and Rockets," "Small Combatants for Export," and "Fates and Other Notes." The last lists the names of the small combatants, name of the building yard, dates of manufacture, launching, completion, and the fate of the vessel, making it the most complete list of small combatants available. "Norman Fiedman is one of this country's leading authorities on navy ships and weapons." He has 11 other books covering other aspects of naval history, ships, and weapons. For the military historian, this book is a must. Public and academic and special libraries will want it for its historical narrative, lists of ships, and excellent photographs.   *H. ROBERT MALINOWSKY*

Indexes

592. The Finding Guide to AIAA Meeting Papers, 1986 Annual Edition.
New York, NY: American Institute of Aeronautics and Astronautics, Technical Information Service, 1987. 104p., index. $25.00. 0894-3818.

The Finding Guide to AIAA Meeting Papers provides a basic index to all available papers from the American Institute of Aeronautics and Astronautics meetings held during the time period covered by each issue of the *Guide*. The first issue is a five-year cumulation covering papers from meetings held in 1981 to 1985, inclusive (available for $50.00). Thereafter, annual editions have been planned, each covering a single year, with the 1986 issue currently available. Each issue is divided into four sections. The first is a numerical/chronological index listed in sequential order by AIAA paper number, covering all the available numbered papers appearing in either preprint format or in a printed proceedings volume. The paper number, title, and author name(s) are given. There are no abstracts. The second section is an alphabetical title index followed by each title's AIAA paper number. The third section is an alphabetical author listing with associated AIAA paper numbers. The final section is an alphabetical list of AIAA meetings, showing where and when each was held and the range of AIAA paper numbers assigned to each

meeting. The total number of papers available from each meeting is also shown. For libraries subscribing to ISI's *Index to Scientific and Technical Proceedings*, the *Guide* is not an essential purchase, because ISI's *Index* provides the same sort of information plus subject access. However, where there is just the need for information about AIAA papers, the *Guide* will provide adequate information at less overall expense.

FRED O'BRYANT

Manuals

593. The Aviator's Guide to Modern Navigation.
Donald J. Clausing. Blue Ridge Summit, PA: TAB Books, Inc., 1987. 248p., illus., index. $26.95, $17.95pbk. 0-8306-0208-9, 0-8306-2408-2pbk.

Written by a professional aviator, this reference manual is primarily directed to student pilots. The author examines several techniques of navigation all of which are not always included in books written for this audience. *The Aviator's Guide to Modern Navigation* "looks at every aspect of modern instrument navigation, from the basics of dead reckoning and VOR navigation, to current instrument airway and approach navigation and advanced techniques of area navigation." Topics that are covered in some detail are ground-based radar navigation, LORAN-C navigation, inertial navigation, radar terrain mapping, electronic flight information systems, and navigation and flight management systems. Each chapter contains appropriate photographs, charts, and diagrams. Acronyms and abbreviations are clearly defined in the text and are again listed in an appendix. Other reference sections are devoted to lists of where to obtain aeronautical charts and other documents from federal agencies, commercial firms, and associations. In addition, there are lists of abbreviations and navigation symbols. Libraries with materials on aeronautics will find this a useful source of navigation information.

CHARLES R. LORD

594. Bicycle Mechanics in Workshop and Competition.
Steve Snowling; Ken Evans. Champaign, IL: Leisure Press, 1986. 160p., illus., index. $19.95. 0-88011-294-8.

"This is the first book to deal comprehensively, and in meticulous detail, with the care, preparation and repair of the racing machine." For those who own a racing bike, this book should be invaluable. It is written in a clear and concise manner with excellent illustrations. It is not a manual of basic mechanics; that knowledge is assumed. It does include some good commonsense instructions with examples of how they can benefit the reader. The first chapter covers the "Tools of the Trade" and has a checklist of what should be in the home workshop, the traveling tool chest, and the service toolbox. The following chapters cover specific parts of the bicycle or some aspect of servicing the bicycle: "Cleaning the Bike," "The Bike Check," "Race Service,"

"The Main Bearings," "Frame Problems," "The Wheels," "The Transmission," "Freewheels and Gears," "Points of Contact," "The Brakes," "My Life as a Professional Mechanic," "Track Bikes," "Transferring the Rider Position," "Equipment and Parts I would Recommend."

The text is concise and speaks to the reader on a one-to-one basis. Excellent illustrations aid in showing what is being discussed. They not only show the technique but include the hands of the author so that one can readily understand how to grasp a tool or part so that maintenance, repair, or replacement is as easy as possible. Snowling is an experienced cycle race mechanic and has worked for the national teams of the West German Federation. He understands and sympathizes with the problems of racing cyclists. Ken is a former racing cyclist and now a free-lance cycling writer. This is a must book for any racing cyclist and an excellent reference book for school and public libraries.

H. ROBERT MALINOWSKY

595. How to Buy and Maintain a Used Car: For the Non-Mechanical Person.
Brad Crouch. Van Nuys, CA: American Pacific Publishing Co., 1986. 159p., illus., index. $9.95pbk. 0-940057-00-Xpbk.

With 20 years of experience as an auto mechanic and currently operating a house call maintenance service for Mercedes Benz automobiles, Brad Couch has written an extremely useful guide and checklist "to help people who are not mechanically oriented to buy and maintain a used car." The book is in the form of a checklist that permits the reader to systematically go over a car and determine its true worth. The text is well written, straight forward, and entertaining with many tips to watch for. After the text are the actual checklists that one can use when looking at the car in question. Obviously this is a book for the consumer and may not have all that much reference value, but libraries will want to have a copy at hand for that patron who may not be aware that such a book exists.

H. ROBERT MALINOWSKY

596. The Loran, RNAV and Nav/Comm Guide.
Keith Connes. San Angelmo, CA: Butterfield Press, 1987. 128p., illus., bibliog., glossary, index. $14.95pbk. 0-932579-02-7pbk.

The author of *The Loran, RNAV and Nav/ Comm Guide*, a freelance aviation writer, has been the associate editor of *Flying* and editor-in-chief of *Air Progress*. Connes is also the author of another work, entitled *Know Your Airplane*. The cover of his guide describes its contents quite well: "How to Choose and Use Loran, RNAV, Nav Comm, and Handheld Radios." The 14 chapters in the book include "Choosing Your Equipment," "How the Loran System Works," "How the VOR-DME-RNAV System Works," "RNAV vs Loran," "Features to Look for in a Loran Receiver," "Should You Buy an IFR Certified Loran?" "Loran Models" (features dimensions and prices of 22 Lorans), "Equipment that

Works with Loran," "Features to Look for in VHF Nav and Comm Radios," "VHF Nav and Comm Models," "The HSI Explained," "The RNAV Models," "The Handheld Models," and "How to Choose a Radio Shop." The author has taken great pains to be as comprehensive and up to date as possible about this equipment. He even has a newsletter, *Avionics Update*, which provides the latest information to those who purchase his book through the Butterfield Press. The appendices are as useful as the information in the chapters. There are six, including "Understanding Latitude and Longitude," "A Guide to Guides," "Loran Antenna Installation," "Calling the Coast Guard," "Directory of Manufacturers," and "Glossary of Abbreviations, Acronyms and Terms." Libraries strong in aeronautical literature will want a copy.

JO BUTTERWORTH

597. The Private Pilot's Handy Reference Manual.
2nd. Joe Christy. Blue Ridge Summit, PA: TAB Books, Inc., 1987. 261p., illus., index. $21.95, $14.60pbk. 0-8306-2167-9, 0-8306-1867-8pbk.

This manual provides practical information about "procedures, methods, facts, and techniques" for the private pilot who must know how to operate an aircraft in the United States. The contents range from operating basics and aviation weather to cross-country flying and survival considerations. It also includes useful aids for ownership and maintenance of private aircraft. The second edition features new techniques developed in the 1980s. The style is easy to understand, and a detailed 92-page glossary explains the jargon used by air traffic controllers and pilots. Access to information is not as easy as it could be with the limited index. The typography, use of boldface, numerous photographs, and useful illustrations add much to the book. A directory of federal offices, private firms, and organizations in the field is included. The author, a private pilot since 1937, has written on this topic for 30 years. In addition to private pilots, this manual would be useful for students and teachers in aviation courses. The glossary makes it a useful reference source for public libraries.

AZIZ RHAZAOUI

598. The Proficient Pilot.
Rev. and enl. Barry Schiff. New York, NY: Macmillan Publishing Co., 1985. 337p. illus., index. $18.95. "An Eleanor Friede Book." 0-02-607150-9.

The Proficient Pilot, a revision and enlargement of the 1980 edition, is a collection of practical essays centered around the following topics: dynamics of flight, flying proficiency and techniques, flightworthy considerations, navigation techniques, emergency tactics, and multiengine flying. The author is an experienced pilot who has generously incorporated personal insights from his own flying stories, knowledge, and techniques.

It is the personal aspect of these writings that makes this book different from other books that are specifically concerned with techniques. Although this material is written for all levels of

flying experience, readers will learn in the first essay in the section on the dynamics of flight "that most pilots do not really know how lift is created; they only think they do." In response to this observation, the author explains "lift" with a series of simple examples diagrams, and notes, and his sense of humor makes this very enjoyable reading. The purpose of these articles is not to outline step-by-step instructions on how to fly, but it is to spark one's thinking on the concepts and principles which are inherent in the art of flying. Remember, "although water skiing, ice skating, and bobsledding can be a lot of fun, doing the same on a set of aircraft tires can produce thrills you are likely never to forget." Public libraries will find this an interesting reference source. *CHARLES R. LORD*

599. The Proficient Pilot II. Barry Schiff. New York, NY: Macmillan Publishing Co., 1987. 337p., illus., index. $19.95. "An AOPA Book"; "An Eleanor Friede Book." 0-02-607151-7.

The Proficient Pilot II is a second volume of essays written by Barry Schiff who is a contributing editor to the *AOPA Journal*. The topics covered are of interest to both student and professional aviators. Although the primary topical headings are similar to the first volume, e.g., flight dynamics, flying techniques and proficiency instruments, and systems management, he also covers other significant areas. Under the heading of "Powerful Topics," Schiff discusses propellers, power management, turbocharging, hot starts, and engine fires. The last chapter includes flying to Europe by yourself, ditching the plane, formation flying, and how to set a world aviation record.

Each article contains very practical information and techniques. Helpful charts and diagrams are included to assist in the explanation of technical information. For example, the article on "Pressurization" includes a chart of standard atmospheric readings beginning at sea level. Because of the vast aviation experience of the author, the articles include information appropriate from light aircraft to jets. The articles are short but well written and should be of interest even to people who are not pilots but just interested in aviation in general. Although readers will not learn all the aviation techniques necessary to fly, the book gives some insight into how experienced aviators view the world both from the ground and from the air. And if the reader wants to take on another genre, there is a collection of brainteasers in the last chapter. Recommended for public libraries. *CHARLES R. LORD*

600. Snowmobile Service Manual. 10th. Overland Park, KS: Intertec Publishing Corp, 1986. 488p., illus. $16.95pbk. 0-87288-236-5pbk.

Over 70 makes and models of snowmobiles are included in this service manual. It is divided into six sections the largest of which are the vehicle service and the engine service fundamentals sections. Each of these two sections is arranged by make of snowmobile or engine. The vehicle service section has a table of contents that indicates which makes are covered, but the service fundamentals section lacks this aid. The other four sections cover converter (belt drive) section, track drive, track and suspension, and skis and steering. Most of the makes were manufactured in the 1960s and 1970s, but there are a few from the 1980s. Each entry gives a list of the engine carburetor, sprocket ration, chain size, and clutch data, followed with such information as lubrication, adjustments, and overhaul. The schematics are clear, but many of the photographs are dark, making it difficult to see what is being illustrated. The textual instructions are clear and easy to follow. There are numerous "exploded views" of the various drive trains and other parts of the snowmobile with corresponding lists of the various parts to aid in ordering parts. With 40 years of expertise in maintenance and repair information, Intertec has published a manual that should be very useful for shop personnel and also for reference in a public library. *H. ROBERT MALINOWSKY*

601. Volkswagen GTI, Golf, and Jetta: Official Factory Repair Manual, 1985, 1986, 1987, Gasoline, Diesel, and Turbo Diesel Including Golf GT, Jetta GLI, and 16V Models. Cambridge, MA: Robert Bentley, Inc., 1987. 1v. various paging, illus., index. $39.95pbk. 0-8376-0337-4pbk.

This factory repair manual by Robert Bentley, seasoned author of many Volkswagen and Audi repair manuals, has been developed primarily with the professional automotive technician in mind, though anyone familiar with basic auto repair can easily follow the step-by-step procedures and accurate specifications. It covers U.S. and Canadian models only.

The topics are clearly and thoroughly covered. Not only is this manual informative and easy-to-use, the format is excellent. There are numerous access points. An eight-page index in the front of the manual clearly outlines the contents of the 97 chapters by alphabetically arranging subheadings under the boldface headings of General Engine, Fuel Supply/Exhaust/Alternator, Transmissions, Suspension/Brakes, Body, and Heater/Electrical System. Unique to this manual is that the chapter numbers align with the first two numbers of the Volkswagen repair time and warranty codes, which is extremely helpful when ordering replacement parts and dealing with dealers' parts departments. The pages of each chapter are margin indexed for quick location with the first page of each chapter containing an index of that chapter. Pagination appears in boldface type on the bottom of each page and consists of a chapter number followed by a page number. On the back cover is a "Quick Data Chart."

Though lacking in narrative style, this manual goes far beyond the other automotive repair manuals in its clarity and straightforward approach. The sparse narrative is accompanied by a multitude of black-and-white photographs, line

drawings, technical data tables, and diagrams, clearly illustrating set-ups and procedures. The book lies flat which will work well in the mechanic's workspace, and the binding appears stable enough to hold up to frequent use. This repair manual is an excellent source for do-it-yourself auto repair or for someone who wants to be more knowledgeable when taking a Volkswagen into a shop for professional repair. *LINDA VINCENT*

Treatises

602. The Seventy-Four Gun Ship: A Practical Treatise on the Art of Naval Architecture—v. 3, Masts, Sails, Rigging. Jean Boudriot. Annapolis, MD: Naval Institute Press, 1987. 280p., illus. (part col.), index. $59.36. 0-87021-618-X.

This is a beautiful oversized volume of a French work translated by David H. Roberts. It is the third of a four-volume set with volume 1 covering hull construction; volume 2 covering fitting out the hull; volume 3 (this volume) covering masts, sails, and rigging; and volume 4 covering manning and shiphandling. This is a fascinating book to look through, read, and study.

The text is well written, descriptive, and accurate, but it is the extremely detailed illustrations that make this a highly recommended book for the ship enthusiast and the historian. The detail of the line drawings is remarkable. Each illustration is thoroughly labelled with detailed margin notations. Also, throughout the book are marginal definitions of terms used in the construction of masts, sails, and rigging. In addition, to almost every page containing these detailed drawings, there are large fold-out illustrations showing details on a much larger scale.

In addition to the separate sections on masts, sails, and rigging, there is a section on maintenance. The volume concludes with a "Navy list 1780" giving the names of the ships, when launched, where built, the designer, length, breadth, depth-in-hold, guns, and remarks. Another very interesting section is a section that gives the "cost of building, rigging, and fitting out with powder and shot, victuals, and other stores, cost of commissioning and de-commissioning. (1778 values)." Every possible expense has been listed from timber, ironwork, nails, and labor to anchors, boats, casks, ballast, and victuals. Also included is a table that gives the labor required to build a 74-gun ship, listing the various tradespeople, the number of days worked, and the cost per day (1778).

Finally, there is a table that gives the prices of various supplies. Using all of this information, the author has calculated the cost of various sizes of naval vessels in 1778 and listed them in a table. Throughout the book, nautical terms are defined in the margins. A special index of these terms is included with the page number on which the definition may be found. In addition there is a general index and indexes to tables, plates, and illustrations.

This is a highly recommended book. Public libraries will find it quite useful, especially those that are located in and around harbors that may have museums of sailing vessels.

H. ROBERT MALINOWSKY

INDEXES

TITLE INDEX

NAME INDEX

SUBJECT INDEX

TYPE OF LIBRARY INDEX

Public

School

University

CPSIA information can be obtained at www.ICGtesting.com
Printed in the USA
BVOW11s0525061115

425637BV00011B/47/P